recent advances in phytochemistry

volume 34

Evolution of Metabolic Pathways

RECENT ADVANCES IN PHYTOCHEMISTRY

Proceedings of the Phytochemical Society of North America
General Editor: John T. Romeo, *University of South Florida, Tampa, Florida*

Recent Volumes in the Series:

Volume 26 **Phenolic Metabolism in Plants**
Proceedings of the Thirty-first Annual Meeting of the Phytochemical Society of North America, Fort Collins, Colorado, June, 1991

Volume 27 **Phytochemical Potential of Tropical Plants**
Proceedings of the Thirty-second Annual Meeting of the Phytochemical Society of North America, Miami Beach, Florida, June, 1992

Volume 28 **Genetic Engineering of Plant Secondary Metabolism**
Proceedings of the Thirty-third Annual Meeting of the Phytochemical Society of North America, Pacific Grove, California, June-July, 1993

Volume 29 **Phytochemistry of Medicinal Plants**
Proceedings of the Thirty-fourth Annual Meeting of the Phytochemical Society of North America, Mexico City, Mexico, August, 1994

Volume 30 **Phytochemical Diversity and Redundancy in Ecological Interactions**
Proceedings of the Thirty-fifth Annual Meeting of the Phytochemical Society of North America, Sault Ste. Marie, Ontario, Canada, August, 1995

Volume 31 **Functionality of Food Phytochemicals**
Proceedings of the Thirty-sixth Annual Meeting of the Phytochemical Society of North America, New Orleans, Louisiana, August, 1996

Volume 32 **Phytochemical Signals and Plant-Microbe Interactions**
Proceedings of a joint meeting of the Phytochemical Society of North America and the Phytochemical Society of Europe, Noordwijkerhout, The Netherlands, April, 1997

Volume 33 **Phytochemicals in Human Health Protection, Nutrition, and Plant Defense**
Proceedings of the Thirty-eighth Annual Meeting of the Phytochemical Society of North America, Pullman, Washington, July, 1998

Volume 34 **Evolution of Metabolic Pathways**
Proceedings of the Thirty-ninth Annual Meeting of the Phytochemical Society of North America, Montreal, Canada, July, 1999

recent advances in phytochemistry

volume 34

Evolution of Metabolic Pathways

Edited by

John T. Romeo
University of South Florida
Tampa, Florida

Ragai Ibrahim
Concordia University
Montreal, Canada

Luc Varin
Concordia University
Montreal, Canada

and

Vincenzo De Luca
University of Montreal
Montreal, Canada

2000

PERGAMON
An Imprint of Elsevier Science

Amsterdam - Lausanne - New York - Oxford - Shannon - Singapore - Tokyo

ELSEVIER SCIENCE Ltd
The Boulevard, Langford Lane
Kidlington, Oxford OX5 1GB, UK

First edition 2000

Library of Congress Cataloging in Publication Data
A catalog record from the Library of Congress has been applied for.

British Library Cataloguing in Publication Data
A catalogue record from the British Library has been applied for.
Transferred to digital printing 2005
Printed and bound by Antony Rowe Ltd, Eastbourne

ISBN: 0 08 043860 1

⊗The paper used in this publication meets the requirements of ANSI/NISO Z39.48-1992 (Permanence of Paper).

Dedicated to the memory of

Dr. Arthur C. Neish (1916-1972)

PREFACE

Over the past few years the Phytochemical Society of North America has devoted its annual symposia to topics dealing with the biochemistry, molecular biology, chemical ecology, and the nutritional and medicinal value of natural plant products. Recent developments in the cloning of genes that encode enzymes of secondary metabolism and the ongoing efforts in sequencing of the *Arabidopsis* genome provided an incentive to devote the 1999 meeting to the functional diversity and evolution of these metabolites and the pathways involved in their biosynthesis. This symposium, entitled "*Evolution of Metabolic Pathways*", was organized by Vincenzo De Luca, Ragai Ibrahim, and Luc Varin, and was held at Concordia University, Montreal, July 10-13, 1999.

This volume is dedicated to the memory of Dr. Arthur C. Neish (1916-1972), a Canadian scientist who in the early 1950's and 1960's pioneered in the synthesis and use of radiolabeled precursors, and laid the foundations of our contemporary knowledge of the biosynthetic pathways of plant natural products. Additionally, his legacy is being honored by the inauguration of a "Young Investigators Minisymposium". The subject of the first of these, "Biochemistry and Molecular Biology of Brassinosteroid Hormones", has been integrated into this volume. The papers presented here are grouped under five major topics: 1. Role of secondary metabolites in evolution; 2. Evolutionary origins of polyketides and terpenes; 3. Roles of oxidative reactions in the evolution of secondary metabolism; 4. Evolutionary origin of substitution reactions: acylation, glycosylation, and methylation; and 5. Biochemistry and molecular biology of brassinosteroids.

The role of natural plant products in evolution is discussed in a broad perspective. Jarvis focuses on polyketide antibiotics, terpenoids, and polyphenols as examples. He hypothesizes that transposition events proximal to gene clusters that encode biosynthetic enzymes played a fundamental role in the evolution of plant secondary metabolism. Stafford deals primarily with a review of phenolic compounds, including soluble esters, hydrolyzable and condensed tannins, and pigments and lignans as biochemical end products of evolutionary trends, and their distribution among plant groups. However, it is difficult to pinpoint the missing links among these taxa due to our scant knowledge of the biochemistry and molecular biology of lower plants.

The evolutionary origin of chalcone synthase, the prototype of plant polyketide synthases, is discussed by Schröder in relation to other members of this enzyme family that includes stilbene synthase, acridone synthase, and a pyrone synthase found among a few other chalcone synthase-related sequences in some bacteria. This ubiquitous enzyme reaction predated the evolution of land plants, where gene duplication and mutation have resulted in much functional diversity. Tsantrizos and Yang show how microorganisms can modulate their multifunctional polyketide synthases and peptide synthetases to produce structurally-diverse, value-added pharmaceuticals. They discuss the implication of the use of genetically engineered mutants in altering the biosynthetic pathways and structure of the metabolites produced. Bohlmann *et al.* discuss the evolution of terpenoid biosynthesis in plants in relation to several factors: biosynthesis of isoprenoid precursors in plastids and cytosol; sequence relatedness of linear prenyltransferases and the extensive terpene synthase (cyclase) family; constitutive vs. inducible synthesis of these metabolites; and the accumulation of compounds in specialized anatomical structures *vs.* their emission as volatiles. Such perspectives shed new light on the evolutionary origin of these enzymes and their metabolites.

Several steps in the biosynthesis of secondary metabolites involve oxidative reactions that are catalyzed by a large, versatile group of cytochrome P450 enzymes. The diversity of types of reactions they catalyze and the multiplicity of the substrates they recognize are reviewed by Kahn and Durst. They emphasize that the structural and functional diversity of the 240 known members of the superfamily are phylogenetically related to those already cloned from bacteria, fungi, and animals. In the following paper, Møller *et al.* review the isolation, reconstitution, and cloning of two, novel multifunctional P450s from manihot that are involved in the biosynthesis of dhurrin, the 'cyanide-bomb' produced by cyanophoric plants. These clones, together with one encoding the cytosolic UDP-glucosyltransferase, were used to reconstitute the complete pathway for the biosynthesis of the cyanogenic glucoside. Both transgenic tobacco and *Arabidopsis*, containing the P450 genes, are cyanophoric, indicating their capability to synthesize and accumulate "un-natural" natural metabolites. The other class of oxidative reactions, catalyzed by the 2-oxoglutarate-dependent dioxygenases, involves the hydroxylation/desaturation of several types of metabolites, including flavonoids, alkaloids, gibberellins, and ethylene. Prescott reviews these enzymes and points out their flexibility, which may have aided the elaboration of various biosynthetic pathways. She suggests that dioxygenases are a major driving force in the evolution of metabolic pathways, an attribute that can be harnessed to produce new metabolites by mutant enzymes.

The diversity of natural plant products is the result of a variety of enzymatic substitution reactions. St-Pierre and DeLuca review the acyltransferase superfamily of enzymes that utilizes diverse coenzyme A esters as substrates. This family shares a highly conserved histidine-containing active-site motif. These multifunctional proteins seem to be evolutionarily related to other enzymes, such as those involved in the synthesis of epicuticular wax (*cer2* and *glossy2)*, the chloramphenicol *O*-acetyltransferase family, as well as the dihydrolipoamide acyltransferase and carnitine acyltransferase involved in primary metabolism. Our current knowledge of the biochemistry, molecular biology, and functions of glycosylation is reviewed by Vogt. These enzymes constitute a superfamily of modifying enzymes involved in late steps in the biosynthesis of natural products. While all glycosyltransferases have several consistent properties in common, cloning and expression studies have defined high substrate specificities towards the sugar receptor. Dendrograms deduced from amino acid sequence data reflect this. Ibrahim and Muzac discuss the functional diversity and evolution of methyltransferases, an extensive family of enzymes that catalyze the methyl group transfer from *S*-adenosyl-L-methionine to the *O, C, N,* or *S*-atoms of a vast variety of acceptor molecules. In contrast with the *C, N,* and *S*-methyltransferases, which are not evolutionary related, the *O*-methyltransferases are highly conserved and have probably evolved from a common ancestral gene.
Finally, the phytochemical, biochemical, and molecular aspects of brassinosteroids, the most recently recognized class of plant hormones are discussed. Distribution and methods used for their synthesis and structure elucidation are reviewed by Schmidt. They appear to be ubiquitous in different plant organs and various taxa. However, studies of their biochemistry and their role in plant physiology present a real challenge to plant scientists, mainly because of low concentrations and the difficulty of their isolation from natural sources. Wang and Chory describe a combination of genetic, biochemical, and molecular approaches to identify the components of the complex interaction between light and brassinosteroid signal transduction pathways involved in plant developmental processes. The modulation of brassinosteroid biological activity by enzymatic sulfonation is discussed by Marsolais and Varin. They report on a member of the *Brassica* gene family that encodes an enzyme that catalyzes the *O*-sulfonation of brassinosteroids and mammalian estrogenic steroids. They correlate the similarity between brassinosteroids and estrogen hormone inactivation to the degree of functional conservation between the plant and mammalian homologs.

We greatly enjoyed working with the authors of these chapters and with our new publisher, Elsevier in the production of this volume. The special ambience of Montreal, the French capital of North America, will long be remembered by meeting participants and contributors alike. JTR expresses special appreciation to Dawn McGowan, Darrin King and Alicia Wassenberg for their dedication, patience, and goodwill throughout the long process of bringing the book to fruition.

John T. Romeo, University of South Florida
Ragai Ibrahim, Concordia University
Luc Varin, Concordia University
Vincenzo DeLuca, University of Montreal

CONTENTS

Chapter One

THE ROLE OF NATURAL PRODUCTS IN EVOLUTION

Bruce B. Jarvis

Department of Chemistry and Biochemistry
University of Maryland
College Park, MD 20742

INTRODUCTION

An earlier paper on this subject presented arguments that "natural products are chemicals released within a system by one component which conveys information and instructions to another component(s) within the system. They are a natural outgrowth and consequence of an increase in complexity, and they are a critical part of the chemical "glue" that holds systems together. That natural products usually tend to be small organic molecules is a natural consequence of the functions they serve: messengers that must survive long enough to shuttle between the various components of the system."[1] Much of the basis for the arguments presented in that paper was rooted in chemical ecology and the many roles played by natural products in mediating and modulating interactions among organisms. Numerous examples were provided on the effects of specific natural products in biological systems in support of the hypotheses that: 1) extant natural products

Evolution of Metabolic Pathways – J.T. Romeo et al. (Editors)

arose from mutations of enzymes involved in the synthesis of primary metabolites; 2) the natural products that emerged over time were modified by variants of other enzymes involved in primary metabolism in a gene-for-gene fashion with other biological components of the ecosystem; 3) under the pressure of natural selection, natural products evolved as messages, defined in broad terms, *i.e.* come hither, don't eat me, germinate, divide, attach, disengage, etc.; 4) horizontal transfer of genes (that code for biosynthetic pathways) between species occurred and conferred selective advantages to organisms; 5) the actions of natural products in human physiology are often not relevant to their function(s) in the producing organisms.

This paper will delve more deeply into the origins of natural products, with an emphasis on the fundamental problem of just how such an extraordinary and bewildering array of structure types has arisen through evolutionary processes. Much of the discussion will be of a general nature but may provide background for the other papers of this volume. Throughout, an emphasis will be placed on the flow of information and how the machinery for the biosynthesis of natural products supports that information flow. In fact, sound arguments can be made that information flow is a prime mover in life.[2]

Another aspect to be emphasized is the contrast between primary metabolism and what some call secondary metabolism. This distinction is often blurred and difficult to make,[3] and further, such usage implies that "secondary metabolites" (natural products) are not of importance to the producing organism. In fact, natural products are quite important, and use of the term, secondary metabolite, will be avoided. Having said this, data will be presented to show that the gene organization of these metabolic pathways in general is usually not ordered in the same manner as the primary metabolic pathways in eucaryotes, *i.e.* the pathway genes for natural products occur in clusters; whereas, the genes involved in primary metabolic pathways are spread about the genome (even on different chromosomes) and appear to have no spatial order.

Background

All processes are driven by changes in free energy, which in turn is made up of two parts, enthalpy and entropy. Since all systems tend to go to their highest states of randomness (increase in entropy), naturally occurring ordered states must arise through enthalpic compensation. Structures that self-organize (crystals, double helical DNA, etc.) do so because of compensating energy factors (*e.g.* favorable intramolecular attractions such as hydrogen bonding); whereas, systems that appear to "spontaneously" self-organize (*e.g.* life) may well involve self-ordering components but also require an external source of energy to compensate for unfavorable entropy.

Over fifty years ago, information theory showed that entropy and information are inversely related such that an increase in information is equivalent to a decrease in entropy.[2] Thus, the agents of information (in our case, natural products) are employed by nature to overcome the natural tendency toward disorder, and, fueled by the energy from the sun (or other less common thermal sources of energy here on earth), these small molecules have become indispensable for life on this planet.

In attempting to evaluate the role of natural products in living systems today and then using that knowledge as a basis for understanding the roles played in evolution, we face formidable obstacles. Central to the problem is the complexity of the cell -and here complex systems are those that arise spontaneously through the interconnection of cycles whose processes are controlled by a complicated series of feedback mechanisms, both positive and negative.[1] The hallmark of these systems (*e.g.* developmental embryology and neural networks) is that they are not fully explained by an understanding of their component parts, *i.e.* they seem not to be readily analyzable by reductionist processes. If the systems become extremely complex (*e.g.* life itself), identifying all the components becomes near impossible. To further complicate matters, some complex systems appear to require initial conditions or agents that have long disappeared from the system but were necessary for the system to evolve to its present state -such situations are common for ecosystems. Current interest in the science of complexity is evident from the numerous recent books on the subject, and *Science* devoted a significant portion of the April 2, 1999 issue to this subject.

The central unit of life is the cell. Cells first appeared on earth over 3.5 billion years ago and remained in the unicellular state for some 2.5 billion years. Much of the cellular machinery of today's procaryotes is believed to have been in place very early, although we have no direct knowledge of this. What is clear, however, is that there must have been a substantial barrier to the formation of multicellular organisms since these "cooperating" cellular systems did not appear until about a billion years ago. At the heart of this resistance to cooperate were various barriers, including cell walls and membranes and, perhaps most challenging, the specific cellular information engines, the transcription-translation machinery of the cell that required robust protection against external factors. Even today, we can see a reluctance to communicate between cells, manifested in the horizontal transfer of genes among the procaryotes. Such transfers have occurred and continue to occur with regularity for operational ("house keeping") genes, but such horizontal gene transfers are far less common with the genes that code for the elements involved in the transcription-translation machinery.[4]

THE MACHINE

An inherent problem with analyzing the role(s) played by natural products in evolution is their involvement in many interconnecting processes -it is difficult to know where to start deconstructing the system in a reductionist manner. One place to start is at the level of molecular interactions of natural products with the cellular machinery. At first glance, this would appear hopeless, since there are countless natural products produced by a myriad of organisms that seem to elicit any number of unending physiological responses. If one takes the view that natural products are bits of information, however, then an analysis becomes more manageable.

Although the panorama of molecular interactions of natural products with cellular targets is vast, the number of fundamental mechanisms for these interactions is limited. There are a few basic mechanisms for intercellular communication, the most relevant to natural products being 1) binding to intracellular receptors after dissolving in and passing through the lipophilic membrane, 2) binding to and affecting the function of protein-lined membrane pores (*e.g.* sodium channels), and 3) binding to a cell surface receptor initiating the transduction of an intracellular signal.[3]

Here the term "mechanism" is not meant in the sense of describing the detailed molecular behavior of these compounds as they affect biological systems (which collectively are far beyond the means of anyone to describe in this limited space). Rather, "mechanism" relates to the general properties of the information system and how small molecules can effect such dramatic responses in cells. We will use steroids as the example, noting that much of what can be ascribed to steroid hormones can also be said about many natural products. For example, cortisone (produced in the adrenal glands) elicits production of different sets of proteins, depending upon which type of cell it enters. This is exactly what some flavonoids do, depending upon which tissue types they interact with in plants.[5] The fact that we may view cortisone as an important physiological hormone and the plant-derived flavonoids as natural products does not alter the basic fact that they are both operating functionally as units of information to turn on specific biochemical circuits in specific cells. The use of the term "circuits" was not used in an off-hand way. Steroid hormones (as well as many natural products) function as on-off switches -like transistors in integrated circuits that are yes/no (1/0 in the parlance of digital computers) answers. But, what is the molecular question -the circuit, as it were? The ultimate question for the circuit is do you want to produce the gene product(s), or perhaps stop producing the product? Do you wish the transcription-translation machinery to be put into motion (or not, as the case may be)? Because of the complexity of the transcription-translation machinery, this would seem to be a difficult question to address. Here, however, is the beauty of the transcription-translation machine; it is a modular arrangement of proteins whose core functioning units are repeated over and over again in various transcription factors, each of which has specific receptor sites that recognize one another and recognize the steroid-bound protein that initiates the transcription at the specific gene on the DNA (Fig. 1).

The lipophilic steroid passes easily through the lipid membrane and into the cytosol where it encounters its receptor protein. This protein has three important regions: a steroid binding site, a zinc finger region that has the potential to interact directly with the DNA double strand, and a third recognition site that will eventually bind this protein to another protein, a translational factor located at the gene that will be transcribed.[2] However, in the absence of the steroid, the zinc finger, which is destined for intimate contact with the DNA, is blocked by yet another protein. The binding of the steroid to its receptor alters the conformation of the steroid receptor (an allosteric effect) such that the blocking protein falls off. The steroid-bound activated receptor molecule then travels to the nucleus where one site recognizes a DNA-bound translation factor, and the zinc finger site insinuates itself into the DNA double helix and helps twist apart the strands so that the RNA polymerase machinery can begin operation (Fig. 1).[2]

Although what follows is certainly a complicated process leading to

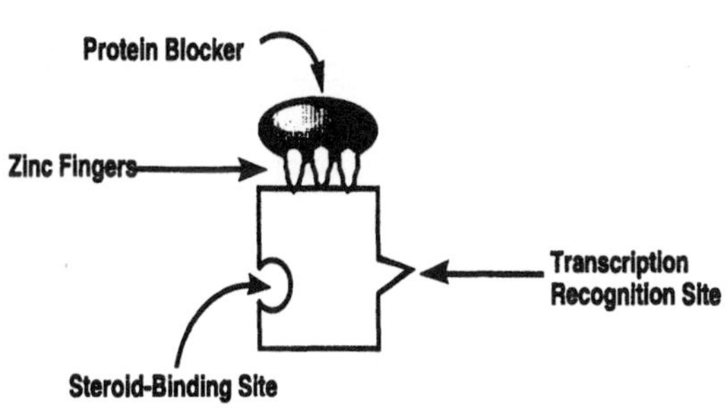

Figure 1. Schematic Representation of a Steroid-binding Receptor Protein.

eventual translation into the protein gene product, the role played by the steroid is informationally quite straightforward. It also accounts for the versatility of hormonal action as a function of tissue type: different tissues produce different steroid receptors that in turn recognize alternate translational factors located at different gene sites. Thus, cortisone receptor molecules from different tissues have similar steroid receptor sites but, upon loss of the zinc finger blocking protein (Fig. 1), plug into different transcription factors. Also built into the system is great plasticity with respect to cell differentiation and evolution. Gene duplication followed by mutation allows for all sorts of combinations and permutations such that different steroids turn on different genes.

This discussion of steroids offers the opportunity to comment on another characteristic of some natural products: the recurrence of certain structure types. Although we marvel at the seemingly endless array of natural product structure types, note should be taken that many structure types, *e.g.* steroids, are used commonly across life. Steroids and the closely related hopanes (found commonly in bacteria) are found throughout the procaryotes and eucaryotes. A few animals are unable to synthesize steroids *de novo* but can readily process food-derived steroids and even make use of such processed steroids for their own use, *e.g.* the ecdysone molting insect hormones.[6]

The triterpenoid steroids are complex molecules requiring significant investment by organisms to synthesize and process. Steroids appear to be evolutionarily ancient molecules whose fundamental structures have been maintained over a long period of time. Steroids and related hopanes perhaps arose first as compounds to provide rigidity to membranes. Their molecular lengths are about one-half the distance of the lipid bilayer, and their lipophilic nature ensures a good fit with the membrane environment. The rigidity of steroids is due in large measure to the *trans*-fused B/C ring junction found typically in steroids and hopanes (Fig. 2). Although the outer rings (A and D) may be either *trans*- or *cis*-fused, the central core of these compounds remains rigid. Further versatility arises upon elaboration of these ring systems with polar groups that are often found arrayed along one face of the molecule, conferring an intrinsic amphiphilic nature to the compounds. One can well imagine these versatile and plentiful molecules being modified and selected by nature as carriers of information for various metabolic pathways.

Biosynthesis of Natural Products

Any treatment of the biosynthesis of natural products herein must be

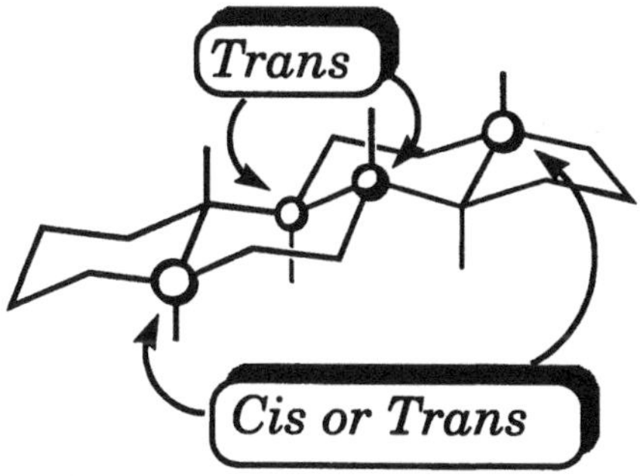

Figure 2. Stereochemical View of the Steroid Ring System.

limited because of the sheer volume of information on the subject.[7] Rather than dealing with the particulars as to how natural products are biosynthesized from a chemical point of view, this section will focus on how the enzymatic systems are ordered so as to result in the biosyntheses of large numbers of different natural products. It has long been recognized that many (but certainly not all) natural products fall into a few general classes of compounds (*e.g.* polyketides, non-ribosomal peptides (NRPs), isoprenoids, alkaloids, etc.) that arise from the condensations of small organic units. However, unlike the oligiomerization products of nucleotides (DNA and RNA) and amino acids (proteins), natural products are further elaborated (tailored) by numerous reactions: oxidations, reductions, cyclizations, eliminations, rearrangements, etc. Rather than being concerned with the specific chemistry involved in these processes, I want to focus on how the biosynthetic machinery is organized to carry out these highly coordinated syntheses. Since all (or virtually all) of the biosynthetic steps are catalyzed by enzymes, we eventually will need to turn our attention to the structural arrangements of the biosynthetic genes and how the organizational properties of these genes support the biosynthesis process. We will consider only a few classes of natural products, but the underlying principles described will be broadly applicable to the biosynthesis of natural products in general.

Polyketides (and related fatty acid derivatives) present an enormous variety of compounds, *e.g.* fatty acids, prostaglandins, macrolide antibiotics, anthroquinones, polyethers, numerous phenolics, flavonoids, etc. It came as something of a surprise to find that the principle of "one reaction - one enzyme" does not typically apply to the biosynthesis of polyketides.[8] Rather, in bacteria, polyketide syntheses can be carried out either by an enzyme that performs multiple enzymatic steps [type I polyketide synthases (PKS)] or by a few enzymes that are able to repeat catalytic steps iteratively on a growing polyketide chain (type II PKS). In fungi, the PKS systems often appear to be hybrids: iterative type I PKSs. These fungal PKS systems can give rise to polyketides similar to those produced by bacteria (*e.g.* polycyclic aromatic compounds) or more complex polyketides such as lovastatin. The type I PKSs are large multi-domain enzymes that carry out a number of enzymatic processes, including acyl transfers, keto condensations (*via* an acyl carrier domain), dehydrations, enol and keto reductions, and upon departing the PKS, the chain is often cyclized.[9] The process can be compared with a modern assembly line where the individual domains of the PKS each have specific functions that are dictated by the structural nature of each domain. Within a given PKS, there may be many modules each containing their own subset of domains. Further details of this can be found in the chapter by Tsantrizos and Yang.

The non-ribosomal peptides (*e.g.* penicillins and cyclosporins) are biosynthesized in a manner reminiscent of the polyketides.[8,9] Microorganisms employ non-ribosomal peptide synthases (NRPSs) to construct these polypeptides. The NRPSs have modular units within which reside various catalytic domains that have functions analogous to those found in the PKSs. The NRPSs have peptidyl carrier protein domains (in place of the acyl carrier domains) but lack the reduction and dehydration domains. However, the NRPSs typically have epimerase domains (responsible for the presence of many of the D-amino acids in antibiotics) and methyl transferase domains (for N-methylation) that are missing in the PKSs.[8,9] Further details can be found in the chapter by Tsantrizos and Yang.

The polyketide lovastatin (mevinolin) is a medically important antihypercholesterolemic produced by the fungus *Aspergillus terreus*. The molecule consists of a methylated nonaketide esterified by a methylated diketide unit at C-8 (see Fig. 3). As is typical for natural product biosynthetic genes in microorganisms, the lovastatin biosynthetic genes are clustered.[10] These genes (as many as 18, although functions have not been assigned to all) are found in a 64 kb region and have within the cluster 3 PKS genes. The Vederas-Hutchinson groups showed that two of these PKSs were sufficient to produce the two polyketide units of lovastatin, a nonaketide synthase, LNKS (from the lovB gene), and diketide synthase, LDKS (from the lovF gene), but that a third, much smaller PKS (from the lovC gene) was required for the correct processing of the nonaketide unit produced by the lovB PKS (LNKS).[10] Together, LNKS and the lovC PKS are sufficient to produce dihydromonacolin L in a transformed culture of *A. nidulans* (Fig. 3). The requirement for both LNKS and the lovC PKS to produce the polyketide is an unusual finding since PKSs are normally self-sufficient. The 363-amino acid protein encoded by the lovC gene shows high similarity to the enol reductase (ER) domains of PKSs. Furthermore, the PKS encoded by lovB (LNKS) seems to lack ER activity, which appears to be assumed by the lovC PKS, interacting closely with LNKS during the construction of the nonaketide unit.

There are several unusual features of this system in addition to the one noted above. Both LNKS and LDKS PKSs have methyl transferase domains, structurally related to those found in the non-ribosomal peptide synthases (NRPSs). Furthermore, in LNKS there is a condensation domain located at the COOH-terminus that is similar to ones found in NRPSs and heretofore reported in only one other PKS.[11] Finally, LNKS appears to have a domain that catalyzes an intramolecular Diels-Alder reaction, a cyclization not known to be catalyzed by any other PKS.[10]

LovC may well have arisen through duplication of a segment of an ancient lovB, but that does not account for the development of the methyl transferase and condensation domains of LNKS. From an evolutionary standpoint, the lovB and lovC genes appear to have acquired elements of an NRPS gene(s). The process by which this acquisition may have occurred will be discussed later.

One of the theses presented earlier was that the enzymes employed in the biosynthesis of natural products can trace their origins to primary metabolic enzymes.[1] Fungal PKSs are closely related to eucaryotic fatty acid synthases (F ASs),[8] and, on rare occasions, FASs are found within the gene clusters of polyketide natural products.[12] However, the lovastatin biosynthetic pathway appears to be highly evolved, presumably from a much simpler pathway in the distant past, and we can only speculate on the origins of this complex pathway. Are there natural products whose biosynthetic pathways are considerably less complex and of more recent origin? A good candidate is the biosynthetic pathway that leads to the DIBOA and DIMBOA hydroxamic acids (Fig. 4) that are employed by grasses (Gramineae) in their defense against pathogens.[13] The pathway to DIBOA starts with the conversion of indole-3-glycerol phosphate to indole, followed by the oxidation of indole by a series of cytochrome P-450-dependent monooxygenases. The first step is catalyzed by the BX1 enzyme, and the oxidation of indole to DIBOA is catalyzed by enzymes BX2-BX5; the genes that encode the BX1-BX5 enzymes in maize are clustered.[13] The exon sequence in the BX1 gene is homologous with that of tryptophan synthase α (TSA). Tryptophan synthesis in bacteria (and most likely in plants as well) is carried out by an $\alpha\alpha\beta\beta$ complex of tryptophan synthase α (TSA) and tryptophan synthase β (TSB) acting on indole-3-glycerol phosphate (Fig.4). This complex does not accept

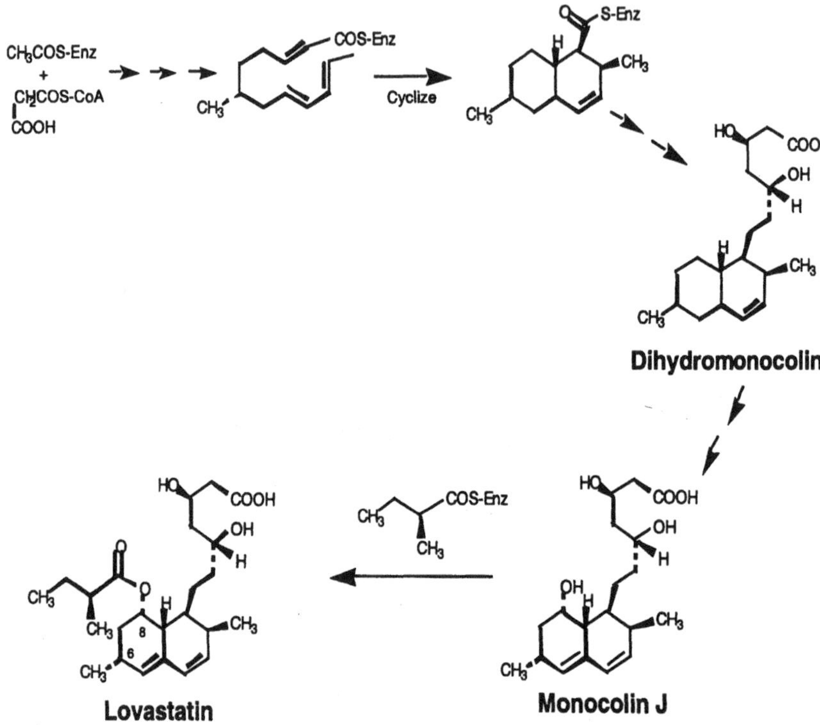

Figure 3. The Lovastatin Biosynthetic Pathway.

exogenous indole, whereas, the BX enzymes efficiently convert exogenous indole to DIBOA. Thus, maize appears to have two independent indole-generating systems employing two independent TSA genes, one committed to tryptophan synthesis, the other to DIBOA synthesis. Remarkably, the level of DIBOA in maize seedlings is 20-30 times that of tryptophan.[13]

The BX2-BX5 enzymes share 45-60% homology and yet are highly specific in that they accept as substrates only the compounds indicated in Figure 4.[13] The BX cluster can easily be envisaged as arising through duplication and movement of the TSA gene to a site adjacent to an ancestral cytochrome P-450-monooxygenase. This cytochrome P-450-monooxygenase gene then underwent a series of duplications that were selected for their phenotypic characteristics. It would be interesting to compare the amino acid sequence relationships of the genes in the BX clusters of the more distantly related grasses to those of rye, wheat, and maize.

Perhaps the most abundant of the natural products are the terpenoids and related compounds that constitute 20% or more of the reported >100,000 natural products reported in the literature. Although many terpenoid compounds are known to serve diverse roles in the life of the producing organisms, many of which are essential, overall, we know little about the functions of most of these metabolites.[14]

Terpene biosynthesis starts first with the generation of isopentenyl diphosphate (isopentenyl pyrophosphate, IPP), followed by the condensation of IPP with its allylic isomer, dimethylallyl diphosphate (DMAPP, formed via IPP isomerase), to give geranyl diphosphate (GPP),[15] the precursor to the monoterpenes. Further prenylation of GPP followed by elongation/cyclization reactions gives rise to the more complex terpenoids and carotenoids (Fig. 5). Two of the basic processes involved in terpenoid biosynthesis are chain elongation (prenylation, catalyzed by terpene transferases) and cyclization (catalyzed by terpene cyclases, usually referred to as terpene synthases). The bewildering array of structure types resulting from these processes (especially cyclizations) can be traced to the versatility of the various enzyme systems operating through reactive intermediate carbocations.[16]

From an evolutionary view, how did such diversity arise? This question is especially provocative when ones considers that there is little similarity in the amino acid sequences of the various terpene synthases, except within the plant monoterpene synthases.[16,17] However, recent X-ray crystal data for three terpene cyclases have shown that, even though there is little homology among them, they share a common molecular topology.[18-20] The terpene synthase enzymes have similar structure motifs, consisting of 10 to 12 mostly antiparallel a helices that form a large active site cavity (isoprenoid synthase fold).[21] Facing into the cavity is a conserved AspAspXXAsp motif that appears to align a Mg^{2+} cation for complexation with the diphosphate leaving group. Farnesyl diphosphate synthase has two of these aspartate-rich motifs, apparently one for each of the diphosphate units of the molecules involved in the elongation reaction.[21] A further common feature in these enzymes is the presence of a high density of aromatic amino acids that are believed to help stabilize (and thus help direct the formation of) the intermediate carbocations. These data support the suggestion that terpene synthases share a common ancestral origin.[20]

A second recent (and quite unexpected) finding in terpene chemistry has been the discovery that there is a second pathway to IPP, in addition to the mevalonate pathway, the deoxyxylulose pathway.[22] This was reported first in

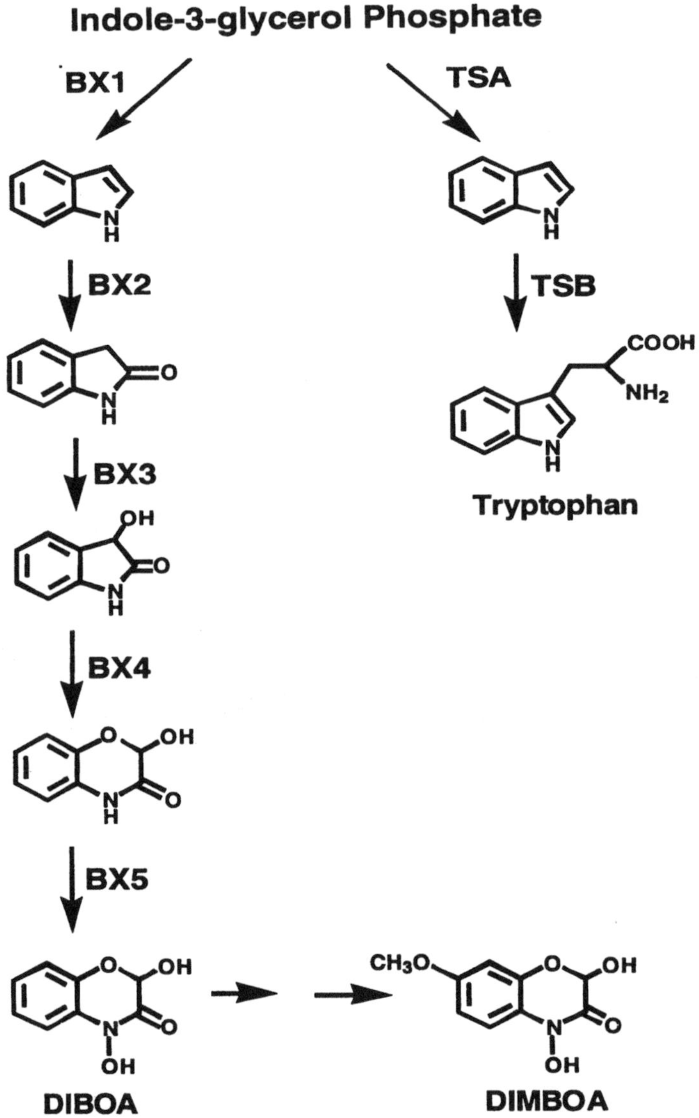

Figure 4. The DIBOA and DIMBOA Biosynthetic Pathways.

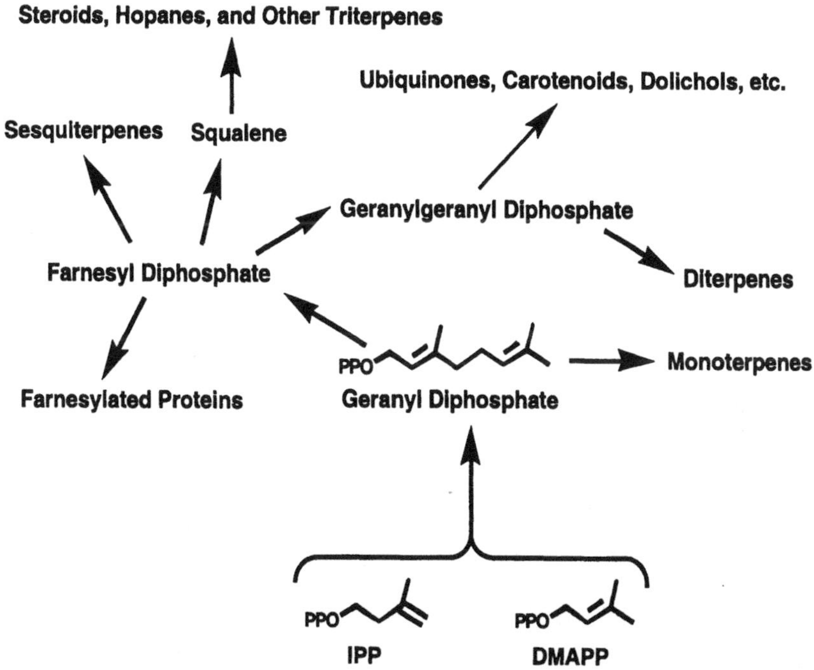

Figure 5. The Biosynthesis of Terpenes.

bacteria and green algae, but subsequently has been shown to be operative in higher plants,[17] but apparently not in fungi and yeast.[23] This pathway is discussed in detail elsewhere in this book (Bohlmann, Chapter 5). From an evolutionary standpoint, terpene synthesis in higher plants is the most interesting since there appears to be a clear division of labor; monoterpene and diterpene syntheses occur in plastids, whereas sesquiterpene and triterpene syntheses occur in the cytosol, although this division of labor is not always strictly followed.[17] The monoterpene and diterpene synthases, encoded by nuclear genes, are metabolically channeled (employing amino-terminal plastid-targeting sequences in the preterpene synthase) to the plastids where they use IPP produced by the alternate deoxyxylulose pathway. This behavior is consistent with the evolutionary history of plastids in plant cells that are believed to have arisen from capture of a cyanobacterium cell by an ancient eucaryotic cell.[24]

Organization of the Genome

In procaryotes, the genes for metabolic pathways are typically organized into operons and, upon transcription, give polycistronic m-RNA. Such unidirectional, organized, and clustered genes would appear to make good engineering sense. One starts first with the control genes that promote the transcription of the first gene in the metabolic pathway, followed by transcription of the second gene, then the third, etc., to yield the polycistronic transcript. So the genes coding for the enzymes of the metabolic pathways in procaryotes are often transcribed into a long string of connected m-RNA that is then translated sequentially into the various enzymes that catalyze the steps of the metabolic pathway. Not only are the genes aligned sequentially (product of gene #1 catalyzes step # 1, gene #5 catalyzes step #5, etc.), but they also are usually aligned in the same direction along the DNA strand. One of the striking differences between this arrangement in procaryotes and that found in eucaryotes is that typically, not only are the genes of the metabolic pathways in eucaryotes not organized in clusters, but they often are located on different chromosomes. Thus, unlike the metabolic genes of procaryotes, those of eucaryotes are not organized spatially, raising interesting questions about the control of gene expression in eucaryotes, a discussion that is beyond the scope of this review.

As noted earlier, a distinction has been made between "primary" and "secondary" metabolism, although in practice, this distinction is often difficult to make.[3] Perhaps a better usage would be dispensable metabolic pathway, rather than secondary metabolic pathway.[25] Dispensable metabolic pathways lead to products that are not required for the internal economy of the organism,[26] and are expressed usually under suboptimal growing conditions. These pathways are usually not required for growth or if required, only under a limited range of conditions.[25]

What is at issue here is not the differences between genomic arrangements in procaryotes vs those found in eucaryotes but whether the primary metabolic genes are organized differently from those genes employed in dispensable metabolic pathways -the limited data suggest that they often are, at least in eucaryotes. In fungi, the biosynthetic genes coding for the production of natural products are typically clustered.[25] However, although this is the common situation, it is by no means universal. For example, the genes coding for the penicillin-cephalosporin antibiotic complex, the trichothecenes, and the aflatoxin mycotoxins are clustered.[25,27] However, the *mel* genes for melanin production[28] in *Alternaria alternata* are clustered,[29] but the corresponding genes in *Colletotrichum lagenarium*[30] and *Magnaporthe grisea*[31] are dispersed. Interestingly, if the clustered *mel* genes in *A. alternata* are disrupted, the fungus remains pathogenic, but disruption of the *mel* genes in *C. lagenarium* and *M. grisea* results in fungi that are no longer pathogenic.

The penicillin-cephalosporin antibiotics are particularly interesting because they are the only natural products synthesized by both procaryotes and eucaryotes for which we have genomic information. The biosyntheses of these antibiotics start with the synthesis of the tripeptide, d-(L-a-aminoadipyl)-L-cysteinyl-D-valine (ACV) that is then cyclized to isopenicillin N (IPN), at which point the penicillin and cephalosporin biosynthetic pathways diverge. In *Penicillium chrysogenum*[32] and *Aspergillus nidulans*,[33] the penicillin pathway genes are clustered, as are the cephalosporin pathway genes in *Acremonium chrysogenum*. The ACV synthase and IPN synthase genes for these pathways are closely linked and have the same orientation. However, in *Acremonium chrysogenum*, although the ACV synthase and IPN synthase genes are linked and oriented as in the fungi above, the expandase/hydroxylase and acetyltransferase genes involved in the latter stages of cephalosporin biosynthesis are linked to one another but on a different chromosome from that which contains the ACV synthase and IPN synthase genes.[34]

As expected, these genes in procaryotes are clustered, and, based on the high degree (ca. 60%) of homology between the IPN synthase genes in the above fungi and those found in procaryotes, it has been suggested that these pathway genes arrived in a eucaryote *via* horizontal gene transfer from a procaryote.[35,36] However, this mechanism has been challenged because of the lack of additional data.[37]

To further complicate matters, the cluster of genes in the procaryotes that code for erythromycin biosynthesis are not transcribed into polycistronic m-RNA. The genes give monocistronic m-RNA, and in fact, the erythromycin genes are transcribed in opposite directions in the cluster, *i.e.* the genes are located on different strands of the genomic DNA.[38] This sort of variation of the genes in a cluster from one strand of the DNA to the other is characteristic of eucaryotes but much less so of procaryotes.

A typical difference between the genomes of procaryotes and those of eucaryotes is the apparent lack of organization in the latter, *i.e.* the various genes for a given metabolic pathway are usually dispersed about, even to the extent of being located on different chromosomes. Even in the cases of natural product metabolic pathways, however, where the genes are clustered, there may be involvement of pathway genes far removed from the loci of the cluster. Within antibiotic/toxin gene clusters are found genes responsible for resistance to the toxic effects of the natural products to the producing organism. However, in *Fusarium graminearium*, one of the resistance genes, 3-O-acetyltransferase gene (*tri*101),[39] is located far from the trichothecene gene cluster.[40] Present evidence suggests that this gene was acquired by *Fusarium via* horizontal gene transfer and recruited to participate in trichothecene biosynthesis.[40] In *F. graminearium* where the intracellular trichothecenes are 3-O-acetylated (and thus rendered relatively non-toxic), the expression of *tri*101 is triggered by sublethal levels of 3-hydroxylated trichothecene toxins. In *F. sporotrichioides*, which also has a functioning *tri*101 gene located in the same genomic position as is the *tri*101 gene in *F. graminearium*, the *tri*101 gene appears not to be up-regulated by the 3-hydroxylated trichothecene toxins, *e.g.* T-2 toxin, the principal trichothecene produced by *F. sporotrichioides*,[40] although the possible role of esterase(s) in the biosynthetic pathway complicates the interpretation of these data.[41] Recent work has shown that the disruption of the *tri*101 gene in *F. sporotrichioides* blocked the biosynthesis of T-2 toxin at the isotrichodermol to acetylisotrichodermol (isotrichodermin) step; furthermore, the resulting strains remained insensitive to the toxic effects of the trichothecenes.[41] Thus, in the trichothecene-producing *Fusarium* species, 3-O-acetyltransferase (from *tri*101) can serve roles in both resistance and as a catalyst in a biosynthetic step.

The above example is unusual in having a resistance gene located outside the cluster. Even with all the constant reshuffling of genes as a result of persistent breaking and rejoining of DNA, natural product metabolic clusters stay together; there must be a natural advantage to maintaining these clusters.[42]

The natural product biosynthetic enzyme systems are highly variable, ranging from synthases that are substrate-specific to ones that will accept a wide range of structurally related substrates. Some enzymes convert a single substrate into several products (*e.g.* monoterpene synthases).[17] In some cases, small changes in amino acid sequences in an enzyme can result in the substrate undergoing a significantly different type of reaction.[43] An example of this is found in the plant fatty acid synthases where hydroxylase and dehydrase activities can be exchanged in an enzyme by changing only a few of the amino acids of the enzymes.[44] Another variation is found in the terpene synthases where the enzymes may vary greatly in amino acid sequences, but where many terpene synthases convert the same substrate into quite different products. Here, the overall molecular topology of the various terpene synthases is conserved, but the extensive variation in the primary protein structures (through mutation / selection) leads to enzymes that give rise to different products from the same substrates.[18-21] Another variation is the modular system found in the polyketide and non-ribosomal peptide synthases. Single giant proteins with multiple catalytic domains, repeated in modules, oligiomerize and transform small units to build up complex natural products (see chapter by Tsantrizos and Yang). These varied enzymatic properties ensure overall catalytic plasticity in the world of natural products.

Even within metabolic pathways that are dedicated to producing natural products of a single class, however, two (or more) separate gene clusters can act in concert, *e.g.* the cephalosporin pathway in *Acremonium chrysogenum*.[34] Another recently uncovered example of this is the biosynthesis of gibberellins in the fungus, *Gibberella fujikuroi*. Gibberellins are particularly interesting since they are produced both by fungi (where they were first discovered) and plants, where they function as plant growth hormones.[45] Although it has been suggested that plants acquired the gibberellin biosynthetic genes from a fungus (or *vice versa*),[1] emerging evidence is suggesting that convergence is a more likely explanation.[46] What is clear, however, is that there is a cluster of five (or more) linked genes in *G. fujikuroi* whose expression patterns suggest that they are under control of a positive acting pathway-specific regulator that is presumably also a member of this cluster.[47] The five linked genes consist of three cytochrome P450 monooxygenases, a geranylgeranyl cyclase (*cps*), and a geranylgeranyl diphosphate synthase (*ggs2*).[47] In fungi, a single cyclase (*cps*, in the case of *G. fujikuroi*) transforms geranylgeranyl diphosphate to *ent*-kaurene on the way to the gibberellins; whereas, in higher plants, this process involves a two-step cyclization catalyzed by two separate enzymes.[48] The *ggs2* gene is particularly interesting since it is the second geranylgeranyl diphosphate synthase gene characterized in *G. fujikuroi*,[49] but the first gene, *ggs1* is expressed constitutively. Thus, *G. fujikuroi* appears to have evolved a geranylgeranyl diphosphate synthase gene (*gg2*) dedicated to the gibberellin biosynthetic pathway. This gibberellin gene cluster is responsible for the biosynthesis of the gibberellin G4, G7, G14 series,[46,47] while the oxidases that transform G7 to the gibberellin G1/G3 series are not linked to this cluster.[47] It is concluded from this,[47] and related work,[50] that the early part of the biosynthesis of the gibberellins in *G. fujikuroi* occurs in special compartments (microbodies) prior to export to the cytosol where further oxidative transformations occur.

Higher Organization

The genomes of several procaryotes and eucaryotes have been completely sequenced, and, within a few years, that of *Homo sapiens* will be completed. It seems likely that hundreds of genomes will be sequenced by the end of the first decade of the new century. What will be done with all these data when it comes to natural products? One outcome will be our ability to home in on the biosynthetic genes for natural products much more easily than we do today. At the moment, we make educated guesses as to which primers to select to pull out DNA for PCR amplification. With the complete genomic map of the organism before us, and based on homologies with known relevant biosynthetic genes from other related organisms, we will be better able to pick out these gene clusters. These "gene maps" may reveal much about genomic organizations that at the moment is obscure.

Many natural products are formed as the result of mixed pathway biosynthesis, *e.g.* terpene-polyketide, polyketide-non-ribosomal peptide, etc. The natural product systems about which we have the best genomic understanding are the result of single biosynthetic pathways (aflatoxins, penicillins, and the trichothecenes). What of the many natural products that arise through a mixing of biosynthetic pathways? What are the genomic arrangements in these systems? The

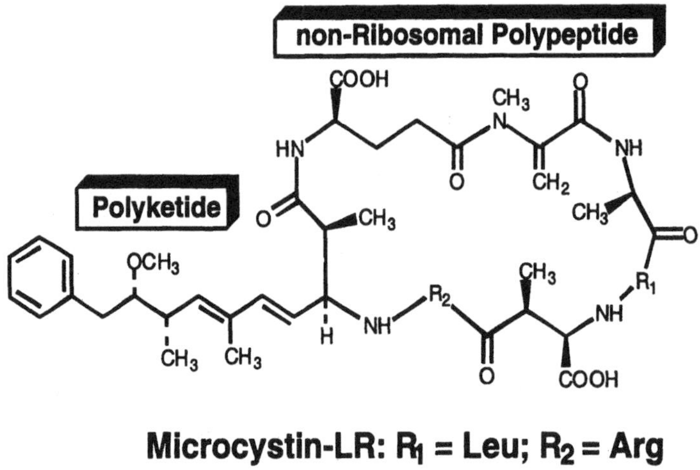

Figure 6. Structure of Microcystin-LR.

data are scant, but what is available is suggestive of yet a higher state of organization -a super cluster brought about through the action of movable genomic elements, transposons.

Microcystins are cyanobacterial toxins of mixed biosynthetic origin: non-ribosomal peptide-polyketide.[51] Recent work has shown that microcystin-LR (Fig. 6) results from the cooperation of two closely linked clusters, one that codes for non-ribosomal peptide synthase (NPS)[52] and the other that codes for a polyketide synthase (PKS). The latter is flanked by a transposase gene (Borner, personal communication), suggesting that the PKS may have been imported in the past by a transposon, which may then have lost part of its element, thus fixing the PKS and NPS clusters together. Although this is speculation, it does suggest a mechanism for the coupling of biosynthetic pathways that lead to natural products of mixed biosynthetic origin.

CONCLUSIONS

The molecular circuits of life that recognize or give rise to natural products are highly complex and interconnected with the circuits that are responsive to developmental and environmental controls.[3] Natural products often function as "transistors" to switch on or switch off these circuits.

Natural product biosynthetic pathway genes, especially those in microorganisms, are usually clustered, and their expression is often under developmental and environmental control.[3,25] Nature has an ability to generate many natural products of a mixed biosynthetic origin, but we currently know little of their biosynthetic genomic organization. Some tantalizing data suggest that the biosynthetic genes of these mixed pathway natural products are organized in "super clusters." How such clusters evolved is an interesting question. Nature being the ever-inventive agent has no doubt evolved more than one means to accomplish this. Preliminary data suggest that one organizational mechanism makes use of transposons to duplicate and move DNA segments (that code for protein domains)[53] into existing biosynthetic genes. This mechanism could move new genes into exisiting natural product gene clusters or could move an entire gene cluster of one natural product class (*e.g.*, a polyketide) close to that of another class (*e.g.*, a terpene), such that the expression of the "supercluster" becomes coupled and gives rise to a hybrid-type natural product.[53,54]

Another recurring theme that appears common is the recruitment of (or the seeding by) genes originally involved in primary metabolism for the natural product biosynthesis gene clusters. Examples noted above include:

a) FAS genes in *Aspergillus*, one dedicated to sterigmatocystin biosynthesis and the other(s) to primary metabolism.

b) The TSA genes for indole biosynthesis in maize, one for DIBOA biosynthesis and the other for tryptophan biosynthesis.

c) the two diterpene synthases *ggs* genes in *Gibberellin fujikuroi*, one expressed constitutively, the other dedicated to gibberellin biosynthesis.

These are likely but the tip of the iceberg.

The clustering of natural products biosynthetic pathway genes is an obvious organizational feature. However, on closer examination, many of these clusters and their gene products appear to have further pathway connections that lie outside the cluster but nonetheless are coordinately linked in the biosynthesis process. How all these elements are organized and managed will no doubt be addressed in the coming years.

A central challenge with the gene sequence data that are being generated for whole organisms is to understand the functions of the gene products. Significant strides have been made recently in this area that give hope that this daunting problem may be manageable.[55] The functional characterization of unknown (but apparent) natural product gene clusters may be within our reach within a few years. Further impressive advances in the identification and characterization of the protein receptors for natural products also are being reported.[56] Related to these issues are the problems associated with the characterization of natural products genes from "unculturable" microorganisms. Since >99% of soil microorganisms are believed to be unculturable,[57] this represents a great potential store of new natural products. Recent efforts to characterize natural product biosynthetic genes from these organisms hold great promise for generating a multitude of new and interesting natural products.[57]

REFERENCES

1. JARVIS, B.B., MILLER, J.D. 1996. Natural products, complexity, and evolution. pp. 265-293 *in* Recent Advances in Phytochemistry: Phytochemical Diversity and Redundancy in Ecological Systems, Vol. 30, (J. T. Romeo, J. A. Saunders, and P. Barbosa, eds.), Plenum, New York.
2. LOEWENSTEIN, W.R. 1999. The Touchstone of Life. Oxford University Press, 340 p.
3. CHRISTOPHERSEN, C. 1996. Theory of the origin, function, and evolution of secondary metabolites. pp. 677-737 *in* Studies in Natural Products, Vol. 18 (Atta-ur-Rahman, ed.), Elsevier Science, Amsterdam.
4. JAIN, R., RIVERA, M.C., LAKE, J.A., 1999. Horizontal gene transfer among genomes: The complexity hypothesis. Proc. Natl. Acad. Sci. USA 96: 3801-3806.
5. BOHM, B.A. 1998. Introduction to Flavonoids. Harwood Academic Publishers, Amsterdam, pp. 351-353.
6. WARREN, J.T., DAI, J.D., GILBERT, L.I., 1999. Can the insect nervous

system synthesize ecdysteroids? Insect Biochem. Mol. Biol. 29: 571-579.
7. BARTON, D., NAKANISHI, K., METH-COHN, O. (eds.) 1999. Comprehensive Natural Products Vol. 1-9, Pergamon Press, New York.
8. CANE, D.E., WALSH, C.T., KHOSLA, C., 1998. Harnessing the biosynthetic code: Combinations, permutations, and mutations. Science 282: 63-68.
9. KONZ, D., MARAHIEL M.A., 1999. How do peptide synthases generate structural diversity? Chem. Biol. 6: R39-R48.
10. KENNEDY, J., AUCLAIR, K., KENDREW, S.G., PARK, C., VEDERAS, J.C., HUTCHINSON, C.R., 1999. Modulation of polyketide synthase activity by accessory proteins during lovastatin biosynthesis. Science 284: 1368-1372.
11. FITZMAURICE, A.M., KOLATTUKUDY, P.E., 1998. An acyl-CoA synthase (acoas) gene adjacent to the mycocerosic acid synthase (mas) locus is necessary for mycocerosyl lipid synthesis in *Mycobacterium tuberculosis* var. bovis BCG. J. Biol. Chem. 273: 8033-8039.
12. BROWN, D.W., ADAMS, T.H., KELLER, N.P., 1996. *Aspergillus* has a distinct fatty acid synthase for primary and secondary metabolites. Proc. Natl. Acad. Sci. USA 93: 14873-14877.
13. FREY, M., CHOMET, P., GLAWISCHNIG, E., STETTNER, C., GRÜN, S., WINKLMAIR, A., EISENREICH, W., BACHER, A., MEELEY, R.B., BRIGGS, S.P., SIMCOX, K., GIERL, A., 1997. Analysis of a chemical plant defense mechanism in grasses. Science 277: 696-699.
14. CANE, D.E. (ed.) 1999. Comprehensive Natural Products: Isoprenoids Including Carotenoids and Steroids, Vol. 2, Pergamon Press, New York, 446 p.
15. KOYAMA, T., OGURA, K. 1999. Isopentenyl diphosphate isomerase and phenyltransferases. pp. 69-96 *in* Comprehensive Natural Products: Isoprenoids Including Carotenoids and Steroids, Vol. 2 (D. E. Cane, ed.), Pergamon Press, New York.
16. CANE, D.E. 1999. Isoprenoid biosynthesis: overview. pp. 1-13 *in* Comprehensive Natural Products: Isoprenoids Including Carotenoids and Steroids, Vol. 2 (D. E. Cane, ed.), Pergamon Press, New York.
17. WISE, M.L., CROTEAU, R. 1999. Monoterpene biosynthesis. pp. 97-153 *in* Comprehensive Natural Products: Isoprenoids Including Carotenoids and Steroids, Vol. 2 (D. E. Cane, ed.), Pergamon Press, New York.
18. WENDT, K.U., PORALLA, K., SCHULZ, G.E., 1997. Structure and function of a squalene cyclase. Science 277: 1811-1815.
19. STARKS, C.M., BACK, K., CHAPPELL, J. NOEL, J.P., 1997. Structural basis for cyclic terpene biosynthesis by tobacco 5-epi-aristolochene synthase.

Science 277: 1815-1820.
20. LESBURG, C. A., ZHAI, G., CANE, D.E., CHRISTIANSON, D.W., 1997. Crystal structure of pentalenene synthase: mechanistic insights on terpenoid cyclization reaction in biology. Science 277: 1820-1824.
21. SACCHETTINI, J.C., POULTER, C.D. 1997. Creating isoprenoid diversity. Science 277: 1788-1789.
22. ROHMER, M. 1999. A mevalonate-independent route to isopentenyl diphosphate. pp. 45-67 *in* Comprehensive Natural Products: Isoprenoids Including Carotenoids and Steroids, Vol. 2 (D. E. Cane, ed.), Pergamon Press, New York.
23. DISCH, A., ROHMER, M., 1998. On the absence of the glyceraldehyde 3-phosphate pyruvate pathway for isoprenoid biosynthesis in fungi and yeasts. FEMS Microbiol. Lett. 168: 201-208.
24. MARGULIS, L., FESTER, R. (eds.) 1991. Symbiosis as a Source of Evolutionary Innovation, The MIT Press, Cambridge, MA, 435 pp.
25. KELLER, N.P., HOHN, T.M., 1997. Metabolic pathway gene clusters in filamentous fungi. Fungal Genet. Biol. 21: 17-29.
26. MAPLESTONE, R.A., STONE, M.J., WILLIAMS, D.H., 1992. The evolutionary role of secondary metabolites - a review. Gene 115: 151-157.
27. BROWN, M.P., BROWN-JENCO, C.S., and PAYNE, G.A., 1999. Genetic and molecular analysis of aflatoxin biosynthesis. Fungal Genet. Biol. 26: 99-117.
28. KUBO, Y., 1998. Melanin biosynthesis genes in fungi. Nippon Nogcik Kaishi 72: 60-62.
29. KIMURA, N., TSUGE, T., 1993. Gene cluster involved in melanin biosynthesis of the filamentous fungus *Alternaria alternata*. J. Bacteriol. 175: 4425-4435.
30. KUBO, Y., TAKANO, Y., ENDO, N., YASUDA, N., TAJMA, S., FURUSAWA, I., 1996. Cloning and structural analysis of the melanin biosynthesis Gene SCD1 encoding scytalone dehydratase in *Colletotrichum lagenarium*. Appl. Environ. Microbiol. 62: 4340-4344.
31. CHUMLEY, F.G., VALENT, B., 1990. Genetic analysis of melanin-deficient non-pathogenic mutants of *Magnaporthe grisea*. Mol. Plant-Microbe Interact. 3: 135-143.
32. DIEZ, B., GUTIERREZ, S., BARREDO, J.I., VAN SOLINGER, P., VAN DER VOORT, L.H.M., MARTIN, J.F., 1990. The cluster of penicillin biosynthetic genes. Identification and characterization of the *pcb*AB gene encoding the a-aminoadipyl-cysteinyl-valine synthase and linkage to the *pcb*C

and *pcb*DE gene. J. Biol. Chem. 265: 16358-16365.
33. MACCABE, A.P., RIACH, M.B.R., UNCLES, S.E., KINGHORN, J.R., 1990. The *Aspergillus nidulans npe*A locus consists of three contiguous genes required for penicillin biosynthesis. EMBO J. 9: 279-285.
34. MATHISON, L., SOLIDAY, C., STEPAN, T., ALDRICH, T., RAMBOSEK, J., 1993. Cloning, characterization, and use in strain improvement of the *Cephalosporium acremonium* gene *cef*G encoding acetyl transferase. Curr. Genet. 23: 33-41.
35. LANDAN, G., COHEN, F., AHRONOWITZ, Y., SHUALI, Y., GRAUR, D., SHIFFMAN, D., 1990. Evolution of isopenicillin N synthase genes may have involved horizontal gene transfer. Mol. Biol. Evol. 7: 399-406.
36. PENALVA, M.A., MOYA, A., DOPAZO, J., RAMON, D., 1990. Sequences of isopenicillin N synthase genes suggest horizontal gene transfer from procaryotes to eucaryotes. Proc. Royal Soc. London 241: 164-169.
37. SMITH, M.W., FENG, D.F., DOOLITTLE, R.F., 1992. Evolution by acquisition: The case for horizontal gene transfers. Trends Biochem. Sci. 17: 489-493.
38. XUE, Y.Q., ZHAO, L.S., LIU, H.W., SHERMAN, D.H., 1998. A gene cluster for macrolide antibiotic biosynthesis in *Streptomyces venezuelae*: architecture of metabolic diversity. Proc. Natl. Acad. Sci. USA 95: 12111-12116.
39. KIMURA, M., KANEKO, I., KOMIYAMA, M., TAKATSUKI, A., KOSHINO, H., YONEYAMA, K., YAMAGUCHI, I., 1998. Trichothecene 3-O-acetyltransferase protects both the producing organism and transformed yeast from related mycotoxins. Cloning and characterization of Tri101. J. Biol. Chem. 273: 1654-1661.
40. KIMURA, M., MATSUMOTO, G., SHINGU, Y., YONEYAMA, K., YAMAGUCHI, I., 1998. The mystery of the trichothecene 3-O-acetyltransferase gene - analysis of the region around Tri101 and characterization of its homologue from *Fusarium sporotrichioides*. FEBS Lett. 435: 163-168.
41. MCCORMICK, S.P., ALEXANDER, N.J., HOHN, T.M., TRAPP, S.C., 1999. The 3-O-acetyltransferase involved in trichothecene resistance is required for T-2 toxin production in *Fusarium sporotrichioides*. Appl. Environ. Microbiol., 65: 5252-5256.
42. DOOLITTLE, R.F., 1997. Microbial genomes opened up. Nature 392: 339-342.
43. O'BRIEN, P.J., HERSCHLAG, D., 1999. Catalytic promiscuity and the evolution of new enzymatic activities. Chem. Biol. 6: R91-R118.

44. BROUN, P., SHANKLIN, J., WHITTLE, E., SOMERVILLE, C., 1998. Catalytic plasticity of fatty acid modification enzymes underlying chemical diversity of plant lipids. Science 282: 1315-1317.
45. MANDER, L.N., 1992. The chemistry of gibberellins: An overview. Chem. Rev. 92: 573-612.
46. MACMILLAN, J., BEALE, M.H., 1999. Diterpene biosynthesis. pp. 217-243 *in* Comprehensive Natural Products: Isoprenoids Including Carotenoids and Steroids, Vol. 2 (D. E. Cane, ed.), Pergamon Press, New York.
47. TUDZYNSKI, B., HÖLTER, K., 1998. Gibberellin biosynthetic pathway in *Gibberella fujikuroi*: Evidence for a gene cluster. Fungal Genet. Biol. 25: 157-170.
48. HEDDEN, P., KAMIYA, Y., 1997. Gibberellin biosynthesis: Enzymes, genes, and their regulation. Annu. Rev. Plant Physiol. Plant Mol. Biol. 48: 431-460.
49. MENDE, K., HOMANN, V., TUDZYNSKI, B., 1997. The geranylgeranyl diphosphate synthase gene of *Gibberella fujikuroi*: Isolation and expression. Mol. Gen. Genet. 255: 96-103.
50. DOMENECH, C.E., GIORDANO, W., AVALOS, J., CERDÁ-OLMEDO, E., 1996. Separate compartments for the production of sterols, carotenoids, and gibberellins in *Gibberella fujikuroi*. Eur. J. Biochem. 239: 720-725.
51. DAWSON, R.M., 1998. The toxicology of microcystins. Toxicon 36: 953-962.
52. DITTMANN, E., NEILAN, B.A., ERHARD, M., DOHREN, H. von; BORNER, T., 1997. Insertional mutagenesis of a peptide synthetase gene which is responsible for hepatotoxin production in the cyanobacterium *Microcyctis aeruginosa* PCC 7806. Mol. Microbiol. 26: 779-788.
53. DOOLITTLE, R.F., 1995. The multiplicity of domains in proteins. Annu. Rev. Biochem. 64: 287-314.
54. SUNDSTROM, L., 1998. The potential of integrons and connected programmed rearrangements for mediating horizontal gene transfer. APMIS 106: 37-42, Suppl. 84.
55. WINZELER, E.A., SHOEMAKER, D.D., ASTROMOFF, A., *et al.*, 1999. Functional characterization of the *S. cerevisiae* genome by gene deletion and parallel analysis. Science 285: 901-906.
56. SCHE, P.P., MCKENZIE, K.M., WHITE, J.D., AUSTIN, D.J., 1999. Display cloning: Functional identification of natural product receptors using cDNA-phage display. Chem. Biol. 6: 707-716.
57. HANDELMAN, J., RONDOD, M.R., BRADY, S.F., CLARDY, J., GOODMAN, R.M., 1998. Molecular biological access to the chemistry of unknown soil microbes. Chem. Biol. 5: R245-R256.

Chapter Two

THE EVOLUTION OF PHENOLICS IN PLANTS

Helen A. Stafford

Reed College
Biology Department
Portland, OR 97202-8199

INTRODUCTION

The evolution of the basic compounds to be emphasized include the so-called 'tannins' that form brown, black, red, and red-brown colored polymers upon senescence: phloroglucinol polymers, caffeoyl-esters (caffeoyltannins), galloyl-esters (hydrolyzable tannins), proanthocyanidins (condensed tannins), the lignin-lignan group, and anthocyanidins (Fig. 1). Both 3-deoxy vs 3-hydroxy forms of the flavonoids, proanthocyanidins, and anthocyanidins will be emphasized. All of these are called micrometabolites, except for the macrometabolite, lignin.[1,2] Betalains such as betacyanin (alkaloid derived from tyrosine) are included because of their effective substitution for anthocyanins in one plant group.

Evolution of Metabolic Pathways – J.T. Romeo et al. (Editors)

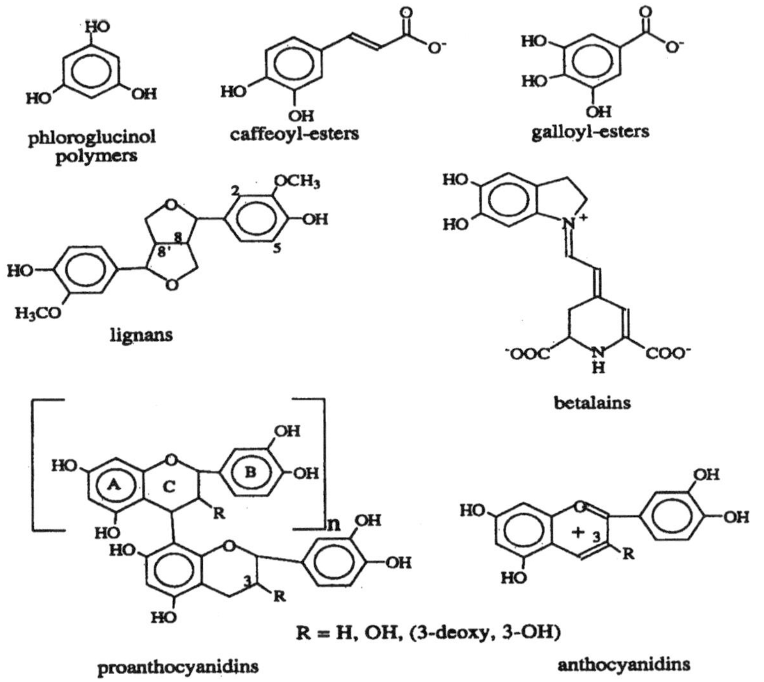

Figure 1. Basic plant phenolic compounds discussed: phloroglucinol polymers (phlorotannins); C^6-C^3 caffeoyl-esters (caffeoyltannins); C^6-C^1 galloyl-esters (hydrolyzable tannins); lignans (optically active dimers of C^6-C^3 units, bonded at various positions such as 8 - 8'); betalains (alkaloids); proanthocyanidins (C^6-C^3- C^6 flavonoid polymers, condensed tannins); anthocyanidins (C^6-C^3- C^3 flavonoids).

Basic functions of the above phenolics involve support, transport, and protection within the individual plant by lignin, and attraction of insects and animals, including pollination of flowers and seed dispersal, as well as defense against microbial pathogens and animals by flavonoids and lignans. Signals between the organism and the environment, including the microbe, animal, and the host, are important aspects in both attraction and defense functions of phenolics. Functions of these are variable and include actions as phytoalexins, toxins, 'tannins', resins, antioxidants, protein precipitation capabilities, and UV and visible light screening for specific wave length effects as signals.[3-5]

Major phenolic pathways involved are the shikimate-aerogenate pathway to the aromatic amino acids, phenylalanine and tyrosine, the phenylpropanoid pathway that forms C_6-C_3 units from phenylalanine or sometimes tyrosine by hydroxylation of cinnamate, and the acetate pathway that forms C_2 units from malonyl-CoA.[6] Flavonoids are synthesized by a combination of three of the above C_2 units to form the C_6 A-ring (minus 3 carbons as CO_2) with one C_6-C_3 unit to form the B- and C-rings of a flavonoid such as naringenin.[4,5] C_6-C_1 units involve a sequence of decarboxylations. Lignin (an insoluble polymer) and lignans (soluble dimers) are formed from coumaryl alcohols produced from phenylpropanoids with varied hydroxylation or methylation patterns of the B-ring[7].

EVOLUTIONARY TRENDS OF PHENOLIC PRODUCTS AND ENZYMATIC PATHWAYS

Algae

General evolutionary trends in both the production of the basic phenolic products and the major taxonomic groups based mainly on morphological data, are shown in Figure 2. Whereas brown algae (Phaeophyta) produce phlorotannins composed of phloroglucinol polymers (acetate-malonate derived), members of the green algae (Charophyceae of the Chlorophyta) are considered the most likely precursor aquatic group of land plants.[8-10] A cladogram of the green algae lineage based on rDNA sequences shows the relationships between charophyte green algae and bryophytes.[11] Sporopollenin, an acetolysis-resistant material in the algal walls, is a key piece of evidence, but its composition is controversial and difficult to determine because of its insolubility and because of possible fusion with various compounds, possibly with phenolics. Lignin residues have been hard to analyze for the same reason, plus confusion with the soluble dimers called lignans that are sometimes found in walls. Sporopollenin appears to be a cross-linked, aliphatic polymer with saturated and unsaturated hydrocarbons, probably with associated phenolics. An important function is in UV-protection.[12] Similar acetolysis-resistant compounds have now been identified in the exine layer of all pollen types, including those in Gymnosperms and Angiosperms.[9,13-16]

Bryophytes

Bryophytes are an important group for the study of the early evolution of phenolic compounds in land plants, including flavonoids and lignans and possibly lignin (Fig. 2). The flavonoids (C6-C_3-C_6 structures) identified in bryophytes are widely distributed in mosses and liverworts, but the information is probably still incomplete.[17] Most of the identified phenolics are presumably found in the gametophyte generation (haploid), although occasionally the small sporophyte generation (diploid) is cited as the source of a compound. Many species have not been investigated biochemically. Flavonoids, mainly 3-deoxy-forms such as flavone glycosides, 3-deoxyanthocyanins, and aurones, are found in about 60% of the mosses (*Bryopsida*) and liverworts (*Hepaticeae*) of this taxonomic group. None has been identified in hornworts (*Anthocerotae*), the third bryophyte group. Flavones are the most abundant group in both liverworts and mosses. They play an important role in protection against UV-B, due to their absorption spectra between 310-350 nm, especially the glycosides at about 320 nm.[18] Apigenin and luteolin glycosides are major flavonoids in mosses. Visible reddish pigments are widespead in mosses and liverworts. Sap-soluble 'red' pigments in *Bryum* spp. of mosses are 3-deoxy-anthocyanidins, such as luteolinidin. Since 3-deoxy-anthocyanidins have absorption peaks around 480-495 nm, the origin of the red color rather than yellow to orange is unclear.[3,19] Violet coloring in cell walls of a liverwort has been identified as being due to a new type of anthocyanidin (riccionidin A and B), with a pattern formed by reversing the additions of the phenylpropanoid and malonyl-CoA moieties to the left (A-) and right (B-) rings (a retro-flavonoid) to produce an apparent 3-deoxy anthocyanidin, plus an additional connection between the right and central (C-) rings.[20] The reddish-violet pigments in cell walls of mosses (sphagnorubins) are 3-deoxyanthocyanidins with a caffeic acid moiety. Luteolinidin glycosides (3-deoxy-anthocyanidins) are dominant in liverworts. No 3-hydroxy-anthocyanidins have been identified. The more enzymically-advanced flavonols have possibly been identified in liverworts, but not mosses. Another enzymically-advanced group, isoflavones or isoflavone-related compounds, has been identified occasionally in mosses.[21] Neither 3-deoxy- or 3-hydroxy-proanthocyanidins, the latter involving an additional complex set of enzymic reactions, has been convincingly shown to be present in either mosses or liverworts. Biflavonoids, flavone dimers common in pteridophytes and gymnosperms, are also found in mosses, probably only in the walls.[22] Bibenzyls (dihydrostilbenes) are characteristic of liverworts.[23] A variety of C_6-C_3 phenolics, including caffeic acid 'dimers' such as rosmarinic acid, and a novel lignan are examples of the few phenolics identified so far in hornworts. No flavonoids have been found.[4,17,24-27]

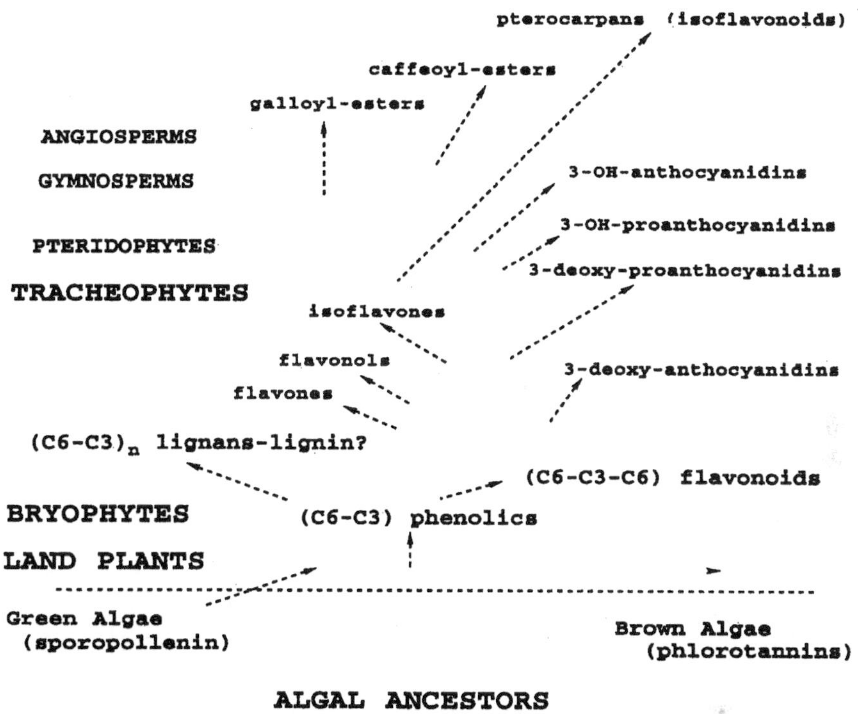

Figure 2. Evolutionary scheme of phenolic end-products and major taxonomic groups from algal ancestors. Lignin probably arose in pteridophytes.

Lignans and possibly lignin-like components, C_6-C_3 biochemical structures, have been identified in bryophytes (mosses and liverworts). Lignans are present in hornworts,[28]and lignins may be present in stalks of mosses.[24] Some authors state, however, that the presence or non-presence of lignin in bryophytes is still an unanswered question.[22,23] Much more needs to be learned about the biochemistry, physiology, and molecular genetics of bryophytes.

A recent review of the phylogeny of early land plants, based on insights from genes, genomes, and cladistic analyses, is presented by Qiu et al.[29] *Chara* and *Coleochaete* of the charophytes are considered the closest relatives to land plants. Some genomic structural information and some genetic sequence data indicate that liverworts of the bryophytes are probably the earliest land plants. Based on chloroplast DNA, lycopods of the pteridophytes have been considered the first lineage of vascular plants (tracheophytes), but subsequent sequence data do not support this. Evolutionary relationships among the spore bearing groups of the early tracheophytes (pteridophytes) and those of early seed plants (gymnosperms and angiosperms) remain unresolved. A summary of various cladistic analyses of land plants, including bryophytes, is now available and will be discussed later under future research.[30]

Tracheophytes

Flavonoids in tracheophytes: Flavonoid evolution in bryophytes and seed-bearing tracheophytes, including the structural enzymes involved in their synthesis, is summarized in Figures 2 and 3.[31-34] Bryophytes contain the basic types of flavonols and isoflavones as discussed above. Within tracheophytes, both 3-deoxy- and 3-hydroxy-proanthocyanidins appear in the pteridophytes (lycopods, ferns). Pterocarpans, the second largest group of natural flavonoids (isoflavones are the largest), are found as both endogenous compounds and induced as phytoalexins in legumes. Proanthocyanidins (3-hydroxy type, condensed tannins) are widespread in gymnosperms and woody angiosperms. They are present in a variety of cells of the cortex, periderm (bark), and wood. The highest concentrations are in the periderm. Histochemical tests for proanthocyanidins are available.[32,33] They are frequently called 'condensed tannins' and will be compared with C_6-C_1 galloyl-esters (hydrolyzable tannins) formed from the C_6-C_3 pathway, which are found mainly in several major taxonomic groups of the dicotyledons. Replacement of both hydolyzable tannins and proanthocyanidins by the C_6-C_3 caffeoyl-esters, also called 'caffeoyltannins',[1] will be discussed later. 3-Deoxy-proanthocyanidins have been identified in the pericarp of sorghum fruits and possibly maize, monocots of the Poaceae.[3,35-37] They function as defense compounds (phytoalexins) in sorghum seedlings.[38] There is still unresolved confusion about the pericarp of maize with its production of so-called 'phlobaphenes' (red-brown complexes) as to whether flavan-4-ols or 3-deoxy-proanthocyanidins are involved.[3,35-37] 3-Hydroxy-anthocyanidins are found in gymnosperms and angiosperms.[37] The 3-deoxy-anthocyanidins, more common in bryophytes, are found only rarely in angiosperms (Figs. 3, 4). One example is in the Gesneriaceae (Asteridae, Lamiales), a group close to Scrophulariaceae (*Antirrhinum* , snapdragon).[34]

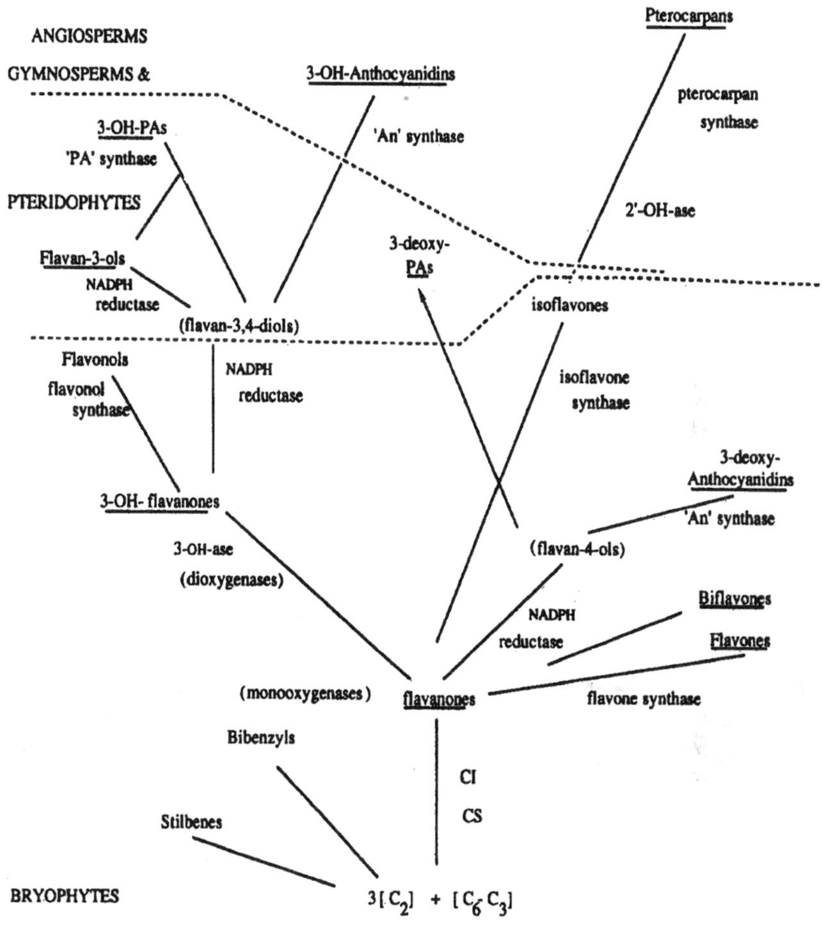

Figure 3. Evolutionary scheme of enzymic steps of the major subgroups of flavonoids in bryophytes, pteridophytes, gymnosperms, and angiosperms. CS = chalcone synthase; CI = chalcone isomerase; PA = proanthocyanidin; An = anthocyanin.

The isoflavonoid pathway is of particular interest in that it can be the longest pathway of structural enzymes in the flavonoid group when pterocarpans are formed from isoflavones, and because of its involvement in root nodules and nitrogen fixation (Fig.3).[39,40] This pathway can be divided into at least two sections, the first leading to 5-hydroxy isoflavones (mainly endogenous signal molecules), the second to the more complex forms such as 5-deoxy pterocarpans (both endogenous and inducible defense compounds).[40-42] Pterocarpans might be considered the most advanced form of a phenolic end-product. Whereas isoflavonoids such as pterocarpans are found predominantly in the leguminosae (mainly in the Papilionoideae), the simpler isoflavones have also been identified in a variety of other taxa, including a few identifications in the bryophytes.[41]

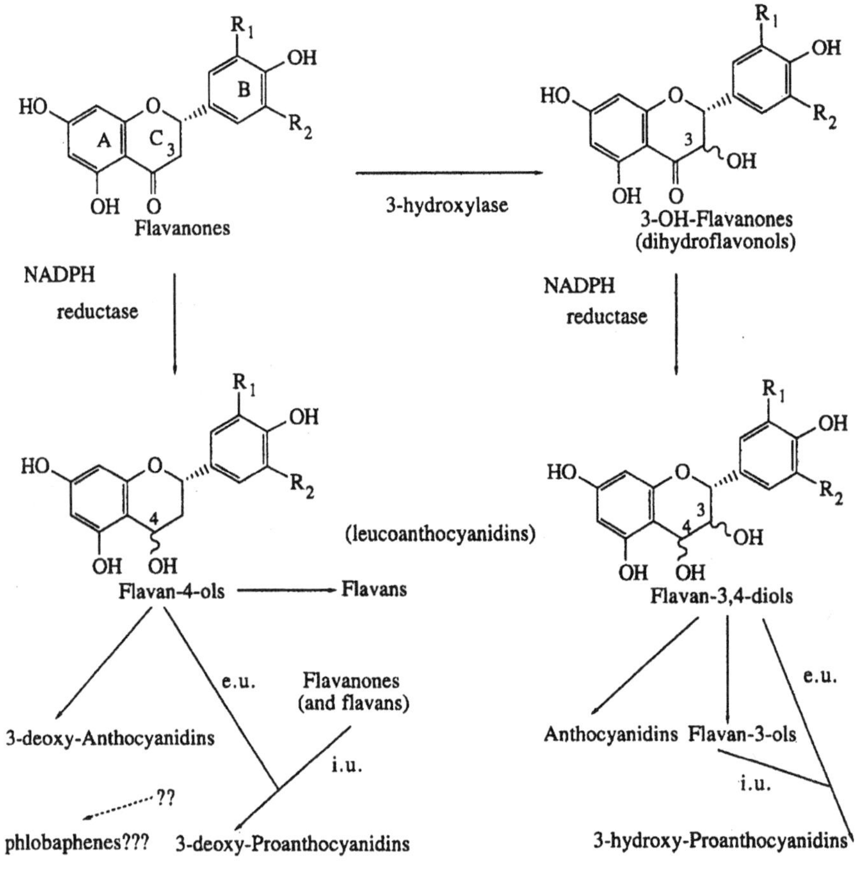

Figure 4. Biosynthesis of 3-deoxy- and 3-hydroxy- anthocyanidins and proanthocyanidins.[3,30-32]

Structural and regulatory genes and their proteins of the phenolic pathways, including evolutionary relationships. The biosynthesis of many of the phenolics to be discussed in this chapter are regulated endogenously, a type of regulation that has been poorly studied. It varies in different tissues. The biosynthesis of some flavonoids involved in defense that are induced by exogenous agents, such as pathogens and light, have been more extensively studied. The majority of the structural enzymes involved in the phenylpropanoid and flavonoid pathways have been identified and isolated. However, the presence of a specific, soluble 3-hydroxylase in the production of caffeic acid and its caffeoyl-esters is still unclear,[43] and the origin of gallic acid (C_6-C_1) from cinnamic acid (C_6-C_3), leading to the complex esters (gallotannins, hydrolyzable tannins) is still being studied.[44-46] Most of the genes of the flavonoid pathway are known.

Evolutionary similarities between flavonoid structural enzymes and other metabolic pathways are being studied (Figs. 3, 4). Similarities have been found among chalcone, stilbene, and fatty acid synthases, all of which are classified as polyketide synthases.[47,48] Flavonoid hydroxylases are related to other cytochrome P450-dependent monooxygenases.[49-52] The two synthases leading to 3-hydroxy proanthocyanidins and anthocyanidins have not been isolated. Recently, however, a molecular genetics study has produced a recombinant anthocyanidin synthase that produces a colorless intermediate(s) *in vitro* that can subsequently be converted non-enzymically to an anthocyanidin by the addition of HCl.[53] This anthocyanidin synthase-catalyzed reaction occurs in the presence of ferrous ion, 2-oxoglutarate, and ascorbate and is considered to be part of a family of four 2-oxoglutarate-dependent oxygenases involved in flavonoid pathways. A common acestor has been postulated for mammalian 3-β-hydroxysteroid dehydrogenase and the plant NADPH-dependent dihydroflavonol reductase.[54] NADPH reductases involved in lignan and isoflavone biosyntheses are considered related to a variety of 'homologs' found in both woody and non-woody gymnosperms and angiosperms.[55] A cladistic analysis of the amino acids of plant *0*-methyltransferases places them within a monophyletic group derived from non-plant genes, indicating a divergence through gene duplication and mutation to yield the various functional enzyme groups currently recognized in plants.[56]

A variety of regulatory genes and their proteins have been identified that control flavonoid structural genes. These transcription factors can be broadly classified on the basis of the DNA-binding domain. Two distinct families of such factors involve Myb-related regulatory proteins in the C_1 family and the basic helix-loop-helix motif of the R family of the monocot maize to regulate anthocyanin (3-hydroxy) biosynthesis as one coordinate system (*via* metabolic channelling) at the C_{15} level.[3,37,57] However, there appears to be two separate systems, an 'early' and 'late' one, for other plants such as petunia and antirrhinum.[5,51,58-60] Whereas the regulatory anthocyanin loci are known for the second set, those for the first set are unknown, but the differences may lie in the promoters of the structural genes.[58] Enzymes of the 3-deoxy and 3-hydroxy pathways to anthocyanidins in the pericarp, cob, and parts of the floral tissue of maize are controlled independently by the *P* gene of the P family.[61-63] A good example of a coordinated regulation of a pathway is also found in the pterocarpan pathway in isoflavonoids of legumes.[40] From the level of phenylalanine, 11 enzymatic steps are involved after exposure to exogenous elicitor signals in the phytoallexin response to fungal pathogens (Fig. 2). This is in contrast to endogenous signnals for beneficial fungal interactions in which the early genes of the isoflavone pathway are uncoupled from that of the later phytoalexin specific genes.

The betalain-anthocyanin substitution. In at least one instance, betacyanin, the red-colored betalain, an alkaloid that is found only in the Caryophyllales taxonomic group, completely substitutes for the ubiquitous flavonoid compounds called anthocyanins, and apparently takes over the function and tissue distribution of this visually pigmented flavonoid (Fig. 1). The evolutionary significance of the apparent mutual exclusion of the rare betalains and the ubiquitous anthocyanins, or the sequence in which one pathway is lost and the other gained, is still basically unknown. See Clement et al.[64] for a list of the numerous Caryophyllean genera of 15 families known to contain betalains, and the smaller number of genera in two families (Caryophyllaceae and Molluginaceae) that are known to contain anthocyanins. Red colored betalains found in plants such as *Beta vulgaris* are derived from tyrosine that forms the intermediates cyclo-dopa and betalamic acid *via* L-DOPA. Although they are alkaloids, they contain hydroxyl units attached to an aromatic ring similar to phenolic hydroxyls.[65,66] A specific tyrosinase catalyzes the formation of both Dopa and *cyclo*-Dopa, precursors of intermediates that condense to form betanidin, a red colored betalain.[67] Surprisingly, this condensation step to form a yellow colored betalain has been shown to occur non-enzymically, based on external feeding experiments in hairy root cultures of yellow beet and in young fodder beet plants.[68] The function of the yellow pigments compared with the red ones is unknown. There is also the problem of unpigmented taxa in the Molluginaceae.[64] The red colored betalain end-products serve as a complete functional substitute for anthocyanins, and so far, no plant has been found to contain both types of red pigments. Further work is needed to determine the subcellular compartment in which this pathway occurs. As in the case of anthocyanidin synthase, cell-free evidence of the betalain 'synthases' (if enzymatic) has not been demonstrated. Some information is available concerning genetic control of this pathway.[69] Molecular genetic studies will be necessary to understand the basis for this taxonomic distribution of the anthocyanidin and betalain synthase genes within the Caryophyllales.[64]

Lignans and lignins. Lignans, soluble C_6-C_3 dimers formed from the phenylpropanoid pathway, function in defense. Some lignans have higher molecular weights than just dimeric forms and can be found in cell walls of pre-lignified cells of heartwood in gymnosperms.[7,55,71] The 8-8'-linkages are common, but a variety of other linkages are known and these are sometimes grouped as neolignans. Most are optically active. In early land plants, these soluble lignan dimers probably preceded and were more widespread than lignins. They are now present in bryophytes (in hornworts) and ferns as well as in both advanced groups, gymnosperms and angiosperms (both dicots and monocots).[28] In monocots, they are found in roots, stems, leaves, and fruits.[70] Since their chemistry is similar to that of lignin components, the identifications of these phenolic compounds are often confused; this has caused discord between two major groups.[7,71]

Lignins, insoluble components of certain cell walls, are the most energetically expensive metabolic end-products of the phenylpropanoid pathway in vascular plants (Fig. 5). In combination with lignans, they collectively account for more than 25% of the dry weight of woody plants; only cellulose is present in larger amounts.[7] Lignins are present in xylem cells of all herbaceous and woody terrestrial plants of tracheophytes, and also in the cortex and bark of woody plants. Histochemical tests are available.[31,32] Lignins function in both support and water conduction in tracheids and vessels of the xylem, in dryness protection of internal tissues of the sclerenchyma, and in cell walls of the periderm of secondary growth that are produced upon senescence or programmed cell death.[7,72] They are frequently considered to be defense compounds against pathogens, formed upon wounding in both herbaceous and woody plants.[73] However, the distinction between lignin and lignans in the analytical methods used has never been made clear.[32] This has recently been discussed by Lewis and Sarkanen.[7] The primitive function of lignin-like (lignans?) products in early bryophytes was probably in defense. The chemical composition of lignin components is an important taxonomic character. Three basic units are formed that vary in the number of methyl groups in the aromatic ring, 4-coumaryl-alcohol, coniferyl-alcohol, and sinapyl-alcohol (Fig. 5). Gymnosperms contain just the coniferyl type, angiosperms both coniferyl- and sinapyl-forms, whereas grasses contain 4-coumaryl-alcohol in addition. Ferulic acid is involved in cross-linking cell wall carbohydrates to lignin in monocots and occasionally in dicots.[74] Pathways to monolignols have been recently studied.[75,76] Formation of heartwood, the non-functional dark-colored portions of the inner wood that is external to the sapwood of a tree trunk, has been described by Gang et al.[77]

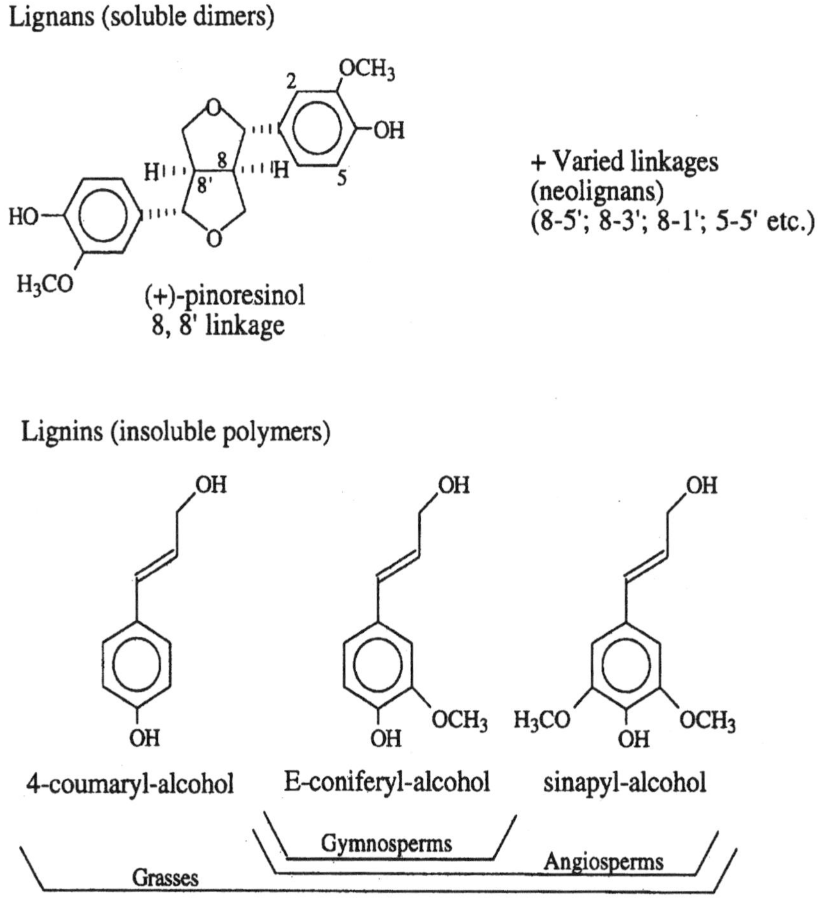

Figure 5. Lignans-lignins in vascular, land plants.

The insolubility of lignins has made it difficult to determine the actual linkages among the different C_6-C_3 components. Whereas much work has been done with *in vitro* or cell-free synthesis of lignins, the *in vivo* or *in plantae* biosynthesis appears to be quite different according to recent data, and the involvement of dirigent proteins in directing specific types is being actively studied.[7,70,71,78] There is considerable disagreement even about the definition as well as the description of lignin biosynthesis between two major groups.[7,78,79] Genetic manipulation of lignin that functions in defense interactions with microorganisms has involved a series of target enzymes to reduce lignin contents in cereals.[80] Molecular genetics and genetic engineering will make vital, future contributions to our knowledge of both insoluble polymers of lignin and soluble dimers of lignans.[72,81] The lignin story is still incomplete.

EVOLUTIONARY SCHEMES OF DICOTYLEDONOUS PLANTS

Morphological vs. Molecular Schemes

The Magnoliaceae (Magnolidae) are generally considered to be relatively primitive families, with woody members (shrubs or small trees) of the Magnolialean group as the earliest or ancestral forms.[1,2,82] Recent arguments, however, have been made for a 'paleoherb' hypothesis in which herbaceous plants with a rhizomatous perennial habit and simple flowers as found in Nymphaeales or Piperales (Chloranthaceae and Piperaceae) within the Magnoliidae are considered the primitive base.[83-86] (see Figure 6 based on Cronquist[87]). This placement also suggests that monocots are early derivatives of the paleoherbs, based on *rbcL* and 18S rDNA sequence-based cladograms.[88-90] Unfortunately, the terms woody and herbaceous are not rigidly defined in any of the above schemes. Presumably, woody plants contain secondary vascular tissue due to vascular cambial layers, whereas herbaceous plants have only primary growth.[91] In the condensed diagram of Cronquist shown in Figure 6,[81] the Asteridae order that is considered advanced contains about forty herbaceous families (Table 1 in Gottlieb[1]). In the more highly dissected groups of dahlgrenogram,[86] the Asteridae order is subdivided into at least 4 major groups, the largest being that of Asterales and Lamiales.[87] This permits a more detailed diagram of end-product localization in the Asteridae order, as used by Molgaard and Raven[92] for their survey of caffeoyl esters and other related C_6-C_3 compounds such as ferulic and chlorogenic acids in dicotyledons.

Diagrams of two major taxonomic evolutionary schemes of dicotyledonous plants, based mainly on morphological and developmental characteristics, have been constructed by Dahlgren[93] and Cronquist.[87] Unfortunately, the taxonomic nomenclature used for the sub-classes, orders, and

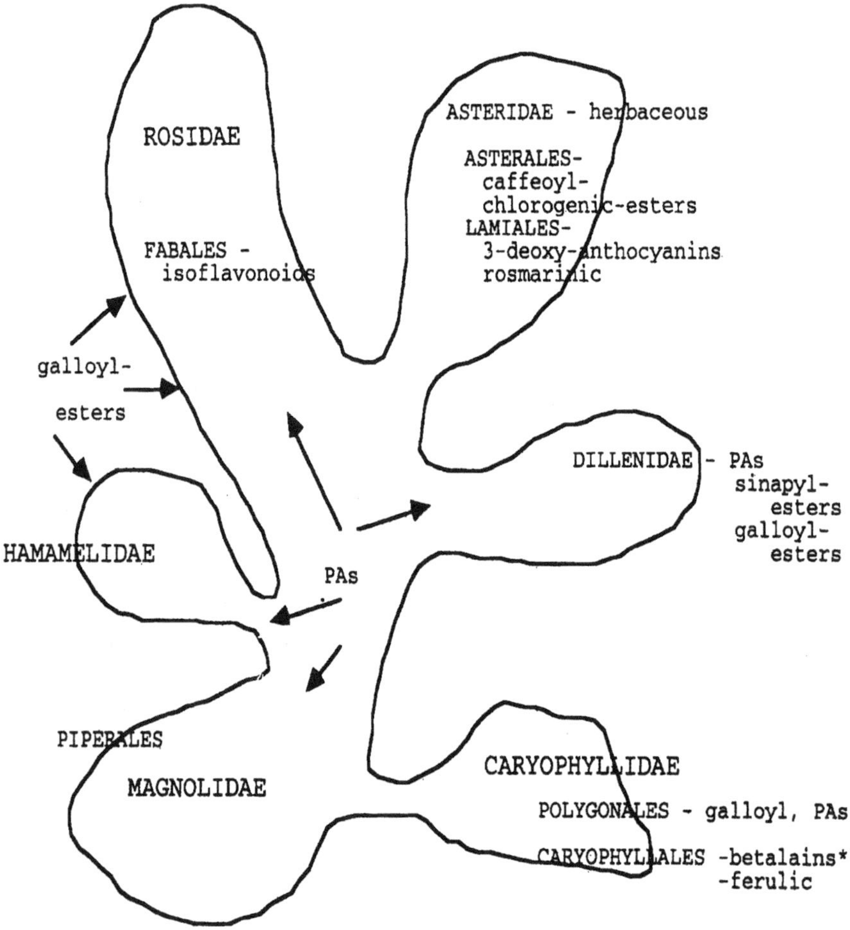

Figure 6. Diagram of six major orders of dicotyledons and their evolutionary relationships, based on the morphological groupings of Cronquist,[82] with indications of major phenolic end-products discussed. (Fabales also known as Leguminosae). *betalains are found in all groups of Caryophyllales, except for the Molluginaceae and Caryophyllaceae.[62]

families differs in the two systems. Floral morphology and pollination mechanisms form the bases of many of these morphological taxonomic divisions. The simpler Cronquist taxonomic model is summarized in diagramatic form in Figure 6, and is the one used by Gottlieb.[1,2] Comparisons can be made with the more complex diagram of orders and families in the 'dahlgrenograms' of

Dahlgren,[93] in which the dicots are divided into a much larger number of taxonomic groups of different sizes.[92-95] Both of these taxonomic systems are based on morphological data,[93] rather than on molecular data such as *rbcL* and 18S rDNA sequences, which are the basis of the morphological data, rather than on molecular data such as *rbcL* and 18S rDNA sequences, which are the basis of cladograms of modern molecular systematics.[88,89] A basic criticism of this more modern system of cladograms was presented by Cronquist.[96] A cladogram based on *rbcL* sequence data showing the phylogeny of nodulating flowering plants of the Leguminosae (Fabaceae) can be compared with a classical scheme based on morphological data.[88,97] The traditional and the molecular phylogenetic classification schemes are compared in Figure 1 of Doyle *et al.*[97] In the *rbcL* cladogram analysis, practically all nodulating and nitrogen-fixing angiosperms belong to a single extensive lineage (Rosid 1) that might be due to a single origin of a 'predisposition' for nodulation, but still with a scattered distribution within the Rosid I lineage. This concept, however, is consistent with multiple origins of symbioses within this Rosid 1 group. In contrast, in the morphologically-based scheme, the families involved in nodulation are scattered throughout most of the major taxonomic sub-groups, and evolved independently.

Phytochemical Distributions

The evolutionary 'picture' of the distributions of biochemical end-products among taxonomic groups can be divided into end-products that are widespread in several major taxonomic groups and those that have a 'spotty' or sporadic distribution.[1,2,98] An attempt has been made in Figure 6 to label some of the different types of phenolic compounds found in each major order of the dicotyledonous group, based on Cronquist's morphological groupings.[87] Terminology problems are present in these taxonomic schemes. In the character correlations of Sporne,[82] Chapman,[85] and Cronquist,[87] the term 'leucoanthocyanin' is used for one of the biochemical compounds characteristic of woody plants, but this term was formerly used for both the intermediate, a flavan-3,4-diol, and for the end-product, because both give rise to anthocyanidins under HCl plus heat conditions. According to modern terminology, the term proanthocyanidin (condensed tannins) should now be substituted for the end-product, leaving the intermediate 3,4-diols as leucoanthocyanins.[3] The second biochemical term used in these schemes is ellagitannins, one group of hydrolyzable tannins that also include gallotannins. The term galloyl-esters to include both types will be used in this chapter. The term 'tannin' will be discussed later.

Some examples of sporadic or unusual phenolic accumulations are shown in Figure 6. The Lamiales group of the Asteridae contains one species (*Sinningia cardinalis* in the Gesneriaceae) in which a 3-deoxy-anthocyanidin has been identified.[34] The disaccharide ester becomes restricted to the Scrophulariaceae and Oleaceae families within that order. Sinapyl esters are found in some of the Dillenidae, and ferulic acid in the Caryophyllidae, associated with cell walls as in monocotyledons. Betalains (alkaloids with phenolic groups on an aromatic ring) are present in ten of the Caryophyllales orders of the Caryophyllidae, but not in the Molluginaceae or Caryophyllaceae in which anthocyanins are found instead.[64,87] Circumstances regarding the gain and loss of these two independent pathways within one taxonomic group are still unknown. Caffeoyl ester conjugates with quinic acids to form chlorogenic acid (caffeoyl dimer) are common in Asteraceae, Solanaceae, and Rubiaceae families, whereas rosmarinic acid is restricted to the Lamiaceae and Boraginaceae families of the Asteridae.[92] Isoflavonoids are found mainly in the subfamily Papilionoideae in Fabales (Leguminosae) in the Rosidae sub-class, but limited amounts have also been identified in a variety of other isolated groups in non-legumes, including other dicotyledons, monocotyledons, gymnosperms, and bryophytes.[41] Why are isoflavonoids more narrowly distributed than other flavonoids? Isoflavones function in nodular nitrogen fixation in legumes, but the more widely distributed flavonoid groups other than isoflavonoids are also involved in nodulation.[99]

Lignans and lignin are not noted on Figure 6, but are found in all of the groups. Lignans, mainly soluble dimers with a C-8-8'- linkage, that function in defense are widespread in vascular plants.[7] Lignin as a structural and transport component is the major phenolic constituent in woody forms of vascular land plants, but although present in much smaller amounts, lignin is still a vital constituent of tracheids and vessels involved in transport and support in all herbaceous vascular plants. Histochemical studies of both herbaceous and woody (plus Klason analyses) species indicate that a rapid lignin deposition at the infection site due to pathogens is closely associated with a resistance response.[7] However, lignans rather than lignin may be involved in these defense responses.[7]

Evolution of 'Tannins'

Phenolics called 'tannins', galloyl-esters (hydrolyzable tannins or gallotannins), and proanthocyanidins (condensed tannins) ultimately produce brown to red-brown colors in mature or senescent tissues.[36] Caffeoyl-esters (called caffeoyltannins by Gottlieb[1]) also turn brown. However, all 'tannins' are originally colorless. Phloroglucinol polymers (phlorotannins) in brown algae were discussed briefly earlier. Both gallotannins and proanthocyanidins are endogenous defense compounds, acting as feeding deterrents or digestion reducing agents due to protein precipitation and action as antioxidants.[100,101] Caffeoyl-esters (caffeoyltannins) are also protective substances.[93] Galloyl-esters (hydrolyzable tannins, including both gallo- and ellagitannins) are found mainly in woody species of dicotyledons such as the Hamamelidae, Rosidae, and Dillenidae (Fig. 6).[1] They are missing in the Magnolidae, most Caryophyllidae, except for the Polygonales order, and in the predominantly herbaceous Asteridae.[44,87,98] They occur, however, in herbaceous species such as the Geraniaceae and fruit pods of Leguminosae.[41] Galloyl-esters are produced from caffeic acid, a C_6-C_3 intermediate, but are chain-shortened to a C_6-C_1 intermediate[45,46] Proanthocyanidins (condensed tannins) are major constituents in some of the same groups as gallotannins, but are missing or low in other examples. They are found in the Polygonales, but not in the remainder of the Caryophyllidae (Fig. 6).[1] Both sometimes exist in the same family, or one type may exceed the other. It is not clear what these differences in distribution are due to, as they have similar, but not identical, protein precipitation and antioxidant characteristics.[100,101] Further studies are necessary to compare the roles of galloyl-esters (gallotannins), proanthocyanidins (condensed tannins), and caffeoyl-esters (caffeoyltannins) in providing these functions. In contrast to galloyl-esters (gallotannins or hydrolyzable tannins) found mainly only in dicots, proanthocyanidins are also found in gymnosperms and pteridophytes.

Caffeoyl- esters are found in all dicot orders as monosaccharide conjugates, but di- and trisaccharides have 'spotty' occurrences in the Asterales and Lamiales, both in the Asteridae (see taxonomic distributions in Table 1, Molgaard et al.[92] The other hydroxycinnamic acids involved are *p*-coumaric-, ferulic-, and sinapic- acids. The same two groups of the Asteridae, Asterales and Lamiales, show the highest number of occurrences of caffeoyltannins, 239 and 219, respectively, with negligible amounts of galloyl-esters and proanthocyanidins (Table 1 of Gottlieb et al.[1]). The only other high occurrence number for caffeoyltannins (102) is in the Rosaceae (Rosidae). Unfortunately, there is no good histochemical assay for either caffeoyl- or galloyl-esters comparable to the ones for proanthocyanidins. Whereas caffeoyl-esters are widespread in dicotyledons, especially in the Asteridae, ferulic acid is restricted to the Caryophyllaceae, and sinapic acid to the Capparales .[93] A high caffeoyl-ester content is indicated in members of the Asteridae, with negligible amounts of galloyl esters and proanthocyanidins.[1]

The use of the term 'tannin' is unfortunate because the compounds initially are colorless (absorption below 350 mm) and are generally soluble constituents of central vacuoles. Upon oxidation during senescence or programmed cell death, however, they are complexed with other compounds and are converted to a variety of brown, red-brown, and even black insoluble mixtures. Molecular weights reported vary from about 500 to 3000 or even greater. Gallotannins such as pentagalloyl glucose consist of 5 gallic acid units attached to glucose, with a molecular weight of about 1030 to about 1350 and an average of about seven units for all gallotannins.[45,46] The related elagitannins after oxidative coupling of adjacent galloyl residues to glucose have similar molecular weights. Proanthocyanidin polymers consist of a variable number of flavan-3,4-diols condensed to a basal flavan-3-ol, generally either catechin or epicatechin.[32,33,39] The sizes are quite variable, but a 4-unit structure would have a molecular weight of about 1088. The molecular weight of a B-1 type procyanidin with 16 epicatechin extender units is about 4930.[100,101] Caffeoyl-esters as mono-or trisaccharides are only in the smaller range from 330 to about 700. Lignins (a large insoluble polymer) and lignans (soluble dimers), however, can also form brown, red-brown, and black colors in mutants.[102,36]

Gottlieb and his associates accept the hypothesis of earlier evolutionists that woody plants in the dicotyledonous group have been replaced during evolution by herbaceous plants in all major taxonomic groups.[1,2,98] This is accompanied by the disappearance of galloyltannins and proanthocyanidin polymers typical of woody plants, followed by replacement with caffeoyl-esters, aliphatics, alkaloids, and isoprenoids, especially in the Asteridae. This involved the strengthening of the acetate pathway leading to aliphatics (no aromatic ring), in contrast to that of the shikimate route to aromatics, according to their argument. However, the shikimate pathway is fundamental for the biosynthesis of aromatic amino acids such as phenylalanine and tyrosine that are precursors of a variety of C_6-C_3 secondary metabolites involving *O*-methyltransferases.[6,56] In addition, whereas there is ample evidence that both galloyl-esters and proanthocyanidins serve as a means of defense due to their abilities to complex with proteins and antioxidant properties,[100,101] no comparable experimental work has been done with caffeoyl-esters to indicate a similar function within the herbaceous forms of the Asteridae, although these esters are considered to be protective compounds.[98] It should also be remembered that the aromatic pathway-derived flavonoids present in all dicots also possess a function as antioxidants, an attribute of the benefits of the above galloyl-esters, proanthocyanidins, and caffeoyl-esters.[4,103]

In the studies of Gottlieb and co-workers, the distribution of secondary compounds in dicots was recorded in a series of Tables in which the number of 'occurrences' of a particular type of compound in different families was recorded.

'Occurrences' were based on the total number or frequencies of compounds registered for a particular species (fossil evidence appears to be included).[1] No information is given concerning the amounts of these compounds based on micrograms per amount of tissue or per cell, the standard or classical technique in biochemical studies. Unfortunately, the only flavonoids shown are proanthocyanidins. Gottlieb does not distinguish effectively between strictly phenylpropanoid pathways leading to products such as caffeoyl-esters and conbined acetate-phenylpropanoid pathways to products such as flavonoids. He apparently uses the term flavonoid in a narrow sense to include only flavones and flavonols, so that proanthocyanidins and anthocyanidins, the latter a major constituent of flowers in the Asteridae and other groups, are not included. All C_{15} products resulting from the initial chalcone synthase step, as shown in Figure 3, should be called flavonoids.

Woody vs. Herbaceous Evolution

A basic problem in comparing the evolution of different 'tannins' or secondary products in woody vs. herbaceous plants is that the ecological and physiological backgrounds of woody vs. herbaceous types are different. Woody plants are specialized for a long-life existence, whereas herbaceous forms generally have shorter life spans. Woody trees extend high into the aerial environment with large roots in the soil. Highly lignified walls in different tissues are required for support, protection, and transport. Light energy requirements are high for woody plants. Herbaceous annuals and perennials, on the other hand, produced mainly by primary tissues only, live in a more limited immediate environment, even though sometimes on tree trunks. Survival in a particular environment by woody and herbaceous plants may require not only variable amounts of secondary products, but different types, depending on the size of the organism. Wind vs. insect pollination is a major difference between the two groups, although both wind and insect pollination occur in woody plants. Volatile end-products, such as isoprenoids and aromatics, or the visual color of flavonoids, such as flavonols and anthocyanins in flowers, are necessary for insect pollination.

Both primary tissues, produced by the terminal meristem, and secondary tissues, produced by the cambium, are present in stems of woody plants. Secondary tissues presumably contain larger amounts of lignin, galloyl-esters ('tannins'), and proanthocyanidins (condensed 'tannins') than primary tissues of the same plant, but comparative data are limited and need to be expanded. Phenolics in primary tissues of both woody and herbaceous plants should be compared qualitatively and quantitatively. The economy of a smaller size may be a key element in any differences. Phenolic contents of leaves and seeds in both types of plants need to be compared. Comparisons need to be made with gymnosperms such as Douglas-fir, where total proanthocyanidins on a dry weight basis have been compared in young stems, needles, and bark of young and old trees.[32]

Whereas both woody and herbaceous forms are found within all major sub-classes shown in Figure 6, and although the trend according to phylogenetic cladograms based on the chloroplast gene *rbcL* and the ribosomal RNA 18S gene is from woody to herbaceous forms in dicotyledons,[89,97] one wonders whether such an evolutionary trend has always occurred and will continue into the future. Both forms have been maintained in evolution. Studies of the relative amounts of phenolic types and their molecular genetic controls should be made with both 'early' herbaceous and 'early' woody forms of dicotyledons, because both types have been considered the first ancestors of dicotyledons. Both the paleoherb hypothesis and the magnolialean hypothesis are based mainly on morphological and developmental characteristics.[82,87,89] The biochemistry, enzymology, and molecular genetics of the Piperales, a possible primitive family according to the paleoherb theory, should be compared with that of a primitive family according to the magnolialean hypothesis (Fig. 6). Monocots, an herbaceous group, considered genetically close to herbaceous forms in the Magnolidae, should be included. The cladogram, based on *rbcL* sequence data, as shown by Doyle in his Figure 1 for nodulating plants, can serve as a model.[97] Also, how similar are the phenolics in herbaceous plants that have evolved at different times? Is there any evidence of either galloyl-esters or proanthocyanidins in 'early' herbaceous forms. The evolution of betalains, tyrosine derived alkaloids, as a substitute for anthocyanins in the Caryophyllidae, consisting of mainly herbaceous plants and some woody shrubs with anomalous secondary growth, is still a fascinating area that needs further research, especially as the common ancestor of the taxonomic groups was an herbaceous plant.[64,65,87]

FUTURE RESEARCH ABOUT PHENOLICS AND PHYLOGENETIC RELATIONSHIPS IN BRYOPHYTES AND EARLY LAND PLANTS

A recent review by Duff et al.[30] on the phylogenetic relationships of land plants compares cladistic analyses based on rDNA sequences with those of other analyses based on both molecular and morphological data. Both monophyletic and paraphyletic taxonomic groups are shown. Major groups of bryophytes are included in all of these as well as tracheophytes. Some detail of genera within the bryophyte three groups is shown, but no biochemical data are included. As shown in Figure 2, flavonoids from flavones to isoflavones (only one identification) have been identified in liverworts and mosses of the bryophytes. Flavanones apparently either do not accumulate or they have just not been identified. No flavonoids have been found in hornworts, but phenolics such as rosmarinic acid (caffeic acid-dihydroxyphenyl-lactic dimer) and lignans are present. Except for the minor appearances of flavonols, isoflavones (only one occurrence), and aurones, flavones, and to a lesser extent, 3-deoxy anthocyanidins can be considered the initial major forms. They are both 3-deoxy-compounds with no substitution at C3 of the C-ring. The vegetative functions of these two types of compounds in bryophytes are unknown, but defense involving protection due to UV-light absorption is a strong possibility for flavones with their 310-350 nm long-wave-length absorption peaks in both the UV B and A areas.[19] This was probably initially important because the shorter wave lengths of UV light were more intense at the time of origin of bryophytes. Flavanones, if accumulated, would also protect by absorption of short UV-light. It is unclear what advantage 3-deoxy-anthocyanidins with major absorption peaks just below 500 nm in the green portion of the spectrum might have in protection of cells below the surface.[3,18] Further work is needed in the area of green light effects.

Chalcone synthase to produce chalcones and chalcone isomerase (initially perhaps only a non-enzymic reaction) to convert to flavanones would be the expected first steps to flavonoids. Then, the origin of a flavone synthase would have permitted the accumulation of flavones, whereas both a NADPH reductase and 3-deoxy-anthocyanidin synthase would be necessary to produce the pigmented 3-deoxy-flavonoids. Markham argues that the lack of flavonoids within some flavonoid producing taxa of existing liverworts is a derived condition.[17] The cladistic analyses as summarized by Duff et al.[30] need to be enlarged to include the genera and species in which flavones and deoxy-anthocyanidins have been identified and those that have been shown to be missing these products. Furthermore, more phytochemical identifications need to be made in bryophytes, including quantitative determinations. Mosses might be a good, initial group to study as they have been considered model systems in developmental studies.[104] Pteridophytes, the first land plant group, also need to be studied more thoroughly.

CONCLUSION

Flavonoids (C_6-C_3-C_6 monomers and polymers) and lignin-lignans (C_6-C_3 polymers) are the major phenolics in the bryophyte-tracheophyte complex of land plants. Flavonoids evolved into the greatest variety of forms as end-products in all groups, whereas lignin, although present in all tracheophytes, became quantitatively the major end-product in woody plants of the gymnosperm and angiosperm groups. Galloyl-esters were dominant in several major angiosperm woody orders, serving a defense function similar to that of the proanthocyanidins (a flavonoid), whereas various forms of caffeoyl-esters of various types were widespread or sometimes present in sporadic groups. Within the dicotyledons of the Angiosperms, the trend is generally believed to be from woody to herbaceous forms. However, since woody forms still dominate large taxonomic orders, the trend might best be considered to be two-directional, leading to woody forms on the one hand, and herbaceous groups on the other. Since woody forms contain secondary tissues in addition to primary tissues, the concentrations of phenolics per plant are much larger, and the types of phenolics may vary in the two groups. For instance, procyanidins and galloyl-esters (both sometimes called 'tannins) are probably more abundant in secondary tissues, whereas the simpler caffeoyl-esters may be more characteristic of and more concentrated in herbaceous forms that have only primary tissues.

Much biochemical and enzymatic work still needs to be done and then must be related to both structural and regulatory genes being studied with molecular techniques. The recent use of cladograms based on molecular data may now be the best method to portray phylogenetic relationships among tracheophyte groups, but the classifications of Cronquist or Dahlgren based mainly on morphology are still useful and are easier to visualize. A comparison of the evolution of phenolics and their functions in all types of land plants is still a primary need.

REFERENCES

1. GOTTLIEB, O. R., BORIN, M. R. D. M. B., KAPLAN, M. A. C. 1995. Biosynthetic interdependence of lignins and secondary metabolites in angiosperms. Phytochemistry 40:99-133.
2. GOTTLIEB, O. R., BORIN, M. R. D. M. B. 1998. Evolution of angiosperms *via* modulation of antagonisms. Phytochemistry 49:1-15.
3. STAFFORD H. A. 1995. Metabolism and regulation of phenolics: Gaps in our knowledge. *in* Phytochemicals and Health, pp. 15-30 *in* Current Topics in Plant Physiology, (DL Gustine and HF Flores, eds). ASPP Series: Vol 15. Rockville, Maryland.
4. COOPER-DRIVER, G. A, BHATTACHARYA, D. 1998. Role of phenolics in plant evolution. Phytochemistry 49:1165-1174.

5. KOES, R. E., QUATTROCCHIO, F., MOL, J. N. M. 1994. The flavonoid biosynthetic pathway in plants: Function and evolution. Bioessays 16:123-132.
6. WEAVER, L. M., HERRMANN, K. L. 1997. Dynamics of the shikimate pathway. Trends Plant Sci. 2:346-351.
7. LEWIS, N.G., DAVIN, L. B., SARKANEN, S. 1998. Lignin and lignan biosynthesis: Distinctions and reconciliations. pp. 1-27 *in* Lignin and Lignan Biosynthesis (N. G. Lewis, and S. Sarkanen, eds.). American Chemical Society Symposium Series 697. 435p.
8. STERN ,J. L., HAGERMAN, A. E., STEINBERG, P. D. MASON, P. K. 1996. Phlorotannin-protein interactions. J. Chem. Ecol. 22:1877-1999.
9. GRAHAM, L. E. 1993.Origin of Land Plants. Wiley & Sons. N.Y. 287p.
10. ARNOLD, T.M., TARGETT, N. M. 1998. Quantifying *in situ* rates of phlorotannin synthesis and olymerization in marine brown algae. J. Chem. Ecol. 24:577-595.
11. BHATTACHARYA, D., MEDLIN, L. 1998. Algal phylogeny and the origin of land plants. Plant Physiol. 116:9-15.
12. XIONG F, KOMENDA, J., KOPECKY, J., NEDBAL, L. 1997. Strategies of ultraviolet-B protection in microscopic algae. Physiol. Plant. 100:378-388.
13. DELWICHE, C. F. 1989. Lignin-like compounds and sporopollenin in *Coleochaete*, an algal model for land plant ancestry. Science 245:399-401.
14. WIERMANN, R., GUBATZ, S. 1992. Pollen wall and sporopollenin 1992 . Intern. Rev. Cytol. 140: 35-72.
15. KAWASE, M., TAKAHASHI, M. 1996. Gas chromatography-mass spectrophotometric analysis of oxidative degradation products of sporopollenin in *Magnolia grandiflora* (Magnoliaceae) and *Hibiscus yriacua* (Malvaceae). J. Plant Res. 109:297-299.
16. KENRICK P., CRANE, P.R. 1997. The origin and early evolution of plants on land. Nature 389:33-39.
17. MARKHAM, K.R. 1990. Bryophyte flavonoids, their structures, distribution, and evolutionary significance. pp. 143-159 *in* Bryophytes. Their Chemistry and Chemical Taxonomy. (H.D. Zinsmeister and R. Mues, eds.) Clarendon Press, Oxford.
18. MARKHAM, K.R, RYAN, J. G., BLOOR, S. J., MITCHELL, K. L. 1998. An increase in luteolin:apigenin ratio in *Marchantia polymorpha* on UB-B enhancement. Phytochemistry. 48:791-794.

19. STAFFORD, H. A. 1965. Flavonoids and related phenolic compounds produced in the first internode of *Sorghum vulgare* Pers. in darkness and in light. Plant Physiolo. 40:130-138.
20. KUNZ, S. , BURKHARDT, G., BECKER, H. 1994. Riccionidins A and B, anthocyanidins from the cell walls of the liverwort *Ricciocarpos natans*. Phytochemistry 35:233-235.
21. BRINKMEIR, E., GEIGER, H., ZIINSMEISTER, H. D. 1999 Biflavonoids and 4,2'-epoxy-3-phenylcoumarins from the moss *Mnium hornum*. Phytochemistry 52:297-302.
22. GEIGER H. 1990. Biflavonoids in bryophytes. pp. 161-170 *in* Bryophytes. Their Chemistry and Chemical Taxonomy. (H. D. Zinsmeister and R. Mues, eds.) Clarendon Press, Oxford.
23. GORHAM, J. 1990. Phenolic compounds, other than flavonoids, from bryophytes. pp. 171-200. *in* Bryophytes. Their Chemistry and Chemical Taxonomy. (H. D. Zinsmeister and R. Mues eds.) Clarendon Press, Oxford.
24. MARKHAM, K.R., PORTER, L. J. 1978. Chemical constituents of the bryophytes. Progr. in Phytochemistry. 5:181-272.
25. MARKHAM, K. R.,GIVEN, D. R., 1998. The major flavonoids of an antarctic Bryum. Phytochemistry 27:2843-2845.
26. SWAIN, T. 1975. Evolution of flavonoid compounds. pp. 1096-1129 *in* The Flavonoids. (J.B. Harborne, T.J. Mabry,H. Mabry eds.). Academic Press, N. Y.
27. MUES, R. 1990. The significance of flavonoids for the classification of bryophyte taxa of different taxonomic rank. pp. 421-435. *in* Bryophytes. Their Chemistry and Chemical Taxonomy. (H. D. Zinsmeister and R. Mues eds.). Clarendon Press, Oxford.
28. DINKOVA-KOSTOVA A. T., GANG, D. R., DAVIN, L. B., BEDGAR, D. L., CHU A., LEWIS, N.G. 1996. (+)-Pinoresinol/(+)-lariciresinol reductase from *Forsythia intermedia*. J. Biol. Chem. 271:29473-9482.
29. QIU, Y-L, PALMER, J. D. 1999. Phylogeny of early land plants: Insights from genes and genomes. Trends Plant Sci.4:26-30.
30. DUFF, R. J., NICKRENT, D. L. 1999. Phylogenetic relationships of land plants using mitochondrial small-sub-unit RDNA sequences. Amer. J. Bot. 86:372-386.
31. STAFFORD, H. A. 1991. Flavonoid evolution: An enymic approach. Plant Physiol. 96:680-685.
32. STAFFORD, H. A., SMITH, E. C., WEIDER, R. M. 1989. The development of proanthocyanidins (condensed tannins) and other phenolics in bark of *Pseudotsuga menziesii*. Can. J. Bot. 67:1111-1118.

33. STAFFORD, H. A. 1988. Proanthocyanidins and the lignin connection. Phytochemistry 27:1-6.
34. STICH, K., FORKMAN, G. 1988. Biosynthesis of 3-deoxyanthocyanins with flower extracts from *Sinningia cardinalis*. Phytochemistry 27:785-789.
35. GUJER, R., MAGNOLATO, D., SELF, R. 1986. Glucosylated flavonoids and other phenolic compounds from sorghum. Phytochemistry 25:1431-1436.
36. STAFFORD, H. A. 1998. Recent discoveries about phenolics. pp. 345-361 *in*: Discoveries in Plant Biology, Vol. 1, (S-D Kung and S-F Yang, eds.). World Scientific, Singapore.
37. STAFFORD, H. A. 1998. Teosinte to maize - some aspects of missing biochemical and physiological data concerning regulation of flavonoid pathways. Phytochemistry 49:285-293.
38. NICHOLSON R. L., HIPSKIND, J. D. 1996. Resistance in the Poaceae: Different roles for phenolic compounds. pp. 209-217. *in* Proc. of the Japan-US Scientific Seminar on Molecular Aspects of Pathogenicity and Host Resistance: Requirements for Signal Transduction. (D. Mills, H. Kunoh, N. T. Keen, eds.) Mayamma. PS Press.
39. STAFFORD, H. A. 1990. Flavonoid Metabolism. CRC Press, Inc. Boca Raton, FL. 298p.
40. DIXON, R. A., HARRISON, M. J., PAIVA, N. L. 1995. Isoflavonoid phytoalexin pathway: From enzymes to genes to transcription factors. Physiol. Plant. 93: 385-392.
41. DEWICK P. M. 1994. Isoflavonoids. pp. 117-238 *in* The Flavonoids: Advances in Research Since 1986. (J. B. Harborne, ed.) Chaplin and Hall University Press, Cambridge, Great Britain.
42. GUO, Z-J, LAMB, C., DIXON, R. A. 1997. Release and biological activity of diffusible signal compounds from elicited plant cells. J. Plant Physiol. 151:699-710.
43. WANG, Z-X, Li, S-M., LÖSCHER, R., HEIDE, L. 1997. 4-Coumaryl coenzyme A 3-hydroxylase activity from cell cultures of *Lithospermum erythrorhizon* and its relationship to polyphenol oxidase. Arch. Biochem. Biophys. 347:249-255.
44. HADDOCK, E. A., GUPTA, R. K., SABAH, M. K. , AL-SHAFI, S. M. K., LAYDEN, K., HASLAM, E., MAGNOLATO, D. 1982. The metabolism of gallic acid and hexahydroxydiphenic acid in plants: Biogenetic and molecular taxonomic considerations. Phytochemistry 21:1049-1062.
45. DENZEL, K., GROSS, G. G. 1991. Biosynthesis of gallotannins. Planta 184:285-289.

46. GROSS, G. G. 1992. Enzymatic synthesis of gallotannins and related compounds. Recent Adv. in Phytochemistry 26:297-324.
47. SCHRÖDER, J. 1997. A family of plant-specific polyketide synthases: Facts and predictions. Trends Plant Sci. 2:373-378.
48. ZUURBIER, K. W. M., LESER, J., BERGER, T., HOFTER, A. J. P., SCHRÖDER, G., VERPORTE, R., SCHRÖDER, J. 1998. 4-Hydroxy-2-pyrone formation by chalcone and stilbene synthase with nonphysiological substrates. Phytochemistry 49:1945-1951.
49. KOCHS, G., WERCK-REICHERT, D., GRIESEBACH, H. 1992. Further characterization of cytochrome P450 involved in phytoalexin synthesis in soybean: Cytochrome P450 cinnamate 4-hydroxylase and 3,9-dihydroxypterocarpan 6a-hydroxylase. Arch. Biochem. Biophys. 293:187-194.
50. CHAPPLE C. 1998. Molecular-genetic analysis of plant cytochrome P450-dependent monooxygenases. Annu. Rev. Plant Phys.& Plant Mol. Biol.49:311-43.
51. DE VETTEN, N., HORST, J. T., VAN SCHAIK, J., H-P, DE BOER, A., MOL, J., KOES, R. 1999. A phytochrome b5 is required for full activity of flavonoid 3',5'-hydroxylase, a cytochrome P450 involved in the formation of blue flower colors. Pro. Nat. Acad. Sci.USA 96:778-783.
52. NELSON, D. R. 1999 Cytochrome P450 and the individuality of species. Arch. Biochem. Biophys. 369:1-10.
53. SAITO, K., KOBAYASHI, M., GONG, Z., TANAKA, Y., YAMAZAKI, M. 1999. Direct evidence for anthocyanidin synthase as a 2-oxoglutarate-dependent oxygenase: Molecular cloning and functional expression of cDNA from a red forma of *Perilla frutescens*. Plant J. 17:181-189.
54. BAKER, M. E. LUU-THE, VAN SIMARD, J., LABRIE, F. 1990. A common ancestor for mammalian 3β-hydroxysteroid dehydrogenase and plant dihydroflavonol reductase. Biochem. J. 269:558-559.
55. GANG D. R., KASHARA, H., XIA, Z-Q., MIJNSBRUGGE, K. V., BAUW, G., BOERJAN, W., MONTAGU, M. V., DAVIN, L. B., LEWIS, N. G. 1999. Evolution of plant defense mechanisms. J. Biol. Chem. 274:7516-7527.
56. IBRAHIM, R. K., BRUNEAU, A., BANTIGNIES, B. 1998. Plant *O*-methyltransferases: Molecular analysis, common signature and claddification. Plant Mol. Biol. 36:1-10.
57. SINGH, K. B. 1998. Transcriptional regulation in plants: The importance of combinatorial control. Plant Physiol. 118:1111-1120.

58. GERATS, A.G.M., MARTIN, C. 1992. Flavonoid synthesis in Petunia hybrida; genetics and molecular biology of flower colour. Recent Adv. in Phytochem.26:165-199.
59. WEISSHAAR, B., JENKINS, G. I. 1998. Phenylpropanoid biosynthesis and its regulation. Cur. Opin. in Plant Biol. 1:251-257.
60. QUATTROCCHIO, F., WING, J. F., VAN DER WOUDE, K., MOL, J. N. M., KOES, R. 1998. Analysis of βHLH and MYB domain proteins: Species-specific regulatory differences are caused by divergent evolution of target anthocyanin genes. Plant J. 13:475-488.
61. STYLES, E. D., CESKA, O. 1989. Pericarp flavonoids in genetic strains of *Zea mays.* Maydica 34:227-237.
62 GROTEWOLD, E., DRUMMOND, B. J., BOWEN, B., PETERSON, T. 1994. The myb-homologous *P* gene controls phlobaphene pigmentation in maize floral organs by directly activating a flavonoid biosynthetic gene subset. Cell 76:543-553.
63. GROTEWOLD, E., CHAMBERLAIN, M., SNOOK, M., SIAME, B., BUTLER, L., SWENSON, J., MADDOCK, S., ST. CLAIR, G., BOWEN, B. 1998. Engineering secondary metabolism in maize cells by ectopic expression of transcription factors. Plant Cell 10:721-740.
64. CLEMENT, J. S., MABRY, T. J., WYLER, H., DREIDING, A. S. 1994. Chemical review and evolutionary significance of the betalains. pp. 247-261. *in*: Evolution and Systematics of the Betalains (H-D. Behnke and T. J. Mabry, eds.) Springer-Verlag, N.Y.
65. STAFFORD H. A. 1994. Anthocyanins and betalains: Evolution of the mutually exclusive pathways. Plant Science 101:91-98.
66. VOGT, T., ZIMMERMAN, E., GRIMM, R., MEYER, M. STRACK, D. 1997. Are the characteristics of betanidin glucosyltransferases from cell-suspension cultures of *Dorotheanthus belliformis* indicative of their phylogenetic relationship with flavonoid glucosyltransferases. Planta 203:349-361.
67. STEINER, U., SCHLIEMANN, W., BÖHM, B., STRACK, D. 1999. Tyrosinase involved in betalain biosynthesis of higher plants. Planta 208:114-124.
68. SCHLIEMANN, W., KOBAYASHI, N., STRACK, D. 1999. The decisive step in betaxanthin biosynthesis is a spontaneous reaction. Plant Physiol. 119:1217-1232.
69. WATSON II, J. F., GOLDMAN, I. L. 1997. Inheritance of a gene conditioning blotchy root color in table beet (*Beta vulgaris* L.). J. Hered. 88:540-543.

70. LEWIS, N. G., DAVIN, L. B. 1998. The biochemical control of monolignol coupling and structure during lignan and lignin biosynthesis. pp. 334-361 *in* Lignin and Lignan Biosynthesis (N. G. Lewis and S. Sarkanen, eds.), American Chemical Society Symposium Series 697. 435 p.
71. WHETTEN, R. W., MACKAY, J. J., SEDEROFF, R. R. 1998. Recent advances in understanding lignin biosynthesis. Ann. Rev. Plant Physiol. Plant Mol. Biol. 49:585-609.
72. BOUDET, A-M. 1998. A new view of lignification. Trends in Plant Sci. 3:67-71.
73. BUCCIARELLI, B., JUNG, H. G., OSTRY, M. E., ANDERSON, N. A., VANCE, C. P. 1998. Wound reponse characteristics as related to phenylpropanoid enzyme activity and lignin deposition in resistant and susceptible *Populus tremuloides* inoculated with *Entoleuca mammata* (*Hypoxylon mammatum*). Can. J. Bot. 76:1282-1288.
74. RALPH, J., HATFIELD, R. D., GRABBER, J. H., JUNG, H-J, G., QUIDEAU, S., HELM, R. F. 1998. Cell wall cross-linking in grasses by ferulates and diferulates. pp. 209-236. *in* Lignin and Lignan Biosynthesis (N.G. Lewis and S. Sarkanen, eds.), American Chemical Society Symposium Series 697. 435 p.
75. ZHONG, R., MORRISON III, W. H., NEGREL, J., YE, Z-H. 1998. Dual methylation pathways in lignin biosynthesis. Plant Cell 10:2033-2045.
76. CHEN, F.,YASUDA, S., FUKUSHIMA, K. 1999. Evidence for a novel biosynthetic pathway that regulates the ratio of syringyl to guaiacyl residues in lignin in the differentiating xylem of *Magnolia kobus* DC. Planta 207:597-603.
77. GANG, D. R., FUJITA, M., DAVIN, L. B., LEWIS, N. G. 1998. The 'abnormal' lignins: mapping heartwood formation through the lignan biosynthetic pathway. pp. 389-421 *in* Lignin and Lignan Biosynthesis (N. G. Lewis, and S. Sarkanen, eds.), American Chemical Society Symposium Series 697. 435 p.
78. SEDEROFF, D. R., MACKAY, J. J., RALPH, J., HATFIELD, R. D. 1999. Unexpected variation in lignin. Cur. Opin. in Plant Biol. 2:145-152.
79. RALPH, J, PENG, J., LU, F., HATFIELD, R. D., HELM, R. F. 1999. Are lignins optically active? J. Agric. Food Chem. 47: 2991-2996.
80. DIXON, R. A., MAXWELL, C. A., WEITING N., OOOMEN, A., PAIVA, N. L. 1994. Genetic manipulation of lignin and phenylpropanoid compounds involved in interactions with microorganisms Recent Adv. in Phytochem. 28:153-178.

81. BOUDET, A-M., GOFFNER, D., MARQUE, C., PETTENATI, J. G., 1998. Genes involved in the final steps of monolignol biosynthesis and their manipulation for tailoring new lignin. *in* Lignin and Lignan Biosynthesis (N. G. Lewis and S. Sarkanen, eds.).
82. SPORNE, K. R. 1980. A re-investigation of character correlations among dicotyledons. New Phytol. 85:419-449.
83. TAYLOR T. N. 1988. The origin of land plants: Some answers, more questions. Taxon 37:805-833.
84. TAYLOR, T. N., HICKEY, L. J. 1992. Phylogenetic evidence for the herbaceous origin of angiosperms. Plant Syst. Evol. 180:137-156.
85. CHAPMAN, J. L. 1987. Sporne's advancement index revisited. New Phytol. 106:319-332.
86. BENEVIDES, P. J. C., SARTORELLI, P., KATO, M. J. 1999. Phenylpropanoides and neolignans from *Piper regnellii*. Phytochemistry 52:339-343.
87. CRONQUIST, A. 1993 An Integrated System of Classification of Flowering Plants. Columbia Univ Press, NY. 1262 p.
88. SOLTIS, D.E., SOLTIS, P. S., MORGAN, D. R., SWENSEN, S. M., MULLIN, B. C., DOWD, J. M., MARTIN, P. G., 1995. Chloroplast gene sequence data suggest a single origin of the predisposition for symbiotic nitrogen fixation in angiosperms. Proc. Natl. Acad. Sci. USA 92:2647-2651.
89. SOLTIS, D. E., SOLTIS, P. S., NICKRENS, D. L., JOHNSON, L. A., HAHN, W. J., HOOT, S. B., SWEERE, J. A., KUZOFF, R. K., KRON, K.A., CHAE, M. W., SWENSEN, S. M., ZIMMER, E. A., CHAW, S-M, GILLESPIE, L. J., KRESS, W. J., SYTSMA, K. J. 1997. Angiosperm phylogeny inferred from 18S ribosomal DNA sequences. Ann. Mo. Bot. Garden 84:1-49.
90. SOLTIS, D. E., SOLTIS, P. S., DOYLE, J. J. , (eds.) 1998. Molecular Systematics of Plants II DNA Sequencing, Kluwer Academic Publishers, Boston. 574 p.
91. ESAU, K. 1977. Anatomy of Seed Plants, 2nd edition, Wiley & Sons, New York. 550 p.
92. MOLGAARD, P., RAVEN, H. 1988. Evolutionary aspects of caffeoyl ester distribution in dicotyledons. (Dahlgrenograms of Dicots). Phytochemistry 27:2411-2421.
93. DAHLGREN, G. 1991. Steps toward a natural system of the dicotyledons: Embryological characters. Aliso 13: 107-165.

94. DAHLGREN, G. 1989. The last Dahlgrenogram. System of classification of the dicotyledons. pp. 249-260 *in* The Davis and Hedge Festschrift. Edinburgh University Press, Edinburgh.
95. CUERRIER, A., BROUILLET, L., BARABÉ, D. 1998. Numerical and comparative analyses of the modern systems of classification of the flowering plants. Bot. Rev. 64: 323-355.
96. CRONQUIST, A. 1987. A botanical critique of cladism. Bot. Rev. 53:1-49.
97. DOYLE, J. J. 1998. Phylogenetic perspectives on nodulation: Evolving views of plants and symbiotic bacteria. Trends Plant Sci. 3:473-478.
98. GOTTLIEB, O. R. , KAPLAN, M. A. C., KUBITZKI, K. 1993. A suggested role of galloyl esters in the evolution of dicotyledons. Taxon 42:539-552.
99. PHILLIPS, D. A. 1992. Flavonoids: Plant signals to soil microbes. Recent Adv. in Phytochem. 26:201-231.
100. HAGERMAN, A.E., REIDI, K. M., JONES G. A., SOVIK, K. N., RITCHARD, N. T., HARZFIELD, P. W., REICHEL, T.R.L. 1998. High molelcular weight plant polyphenolics (tannins) as biological antioxidants. J. Agric. Food Chem. 46:1887-1892.
101. HAGERMAN, A.E., RICE, M. E., RITCHARD, N. T. 1998. Mechanisms of protein precipitation for two tannins, pentagalloyl glucose and epicatechin$_{16}$ (4-->8) catechin (procyanidin). J. Agric. Food Chem. 46: 2590-2595.
102. HALPIN, C., HOLT, K., CHOJECKI, J., OLIVER, D., CHABERT, B., MOMIES, B., EDWARDS. K., EDWARDS, E., BARAKATE, A., FOXON, G. A. 1998. Brown-midrib maize (bm1) - a mutation affecting the cinnamyl alcohol dehydrogenase gene. Plant J. 14:545-563.
103. RICE-EVANS, C. A., MILLER, N. J., PAGANGA, G. 1997. Antioxidant properties of phenolic compounds. Trend. Plant Sci. 2:152-159.
104. COVE, D. J., KNIGHT, C. D., LAMPARTER, T. 1997. Mosses as model systems. Trends in Plant Sci. 2:99-105.

Chapter Three

THE FAMILY OF CHALCONE SYNTHASE-RELATED PROTEINS: FUNCTIONAL DIVERSITY AND EVOLUTION

Joachim Schröder

Institut für Biologie II
Universität Freiburg
Schänzlestrasse 1
D-79104 Freiburg, Germany

Evolution of Metabolic Pathways – J.T. Romeo et al. (Editors)

INTRODUCTION

Chalcone synthase (CHS), the prototype of the family, appears to be ubiquitous in higher plants. It synthesizes the precursor for the biosynthesis of thousands of biologically important substances, and several of these probably have interesting and beneficial effects on human health.[1-5] Roles in plants include flower colors, UV protection, defense against pathogens (phytoalexins), interaction with microorganisms (*e.g.* Rhizobia), and fertility.[6-12] Accordingly, the interest in this enzyme has always been high, initially in the protein and its properties, and later in the genes and the regulation of their expression.

One of the consequences of the shift in interest from enzyme activity to DNA sequences and the ease of modern cloning techniques is that the public databases contain more than 140 sequence entries labelled as CHS, and that the identification in the majority of cases is based on sequence similarity rather than on functional evidence. However, there is clear evidence now that CHS is just one member of a family of related proteins that perform condensing reactions in a variety of different pathways, and it seems possible that some of the earlier assignments will need to be changed in the future.

This review focuses on recent advances in understanding the functions of proteins in this family. The members are basically defined by sequence similarity, and, thus, the assignment requires molecular characterization. However, there are several enzymes whose properties suggest that they belong to the family, and precursor feeding experiments suggest several more enzyme activities that would nicely fit into the family. Examples for both will be included in the discussion. The most exciting prospects, however, are in the identification of new functions and finally in the rational design of new enzyme activities. The recent first elucidation of a CHS crystal structure will provide important clues for such approaches.[13,14]

OVERVIEW OF THE REACTIONS

The proteins are condensing enzymes, and Figure 1 summarizes the principle of the reactions, as deduced from functional and structural analyses.[13,15] The reaction begins with the binding of the starter substrate CoA-ester, the covalent attachment of the substrate residue to the active site cysteines, and the release of the free CoASH. The binding of malonyl-CoA is the next step, and chain extension proceeds with an acetyl-CoA anion produced by decarboxylation of malonyl-CoA. The anion is stabilized by various interactions with residues on the protein and by tautomerization between the keto and enol species, but apparently not by covalent binding of the residue to the enzyme protein.[13] The starter residue is transferred from the active site cysteine to the chain extender unit, thus liberating the -SH group of the active site and producing the diketide product bound to CoA. The diketide residue is the substrate for the next chain extension which follows the same principle. When compared to other polyketide synthases, the distinctive features of the reactions are the direct use of CoA-esters (no involvement of acyl carrier proteins or 4'-phosphopantetheine arms), the presence of only one essential cysteine in the protein (covalent binding of the starter residue, or of the intermediate to be used in the next elongation cycle), and the chain extension without covalent binding of the extender unit to the protein.

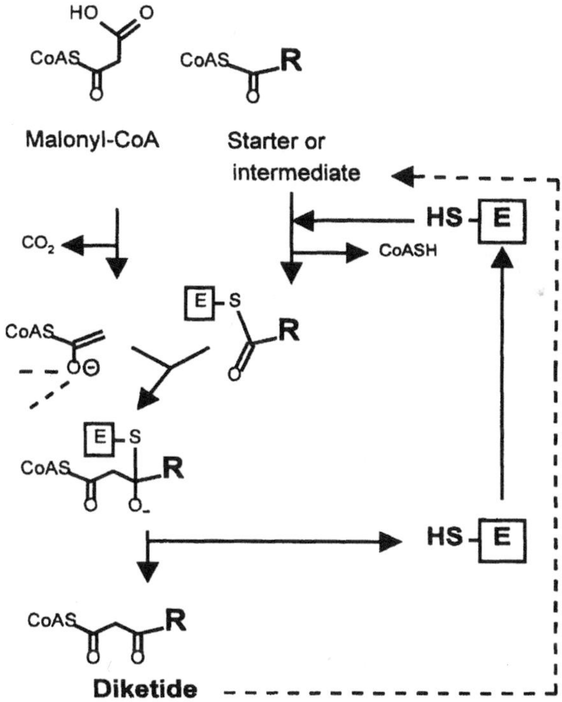

Figure 1. Model of the condensing reaction performed by CHS and related proteins. Only the first condensation is shown; the diketide product is the substrate for the next chain extension. HS-E, the active site cysteine (Cys169)[140] that covalently binds the starter residue prior to the condensing reaction.

There is one report that seemingly contradicts this model; it describes binding experiments that show that the chain extender residue is covalently attached to the active site cysteine.[16] However, there are reasons to argue that the conditions were to some extent artificial and that the results should be interpreted differently. The important points of that work are: a) the binding experiments with malonyl-CoA were carried out in the absence of a starter substrate (*e.g.* coumaroyl-CoA), b) the enzymes decarboxylated malonyl-CoA to acetyl-CoA, and c) the moieties bound to the active site cysteine were predominantly acetyl and not malonyl residues.[16] The detection of malonyl residues might well reflect some transfer in the absence of a starter substrate. The predominance of bound acetyl residues indicates a decarboxylation at some stage. It seems that the interpretation in the paper failed to consider the possibility that the acetyl-CoA produced by decarboxylation of malonyl-CoA could serve as a starter substrate, and that the acetyl residues bound to the active site cysteine actually represented primed starter units for condensing reactions. Experiments in our laboratory (unpublished results) confirm that stilbene synthase (STS) decarboxylates malonyl-CoA, and they also show that the product, acetyl-CoA, is used for condensing reactions with malonyl-CoA, if the physiological starter is not available. The activities are lower than with the physiological substrates, but the products are easily detected if one looks for them. Other data indicate that this is a property shared by all enzymes of the family. The most obvious example of such self-priming by malonyl-CoA decarboxylation is the pyrone synthase (2PS) that will be discussed later in some detail: acetyl-CoA is the physiological substrate, and appreciable reaction rates are obtained in incubations that contain only malonyl-CoA.[17] Taken together, these results suggest that binding experiments with malonyl-CoA in the absence of starter units do not provide convincing evidence that the chain extender unit is covalently bound to the protein in the normal course of the reaction. It is interesting to note that comparable self-priming activities *via* decarboxylation of the chain extender are well-known from other polyketide synthases, and recent data indicate that this may be much more important and complex than previously thought.[18]

The functional diversity of the proteins in the family of CHS-related proteins is achieved by differences in the substrate specificities, in the number of condensation reactions, in the type of ring closure of the released products, and in some cases by modification of reaction intermediates by additional proteins. Figure 2 gives an overview that summarizes the intermediates after one, two, or three condensation reactions, and the released products identified from the various activities of the enzymes. All of the complex reactions are carried out by homodimeric proteins with subunits of 40-44 kDa.

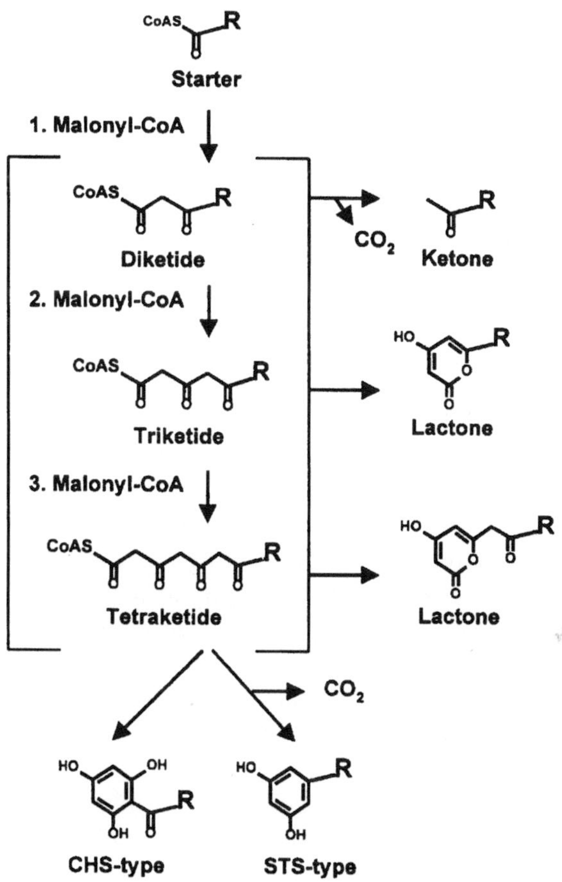

Figure 2. Overview of reaction intermediates (in brackets) and products identified from reactions with CHS-related proteins.

CHALCONE SYNTHASE-TYPE REACTIONS

These enzymes perform three condensation reactions and fold the resulting tetraketide intermediate to a characteristic new aromatic ring system (Fig. 2).

CLONED MEMBERS OF THE FAMILY

The Prototype of the Family: CHS. The enzyme uses starter CoA-esters derived from the phenylpropanoid pathway (Fig. 3). The first demonstration of CHS activity *in vitro* was reported in 1972 with extracts from parsley (*Petroselinum crispum*) cell suspension cultures.[19] The properties of the enzymes from more than 30 plants have been described in the meantime.[20,21] In most plants, 4-coumaroyl-CoA is considered as the physiological substrate, but *in vitro,* the enzymes are also active with other CoA-esters from the phenylpropanoid pathway, *e.g.* cinnamoyl-CoA and caffeoyl-CoA, albeit the activities are usually much lower. Cinnamoyl-CoA is a physiological substrate in some plants because pinocembrin or its derivatives are known from several species. The significance of the activities with caffeoyl-CoA have been discussed repeatedly, but without a clear-cut conclusion.[21,22] Recent data with two CHSs cloned from barley (*Hordeum vulgare*) support that caffeoyl-CoA is a physiological substrate under certain conditions in this plant, because CHS2 (induced by UV stress or pathogen attack) clearly preferred that CoA-ester over 4-coumaroyl-CoA *in vitro*, while CHS1 did not.[23]

The first CHS sequence was published in 1983 for the enzyme from parsley (*Petroselinum crispum*),[24] and many more sequences have been reported in subsequent years. However, as noted before, the primary interest often was in the investigation of transcriptional regulation or of the promoter properties, and the possibility was often ignored that the cloned cDNAs/genes might encode another protein function.

Acridone Synthase (ACS). Acridone alkaloids are mostly found in some genera of the Rutaceae family. Experiments with crude extracts from cell cultures of *Ruta graveolens* revealed an enzyme activity that used *N*-methylanthraniloyl-CoA to synthesize 1,3-dihydroxy-N-methylacridone.[25,26] The product was most easily explained by a CHS-type reaction combined with a second ring formation (Fig. 3). Subsequent experiments with the purified enzyme confirmed the functional similarities to CHS, and the sequences from seven peptides strongly suggested that the protein belonged to the family.[27] The analysis of a cloned cDNA proved that ACS is a member of the protein family (>65% overall identity with CHSs), and the function was unambiguously identified after heterologous expression in *E. coli*.[28] Although the early data suggested that the enzyme is a monomer,[28] a later analysis with the recombinant protein and also with a second ACS cloned from *Ruta graveolens* showed that the enzymes are dimers,[29] like all CHS-related proteins. The ACS does not accept 4-coumaroyl-CoA as substrate.[28] The amino acid residues responsible for the functional differences to CHS (substrate specificity, additional ring closure) are not yet identified, nor has the CHS from *Ruta graveolens* been cloned.

Candidate Enzyme Activities with CHS-Type Reactions

Valerophenone Synthase (VPS). The ripe cones of hop (*Humulus lupulus*, *Cannabinaceae*) contain up to 20% of bitter acids, *e.g.* humulone and cohumulone, which are converted during the brewing process to the isoforms that are important for the flavor and taste of beer. A few years ago, the detection of phlorisovalerophenone and phlorisobutyrophenone and their likely role as intermediates suggested that the synthesis of the aromatic ring could be by a CHS-type reaction that used isovaleryl-CoA and isobutyryl-CoA as substrates, respectively (see Fig. 3 for the reaction with isovaleryl-CoA), and the predicted enzyme activities could be demonstrated *in vitro*.[30,31] All of the properties suggested that the protein could be related to CHS. The enzyme is now called valerophenone synthase (VPS) because of its substrate preference for isovaleryl-CoA. The relationship with CHS was confirmed recently after a protein purification that separated CHS and VPS. The protein had no activity with the CHS substrate, 4-coumaroyl-CoA, and two peptide sequences (35 and 30 amino acids) obtained from the purified protein clearly showed the close relationship to CHS.[32] This was not unexpected because early data had shown that CHS accepts aliphatic substrates,[33] and more recent results indicated that a CHS from a plant not synthesizing the hop-specific products can actually function as VPS.[34] Interestingly, the public databases contain a CHS-related cDNA sequence from hop (accession AB015430) that precisely matches the peptide sequences. The description in the data file strongly suggests that the cDNA encoded protein is VPS, but this needs to be demonstrated directly by heterologous expression and functional analysis.

Benzophenone Synthase (BPS). Xanthones are a group of natural products with interesting pharmaceutical properties, and the majority occur in two plant families (Gentianaceae and Hypericaceae). Precursor feeding experiments indicated that the backbone of the xanthone ring system was synthesized from shikimate-derived precursors and *via* the acetate/malonate route.[35] Cell cultures of *Centaurium erythraea* and *C. littorale* were used to investigate the xanthones and the regulation of their formation.[36,37] The proposed biosynthetic pathway suggested the condensation of a benzoyl-CoA

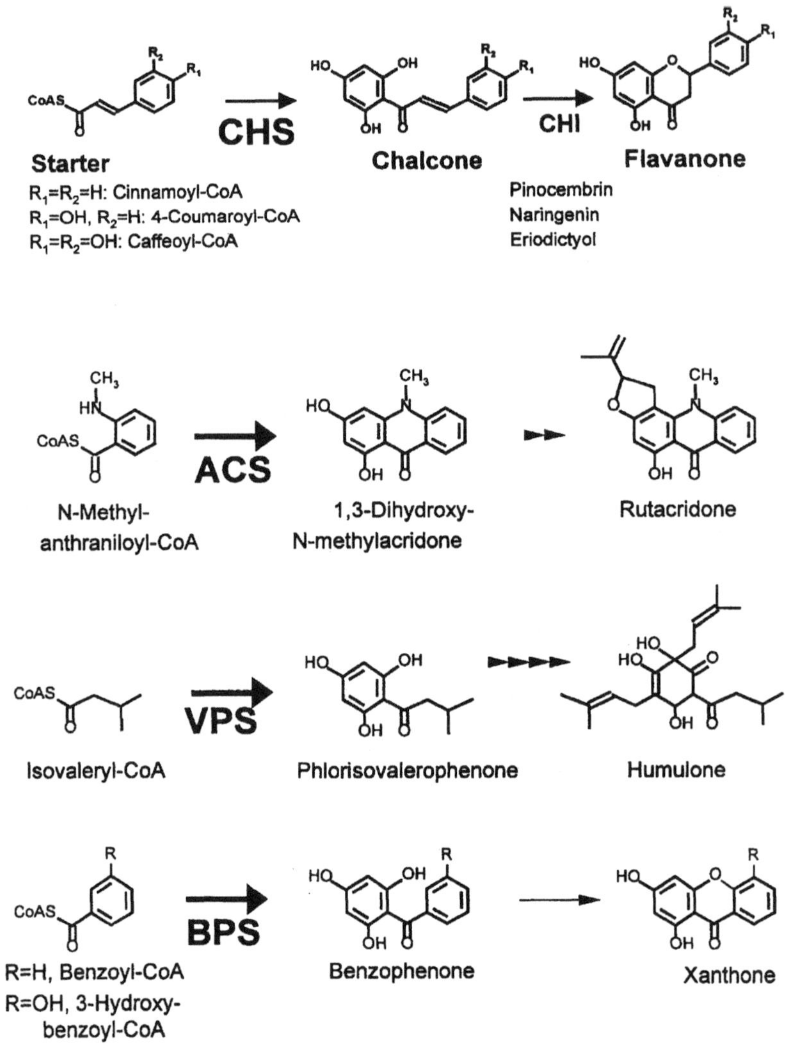

Figure 3. Reactions of CHS-type enzymes. The overview shows the starter substrates, the products, and secondary metabolites containing the backbone synthesized by the CHS-type reaction. CHI, chalcone isomerase; ACS, acridone synthase; VPS, valerophenone synthase; BPS, benzophenone synthase

derivative with three acetate units derived from malonyl-CoA to form the benzophenone, the precursor of the xanthones (Fig. 3).[37,38] The reaction was demonstrated *in vitro* with a partially purified protein preparation from *C. erythraea* incubated with 3-hydroxybenzoyl-CoA and malonyl-CoA.[39] The enzyme also accepted benzoyl-CoA as a starter (44% efficiency), but was completely inactive with 2-hydroxybenzoyl-CoA and 4-hydroxybenzoyl-CoA, indicating a clear specificity for only one position of the hydroxyl group. Later work with extracts from *Hypericum androsaemum* indicated an enzyme preferentially using benzoyl-CoA.[40] The ring formation is clearly of the CHS-type, but the actual relationship of the proteins with CHSs needs to be established. The additional ring closure to the xanthones is a regioselective cyclization (oxidative phenol-coupling) probably catalyzed by a P450-dependent enzyme.[41]

STILBENE SYNTHASE-TYPE REACTIONS

Like CHS, these enzymes are characterized by three condensation reactions, but they are followed by a ring closure type that is clearly different from that of CHS (acylation *versus* aldol condensation). With all characterized STS activities, the formation of the released product involves the removal of the terminal carboxyl group of the tetraketide as CO_2 (Fig. 2).

The STS and CHS ring-folding types appear to be mutually exclusive because enzymes producing both stilbenes and chalcones in comparable amounts are not known, and they were also not detected in experiments converting a CHS into a STS by site-directed mutagenesis,[42] or in attempts to produce hybrids between subunits of CHS and STS.[43] More detailed experiments indicate, however, that small quantities of stilbenes (at most a few percent) are detectable in CHS incubations, and likewise small amounts of chalcones have been recovered from STS reactions (U. Sankawa, personal communication). These are not detected in standard incubations.

Cloned Enzymes: Stilbene (STS) and Bibenzyl (BBS) Synthases

Both STS and BBS use starter CoA-esters originating from the phenylpropanoid pathway (Fig. 4). The first description of a STS demonstration *in vitro* was published in 1978.[44] The activities of STSs and BBS have been characterized from several plants, often after heterologous expression of the cloned proteins. Apart from some differences in the substrate specificities, there seem to be no obvious mechanistic differences between STS and BBS, and, therefore, this review will simply use the name STS for both enzymes. The cDNA or genomic sequences show that the proteins share about 65-73% identity with many CHSs and also with ACS. The STS-type enzymes, thus, belong in the family of CHS-related proteins.[15]

A recent book covers most of the information on stilbenes, bibenzyls, and related substances up to 1994.[45] In contrast to CHS-derived products, stilbenes and their derivatives appear to be rare in higher plants, in particular in important crop plant cultivars, with the exception of wine and peanut. A large number of interesting bibenzyl-derived molecules (stilbenoids) occur in liverworts. However, stilbenes or stilbene-derived substances may be much more widespread in other plants than previously suspected. The literature from the last few years shows, for example, that several monocotyledonous plants contain stilbenes.[46-52]

One of the important functions of stilbenes in plants is probably their role as phytoalexins in disease resistance. Transgenic STS expression has been used several times to introduce the capacity to produce new phytoalexins in species that otherwise do not synthesize stilbenes,[53-57] and these plants now include the important monocots, rice, barley, and wheat.[58,59] STS appears to be one of the simplest possibilities for such experiments because the substrates are the same as for the ubiquitous CHS, and there is no need to pay additional attention to their availability in the transgenic plants. The possible competition of CHS and STS for substrates can apparently lead to male sterility, a side-effect that is interesting for plant breeders.[56] One of the concerns in these strategies has been that stilbenes may be considered as toxic to humans, and that consumption of food containing them would not be desirable. However, it may be possible that some people will consider another aspect more important: considerable attention has been given in the last few years to reports that stilbenes (usually resveratrol and derivatives in wine) could be beneficial to human health because they are antioxidants,[60-63] or because they might be active in prevention of coronary heart disease or even cancer.[64-70] However, many of the issues are still controversial, and it could be that the beneficial effects of red wine consumption are due to alcohol or tannins rather than to stilbenes.[71-73]

Proposals for STS-Type Reactions

STS-type enzyme activities with compounds other than phenylpropanoid substrates have not yet been described. Precursor feeding experiments, however, suggest several candidates. Two will be mentioned briefly here, mainly because the same or similar substrates are known from CHS-type reactions.

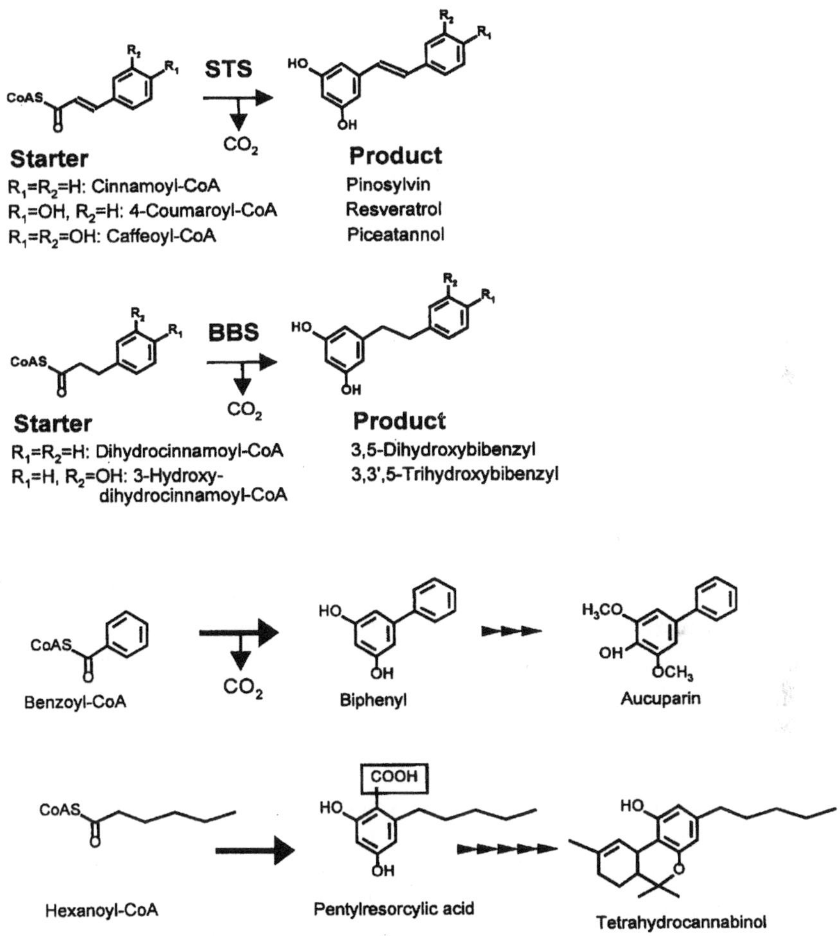

Figure 4. Substrates and products of stilbene (STS) and bibenzyl (BBS) synthase type enzymes. The reactions with benzoyl-CoA and hexanoyl-CoA are proposals derived from precursor feeding studies.

One example is the biosynthesis of the biphenyl backbone in the aucuparin-type secondary products that are phytoalexins in Mountain ash (*Sorbus aucuparia*), apple (*Malus* x *domestica*), and other plants.[74-79] It was proposed a long time ago[35] that the backbone is synthesized from benzoyl-CoA (or a derivative) *via* three condensation reactions with malonyl-CoA, followed by an STS-type ring closure accompanied by removal of the terminal carboxyl group, as shown in Figure 4. The same reaction type, but with an aliphatic starter substrate, has been proposed for the formation of the backbone in tetrahydrocannabinol in Indian hemp (*Cannabis sativa*) (Fig. 4). The reader will find more interesting details and other examples in a book devoted to the biosynthesis of secondary plant products.[80] It is interesting to note that the proposal suggests a STS-type reaction with retention of the carboxyl group that is removed in the reactions of the known STS, but this needs to be demonstrated *in vitro*.

THREE CONDENSATIONS: OTHER PRODUCTS

Modification of Intermediates by Reduction

The CHS reaction does not involve a modification of reaction intermediates prior to the ring closure. The only demonstrated example is the formation of 6'-deoxychalcone, the precursor for a large number of biologically and medically important secondary products.[1,3,7,81] The modification is performed by an enzyme that uses NADPH to reduce a specific carbonyl group of the intermediate either after the second or third condensation reactions. Available evidence indicates that the reduction occurs at the polyketide level, prior to the formation of the new aromatic ring (Fig. 5). Therefore, the name polyketide reductase (PKR) was proposed for the enzyme, but the old name chalcone reductase is still often used although it is to some extent misleading.[21] The enzyme activities have been characterized from various plants.[82-86] Sequences for cDNAs and a functional identification after heterologous enzyme expression were first described for *Glycine max*,[87] and related sequences are known from *Medicago sativa*,[88-90] *Sesbania rostrata*,[91,92] *Pueraria lobata*,[93] *Glycyrrhiza echinata*,[94] *Glycyrrhiza glabra*,[95] and *Fragraria ananassa*.[96] These proteins belong to the aldo/keto-reductase superfamily.[97] The members share three highly conserved protein sequence elements, but the diversity of functions within the family is at least as large as in the family of CHS-related proteins. A functional identification should, therefore, be mandatory, but this has not been done with all sequences cloned from plants. Interestingly, the transgenic expression of the *Medicago sativa* PKR in *Petunia* has led to a new yellow flower color by redirection of the flavonoid biosynthesis.[98]

With STS, a comparable interaction with a reductase has not yet been demonstrated *in vitro*. However, the formation of the backbone in stilbenoids is likely to involve such a co-action, and a reduction of the same carbonyl group as in the CHS/PKR interaction can be postulated. Figure 5 shows an example of the reactions predicted for the formation of hydrangeic acid in the Garden Hortensia (*Hydrangea macrophylla*). The reaction has not yet been demonstrated *in vitro*, nor has its gene been cloned. It is interesting to note that the product of the reaction is a stilbenecarboxylic acid, *i.e.* it retains the carboxyl group that is usually removed during the known STS reactions. The early precursor feeding experiments strongly suggest that stilbenecarboxylic acids and their many derivatives in liverworts, but also in ferns and some higher plants, are derived from phenylpropanoid starters and STS-type reactions.[45]

Three Condensations, but No Ring Closure

A recent publication showed that the family of CHS-related proteins still has some surprises in store with respect to possible products.[99] The work described the characterization of cDNA clones for CHS-related proteins from *Hydrangea macrophylla* var. *thunbergii*. One of the predicted proteins shared about 90% identity with many CHSs, but the other had only 70-74% identity with any other enzyme from the protein family. Heterologous functional expression in *E. coli* showed that the first protein was a CHS, but the second protein synthesized two products from 4-coumaroyl-CoA that were not those expected from CHS (naringenin chalcone), STS (resveratrol), or a stilbenecarboxylic acid synthase (like the STS-type postulated in hydrangeic acid biosynthesis, see Fig. 5). The identification by NMR showed that the products were two lactones: bisnoryangonin and 4-coumaroyltriacetic acid lactone (CTAL) (Fig. 6). Bisnoryangonin is the product of the styrylpyrone synthase in ferns (discussed later), but it is also well-known as a derailment product of CHS reactions under non-optimal assay conditions. CTAL is the corresponding lactone product after three condensation reactions. Bisnoryangonin, CTAL, or their derivatives are seemingly not known in *Hydrangea*, but the linear tetraketide (coumaroyltriacetic acid) could be a precursor of hydramacroside B (Fig. 6), a secondary plant product in *Hydrangea* that contains a secologanin moiety.[100] The authors, therefore, propose coumaroyltriacetic acid synthase (CTAS) activity as a new function in the family of CHS-related proteins: an enzyme performing three condensations but no subsequent cyclization reaction. The CTAL detected *in vitro* is interpreted as a derailment product that is released in absence of the proteins carrying out further reactions.[99]

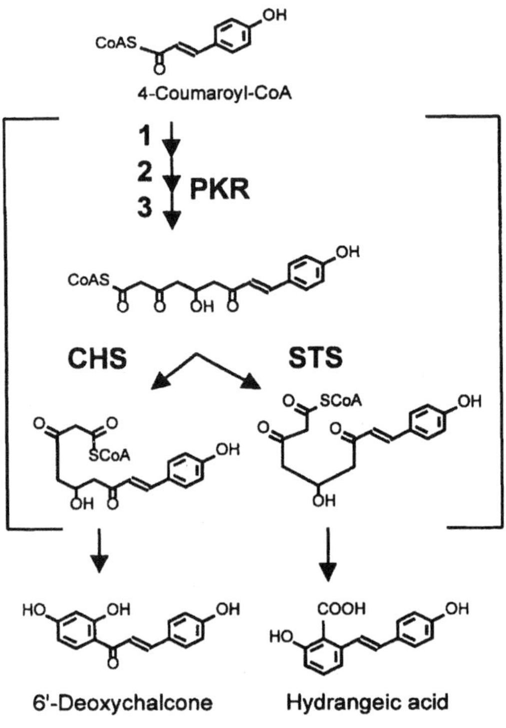

Figure 5. Modification of reaction intermediates. Biosynthesis of 6'-deoxychalcone (CHS-type) and reaction proposed for the biosynthesis of hydrangeic acid (STS-type). PKR, polyketide reductase which reduces with NADPH a specific carbonyl group either at the tri- or tetraketide intermediate stage prior to the ring closure.

Origin of Methyl Groups in C-Methylated Flavonoids

C-methylated flavonoids are not of common occurrence and are considered minor constituents, although more than 50 have been described from different plants. The origin of the methyl group had not been investigated, but recent results suggest that it may be introduced during the biosynthesis of the chalcone backbone.[101] The basis for this investigation was the finding that cDNA libraries from *Pinus strobus* L. (Eastern white pine) contained sequences for two different CHS-related proteins (PStrCHS1 and PStrCHS2, sharing 87.6% identity). Heterologous expression in *E. coli* showed that PStrCHS1 had all the properties of a typical CHS, but PstrCHS2 was completely inactive with any of the substrates from the phenylpropanoid pathway, and it was also inactive with all other substrates that are known for CHS-related enzymes. *P. strobus* is one of the plants containing various C-methylated flavonoids,[102] and the methyl group in all of them could be explained by a CHS reaction in which the second condensation is performed with methylmalonyl-CoA instead of malonyl-CoA. The investigations showed that PstrCHS2 performs this reaction, and this was the only activity that could be demonstrated.[101] If the methyl group is indeed introduced during the biosynthesis of the chalcone backbone, one has to postulate a cooperation of at least two proteins that act as modules, each protein performing a specific part of the complete reaction sequence to the C-methylated chalcone.

ENZYMES WITH ONLY TWO OR ONE CONDENSATION REACTION

The Cloned Enzyme: Pyrone Synthase (2PS)

The cDNA for this protein was originally discovered in *Gerbera hybrida* while screening for CHS-related sequences.[103] Three different types of clones were identified. Functional studies after heterologous expression in *E. coli* showed that two of the proteins (GCHS1 and GCHS3) were typical CHS. The third protein (previously called GCHS2, now renamed to 2PS) was different. It shared 74% identity with the CHS from the same plant and with other CHSs, but was inactive with any of the CHS substrates. Enzyme activity could only be demonstrated with benzoyl-CoA, but the reaction product was not identified in the early studies.[103,104] A reverse genetic approach with antisense cDNA constructs for the CHS-related protein completely suppressed the enzyme activity with benzoyl-CoA, but only partially the formation of flavonoids, indicating that the enzyme functions in another pathway.[105]

Further investigation of the transgenic GCHS2 antisense plants revealed that they lack two compounds that are dominant in the control plants, gerberin and parasorboside (Fig. 7).[17] As suggested by the name, gerberin had been isolated earlier from *Gerbera*.[106] Parasorboside was a new discovery in *Gerbera*,

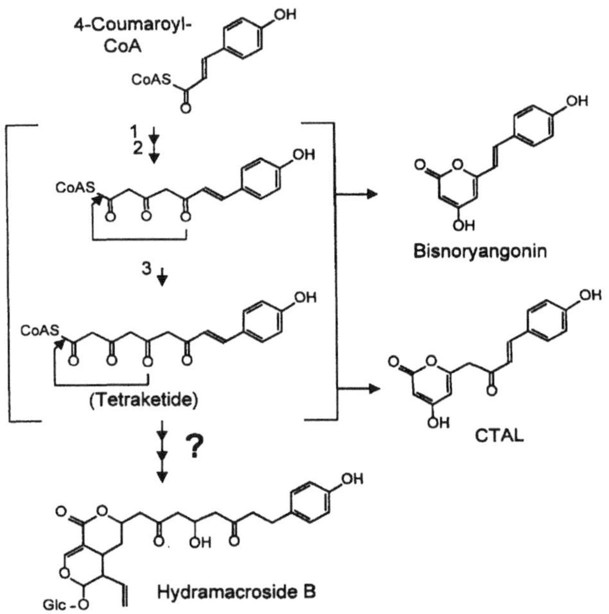

Figure 6. Reactions of coumaroyltriacetic acid synthase (CTAS) and the derailment products detected *in vitro*, bisnoryangonin and coumaroyltriacetic acid lactone (CTAL). It is proposed that the enzyme does not perform a cyclization, and that the linear tetraketide is a precursor of hydramacroside B.[99]

but it had been described from cranberry and the fern *Osmunda japonica*, and it is a major constituent in the berries of mountain ash (*Sorbus aucuparia*) (reviewed in ref. 17). The biosynthesis had not been investigated, but the absence of the secondary products in the antisense plants immediately suggests that the backbone is synthesized by a CHS-related enzyme that uses acetyl-CoA as the starter substrate, performs two condensation reactions with malonyl-CoA, and releases 6-methyl-4-hydroxy-2-pyrone (also known as triacetic acid lactone) as the product (Fig. 7). The prediction was confirmed with the purified enzyme after heterologous expression in *E. coli*.[17]

The other enzymes necessary to form gerberin and parasorboside are not yet characterized.This pyrone synthase is the first proven case for a CHS-related protein that is programmed for only two condensation reactions. It is also the first example that the CHS family contains members that, by using acetyl-CoA as substrate, are functionally more similar to all other types of polyketide synthases than to CHS. It cannot be considered as 'ancient, ancestor-like' CHS, however, because it probably evolved fairly recently from the CHSs in *Gerbera*.[17] As mentioned before, another interesting feature of the enzyme is the self-priming of the reaction sequence catalyzed by the decarboxylation of malonyl-CoA to acetyl-CoA.

Like most other CHS-related proteins, the pyrone synthase is not substrate-specific, and it readily accepts other small hydrophobic substrates. The product from benzoyl-CoA was identified as phenylpyrone.[17] This compound or its derivatives are not known from *Gerbera*, but it contains the aglycone backbone of the psilotins, which are considered as chemical markers of the Psilotaceae (far relatives of ferns) .[107,108] However, precursor feeding studies *in vivo* suggest that the psilotins are not synthesized from benzoyl-CoA, but from phenylpropanoid precursors, with the pyrone ring formed after one condensation reaction.[109] It is noteworthy that the phenylpyrone contains the backbone of substances that became of medical interest as HIV-1 protease inhibitors.[110,111]

Predictions

Two Condensations: Styrylpyrone Synthase (SPS). Styrylpyrones occur in Pteridophytes and in Angiosperms, and they are common constituents in fungi, mainly in the Basidiomycetes.[112] Precursor feeding experiments suggested that they can be synthesized from the precursors used by CHS, but with only two condensation reactions that are followed by a ring closure of the triketide intermediate to the styrylpyrone (Fig. 8).[112-114]

An enzyme from the horsetail fern (*Equisetum arvense*) seems to be the only example where the predicted enzyme activities have been demonstrated *in vitro*.[115,116] The partially purified styrylpyrone synthase (SPS) accepted both 4-coumaroyl-CoA and caffeoyl-CoA, with bisnoryangonin and hispidin as products, respectively. The tissue used in the experiments (gametophytes) does not express CHS activity, and, therefore, SPS must be a different enzyme. There is no clear evidence so far that the SPS belongs to the family of CHS-related

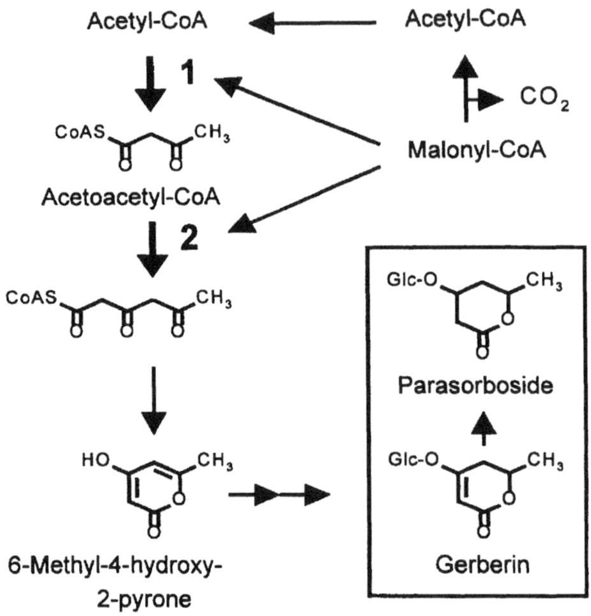

Figure 7. Reactions of the *Gerbera hybrida* pyrone synthase (2PS) in the biosynthesis of the aglycone in gerberin and parasorboside. In absence of the starter substrate, the enzyme decarboxylates malonyl-CoA to acetyl-CoA which is then used as starter.

proteins. *E. arvense* is an interesting plant system to investigate the differential expression and relation of SPS and CHS, because it shows a developmental switch: gametophytes and rhizomes accumulate styrylpyrones as major phenolic constituents, but no flavonoids, while green sprouts (sporophytes) contain various flavonoid glycosides, but no styrylpyrones.[115,117-119]

One Condensation: Benzalacetone Synthase (BAS). The characteristic aroma of raspberries is caused by 4-hydroxyphenylbutan-2-one (pHPB, "raspberry ketone").[120] Two enzymes are involved in its biosynthesis; they were

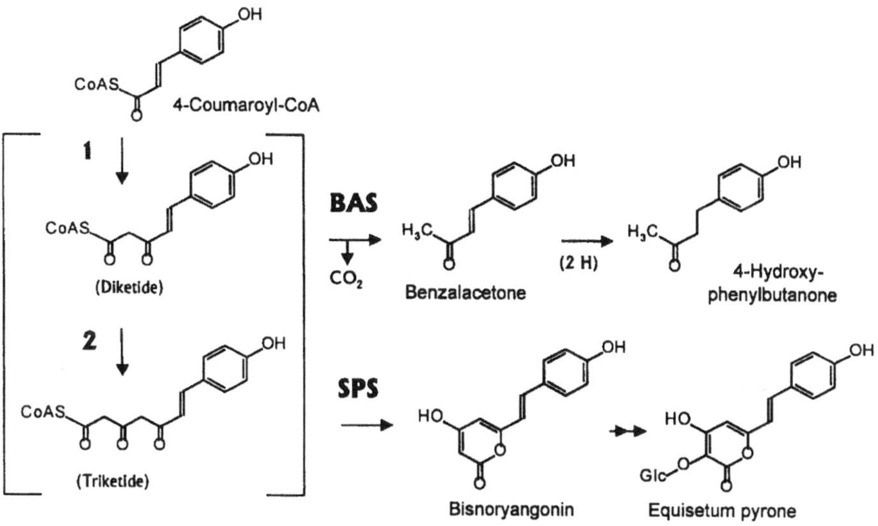

Figure 8. Enzymes with only one or two condensation reactions. BAS, benzalacetone synthase; SPS, styrylpyrone synthase. Only the reactions with 4-coumaroyl-CoA are shown; the enzymes have also high activities with caffeoyl-CoA (see Fig. 3).

investigated in raspberry fruits and tissue cultures.[121] The first, benzalacetone synthase (BAS), uses a phenylpropanoid starter CoA-ester, performs one condensation reaction, and a decarboxylated product is detected *in vitro*.[122] The second enzyme reduces the double bond in the propenoyl moiety with NADPH to form the aroma component (Fig. 8).

BAS was purified and characterized from raspberry fruits. All of its properties and the differential induction argue for a proposal that BAS and CHS are different enzymes.[122] This, however, needs unambiguous proof by cloning and heterologous expression because it is known that purified CHS can produce benzalacetone *in vitro* as a byproduct.[123,124] BAS should be present in other plants as well, because benzalacetone or its glycosides[125] have also been found in several other plants (reviewed in ref. 15). Precursor feeding studies suggest that enzymes performing one condensation reaction to a diketide intermediate may be involved in the formation of other complex substances of considerable interest, *e.g.* gingerol, curcumin, and other secondary products.[126]

RELATIONSHIPS AND EVOLUTION

Correlation of Functions and Sequences

The ease of molecular cloning and sequencing techniques has led in the recent years to many sequences. Both for practical and theoretical reasons, it is an interesting question whether functional differences can be recognized in protein sequences deduced from cDNA or genomic clones. For the workers involved in sequencing projects, or for the researcher finding such sequences by accident, the practical question may be only whether it is possible to predict a probable function from the sequence. It is obvious, however, that the question is much more important for those interested in understanding the correlation between sequence and function: this will be the key to understanding how the diversity of functions is programmed into the proteins, and it will also be the key for the design of new enzyme activities. The recent elucidation of the first crystal structure of a CHS is likely to be of great importance.[13,14] However, it should be realized that it probably will not be possible for some time to use functional predictions from primary sequences with sufficient certainty, and heterologous expression is still the method of choice if one wants to clearly define the function. A more detailed discussion of the problems encountered with functional assignments from sequence data can be found in a previous review.[15]

Evolution

Functional Diversification in Plants. CHS seems to be ubiquitous in plants, and it is thought that it appeared early in land plants, *i.e.* in Charophyceae or in simple Bryophytes.[127-131] It has been argued for the present-day STS[42] and for the 2PS[17] that they evolved fairly recently from CHS by gene duplication and mutation. The main lines of reasoning can also be applied to acridone synthase and CTAS; these enzymes are also of limited distribution and share about 70% identity with each other and with CHS. It seems possible, therefore, that CHSs were the basis for the known functional diversifications. This would not be so easily accepted, however, with sequences for predicted plant proteins of unknown functions that share only about 40% or less identity with the typical CHS. Such cases have been described from several plants (*e.g. Nicotiana sylvestris*, *Arabidopsis thaliana*, *Oryza sativa*, *Pinus radiata*, and *Brassica napus*),[132] and it may be necessary to consider other possibilities. Unfortunately, up to now, no CHS or CHS-related sequences are available from the postulated early land plants and their possible ancestors, and little is known about products of secondary metabolism that could have been synthesized by CHS-related proteins in such plants.

CHS-Related Sequences in Bacteria. Sequences for CHS-related proteins have been identified in several bacteria. The examples include such diverse organisms as *Pseudomonas fluorescens*,[133] *Streptomyces griseus*,[134] *Bacillus subtilis*,[135] *Mycobacterium tuberculosis* (accession nos. Z81011, Z85982, Z95617), *Amycolatopsis orientalis* (accession no. AJ223998), and *Rhodospirillum centenum*.[136] The predicted polypeptides constitute a rather heterogeneous group that shares 25-30% identity with the plant proteins, and certain typical elements are clearly conserved (*e.g.* a cysteine in the position expected for the active site of the condensing reactions). The functions of the proteins are unknown in most cases, but the obvious relation suggests that the evolution of the basic structural and functional properties of these condensing enzymes predated the emergence of land plants.

A recent report identified the function of one of the bacterial proteins.[137] *Streptomyces griseus* contains a gene for a protein that was previously shown to be sufficient for the production in the foreign host *E. coli* of an unknown red-brown pigment.[134] The new results are surprising in several aspects. After correction of the nucleotide sequence, the joining of two small open reading frames yielded a polypeptide (RppA, 40 kDa) that shared about 30% identity with CHSs. An His-tagged protein was expressed in *E. coli*, and the function was tested with the purified enzyme. The assays showed that the protein used malonyl-CoA as the starter substrate, performed four condensation reactions with malonyl-CoA, and released the pentaketide to form 1,3,6,8-tetrahydroxynaphthalene (THN). The final ring closure to THN involved the removal of the carboxyl group of the malonyl-CoA used as the starter substrate (Fig. 9).

This appears to be the first functional identification of a CHS-related bacterial protein. The use of malonyl-CoA as starter and the performance of four condensations are novel and unexpected properties for CHS-related enzymes. Indeed, previous findings with a polyketide synthase cloned from the fungus *Colletotrichum lagenarium* suggested that THN is synthesized by proteins that have no similarity with the family of CHS-related proteins.[138,139] The example of *S. griseus* suggests the possibility that the diversity of the biosynthetic capacities in the protein family may be much larger than previously expected, at least in bacteria. It will be interesting to see whether the same

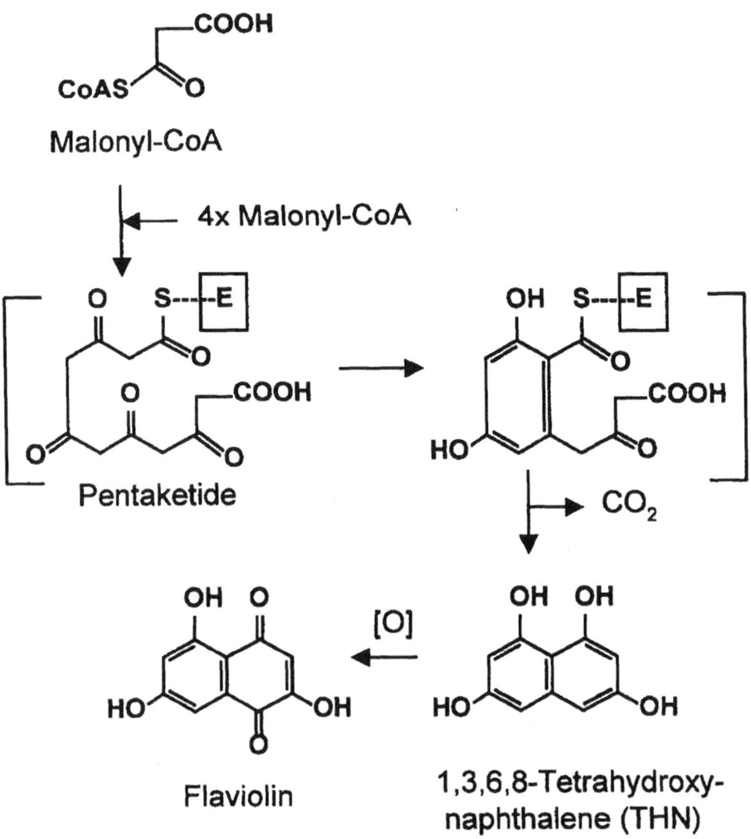

Figure 9. Reaction of the CHS-related enzyme from *Streptomyces griseus*.

applies to plants, or whether the activities in plants represent only a subset of the possibilities realized in bacteria.

CONCLUSIONS

Results in the last few years have shown that the well-known chalcone synthase (CHS) is only one example from a family of plant polyketide synthases. Other members of the family that are identified by function and sequences are the stilbene synthases (STS), acridone synthase (ACS), and a pyrone synthase (2PS); all of these proteins share about 65-70% identity with CHS. The properties of several other enzymes suggest that they are members of the protein family, and precursor feeding studies suggest that the number may be much larger than suspected so far. The diversity of functions is based on different substrate specificities, variations in the number of condensation reactions, folding of intermediates to different products, and the modification of intermediates by other enzymes.

The recently published crystal structure of a CHS raises hopes that it will be possible to understand how the variation in primary sequence can be related to the various functions; this then will facilitate the design of enzymes synthesizing new products.

The understanding of the evolution of the protein family is still rudimentary. The available data suggest that the functional diversity known in present-day plants could be the result of fairly recent modifications of CHS by gene duplication and mutation. The presence of CHS-related sequences in bacteria indicates that the basic function unit predated the evolution of plants. The recent functional identification of such a protein from *Streptomyces griseus* suggests that the functional diversity in bacteria may even be larger than in plants.

ACKNOWLEDGEMENTS

Support by the Deutsche Forschungsgemeinschaft and by Fonds der Chemischen Industrie for many years is gratefully acknowledged.

REFERENCES

1. DIXON, R.A., STEELE, C.L. 1999. Flavonoids and isoflavonoids - a gold mine for metabolic engineering. Trends Plant Sci. 4: 394-400.
2. PAGANGA, G., MILLER, N., RICE-EVANS, C.A. 1999. The polyphenolic content of fruit and vegetables and their antioxidant activities. What does a serving constitute? Free Radic. Res. 30: 153-162.
3. BINGHAM, S.A., ATKINSON, C., LIGGINS, J., BLUCK, L., COWARD, A. 1998. Phyto-oestrogens: Where are we now? Br. J. Nutr. 79: 393-406.
4. WANG, H.K., XIA, Y., YANG, Z.Y., NATSCHKE, S.L.M., LEE, K.H. 1998. Recent advances in the discovery and development of flavonoids and their analogues as antitumor and anti-HIV agents. Adv. Exp. Med. Biol. 439: 191-225.

5. RICE-EVANS, C.A., MILLER, N.J. 1996. Antioxidant activities of flavonoids as bioactive components of food. Biochem. Soc. Transactions 24: 790-794.
6. STAFFORD, H.A. 1997. Roles of flavonoids in symbiotic and defense functions in legume roots. Bot. Rev. 63: 27-39.
7. DIXON, R.A. 1999. Isoflavonoids: biochemistry, molecular biology, and biological functions. pp. 773-823 *in* Polyketides and Other Secondary Metabolites Including Fatty Acids and Their Derivatives (U. Sankawa, ed.), Elsevier, Amsterdam.
8. FORKMANN, G. HELLER, W. 1999. Biosynthesis of flavonoids. pp. 713-748 *in* Polyketides and Other Secondary Metabolites Including Fatty Acids and Their Derivatives (U. Sankawa, ed.), Elsevier, Amsterdam.
9. MOL, J.N.M., GROTEWOLD, E., KOES, R.E. 1998. How genes paint flowers and seeds. Trends Plant Sci. 3: 212-217.
10. COHN, J., DAY, R.B., STACEY, G. 1998. Legume nodule organogenesis. Trends Plant Sci. 3: 105-110.
11. MCKHANN, H.I., PAIVA, N.L., DIXON, R.A., HIRSCH, A.M. 1998. Expression of genes for enzymes of the flavonoid biosynthetic pathway in the early stages of the *Rhizobium*-legume symbiosis. Adv. Exp. Med. Biol. 439: 45-54.
12. GRAHAM, T.L. 1998. Flavonoid and flavonol glycoside metabolism in *Arabidopsis*. Plant Physiol. Biochem. 36: 135-144.
13. FERRER, J.-L., JEZ, J.M., BOWMAN, M.E., DIXON, R.A., NOEL, J.P. 1999. Structure of chalcone synthase and the molecular basis of plant polyketide biosynthesis. Nature Struct. Biol. 6: 775-784.
14. SCHRÖDER, J. 1999. Probing plant polyketide synthesis. Nature Struct. Biol. 6: 714-716.
15. SCHRÖDER, J. 1999. The chalcone/stilbene synthase-type family of condensing enzymes. pp. 749-771 *in* Polyketides and Other Secondary Metabolites Including Fatty Acids and Their Derivatives (U. Sankawa, ed.), Elsevier, Amsterdam.
16. PREISIG-MÜLLER, R. GEHLERT, R., MELCHIOR, F., STIETZ, U., KINDL, H., 1997. Plant polyketide synthases leading to stilbenoids have a domain catalyzing malonyl-CoA:CO_2 exchange, malonyl-CoA decarboxylation, and covalent enzyme modification and a site for chain lengthening. Biochemistry 36: 8349-8358.
17. ECKERMANN, S., SCHRÖDER, G., SCHMIDT, J., STRACK, D., EDRADA, R.A., HELARIUTTA, Y., ELOMAA, P., KOTILAINEN, M., KILPELÄINEN, I., PROKSCH, P., TEERI, T.H., SCHRÖDER, J. 1998.

New pathway to polyketides in plants. Nature (London) 396: 387-390.

18. BISANG, C., LONG, P.F., CORTÉS, J., WESTCOTT, J., CROSBY, J., MATHARU, A.L., COX, R.J., SIMPSON, T.J., STAUNTON, J. LEADLAY, P.F., 1999. A chain initiation factor common to both modular and aromatic polyketide synthases. Nature (London) 401: 502-505.
19. KREUZALER, F., HAHLBROCK, K. 1972. Enzymic synthesis of aromatic compounds in higher plants: formation of naringenin (5,7,4'-trihydroxy-flavanone) from p-coumaroyl-coenzyme A and malonyl-coenzyme A. FEBS Lett. 28: 69-72.
20. MARTIN, C.R. 1993. Structure, function, and regulation of the chalcone synthase. pp. 233-284 *in* International Review of Cytology: A Survey of Cell Biology (K. Jeon and J. Jarvik, eds.), Academic Press, New York.
21. HELLER, W., FORKMANN, G. 1994. Biosynthesis of flavonoids. pp. 499-535 *in* The Flavonoids: Advances in Research Since 1986 (J.B. Harborne, ed.), Chapman & Hall, London.
22. DE COOMAN, L., EVERAERT, E., FACHÉ, P., VANDE CASTEELE, K., VAN SUMERE, C.F. 1993. Flavonoid biosynthesis in petals of *Rhododendron simsii*. Phytochemistry 33: 1419-1426.
23. CHRISTENSEN, A.B., GREGERSEN, P.L., SCHRÖDER, J., COLLINGE, D.B. 1998. A chalcone synthase with an unusual substrate preference is expressed in barley leaves in response to UV-light and pathogen attack. Plant Mol. Biol. 37: 849-857.
24. REIMOLD, U., KROEGER, M., KREUZALER, F., HAHLBROCK, K. 1983. Coding and 3' noncoding nucleotide sequence of chalcone synthase messenger RNA and assignment of amino acid sequence of the enzyme. EMBO J. 2: 1801-1806.
25. BAUMERT, A., PORZEL, A., SCHMIDT, J., GRÖGER, D. 1992. Formation of 1,3-dihydroxy-N-methylacridone from *N*-methylanthraniloyl-CoA and malonyl-CoA by cell-free extracts of *Ruta graveolens*. Z. Naturforsch. 47c: 365-368.
26. MAIER, W., BAUMERT, A., SCHUMANN, B., FURUKAWA, H., GRÖGER, D. 1993. Synthesis of 1,3-dihydroxy-N-methylacridone and its conversion to rutacridone by cell-free extracts of *Ruta graveolens* cell cultures. Phytochemistry 32: 691-698.
27. BAUMERT, A., MAIER, W., GRÖGER, D., DEUTZMANN, R. 1994. Purification and properties of acridone synthase from cell suspension cultures of *Ruta graveolens*. Z. Naturforsch. 49c: 26-32.
28. JUNGHANNS, K.T., KNEUSEL, R.E., BAUMERT, A., MAIER, W., GRÖGER, D., MATERN, U. 1995. Molecular cloning and recombinant

expression of acridone synthase from elicited *Ruta graveolens* L. cell suspension cultures. Plant Mol. Biol. 27: 681-692.
29. LUKACIN, R., SPRINGOB, K., URBANKE, C., ERNWEIN, C., SCHRÖDER, G., SCHRÖDER, J., MATERN, U. 1999. Native acridone synthases I and II from *Ruta graveolens* L. form homodimers. FEBS Lett. 448: 135-140.
30. FUNG, S.-Y., BRUSSEE, J., VAN DER HOEVEN, R.A.M., NIESSEN, W.M.A., SCHEFFER, J.J.C., VERPOORTE, R. 1994. Analysis of proposed aromatic precursors of hop bitter acids. J. Nat. Prod. 57: 452-459.
31. ZUURBIER, K.W.M., FUNG, S.-Y., SCHEFFER, J.J.C., VERPOORTE, R. 1995. Formation of aromatic intermediates in the biosynthesis of bitter acids in *Humulus lupulus*. Phytochemistry 38: 77-82.
32. PANIEGO, N.B., ZUURBIER, K.W.M., FUNG, S.-Y., VAN DER HEIJDEN, R., SCHEFFER, J.J.C., VERPOORTE, R. 1999. Phlorisovalerophenone synthase, a novel polyketide synthase from hop (*Humulus lupulus* L.) cones. Eur. J. Biochem. 262: 612-616.
33. SCHÜZ, R., HELLER, W., HAHLBROCK, K. 1983. Substrate specificity of chalcone synthase from *Petroselinum hortense*. Formation of phloroglucinol derivatives from aliphatic substrates. J. Biol. Chem. 258: 6730-6734.
34. ZUURBIER, K.W.M., LESER, J., BERGER, T., HOFTE, A.J.P., SCHRÖDER, G., VERPOORTE, R., SCHRÖDER, J. 1998. 4-Hydroxy-2-pyrone formation by chalcone and stilbene synthase with nonphysiological substrates. Phytochemistry 49: 1945-1951.
35. SULTANBAWA, M.U.S. 1980. Xanthonoids of tropical plants. Tetrahedron 36: 1465-1506.
36. BEERHUES, L., BERGER, U. 1994. Xanthones in cell suspension cultures of two *Centaurium* species. Phytochemistry 35: 1227-1231.
37. BEERHUES, L., BERGER, U. 1995. Differential accumulation of xanthones in methyl-jasmonate- and yeast-extract-treated cell cultures of *Centaurium erythrea* and *Centaurium littorale*. Planta 197: 608-612.
38. BENNETT, G.J. LEE, H.H., 1989. Xanthones from Guttiferae. Phytochemistry 28: 967-998.
39. BEERHUES, L. 1996. Benzophenone synthase from cultured cells of *Centaurium erythraea*. FEBS Lett. 383: 264-266.
40. SCHMIDT, W., BEERHUES, L. 1997. Alternative pathway of xanthone biosynthesis in cell cultures of *Hypericum androsaemum* L. FEBS Lett. 420: 143-146.

41. PETERS, S., SCHMIDT, W., BEERHUES, L. 1998. Regioselective oxidative phenol coupling of 2,3',4,6-tetrahydroxybenzophenone in cell cultures of *Centaurium erythraea* RAFN and *Hypericum androsaemum* L. Planta 204: 64-69.
42. TROPF, S., LANZ, T., RENSING, S.A., SCHRÖDER, J., SCHRÖDER, G. 1994. Evidence that stilbene synthases have developed from chalcone synthases several times in the course of evolution. J. Mol. Evol. 38: 610-618.
43. TROPF, S., KÄRCHER, B., SCHRÖDER, G., SCHRÖDER, J. 1995. Reaction mechanisms of homodimeric plant polyketide synthases (stilbene and chalcone synthase): a single active site for the condensing reaction is sufficient for synthesis of stilbenes, chalcones, and 6'-deoxychalcones. J. Biol. Chem. 270: 7922-7928.
44. RUPPRICH, N., KINDL, H. 1978. Stilbene synthases and stilbenecarboxylate synthases. I. Enzymatic synthesis of 3,5,4×-trihydroxystilbene from p-coumaroyl coenzyme A and malonyl coenzyme A. Hoppe-Seyler's Z. Physiol. Chem. 359: 165-172.
45. GORHAM, J. 1995. The Biochemistry of the Stilbenoids. Chapman & Hall, London.
46. BRINKER, A.M., SEIGLER, D.S. 1991. Isolation and identification of piceatannol as a phytoalexin from sugarcane. Phytochemistry 30: 3229-3232.
47. HARBORNE, J.B., GREENHAM, J., WILLIAMS, C.A., EAGLES, J., MARKHAM, K.R. 1993. Ten isoprenylated and C-methylated flavonoids from the leaves of three *Vellozia* species. Phytochemistry 34: 219-226.
48. CASABUONO, A.C., POMILIO, A.B. 1994. Lignans and a stilbene from *Festuca argentina*. Phytochemistry 35: 479-483.
49. GRAYER, R.J., HARBORNE, J.B. 1994. A survey of antifungal compounds from higher plants, 1982-1993. Phytochemistry 37: 19-42.
50. POWELL, R.G., TEPASKE, M.R., PLATTNER, R.D., WHITE, J.F., CLEMENT, S.L. 1994. Isolation of resveratrol from *Festuca versuta* and evidence for the widespread occurrence of this stilbene in the Poaceae. Phytochemistry 35: 335-338.
51. HÖLSCHER, D., SCHNEIDER, B. 1996. A resveratrol dimer from *Anigozanthos preissii* and *Musa cavendish*. Phytochemistry 43: 471-473.
52. HÖLSCHER, D., SCHNEIDER, B. 1999. HPLC-NMR analysis of phenylphenalenones and a stilbene from *Anigozanthos flavidus*. Phytochemistry 50: 155-161.

53. HAIN, R., BIESELER, B., KINDL, H., SCHRÖDER, G., STÖCKER, R. 1990. Expression of a stilbene synthase gene in *Nicotiana tabacum* results in synthesis of the phytoalexin resveratrol. Plant Mol. Biol. 15: 325-335.
54. HAIN, R., REIF, H.-J., KRAUSE, E., LANGEBARTELS, R., KINDL, H., VORNAM, B., WIESE, W., SCHMELZER, E., SCHREIER, P.H., STÖCKER, R., STENZEL, K. 1993. Disease resistance results from foreign phytoalexin expression in a novel plant. Nature (London) 361: 153-156.
55. FISCHER, R., HAIN, R. 1994. Plant disease resistance resulting from the expression of foreign phytoalexins. Curr. Opin. Biotechnol. 5: 125-130.
56. FISCHER, R., BUDDE, I., HAIN, R. 1997. Stilbene synthase gene expression causes changes in flower colour and male sterility in tobacco. Plant J. 11: 489-498.
57. THOMZIK, J.E., STENZEL, K., STÖCKER, R., SCHREIER, P.H., HAIN, R., STAHL, D.J. 1997. Synthesis of a grapevine phytoalexin in transgenic tomatoes (*Lycopersicon esculentum* Mill.) conditions resistance against *Phytophthora infestans*. Physiol. Mol. Plant Pathol. 51: 265-278.
58. STARK-LORENZEN, P., NELKE, B., HÄNSSLER, G., MÜHLBACH, H.P., THOMZIK, J.E. 1997. Transfer of a grapevine stilbene synthase gene to rice (*Oryza sativa* L.). Plant Cell Rep. 16: 668-673.
59. LECKBAND, G., LÖRZ, H. 1998. Transformation and expression of a stilbene synthase gene of *Vitis vinifera* L. in barley and wheat for increased fungal resistance. Theor. Appl. Genet. 96: 1004-1012.
60. FRÉMONT, L., BELGUENDOUZ, L., DELPAL, S. 1999. Antioxidant activity of resveratrol and alcohol-free wine polyphenols related to LDL oxidation and polyunsaturated fatty acids. Life Sci. 64: 2511-2521.
61. LAPIDOT, T., HAREL, S., AKIRI, B., GRANIT, R., KANNER, J. 1999. pH-dependent forms of red wine anthocyanins as antioxidants. J. Agric. Food Chem. 47: 67-70.
62. BELGUENDOUZ, L., FREMONT, L., LINARD, A. 1997. Resveratrol inhibits metal ion-dependent and independent peroxidation of porcine low-density lipoproteins. Biochem. Pharmacol. 53: 1347-1355.
63. RICE-EVANS, C.A., MILLER, N.J., PAGANGA, G. 1997. Antioxidant properties of phenolic components. Trends Plant Sci. 2: 152-159.
64. PACE-ASCIAK, C.R., HAHN, S., DIAMANDIS, E.P., SOLEAS, G.J., GOLDBERG, D.M. 1995. The red wine phenolics trans-resveratrol and quercetin block human platelet aggregation and eicosanoid synthesis: implications for protection against coronary heart disease. Clin. Chim. Acta 235: 207-219.

65. CARBO, N., COSTELLI, P., BACCINO, F.M., LOPEZ-SORIANO, F.J., ARGILÉS, J.M. 1999. Resveratrol, a natural product present in wine, decreases tumour growth in a rat tumour model. Biochem. Biophys. Res. Commun. 254: 739-743.

66. HUANG, C.S., MA, W.Y., GORANSON, A., DONG, Z.G. 1999. Resveratrol suppresses cell transformation and induces apoptosis through a p53-dependent pathway. Carcinogenesis 20: 237-242.

67. SURH, Y.J. 1999. Molecular mechanisms of chemopreventive effects of selected dietary and medicinal phenolic substances. Mutat. Res. Fund. Mol. Mech. Mutagen. 428: 305-327.

68. SURH, Y.J., HURH, Y.J., KANG, J.Y., LEE, E., KONG, G., LEE, S.J. 1999. Resveratrol, an antioxidant present in red wine, induces apoptosis in human promyelocytic leukemia (HL-60) cells. Cancer Lett. 140: 1-10.

69. CLÉMENT, M.V., HIRPARA, J.L., CHAWDHURY, S.H., PERVAIZ, S. 1998. Chemopreventive agent resveratrol, a natural product derived from grapes, triggers CD95 signaling-dependent apoptosis in human tumor cells. Blood 92: 996-1002.

70. JANG, M., CAI, L., UDEANI, G.O., SLOWING, K.V., THOMAS, C.F., BEECHER, C.W.W., FONG, H.H.S., FARNSWORTH, N.R., KINGHORN, A.D., MEHTA, R.G., MOON, R.C., PEZZUTO, J.M. 1997. Cancer chemopreventive activity of resveratrol, a natural product derived from grapes. Science 275: 218-220.

71. SOLEAS, G.J., DIAMANDIS, E.P., GOLDBERG, D.M. 1997. Wine as a biological fluid: history, production, and role in disease prevention. J. Clin. Lab. Analysis 11: 287-313.

72. RIMM, E.B., KLATSKY, A., GROBBEE, D., STAMPFER, M.J. 1996. Review of moderate alcohol consumption and reduced risk of coronary heart disease - is the effect due to beer, wine, or spirits. British Med. J. 312: 731-736.

73. SOLEAS, G.J., DIAMANDIS, E.P., GOLDBERG, D.M. 1997. Resveratrol: a molecule whose time has come? And gone? Clin. Biochem. 30: 91-113.

74. KOKUBUN, T., HARBORNE, J.B. 1995. Phytoalexin induction in the sapwood of plants of the Maloideae (Rosaceae): biphenyls or dibenzofurans. Phytochemistry 40: 1649-1654.

75. KOKUBUN, T., HARBORNE, J.B., EAGLES, J., WATERMAN, P.G. 1995. Antifungal biphenyl compounds are the phytoalexins of the sapwood of *Sorbus aucuparia*. Phytochemistry 40: 57-59.

76. KOKUBUN, T., HARBORNE, J.B. 1994. A survey of phytoalexin induction in leaves of the Rosaceae by copper ions. Z. Naturforsch. 49c: 628-634.

77. BURKHARDT, G., SCHILD, W., BECKER, H., GRUBERT, M. 1992. Biphenyls and xanthones from the Podostemaceae. Phytochemistry 31: 543-548.
78. WIDYASTUTI, S.M., NONAKA, F., WATANABE, K., SAKO, N., TANAKA, K. 1992. Isolation and characterization of two aucuparin-related phytoalexins from *Photinia glabra* Maxim. Ann. Phytopath. Soc. Japan 58: 228-233.
79. BOREJSZA-WYSOCKI, W., LESTER, C., ATTYGALLE, A.B., HRAZDINA, G. 1999. Elicited cell suspension cultures of apple (*Malus x domestica*) cv. Liberty produce biphenyl phytoalexins. Phytochemistry 50: 231-235.
80. DEWICK, P.M. 1997. Medicinal Natural Products - A Biosynthetic Approach. John Wiley & Sons, Chichester.
81. SETCHELL, K.D.R., CASSIDY, A. 1999. Dietary isoflavones: biological effects and relevance to human health. J. Nutr. 129: 758S-767S.
82. AYABE, S.-I., UDAGAWA, A., FURUYA, T. 1988. NAD(P)H-dependent 6'-deoxychalcone synthase activity in *Glycyrrhiza echinata* cells induced by yeast extract. Arch. Biochem. Biophys. 261: 458-462.
83. HAKAMATSUKA, T., NOGUCHI, H., EBIZUKA, Y., SANKAWA, U. 1988. Deoxychalcone synthase from cell suspension cultures of *Pueraria lobata*. Chem. Pharm. Bull. (Tokyo) 36: 4225-4228.
84. WELLE, R., GRISEBACH, H. 1988. Isolation of a novel NADPH-dependent reductase which coacts with chalcone synthase in the biosynthesis of 6'-deoxychalcone. FEBS Lett. 236: 221-225.
85. WELLE, R., GRISEBACH, H. 1989. Phytoalexin synthesis in soybean cells: elicitor induction of reductase involved in biosynthesis of 6'-deoxychalcone. Arch. Biochem. Biophys. 272: 97-102.
86. HARANO, K., OKADA, N., FURUNO, T., TAKAHASHI, T., AYABE, S.-I., WELLE, R. 1993. Enzymatic synthesis of 6'-deoxychalcone in cultured *Glycyrrhiza echinata* cells. Plant Cell Rep. 12: 66-69.
87. WELLE, R., SCHRÖDER, G., SCHILTZ, E., GRISEBACH, H., SCHRÖDER, J. 1991. Induced plant responses to pathogen attack: analysis and heterologous expression of the key enzyme in the biosynthesis of phytoalexins in soybean (*Glycine max* L. Merr.cv. Harosoy 63). Eur. J. Biochem. 196: 423-430.
88. BALLANCE, G.M., DIXON, R.A. 1995. *Medicago sativa* cDNAs encoding chalcone reductase. Plant Physiol. 107: 1027-1028.
89. SALLAUD, C., EL-TURK, J., BREDA, C., BUFFARD, D., DE KOZAK, I., ESNAULT, R., KONDOROSI, A. 1995. Differential expression of cDNA

coding for chalcone reductase, a key enzyme of the 5-deoxyflavonoid pathway, under various stress conditions in *Medicago sativa*. Plant Sci. 109: 179-190.

90. SALLAUD, C., EL-TURK, J., BIGARRÉ, L., SEVIN, H., WELLE, R., ESNAULT, R. 1995. Nucleotide sequences of three chalcone reductase genes from alfalfa. Plant Physiol. 108: 869-870.
91. GOORMACHTIG, S., VALERIO-LEPINIEC, M., SZCZYGLOWSKI, K., VAN MONTAGU, M., HOLSTERS, M., DE BRUIJN, F.J. 1995. Use of differential display to identify novel *Sesbania rostrata* genes enhanced by *Azorhizobium caulinodans* infection. Mol. Plant-Microbe Interact. 8: 816-824.
92. GOORMACHTIG, S., LIEVENS, S., HERMAN, S., VAN MONTAGU, M., HOLSTERS, M. 1999. Chalcone reductase-homologous transcripts accumulate during development of stem-borne nodules on the tropical legume *Sesbania rostrata*. Planta 209: 45-52.
93. SANKAWA, U., HAKAMATSUKA, T., SHINKAI, T., YOSHIDA, M., PARK, H.-H., EBIZUKA, Y. 1995. Changes of secondary metabolism by elicitor treatment in *Pueraria lobata* cell cultures. pp. 595-604 *in* Current Issues in Plant Molecular and Cellular Biology (M. Terzi, R. Cella, and A. Falavigna, eds.), Kluwer Academic Publishers, Dordrecht.
94. AKASHI, T., FURUNO, T., FUTAMI, K., HONDA, M., TAKAHASHI, T., WELLE, R., AYABE, S.-I. 1996. A cDNA for polyketide reductase (Accession No. D83718) that catalyzes the formation of 6'-deoxychalcone from cultured *Glycyrrhiza echinata* L. cells. Plant Physiol. 111: 347
95. HAYASHI, H., MURAYAMA, K., HIROAKA, N., IKESHIRO, Y. 1996. Nucleotide sequences of cDNAs (accession nos. D86558 and D86559) for *Glycyrrhiza glabra* polyketide reductase. Plant Physiol. 112: 864
96. MANNING, K. 1998. Isolation of a set of ripening-related genes from strawberry: their identification and possible relationship to fruit quality traits. Planta 205: 622-631.
97. JEZ, J.M., FLYNN, T.G., PENNING, T.M. 1997. A new nomenclature for the aldo-keto reductase superfamily. Biochem. Pharmacol. 54: 639-647.
98. DAVIES, K.M., BLOOR, S.J., SPILLER, G.B., DEROLES, S.C. 1998. Production of yellow colour in flowers: redirection of flavonoid biosynthesis in *Petunia*. Plant J. 13: 259-266.
99. AKIYAMA, T., SHIBUYA, M., LIU, H.M., EBIZUKA, Y. 1999. *p*-Coumaroyltriacetic acid synthase, a new homologue of chalcone synthase, from *Hydrangea macrophylla* var. *thunbergii*. Eur. J. Biochem. 263: 834-839.

100.YOSHIKAWA, M., UEDA, T., MATSUDA, H., YAMAHARA, J., MURAKAMI, N. 1994. Absolute stereostructures of hydramacrosides A and B, new bioactive secoiridoid glucoside complexes from the leaves of *Hydrangea macrophylla* Seringe var. *thunbergii* Makino. Chem. Pharm. Bull. (Tokyo) 42: 1691-1693.

101.SCHRÖDER, J., RAIBER, S., BERGER, T., SCHMIDT, A., SCHMIDT, J., SOARES-SELLO, A.M., BARDSHIRI, E., STRACK, D., SIMPSON, T.J., VEIT, M., SCHRÖDER, G. 1998. Plant polyketide synthases: a chalcone synthase-type enzyme which performs a condensation reaction with methylmalonyl-CoA in the biosynthesis of C-methylated chalcones. Biochemistry 37: 8417-8425.

102.ERDTMAN, H. 1963. Some aspects of chemotaxonomy. pp. 89-125 *in* Chemical Plant Taxonomy (T. Swain, ed.), Academic Press, London.

103.HELARIUTTA, Y., ELOMAA, P., KOTILAINEN, M., GRIESBACH, R.J., SCHRÖDER, J., TEERI, T.H. 1995. Chalcone synthase-like genes active during corolla development are differentially expressed and encode enzymes with different catalytic properties in *Gerbera hybrida* (Asteraceae). Plant Mol. Biol. 28: 47-60.

104.HELARIUTTA, Y., KOTILAINEN, M., ELOMAA, P., KALKKINEN, N., BREMER, K., TEERI, T.H., ALBERT, V.A. 1996. Duplication and functional divergence in the chalcone synthase gene family of Asteraceae: evolution with substrate change and catalytic simplification. Proc. Natl. Acad. Sci. USA 93: 9033-9038.

105.ELOMAA, P., HELARIUTTA, Y., KOTILAINEN, M., TEERI, T.H. 1996. Transformation of antisense constructs of the chalcone synthase gene superfamily into *Gerbera hybrida*: differential effect on the expression of family members. Mol. Breed. 2: 41-50.

106.NAGUMO, S., TOYONAGA, T., INOUE, T., NAGAI, M. 1989. New glucosides of a 4-hydroxy-5-methylcoumarin and a dihydro-a-pyrone from *Gerbera Jamesonii hybrida*. Chem. Pharm. Bull. (Tokyo) 37: 2621-2623.

107.MCINNES, A.G., YOSHIDA, S., TOWERS, G.H.N. 1965. A phenolic glycoside from *Psilotum nudum* (L) Griseb. Tetrahedron 21: 2939-2946.

108.TAKAHASHI, S., NAKAMURA, F., SAHASHI, N., OHMOTO, T., MIZUSHIMA, U., SANKAWA, U., TOWERS, G.H.N. 1990. Chemical markers of the Psilotaceae. Biochem. Syst. Ecol. 18: 11-12.

109.LEETE, E., MUIR, A., TOWERS, G.H.N. 1982. Biosynthesis of psilotin from [2',3'-$^{13}C_2$,1'-^{14}C,4-^{3}H]phenylalanine studied with ^{13}C-NMR. Tetrahedron Lett. 23: 2635-2638.

110.HAGEN, S.E., PRASAD, J.V., BOYER, F.E., DOMAGALA, J.M., ELLSWORTH, E.L., GAJDA, C., HAMILTON, H.W., MARKOSKI, L.J., STEINBAUGH, B.A., TAIT, B.D., LUNNEY, E.A., TUMMINO, P.J., FERGUSON, D., HUPÉ, D.J., NOUHAN, C., GRACHECK, S.J., SAUNDERS, J.M., VAN DER ROEST, S. 1997. Synthesis of 5,6-dihydro-4-hydroxy-2-pyrones as HIV-1 protease inhibitors: the profound effect of polarity on antiviral activity. J. Med. Chem. 40: 3707-3711.

111.TAIT, B.D., HAGEN, S.E., DOMAGALA, J.M., ELLSWORTH, E.L., GAJDA, C., HAMILTON, H.W., PRASAD, J.V., FERGUSON, D., GRAHAM, N., HUPÉ, D.J., NOUHAN, C., TUMMINO, P.J., HUMBLET, C., LUNNEY, E.A., PAVLOVSKY, A., RUBIN, J., GRACHECK, S.J., BALDWIN, E.T., BHAT, T.N., ERICKSON, J.W., GULNIK, S.V., LIU, B.S. 1997. 4-Hydroxy-5,6-dihydropyrones. 2. Potent non-peptide inhibitors of HIV protease. J. Med. Chem. 40: 3781-3792.

112.WAT, C.-K., TOWERS, G.H.N. 1979. Metabolism of the aromatic amino acids by fungi. pp. 371-432 *in* Biochemistry of Plant Phenolics (T. Swain, J.B. Harborne, and C.F. Van Sumere, eds.), Plenum Press, New York.

113.HATFIELD, G.M., BRADY, L.R. 1973. Biosynthesis of hispidin in cultures of *Polyporus schweinitzii*. Lloydia 36: 59-65.

114.PERRIN, P.W., TOWERS, G.H.N. 1973. Hispidin biosynthesis in cultures of *Polyporus hispidus*. Phytochemistry 12: 589-592.

115.BECKERT, C., HORN, C., SCHNITZLER, J.-P., LEHNING, A., HELLER, W., VEIT, M. 1997. Styrylpyrone biosynthesis in *Equisetum arvense* L. Phytochemistry 44: 275-283.

116.HERDERICH, M., BECKERT, C., VEIT, M. 1997. Establishing styrylpyrone synthase activity in cell free extracts obtained from gametophytes of *Equisetum arvense* L. by high performance liquid chromatography-tandem mass spectrometry. Phytochem. Anal. 8: 194-197.

117.VEIT, M., GEIGER, H., WRAY, V., ABOU-MANDOUR, A., ROZDZINSKI, W., WITTE, L., STRACK, D., CZYGAN, F.-C. 1993. Equisetumpyrone, a styrylpyrone glucoside in gametophytes from *Equisetum arvense*. Phytochemistry 32: 1029-1032.

118.VEIT, M., BECKERT, C., HÖHNE, C., BAUER, K., GEIGER, H., HOEHNE, C. 1995. Interspecific and intraspecific variation of phenolics in the genus *Equisetum* subgenus *Equisetum*. Phytochemistry 38: 881-891.

119.VEIT, M., GEIGER, H., KAST, B., BECKERT, C., HORN, C., MARKHAM, K.R., WONG, H., CZYGAN, F.-C. 1995. Styrylpyrone glucosides from *Equisetum*. Phytochemistry 39: 915-917.

120.LARSEN, M., POLL, L., CALLESEN, O., LEWIS, M. 1991. Relations between the aroma compounds and the sensory evaluation of 10 raspberry varieties (*Rubus idaeus*) L. Acta Agric. Scand. 41: 447-454.

121.BOREJSZA-WYSOCKI, W., HRAZDINA, G. 1994. Biosynthesis of p-hydroxyphenylbutan-2-one in raspberry fruits and tissue cultures. Phytochemistry 35: 623-628.

122.BOREJSZA-WYSOCKI, W., HRAZDINA, G. 1996. Aromatic polyketide synthases. Purification, characterization, and antibody development to benzalacetone synthase from raspberry fruits. Plant Physiol. 110: 791-799.

123.HRAZDINA, G., KREUZALER, F., HAHLBROCK, K., GRISEBACH, H. 1976. Substrate specificity of flavanone synthase from cell suspension cultures of parsley and structure of release products *in vitro*. Arch. Biochem. Biophys. 175: 392-399.

124.SALEH, N.A.M., FRITSCH, H., KREUZALER, F., GRISEBACH, H. 1978. Flavanone synthase from cell suspension cultures of *Haplopappus gracilis* and comparison with the synthase from parsley. Phytochemistry 17: 183-186.

125.PABST, A., BARRON, D., ADDA, J., SCHREIER, P. 1990. Phenylbutan-2-one b-D-glucosides from raspberry fruit. Phytochemistry 29: 3853-3858.

126.SCHRÖDER, J. 1997. A family of plant-specific polyketide synthases: facts and predictions. Trends Plant Sci. 2: 373-378.

127.KUBITZKI, K. 1987. Phenylpropanoid metabolism in relation to land plant origin and diversification. J. Plant Physiol. 131: 17-24.

128.MARKHAM, K.R. 1988. Distribution of flavonoids in the lower plants and its evolutionary significance. pp. 427-468 *in* The Flavonoids (J.B. Harborne, ed.), Chapman and Hall, London.

129.MARKHAM, K.R. 1990. Bryophyte flavonoids, their structures, distribution, and evolutionary significance. pp. 143-161 *in* Bryophytes - Their Chemistry and Chemical Taxonomy (H.D. Zinsmeister and R. Mues, eds.), Clarendon Press, Oxford.

130.STAFFORD, H.A. 1991. Flavonoid evolution: an enzymic approach. Plant Physiol. 96: 680-685.

131. KOES, R.E., QUATTROCCHIO, F., MOL, J.N.M. 1994. The flavonoid biosynthetic pathway in plants: function and evolution. BioEssays 16: 123-132.

132.ATANASSOV, I., RUSSINOVA, E., ANTONOV, L., ATANASSOV, A. 1998. Expression of an anther-specfic chalcone synthase-iike gene is correlated with uninucleate microspore development in *Nicotina sylvestris*. Plant Mol. Biol. 38: 1169-1178.

133.BANGERA, M.G., THOMASHOW, L.S. 1996. Characterization of a genomic locus required for synthesis of the antibiotic 2,4-diacetylphloroglucinol by the biological control agent *Pseudomonas fluorescens* Q2-87. Mol. Plant-Microbe Interact. 9: 83-90.

134.UEDA, K., KIM, K.-M., BEPPU, T., HORINOUCHI, S. 1995. Overexpression of a gene cluster encoding a chalcone synthase-like protein confers redbrown pigment production in *Streptomyces griseus*. J. Antibiotics 48: 638-646.

135.CAPUANO, V., GALLERON, N., PUJIC, P., SOROKIN, A., EHRLICH, S.D. 1996. Organization of the *Bacillus subtilis* 168 chromosome between *kdg* and the attachment site of the sp-beta prophage: use of long accurate PCR and yeast artificial chromosomes for sequencing. Microbiol. 142: 3005-3015.

136.JIANG, Z.Y., SWEM, L.R., RUSHING, B.G., DEVANATHAN, S., TOLLIN, G., BAUER, C.E. 1999. Bacterial photoreceptor with similarity to photoactive yellow protein and plant phytochromes. Science 285: 406-409.

137.FUNA, N., OHNISHI, Y., FUJII, I., SHIBUYA, M., EBIZUKA, Y., HORINOUCHI, S. 1999. A new pathway for polyketide synthesis in microorganisms. Nature (London) 400: 897-899.

138.FUJII, I., MORI, Y., WATANABE, A., KUBO, Y., TSUJI, G., EBIZUKA, Y. 1999. Heterologous expression and product identification of *Colletotrichum lagenarium* polyketide synthase encoded by the *PKS1* gene involved in melanin biosynthesis. Biosci. Biotechnol. Biochem. 63: 1445-1452.

139.TAKANO, Y., KUBO, Y., SHIMIZU, K., MISE, K., OKUNO, T., FURUSAWA, I. 1995. Structural analysis of PKS1, a polyketide synthase gene involved in melanin biosynthesis in *Colletotrichum lagenarium*. Mol. Gen. Genet. 249: 162-167.

140.LANZ, T., TROPF, S., MARNER, F.-J., SCHRÖDER, J., SCHRÖDER, G. 1991. The role of cysteines in polyketide synthases: site-directed mutagenesis of resveratrol and chalcone synthases, two key enzymes in different plant-specific pathways. J. Biol. Chem. 266: 9971-9976.

Chapter Four

MACROLIDE AND POLYETHER POLYKETIDES
Biosynthesis and Molecular Diversity

Youla S. Tsantrizos* and Xian-Shu Yang

Department of Chemistry
McGill University
801 Sherbrooke Street West
Montreal,Quebec, Canada H3A 2K6

INTRODUCTION

Secondary Metabolites in Medicinal and Agricultural Chemistry

The healing and killing power of secondary metabolites has been explored by humans since prehistoric times. The very term "Pharmacology" was coined from the Greek word "Pharmakon" having the double meaning of both "Potion" and "Poison". Folklore references to such effects deal almost exclusively with plant extracts. Classic examples include the alkaloid metabolites atropine (**1**) and scopolamine (**2**) from *Atropa belladonna* (deadly nightshade), which are known to induce strong hallucinations, delirium, paralysis, and eventual death. Today, atropine is routinely used in ophthalmology for the dilation of the pupils during eye examinations, whereas scopolamine is used in the treatment of dizziness, gastrointestinal spasm, and motion sickness. Another striking example is that of tubocurarine (**3**), a metabolite of the Brazilian vine *Chondodendron tomentosum* and the key ingredient in *curare,* the extract used by South American Indians to poison their arrows. In modern medicine, tubocurarine is a useful clinical drug employed as a muscle relaxant during surgery.

The existence of microorganisms was recognized only in the 17th century when the Dutch microscopist Anton van Leeuwenhoek turned a prototype microscope to the examination of water and decaying matter. However, the tremendous value of microbial metabolites was not realized until much after the

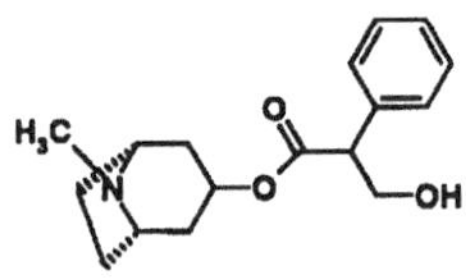

Atropine (1)

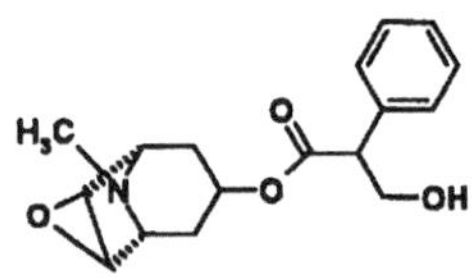

Scopolamine (2)

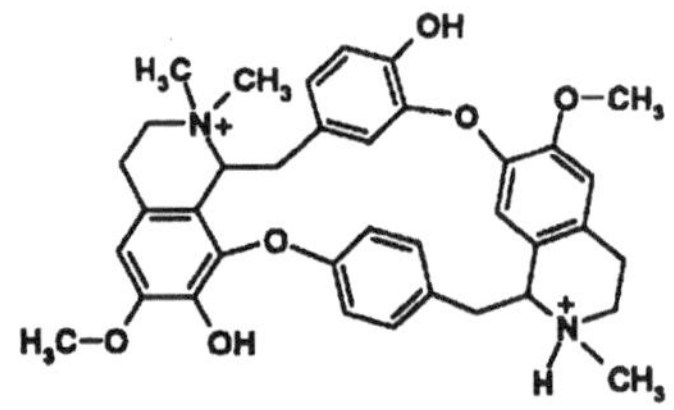

Tubocurarine (3)

antagonistic effects of *Penicillium notatum* against *Staphylococcus aureus* had been recognized by Fleming and the isolation of penicillin G (**4**) had been achieved. This discovery launched a new era in medicine and the chemistry of natural products. In modern times, metabolites isolated from cultures of bacteria, fungi, or marine microorganisms play a pivotal role in the discovery of therapeutic agents. Recent reports indicate that secondary metabolites constitute approximately 60% of all the antitumor and antibiotic drugs on the market, as well as a significant portion of the new compounds undergoing clinical testing or development.[1]

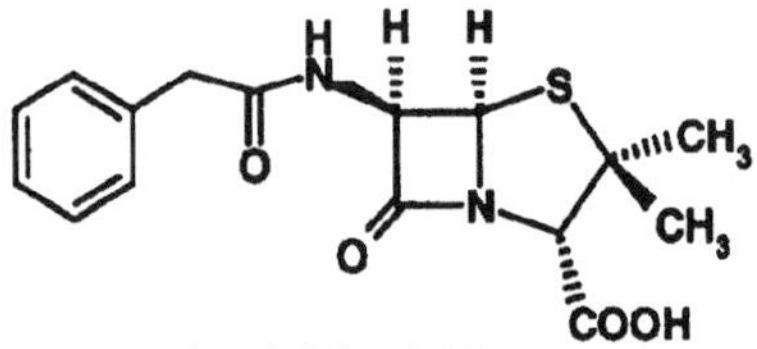

Penicillin G (4)

Polyketides constitute a special class of secondary metabolites, produced primarily by microorganisms and used in agriculture, food sciences, and especially as medicine for both humans and animals. Although these metabolites exhibit remarkable structural diversity, their biogenesis *in vivo* involves, primarily, the successive condensation of simple building blocks such as acetate, propionate, and butyrate. Examples of biologically important microbial polyketides include the fungal aflatoxins, *e.g.* versicolorin A (**5**),[2] as well as numerous metabolites of filamentous bacteria, which include the antibiotics erythromycin A (**6**) and doxorubicin (**7**),[3,4] and the immunosuppressants FK506 (**8**),[5,6] and rapamycin (**9**).[7]

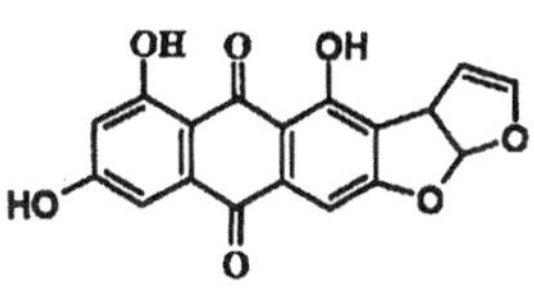

Versicolorin A (5)

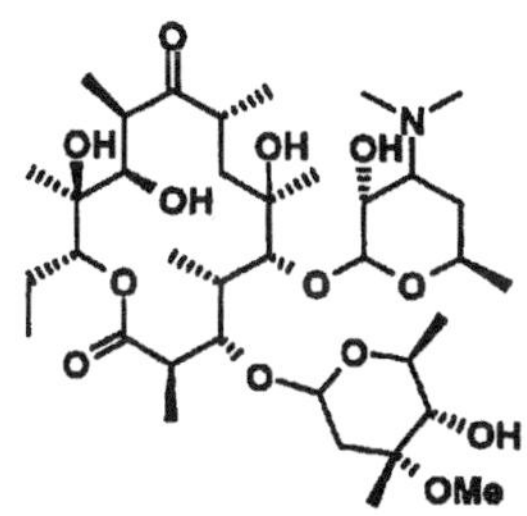

Erythromycin A (6)

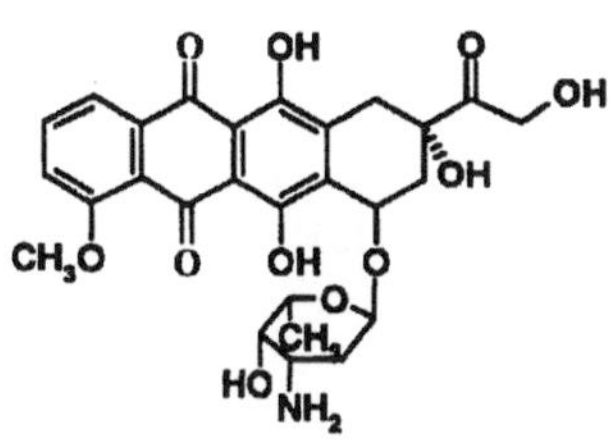

Doxorubicin (7)

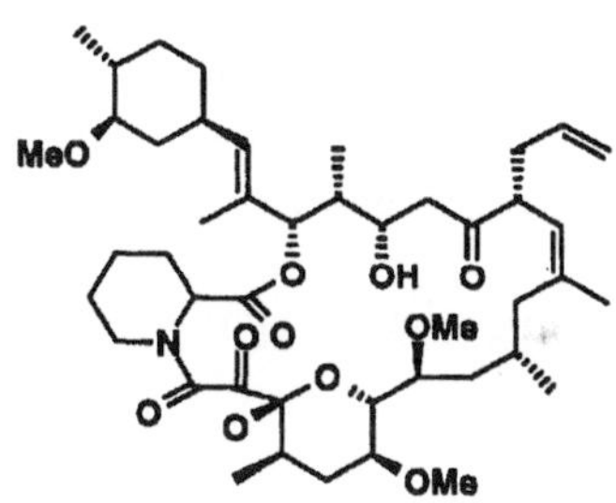

FK506 (8)

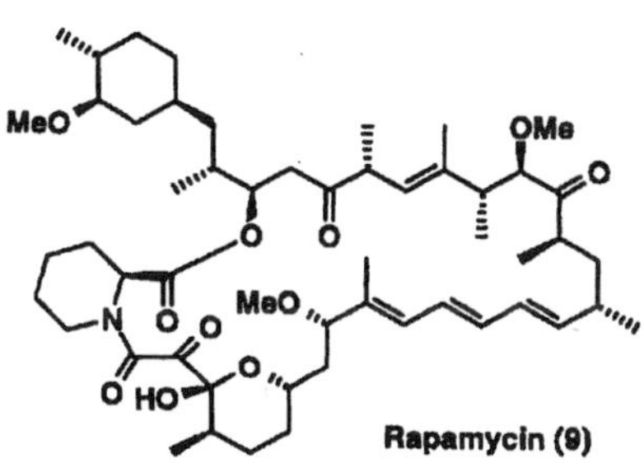

Rapamycin (9)

As with the plant alkaloids **1-3**, toxicity is also common among most clinically useful polyketides, thus compromising their therapeutic value. For example, the third most widely used class of antibiotics worldwide, the erythromycins (*e.g.* metabolite **6**) are often associated with serious gastrointestinal disorders. Similarly, the clinical usefulness of doxorubicin (**7**) is limited by its powerful cardiotoxic, myelosuppresive, and DNA damaging effects.

In this millennium, the key challenges for natural product chemists will be: (a) to fully understand Nature's catalytic processes that are involved in the biogenesis of complex structures, such as those shown above (metabolites **5-9**), and (b) to selectively redesign or reprogram the biosynthetic pathways allowing for the production of new "unnatural" natural products with exquisite efficiency and selectivity.

It is evident that by increasing structural diversity we will improve the chances of discovering new compounds with beneficial biological properties and without undesirable side effects. In the last few years, a number of achievements in this area have provided strong evidence that this goal is realistic and have sparked intense enthusiasm among scientists in the field of natural product biosynthesis and genetics. The main purpose of this chapter is to highlight some examples from this hybrid field of chemistry and biology that are relevant to the author's own research interests. However, it is not intended to be a thorough review of the field; a number of excellent reviews on this topic were published in 1997 as a collective volume in *Chemical Reviews* (volume 97, issue number 7).

BIOSYNTHESIS OF MICROBIAL MACROLIDE AND POLYETHER POLYKETIDES

Biosynthesis, Enzymology, and Genetics

Over the last ten years, extensive investigations into the biosynthesis of microbial polyketides have demonstrated that their formation is catalyzed by a class of enzymes collectively known as the polyketide synthases (PKSs). These enzymes share many mechanistic similarities with those associated with the carbon backbone assembly of fatty acids. Biosynthetic studies and molecular genetics have demonstrated that PKSs catalyze the synthesis of oligoketides from primary precursors *via* repeated decarboxylative Claisen condesnation, analogous to the mechanism of fatty acid synthases (FASs) (Scheme 1).[8,9] However, several catalytic variables are unique to PKS enzymes, such as the choice of starter unit, extender units at each step, control of reductive steps on the ß-keto group of the growing carbon chain, as well as control of stereochemistry of the alkyl substituents (Scheme 1).

Scheme 1: Key catalytic steps in carbon chain elongation of macrolide and polyether polyketides.

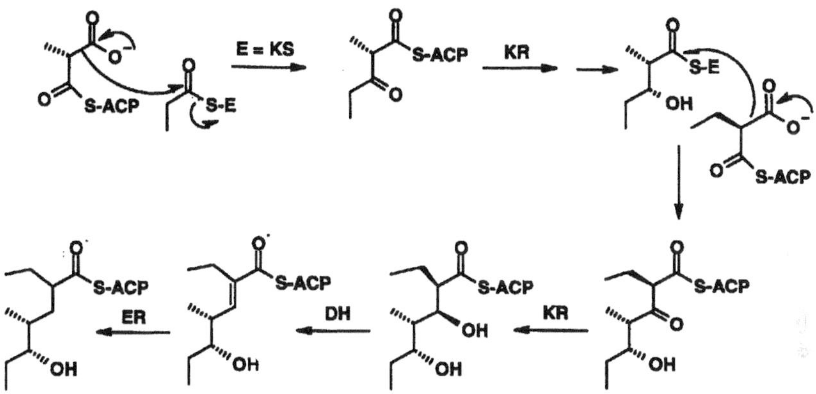

Successful cloning and sequencing of the genes for a variety of PKSs has established the primary protein structures of synthase enzymes from a number of different sources. For example, PKSs have been found that resemble the classical FASs type I, characteristic of fungi and vertebrates, in which all of the catalytic sites, acyl transferase (AT), ketosynthase (KS), acyl carrier protein (ACP), ketoreductase (KR), dehydratase (DH) and enoylreductase (ER) are present as domains along the length of a multifunctional protein. PKS type II enzymes, characteristic of bacterial and plant metabolites, have also been identified. The latter systems are analogous to the type II FASs in which each catalytic site is carried on a separate protein subunit. A number of gene clusters associated with PKS type II enzymes have already been studied. These catalyze primarily the biosynthesis of aromatic polyketides, such as doxorubicin (7),[3,4] tetracenomycin C,[10,11] actinorhodin,[12] and frenolicin.[13] However, detailed discussion of the biosynthesis and enzymology of aromatic polyketides is beyond the scope of this article.

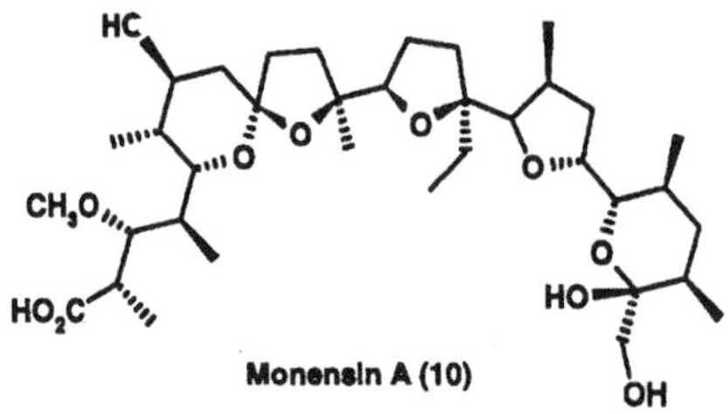

Monensin A (10)

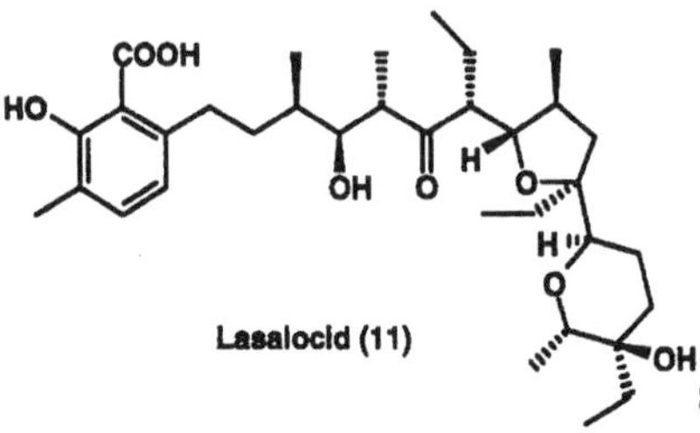

Lasalocid (11)

In this chapter, we focus on the PKS type I enzymes that are associated with the biosynthesis of macrolides, such as erythromycin A (**6**),[14-16] niddamycin,[17] FR-008,[18] FK506 (**8**),[19,20] rapamycin (**9**),[21] polyether polyketides, such as monensin A (**10**), lasalocid (**11**), mutalomycin (**12**), oudenone (**13a**, **13b**), and the family of marine toxins typified by brevetoxin A (**14**).

Erythromycin A (**6**), a metabolite of the mycelia-forming bacterium

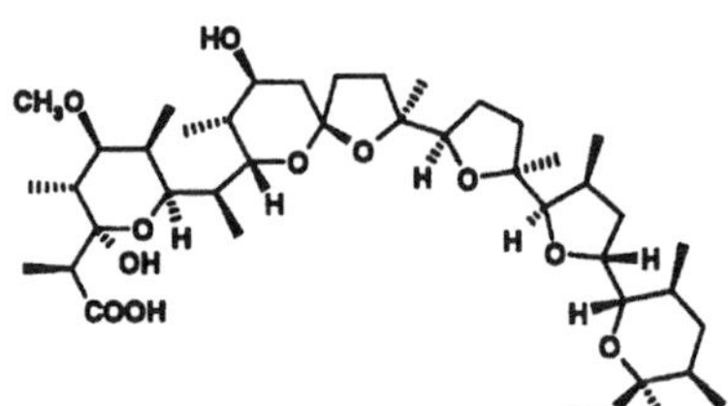

Mutalomycin (12)

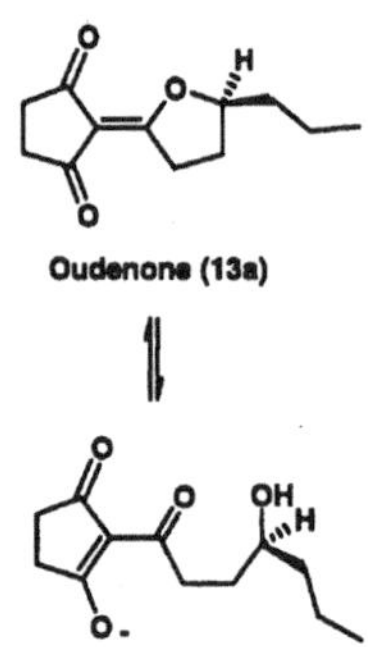

Oudenone (13a)

Oudenone (13b)

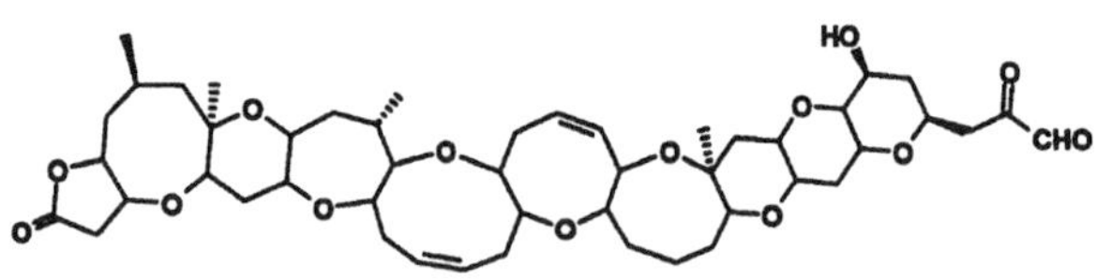

Brevetoxin A (14)

Scheme 2: Incorporation of ^{13}C, ^{18}O-labeled propionate into 6-dEB (**15**).

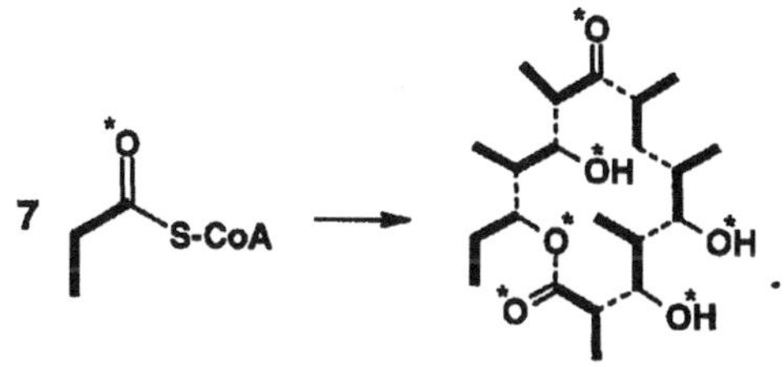

6-dEB (15)

Saccharopolyspora erythraea, is one of the most widely used clinical antibiotics against Gram-positive bacteria. It is also used for many pulmonary infections such as Legionnaire's disease and as an alternative treatment for patients allergic to b-lactam antibiotics (*e.g.* penicillins). Over the last two decades, the biosynthetic origin of this structurally complex molecule has been the subject of numerous studies.[22-24] The 14-membered macrolactone core of erythromycin, 6-deoxyerythronolide B (6-dEB, **15**), is derived from one propionyl CoA starter unit and six methylmalonyl CoA extender units (Scheme 2). Furthermore, all of the oxygen atoms of 6-dEB originate from propionate and not molecular oxygen or water.[25]

The polymerization mechanism leading to the formation of the carbon backbone involves a condensation step of each propionate unit (activated as a methylmalonate unit), which is followed by a reduction to give the required oxidation state and absolute stereochemistry, before the addition of the next propionate unit (Scheme 3). Strong evidence in support of the mechanism outlined in Scheme 3 was first provided by Cane's group in a pioneering experiment showing that an *N*-acetylcysteamine (NAC) thioester derivative of the advanced intermediate diketide **16** could be taken-up by the enzymatic machinery and incorporated *intact* into the final metabolite erythromycin *via* the key intermediate 6-dEB (Scheme 3).[26,27] It is worth noting that the incorporation of the intact carboxylic acid **17** into erythromycin could not be achieved, due to the efficient *in vivo* degradation of fatty acids by β-oxidation. However, NAC thioester derivatives are known to be far better substrates for PKS enzymes than their corresponding carboxylic acids, most likely because they exhibit a high degree of structural homology with the 4'-phosphopantetheine sidearm of the active acyl carrier protein (ACP) of the PKSs, as well as that of coenzyme A.[28,29]

Scheme 3: Model for the biosynthesis of 6-dEB (**15**) catalyzed by the complete 6-deoxyerytheronolide B synthase (three multifuntion proteints: DEB1, 2, and 3)

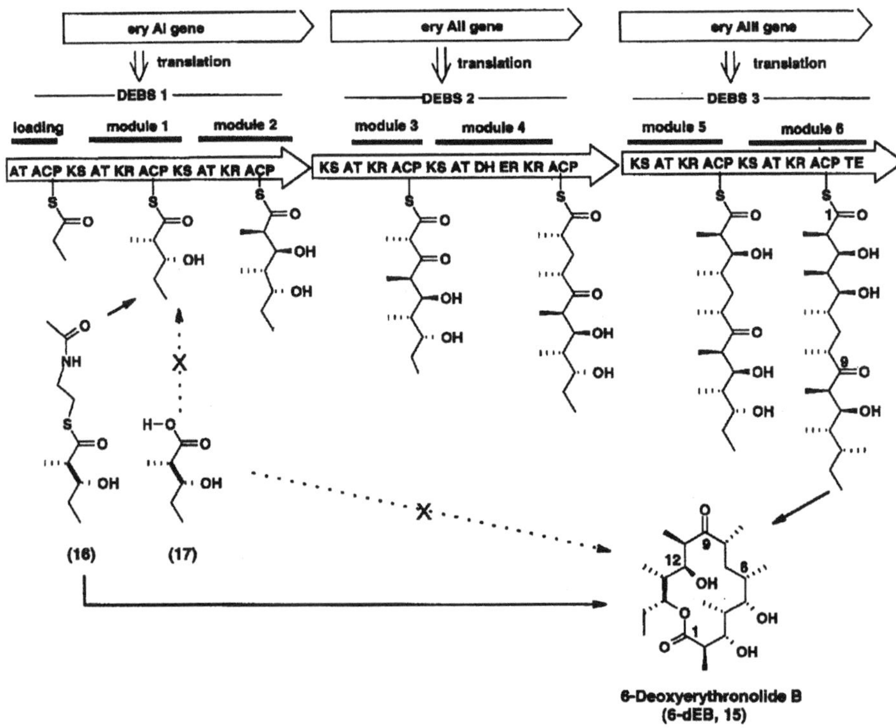

In spite of numerous earlier attempts to isolate the fully active forms of the PKS enzymes involved in the construction of the 6-dEB macrolide (**15**) from *S. erythraea,* it was only in 1995 that Khosla and co-workers reported the first method for the preparation of a recombinant 6-deoxyerythronolide B synthase (DEBS 1, 2, and 3).[30] This enzyme preparation exhibited complete modular function, possessing at least 28 distinct active sites, and was subsequently used in the first cell-free enzymatic synthesis of 6-dEB (**15**).[30]

The availability of this cell-free system allowed the *in vitro* synthesis of 6-dEB (**15**), as well as the synthesis of several smaller lactones (compounds **18-22**) from truncated DEBS mutants (Scheme 4).[31] In addition to altering the macrolide ring size, modifications of its functional groups became possible through reprogramming of the polyketide biosynthetic pathway. Multiple genetic modifications of this system, in a highly selective fashion, have led to the design of mutants that produce novel "unnatural" natural products. These studies have also expanded our knowledge of microbial genetics and enriched the structural diversity of metabolites with potential pharmaceutical or agrochemical value.[32,33]

Although it is generally believed that the biosynthetic steps associated with the skeletal construction of polyether antibiotics are also catalyzed by modular, multifunctional PKS type I enzymes (analogous to those described in the cases of 6-dEB, Scheme 3), little is known about the specific enzymes and genetics associated with this family of natural products. An idea that was initially proposed by Westley for the biogenesis of lasalocid (**11**),[34] and later modified by Cane *et al.* as a unified hypothesis for the biosynthesis of all polyether polyketides [*e.g.*, monesin A (**10**)], is

Scheme 4: Synthesis of novel "unnatural" natural lactones from DEBS mutants.

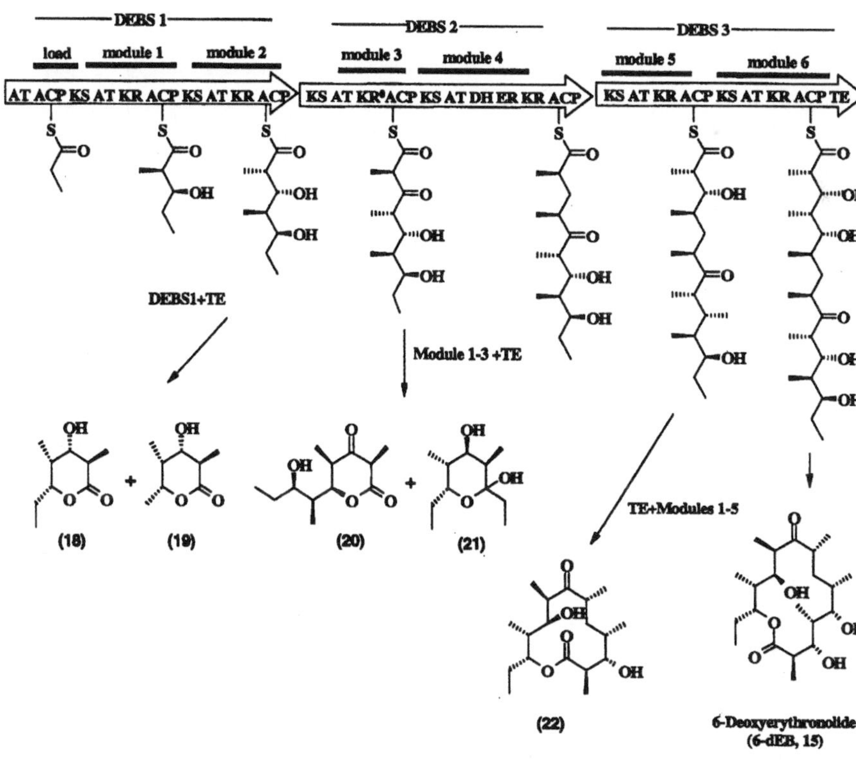

Scheme 5: Models for the biosynthesis of tetrahydrofuranyl and tetrahydropyranyl moieties in monesin A (**10**).

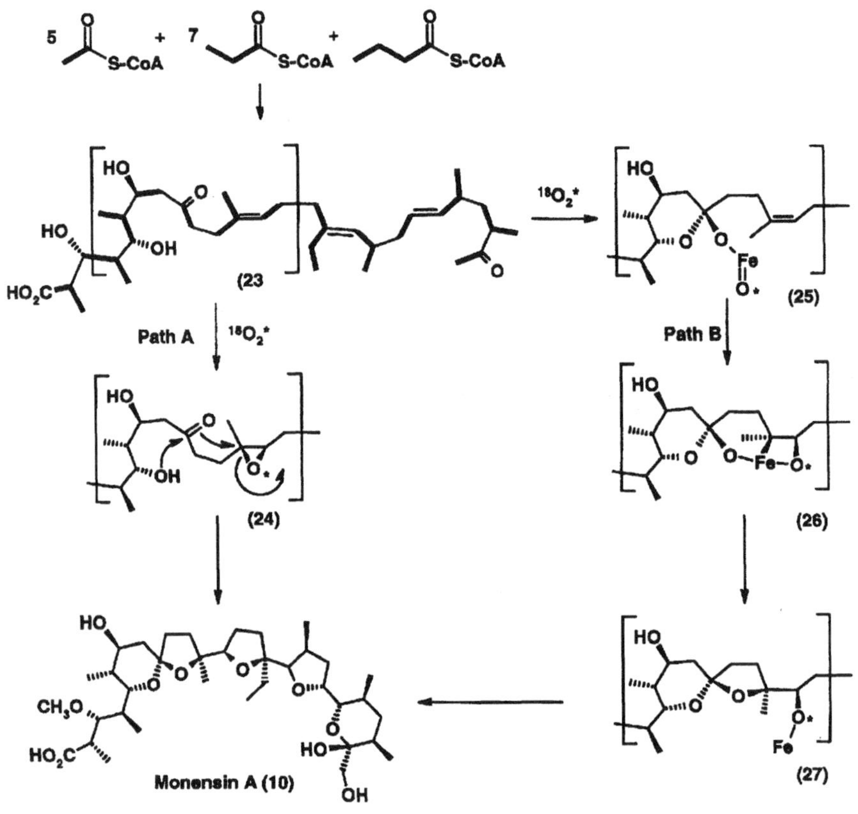

the "polyepoxide cascade" mechanism (Scheme 5, Path A).[35] In this three-step process, the open-chain oligoketide precursor **23** is formed from five acetate, seven propionate, and one butyrate unit *via* repeated decarboxylative Claisen condensations and subsequent modifications of the β-keto group. All of these steps are believed to be catalyzed by the multifunctional PKS type I enzyme(s). The second step is proposed to be an enzymatic polyepoxidation reaction of the acyclic hydroxypolyene precursor **23** to give intermediate **24**, followed by a cascade of intramolecular nucleophilic attacks on the keto and epoxide moieties of metabolite **24**, eventually leading to the formation of the tetrahydrofuran and tetrahydropyran rings of the final product **10** (Scheme 5, Path A).

More recently, the alternative mechanism of "*syn*-oxidative polycyclization" of the hydroxypolyene precursor **23** was proposed by Townsend *et al.* (Scheme 5, Path B).[36-38] In this model, an alkoxy-bound oxo metal derivative **25** is proposed to undergo a series of consecutive intramolecular [2+2] cycloadditions with the double bonds to give the corresponding metallaoxetane intermediate **26**, followed by reductive elimination of the metal and closure of the tetrahydrofuran or tetrahydropyran rings (Scheme 5, Path B).

Each of these models may be relevant in the biogenesis of polyether polyketides from different sources; nonetheless, both proposals remain unproven.

Scheme 6. Incorporation of ^{13}C-labeled acetate into oudenone (**13a** and **13b**)

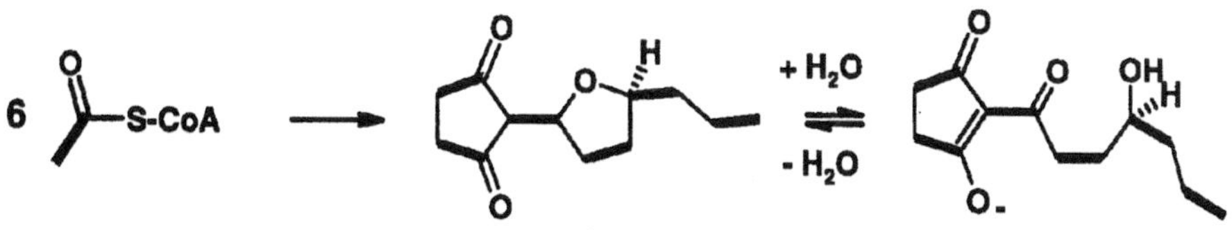

This is mainly due to the chemical complexity and instability of the precursors and intermediates associated with the biosynthesis of such structurally complex molecules as monensin A (**10**) or brevetoxin A (**14**). Recently, we reported our on-going investigations into the biosynthesis and enzymology of the fungal metabolite oudenone (**13a** and **13b**),[39] a structurally unique hexaketide (Scheme 6) characterized by a tetrahydrofuran and a 1,3-cyclopentadione moiety (**13a**). The structure of oudenone in anhydrous organic solvents is that of **13a**. However, in aqueous solvents, **13a** is in dynamic equilibrium with the β-trione anion **13b** *via* the simple addition of water (pHa'=4.1).

We speculated that due to the small molecular structure of oudenone this metabolite could serve as a model for exploration into the enzymology and genetics associated with the biogenesis of tetrahydrofuranyl and tetrahydropyranyl moieties of polyether-type polyketides. We began our investigation by showing that the N-acetylcysteamine (NAC) thioester derivative of (5S)-5-hydroxyoctanoic acid (Scheme 7, NAC-derivative of intermediate **28**) can serve as a substrate of the PKS enzyme(s) catalyzing the biosynthesis of **13** in cultures of *Oudemansiella radicata*. Furthermore, the enzymatic incorporation of the deuterium-labeled NAC-derivative of **28** into the tetrahydrofuran moiety of oudenone (**13a**) was achieved without any change in the absolute stereochemistry of the C9 chiral center. Subsequent studies allowed structure determination of the complete pre-cyclization polyketide precursor of oudenone. Based on our previous experimental data, two plausible hexaketide open-chain precursors were proposed, **29** and **31**, which could be supported by either the "polyepoxide cascade" or the "oxidative polycyclization" models, respectively (Scheme 7).

Recently, we reported the synthesis and successful *in vivo* transformation of the deuterium labeled, NAC-derivative of β-diketone **30** into oudenone (**13a**).[40] This result strongly suggests a biosynthetic mechanism analogous to the "polyepoxide cascade" model (Scheme 7, Path A) and confirms that the linear hexaketide α-diketone **30** is the open-chain precursor of oudenone (**13**). Thus, it is reasonable to assume that the cyclization of **30** proceeds *via* an intramolecular β-addition, followed by a Claisen-type intramolecular condensation and dehydration to give **13** (Scheme 6, Path A).

It should be noted that although the proposed mechanism for the biogenesis of oudenone (**13**) bears many similarities to the "polyepoxide cascade" model (Scheme 7), it is also reminiscent of the catalytic pathway leading to the formation of the 2-amino-3-hydroxycyclopent-2-enone, a unique structural moiety found in several antibiotics, including reductiomycin[41] and asukamycin.[42] A pyridoxal phosphate-dependent cyclization mechanism leading to the formation of this moiety was proposed by Floss in the biosynthesis of reductiomycin.[27]

Further studies on the genetics associated with the biosynthesis of metabolite **13** are currently in progress in our laboratory and hopefully will shed more light on the details of the enzymes and the catalytic mechanisms involved in the formation of oudenone (**13**).

SUMMARY

The biosynthesis of macrolides and polyether polyketides represent only one of the many remarkable examples of Nature's ability to synthesize complex compounds with exquisite efficiency and selectivity. It also presents natural product chemists and genetic engineers with a unique opportunity to harness the catalytic diversity and power of PKS enzymes in mutant microorganisms that can produce novel molecular structures of pharmaceutical or agricultural value. Furthermore, of the estimated 3 million species of bacteria and 1.5 million species of fungi believed to occupy our planet, only 0.1% of the bacteria and 5% of the fungi have been described. An important obstacle in identifying many of the yet unknown microorganisms is our inability to grow these species in the laboratory. However,

Scheme 7: Plausible hexaketide precursors and their cyclization to oudenone (**13a**)

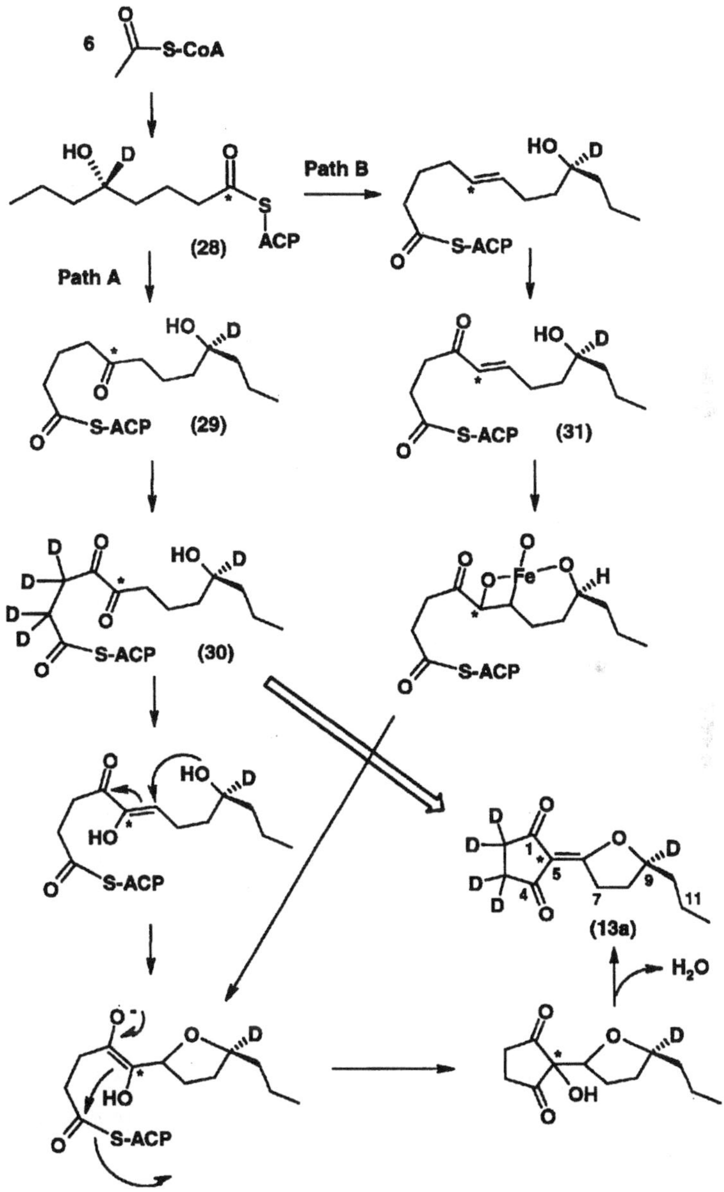

it should be possible to isolate the chromosomal DNA of microorganisms from small samples (some researchers refer to this DNA as "environmental DNA"), and by using sequence homology of known PKSs we should be able to discover gene clusters encoding for new PKS enzymes. Expressing these genes in a heterologous host may subsequently lead to the discovery of new natural products. There is little doubt that a proliferation of studies on the rational manipulation of biosynthetic genes will occur over the next few years. The key results will be the rapid progress in our understanding of complex biosynthetic pathways and a complementary explosion in the molecular diversity of secondary metabolites, or "unnatural" natural products, that will be discovered.

ACKNOWLEDGMENTS

Research on the biosynthesis of oudenone has been supported by the Natural Sciences and Engineering Research Council of Canada. We are grateful to Boehringer Ingelheim (Canada) Ltd., Bio-Méga Research Division for making their facilities available for our academic research, and to Dr. Stephen Kawai for proofreading the manuscript.

REFERENCES

1. CRAGG, G. M., NEWMAN, D. J., SNADER, K. M 1997. Natural products in drug discovery and development. J. Nat. Prod. 60: 52-60.
2. MINTO, R.E., TOWNSEND, C.A. 1997. Enzymology and molecular biology of aflatoxin biosynthesis. Chem Rev. 97: 2537-2555.
3. FUJII, I., EBIZUKA, Y., 1997. Anthracycline biosynthesis in Streptomyces galilaeus. Chem. Rev. 97: 2511-2523.
4. HUTCHINSON, C.R., 1997. Biosynthetic studies of daunorubicin and tetracenomycin C. Chem. Rev. 97: 2525-2535.
5. KINO, T., HATANAKA, H., MIYATA, S., INAMURA, N., NISHIYAMA, N., YAJIMA, T., GOTO, T., OKUHARA, M., KOHSAKA, M., AOKI, H., OCHIAI, T. 1987. FK506, a novel immunosuppressant isolated from a *Streptomyces*. J. Antibiot. 40: 1256-1265.
6. SCHREIBER, S.L., CRABTREE, G.R. 1992. The mechanism of action of cyclosporin A and FK506. Immunol. Today 13: 136-142.
7. CALNE, R.Y., LIM, S., SAMAAN, A., COLLIER, D.St.J., POLLARD, S.G., WHITE, D.J.G. 1989. Rapamycin for immuno suppression in organ

allograftin. Lancet ii 227.
8. HOPWOOD, D.A., SHERMAN, D.H. 1990. Molecular genetics of polyketides and its comparison to fatty acid biosynthesis. Annu. Rev. Genet. 24: 37-66.
9. KATZ, L., DANADIO, S. 1993. Polyketide synthesis: Prospects for hybrid antibiotics. Annu. Rev. Microbiol. 47: 875-912.
10. MOTAMERI, H., HUTCHINSON, C.R. 1987. Cloning and heterologous expression of gene cluster for the biosynthesis of tetracenomycin C, the anthracycline antitumor antibiotic of *Streptomyces glaucescens*. Proc. Natl. Acad. Sci. USA 84: 4445-4449.
11. SHEN, B., SUMMERS, R.G., WENDT-PIENKOWSKI, E., HUTCHINSON, C.R. 1995. The *Streptomyces glaucescens tcmKL* Polyketide synthase and tcmN polyketide cyclase genes govern the size and shape of aromatic polyketides. J. Am. Chem. Soc. 117: 6811-6821.
12. MALPARTIDA, F., HOPWOOD, D.A. 1984. Molecular cloning of the whole biosynthetic pathway of a Streptomyces antibiotic and its expression in a heterologous host. Nature 309: 462-464.
13. BIBB, M.J., SHERMAN, D.H., OMURA, S., HOPWOOD, D.A. 1994. Cloning, sequencing and deduced functions of a cluster of *Streptomyces* genes probably encoding the biosynthesis of the polyketide antibiotic frenolicin. *Gene* 142(1): 31-39.
14. CORTES, J., HAYDOCK, S.F., ROBERTS, G.A., BEVITT, D.J., LEADLAY, P.F. 1990. An unusually large multifunctional polypeptide in the erythromycin-producing polyketidesynthase of *Saccharopolyspora erythraea*. Nature 348: 176-178.
15. CAFFREY, P., BEVITT, D.J., STAUNTON, J., LEADLAY, P.F. 1992. Identification of DEBS 1, DEBS 2 and DEBS 3, the multienzyme polypeptides of the erythromycin-producing polyketide synthase from *Saccharopolyspora erythraea*. *FEBS* Lett. 304: 225-228.
16. DONADIO, S., STAVER, M.J., MCALPINE, J.B., SWANSON, S.J., KATZ, L. 1991. Modular organization of genes required for complex polyketide biosynthesis. Science 252: 675-679.
17. KAKAVAS, S.J., KATZ, L., STASSI, D. 1997. Identification and characterization of the niddamycin polyketide synthase genes from *Streptomyces caelestis*. J. Bacteriol. 179: 7515-7522.
18. HU, Z., BAO, K., ZHOU, X., ZHOU, Q., HOPWOOD, D.A., KIESER, T., DENG, Z. 1994. Repeated polyketide synthase modules involved in the biosynthesis of a heptaene macrolide by *Streptomyces* sp. FR-008. Mol. Microbiol. 14: 163-172.
19. MOTAMEDI, H., SHAFIEE, A. 1998. The biosynthetic gene cluster for the

macrolactone ring of the immunosuppressant FK506. Eur. J. Biochem. 256: 528- 534.

20. MOTAMEDI, H., CAI, S.-J., SHAFIEE, A., ELLISTON, K.O. 1997. Structural organization of a multifunctional polyketide synthase involved in the biosynthesis of the macrolide immunosuppressant FK506. Eur. J. Biochem. 244: 74-80.
21. SCHWECKE, T., APARICIO, J.F., MOLNÁR, I., KÖNIG, A., KHAW, L.E., HAYDOCK, S.F., OLIYNYK, M., CAFFREY, P., CORTÉS, J., LESTER, J.B., BÖHM, G.A., STAUNTON, J., LEADLAY, P.F. 1995. The biosynthetic gene cluster for the polyketide immunosuppressant rapamycin. Proc. Natl. Acad. Sci. USA 92: 78397843.
22. KATZ, L., 1997. Manipulation of modular polyketide synthases. Chem. Rev. 97: 2557-2575.
23. STAUNTON, J., WILKINSON, B., 1997. Biosynthesis of erythromycin and rapamycin. Chem. Rev. 97: 2611-2629.
24. KHOSLA, C., 1997. Harnessing the biosynthetic potential of modular polyketide synthases. Chem. Rev. 97: 2577-2590.
25. CANE, D.E., HASLER, H., LIANG, T.-C. 1981. Macrolide biosynthesis. Origin of the oxygen atoms in the erythromycins. J. Am. Chem. Soc. 103: 5960-5962.
26. CANE, D.E., YANG, C.-C. 1987. Macrolide biosynthesis. 4. Intact incorporation of a chain-elongation intermediate into erythromycin. J. Am. Chem. Soc. 109: 1255-1257.
27. CANE, D.E., PRABHAKARAN, P.C., TAN, W., OTT, W.R. 1991. Macrolide biosynthesis. 6. Mechanism of polyketide chain elongation. Tetrahedron Lett. 32: 5457-5460.
28. YUE, S., DUNCAN, J.S., YAMAMOTO, Y., HUTCHINSON, C.R., 1987. Macrolide biosynthesis. Tylactone formation involves the processive addition of three carbon units. J. Am. Chem. Soc. 109: 1253-1255.
29. YOSHIZAWA, Y., LI, Z., REESE, P.B., VEDERAS, J.C. 1990. Intact incorporation of acetate-derived di- and tetraketides during biosynthesis of dehydrocurvularin, a macrolide phytotoxin from *Allernaria cinerariae*. J. Am. Chem. Soc. 112: 3212-3213.
30. PIEPER, R., LUO, G., CANE, D.E., KHOSLA, C. 1995. Cell-free synthesis of polyketides by recombinant erythromycin polyketide synthases. Nature 378: 263-266.
31. KAO, C.M., LUO, G., KATZ, L., CANE, D.E., KHOSLA, C. 1995. Manipulation of macrolide ring size by directed mutagenesis of a modular polyketide synthase. J. Am. Chem. Soc. 117: 9105-9106.

32. DONADIO, S., McALPINE, J.B., SHELDON, P.J., JACKSON, M., KATZ, L. 1993. An erythromycin analog produced by reprogramming of polyketide synthesis. Proc. Natl. Acad. Sci. USA 90: 7119-7123.
33. McDANIEL, R., THAMCHAIPENET, A., GUSTAFSSON, C., FU, H., BETLACH, M., BETLACH. M., ASHLEY, G. 1999. Multiple genetic modifications of the erythromycin polyketide synthase to produce a library of novel "unnatural" natural products. Proc. Natl. Acad. Sci. USA 96: 1846-1851.
34. WESTLEY, J.W. 1981. pp. 41-73 *in* Antibiotics IV: Biosynthesis. (J.W. Corcoran, ed.) Springer-Verlag, N.Y.
35. CANE, D.E., CELMER, W.D., WESTLEY, J.W. 1983. Unified stereochernical model of polyether antibiotic structure and biogenesis. J. Am. Chem. Soc. 105: 3594-3600.
36. TOWNSEND, C.A., BASAK, A. 1991. Experiments and speculations on the role of oxidative cyclization chemistry in natural product biosynthesis. Tetrahedron 47: 2591-2602.
37. McDONALD, F.E., TOWNE, T.B. 1994. *syn*-Oxidative polycyclization of hydroxy poyenes: a new approach to polyether synthesis. J. Am. Chem. Soc. 116: 7921-7922.
38. KOERT, U. 1995. Oxidative polycyclization versus the "Polyepoxide Cascade": New pathways in polyether (bio)synthesis? Angew. Chem. Int. Ed. Engl. 341:298-300.
39. TSANTRIZOS, Y.S., ZHOU, F., FAMILI, P., YANG, X.-S. 1995. Biosynthesis of the hypotensive metabolite oudenone by *Oudemansiella radicata.* 1. Intact incorporation of a tetraketide chain elongation intermediate. J. Org. Chem. 60: 6922-6929.
40. TSANTRIZOS, Y.S., YANG, X-S., McCLORY, A. 1999. Studies on the biosynthesis of the fungal metabolite oudenone. 2. Synthesis and enzymatic cyclization of an α-diketone, open-chain precursor into oudenone in cultures of *Oudemansiella radicata.* J. Org. Chem. 64: 6609-6614.
41. CHO, H., BEALE, J.M., GRAFF, C., MOCEK, U., NAKAGAWA, A., OMURA, S., FLOSS, H.G. 1993. Studies on the biosynthesis of the antibiotic reductiomycin in *Streptomyces xanthochromogenus.* J. Am. Chem. Soc. 115: 12296-12304.
42. KAKINUMA, K., IKEKAWA, N., NAKAGAWA, A., OMURA, S. 1979. The structure of asukamycin, a possible shunt metabolite from 3-dehydroquinic acid in the shikimate pathway. J. Am. Chem. Soc. 101: 3402-3404.

Chapter Five

BIOCHEMICAL, MOLECULAR GENETIC AND EVOLUTIONARY ASPECTS OF DEFENSE-RELATED TERPENOID METABOLISM IN CONIFERS

Jörg Bohlmann,[1]* Jonathan Gershenzon[1] and Sébastien Aubourg[2]

Max Planck Institute for Chemical Ecology[1]
Carl Zeiss Promenade 10
D-07745 Jena, Germany

Institut de Biotechnologie des Plantes[2]
Laboratoire de Biologie du Dévelopement des Plantes
Université de Paris-Sud, UMR CNRS 8618
F-91405 Orsay Cedex, France

*Author for correspondence.

Evolution of Metabolic Pathways – J.T. Romeo et al. (Editors)

INTRODUCTION

Conifers (order Coniferales) are the largest and most diverse group of living gymnosperms comprising more than 600 species of 60-65 genera in seven families, including the Pinaceae, Podocarpaceae, Araucariaceae, Taxaceae, Cephalotaxaceae, Taxodiaceae, and Cupressaceae.[1,2] Their origin can be traced back to 300 million years (upper Carboniferous) according to fossil records, and their peak of abundance occurred about 150 million years ago at the Jurassic-Cretaceous boundary. Cladistic analyses based on nuclear 28S rRNA sequences and other molecular and morphological data demonstrate that conifers are a monophyletic group with the pine family (Pinaceae) having been the first to diverge from the rest of the assemblage.[2] At present, conifers are dominant members of many forest ecosystems. In the northern hemisphere, conifers are among the most important species for the production of wood and fiber. In addition to their taxonomic abundance, ecological dominance, and evolutionary persistence, conifers are also noted for their ability to form an enormous array of natural products, including soluble and insoluble phenolics, alkaloids of the 2,6-disubstituted piperidine type, and a large variety of terpenoids (isoprenoids).[3] Of these, terpenoids, including hemiterpenes (C_5), monoterpenes (C_{10}), sesquiterpenes (C_{15}), and diterpenes (C_{20}) are perhaps the most characteristic group of conifer metabolites (Fig. 1).

Conifer terpenoids have been extensively studied for their role in plant defense and have been frequently exploited as systematic markers at a variety of taxonomic levels.[4,5] Each species contains a complex mixture of terpenoids that varies among individuals and plant parts. Terpenoid quantity and composition can also vary dramatically during development, and are affected by both the physical and biotic environment.[5-8] The complexity and variability of terpenoid composition characteristic of extant conifers is the product of a wide range of chemical, enzymatic, and genetic factors. This review will focus on the biochemistry and molecular genetics of terpenoid metabolism in members of the pine family (Pinaceae) of conifers and its evolutionary significance. Relevant literature on terpenoid metabolism in angiosperms and species outside the plant kingdom is also reviewed to provide appropriate background and context. Understanding the genetics, biochemistry, and evolution of terpenoids is important not only for appreciating the major role that these substances play in conifer ecology and evolution, but also for developing strategies to genetically engineer terpenoids. Manipulating terpenoids in forest trees and agricultural crops has high potential to improve resistance against herbivores and pathogens and to increase their commercial value.

TERPENOIDS IN CONIFERS

While all plants probably produce some terpenoid natural products, members of the pine family along with other plant families such as the Lamiaceae, Rutaceae, Myrtaceae, Asteraceae, and Apiaceae share the ability to accumulate large amounts of terpenoid secondary products of low molecular weight (C_{10} – C_{20}) as constituents of essential oils, turpentines, and resins.[9] These families have evolved various forms of specialized anatomical structures for storage of volatile and potentially toxic terpenoid products that might otherwise interfere with general cell metabolism.[10] The development of specialized structures for terpenoid biosynthesis and accumulation must be considered an important event in the evolution of these families. In members of the pine family, resin terpenoids are formed in secretory structures, such as resin blisters and resin ducts.[10,11] The extent of such resin systems and their storage capacity varies significantly among genera. For example, short-lived, isolated resin blisters are characteristic of stems of fir (*Abies*), cedar (*Cedrus*), hemlock (*Tsuga*), and golden larch (*Pseudolarix*), while resin ducts are found in larch (*Larix*), Douglas fir (*Pseudotsuga*), pine (*Pinus*), and spruce (*Picea*). Axial and radial resin ducts in the wood and bark of pine and spruce are connected to form a three-dimensional network allowing the accumulation of copious amounts of resin.

The focus of this chapter is on the metabolism of three classes of conifer terpenoids, monoterpenes (C_{10}), sesquiterpenes (C_{15}), and diterpenes (C_{20}), which serve as the major constituents of oleoresin (Fig. 1). These substances are generally believed to serve as physical and chemical defenses against herbivores and fungi.[12-14] Consequently, individual terpenoids and combinations of terpenoids have been proposed as markers for the resistance or susceptibility of certain species and genotypes to herbivores.[15-20] Conifer oleoresin can be separated into a volatile turpentine fraction, which consists mainly of monoterpenes and smaller quantities of sesquiterpenes, and a non-volatile rosin fraction, which contains mainly diterpene resin acids. The low molecular weight components of the turpentine provide a solvent for the mobilization of diterpene resin acids from resin ducts and blisters, where they are stored under pressure. Mechanical injury of the bark and wood results in resin flow to the wound site, where the volatile turpentine evaporates and the diterpene resin acids and other non-volatile resin components harden to form a mechanical barrier sealing the wound site against invading insects and pathogens.[12] In addition to their role in physical defense, monoterpenes and monoterpene-containing mixtures can be deterrent or toxic to stem-boring herbivores and their associated fungal pathogens, as well as to foliage feeders.[21] Despite their toxicity, conifer monoterpenes also serve as attractants for bark

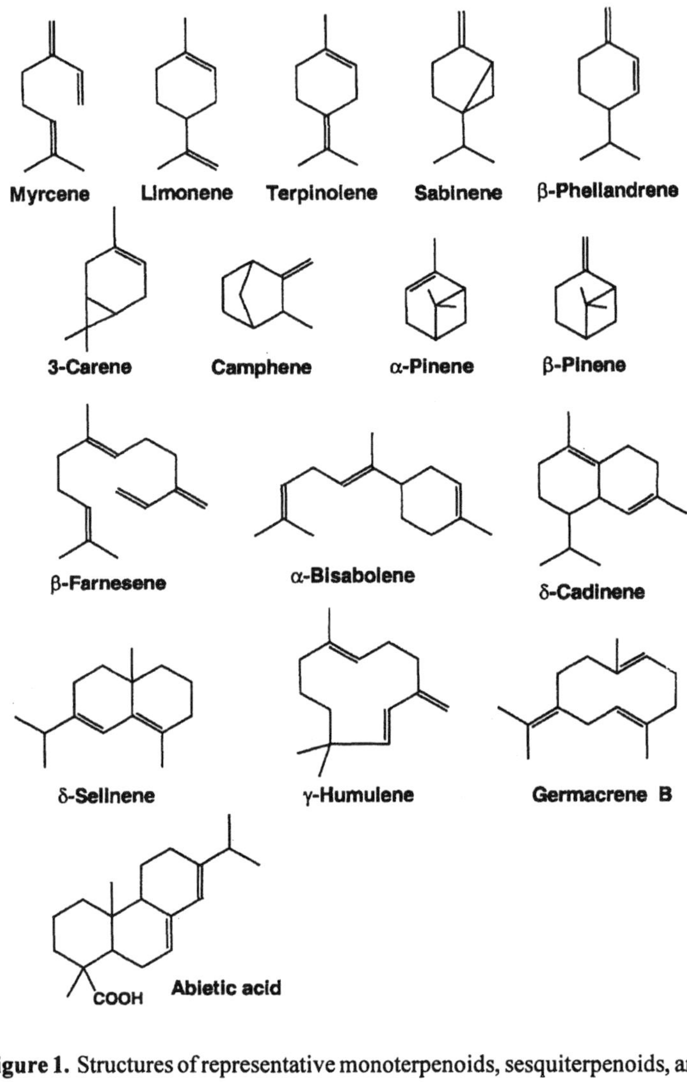

Figure 1. Structures of representative monoterpenoids, sesquiterpenoids, and diterpenoids of conifers.

beetles and other specialist insect herbivores, as precursors for bark beetle pheromones, and as volatile cues for parasitic wasps that attack bark beetles in certain forms of tritrophic plant-insect interactions.[22] The sesquiterpenoid constituents of resins may also function in defense against pathogens, as angiosperm sesquiterpene phytoalexins do.[23] Bisabolene-derived analogs of insect juvenile hormone III are formed in firs (*Abies*) and interfere with insect reproduction and development.[24] Diterpene resin acids also exhibit toxicity and feeding deterrency to herbivores.[25] For example, the survival and development of specialist pine sawfly larvae on *Pinus sylvestris* foliage is significantly reduced by high concentrations of diterpene acids.[26] The many roles of conifer terpenoids as physical and chemical defenses, and as chemical signals for herbivores and herbivore enemies, no doubt reflect the millions of years of conifer - insect coevolution.

THE BASIC PATHWAY OF TERPENOID BIOSYNTHESIS

Formation of the Central C_5-Units of Terpenoids

The general isoprenoid pathway with its many branches and numerous possibilities of chemical variation gives rise to natural products of unsurpassed abundance and structural diversity. For convenience, the formation of monoterpenes, sesquiterpenes, and diterpenes in plants can be divided into several stages (Fig. 2).[27,28] The first stage encompasses the biosynthesis of the central C_5-intermediate, isopentenyl diphosphate (IPP), which originates from two different pathways, the "classical" mevalonic acid pathway and the newly discovered 1-deoxy-D-xylulose 5-phosphate (DXP) pathway. The mevalonic acid pathway and its regulation in plants have been subject of several recent reviews.[28-30] Similar to the situation in animals, 3-hydroxy-3-methylglutaryl-CoA (HMG-CoA) reductase, which converts HMG- CoA to mevalonic acid in a two step reduction, appears to be a rate-limiting enzyme in the biosynthesis of sterols and other terpenoids in plants. Although the mevalonic acid pathway has received scant attention in conifers, HMG-CoA reductase from *Picea abies* appears to have similar properties to those of other plant species.[31,32]

The discovery of a mevalonic acid-independent route to IPP *via* DXP has attracted much attention in recent years.[33-35] Significant progress in elucidating the DXP- pathway has been achieved by a combination of excellent chemical and biochemical studies, along with efforts to sequence the complete genomes of selected species and the transcriptomes (ESTs) of specialized tissues. The first pathway intermediate, DXP, is formed by the condensation of a C_2–unit generated from pyruvate with glyceraldehyde-3- phosphate under the catalysis of DXP synthase, a novel type of transketolase.[36-39] Sequence analyses suggest that DXP synthase-type transketolases from species as distant as *E. coli* and higher plants share a common origin and have evolved as a separate class distinct from the transketolases of the pentose phosphate pathway.[37] In the following step, DXP is converted by means of a reductosiomerase to 2-C-methyl-D-erythritol 4-phosphate

(MEP).[40-42] MEP can then be transformed in a cytidine 5'-triphosphate dependent reaction into 4-diphosphocytidyl-2-C-methylerythritol.[43] Further intermediates in the conversion of MEP to isopentenyl monophosphate (IP) remain to be identified.[44] In the terminal step of the DXP-pathway, IP is converted to IPP *via*

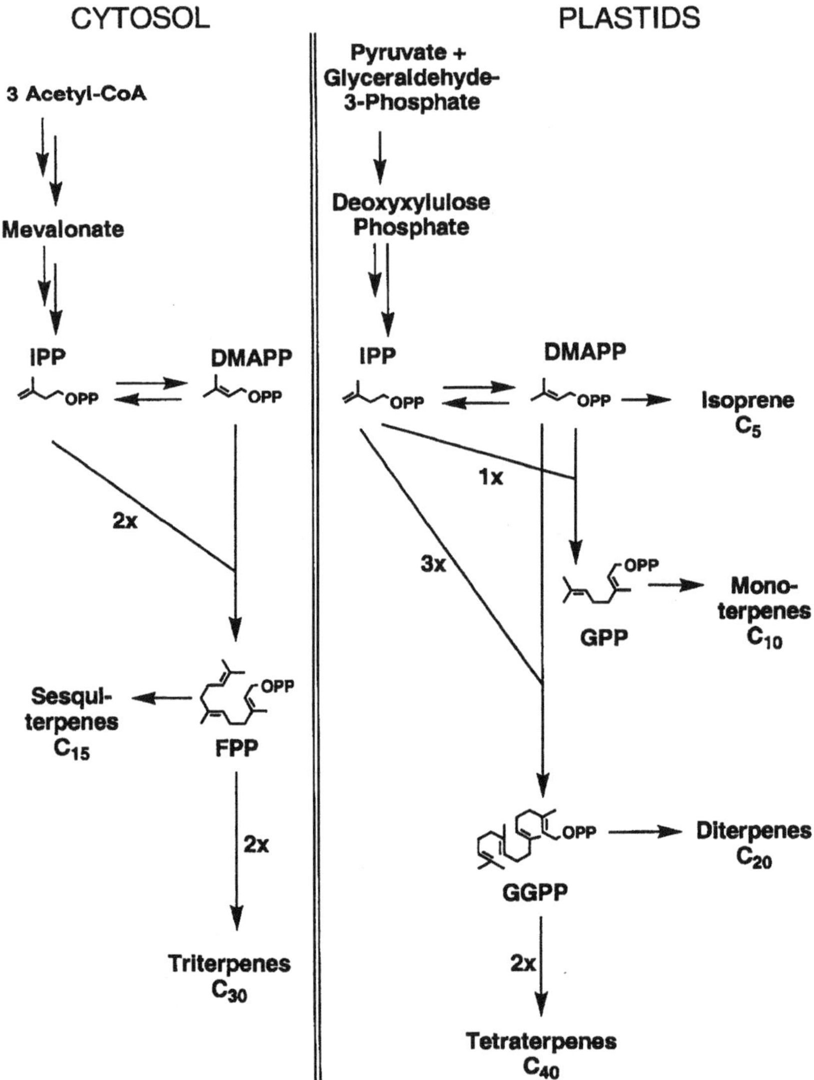

Figure 2. Pathways of plastidial and cytosolic terpenoid biosynthesis in plants.

the action of a monophosphate kinase.[45] Following the formation of IPP *via* either the DXP-pathway or the mevalonic acid pathway, this central intermediate is isomerized to dimethylallyl diphosphate (DMAPP) by IPP isomerase.

The mevalonic acid pathway and the DXP-pathway to IPP are localized in different subcellular compartments of the plant cell and participate in the formation of different groups of terpenoid end products. The mevalonic acid pathway operates in the cytosol and endoplasmic reticulum yielding precursors for the biosyntheses of sesquiterpenes and triterpenes. In contrast, the DXP pathway is active in plastids where it generates IPP for the biosyntheses of the hemiterpene isoprene, and for the formation of monoterpenes, diterpenes, and tetraterpenes. Both pathways appear to contribute to the origin of certain sequiterpenoids, supporting the concept of at least a partial flux of metabolites between terpenoid pathways in the plastids and the cytosol - endoplasmic reticulum.[33,46] The DXP pathway has now been demonstrated in eubacteria and cyanobacteria, in the malaria parasite *Plasmodim falciparum*, and in plastids of various groups of algae and higher plants, but has not been found in archaebacteria, *Euglena*, yeasts, fungi, or mammals.[33,35,45,47,48] Occurrence of the DXP pathway in cyanobacteria and its plastidial localization imply a possible endosymbiotic origin of this pathway in algae and higher plants.[33,35] Genes for the DXP pathway could have been translocated to the nucleus of the eukaryotic cell as part of major rearrangements of plant nuclear and organelle genomes occurring through the course of evolution.[33,49] In conifers, the DXP pathway has so far been demonstrated only in *Taxus chinensis*, where it participates in the biosynthesis of the diterpene paclitaxel (taxol).[50] However, this pathway is undoubtedly of universal occurrence in conifers as in other higher plants.[35] The existence of a mevalonate-independent route to terpenoids helps explain many puzzling observations about conifer terpenoid biosynthesis in the older literature. For example, while radiolabeled mevalonic acid was significantly incorporated into sesquiterpenes in *Pinus pinaster* foliage, this precursor was not incorporated into monoterpenes.[51] These results can be readily accounted for by the fact that monoterpenes arise from the mevalonate-independent, DXP pathway, while sesquiterpenes are formed from the mevalonic acid pathway.

Condensation of C_5-Units by Chain Length-Specific Prenyl Transfearses

In the second stage of terpenoid biosynthesis, specific *E*-isoprenyl diphosphate synthases, known as prenyl transferases, mediate the condensation of the two C_5-units, IPP and DMAPP, to geranyl diphosphate (GPP, C_{10}), and catalyze the subsequent 1'-4 additions of IPP to generate farnesyl diphosphate

Table 1. Cloned short-chain prenyl transferases from plants.

Prenyltransferase	Species	Accession	Reference
GPP Synthase large subunit	*Mentha x piperita*	AF182828	55
GPP Synthase small subunit	*Mentha x piperita*	AF182827	55
FPP Synthase	*Arabidopsis thaliana*	X75789, L46349, L46350	66, 135
	Artemisia annua	U36376	136
	Capsicum annuum	X84695	137
	Gossypium arboreum	Y12072	138
	Hevea brasilinesis	Z49786	139
	Lupinus albus	U15777, U20771	64
	Oryza sativa	D85317	140
	Parthenium argentatum	X82542, X82543	65
	Zea mays	L39789	141
GGPP Synthase	*Arabidopsis thaliana*	D85029, L25813, L40577, U44876, U44877, AB000835	68, 69, 70, 71, 142
	Capsicum annum	X80267	143, 144
	Catharanthus roseus	X92893	145
	Lupinus albus	U15778	146
	Sinapis alba	X98795	147
	Taxus canadensis	AF081514	72

(FPP, C_{15}) and geranylgeranyl diphosphate (GGPP, C_{20}).[52,53] Although little is known about the prenyltransferases of conifers, these activities have been characterized from a number of angiosperm species, and the genes for FPP and GGPP synthase have been frequently cloned (Table 1).[28,54] The first cDNAs encoding GPP synthase were recently isolated from peppermint (*Mentha x piperita*), and the recombinant and native enzymes were shown to be heterodimers.[55] In contrast, FPP synthase and GGPP synthase are homodimeric. GPPSynthase, FPP synthase, and GGPP synthase, the "short chain prenyltransferases", all operate by a common electrophilic reaction mechanism that requires a divalent metal ion cofactor (Mg^{2+}) and two conserved aspartate-rich domains in the active site fold for interaction with the diphosphate substrates.[53,56,57]

Our current knowledge of the structural elements that determine the product specificity of prenyltransferases is based on X-ray crystal structure analyses of wild-type and mutated avian FPP synthase and on functional analyses of a series of mutated archaeal and bacterial prenyltransferases.[56-62] These studies demonstrate that product chain length is determined by a few structural elements in the α-barrel active site fold, the amino acid residues at the fourth and fifth position *N*-terminal of the first aspartate-rich motif (FARM), and the insertion of two amino acids in this motif.[53] These residues may affect the size and shape of the hydrophobic pocket in the active site and, thus, determine the maximum chain-length of the elongated isoprenyl diphosphate product that may be formed. Structure-function analyses of prenyl diphosphate synthases have provided insight on how chain length-specific enzymes could have evolved by mutation of one or a few amino acid residues in the so-called chain-length determination (CLD) domain.[53]

A recent reconstruction of the phylogeny of more than 50 *E*-isoprenyl synthases, placed FPP synthases into a eukaryotic type I class that includes those from plants, and a prokaryotic type II class.[53] GGPP synthases were classified into three different types. GGPP synthases of type I are found in archaea as a sister group of long chain prenyltransferases. Type II GGPP synthases include enzymes from plants and bacteria, and are more closely related to bacterial type II FPP synthases than they are to type I FPP synthases from plants and other eukaryotes. GGPP synthases from plants are most similar to GGPP synthases from cyanobacteria and more distant from GGPP synthases (type III) of other eukaryotes such as yeast and human.[53-63] These findings suggest that plant GGPP synthases, but not FPP synthases, might be of endosymbiotic origin, reminiscent of the proposed origin of the genes of the DXP pathway in plants. Like the enzymes of the DXP pathway, plant GGPP synthases are localized in plastids where they supply precursors for diterpene and tetraterpene formation. FPP synthases, in contrast, are localized in the cytosol where they provide the precursors for sesquiterpene and triterpene formation. The heterodimeric peppermint GPP synthase, whose cDNA was recently isolated, is also plastid localized, like GGPP synthase.[55] Since both subunits of peppermint GPP synthase are more closely related to plant GGPP synthases than to other prenyltransferases, GPP and GGPP synthases may share a close evolutionary relationship.[55]

Prenyltransferases may have an important role in controlling the flux of metabolites at branchpoints in the central terpenoid pathway.[27] However, relatively little is known about the regulation of these enzymes in plants and their patterns of gene expression. Several species of angiosperms express at least two genes for isoforms of FPP synthase.[64,65] *Arabidopsis thaliana* also contains two differentially expressed genes encoding FPP synthase, *FPS1* and *FPS2*.[66] By using alternative transcription start sites, the *FPS1* gene yields two different isoforms of FPP synthase, a cytosolic form and a second form, that is targeted to mitochondria by aid of an *N*-terminal transit peptide specific for this isoform.[67] *A. thaliana* also expresses a family of genes and isoforms for GGPP synthase.[68-71] In light of the recent identification of a GGPP synthase-like protein as a subunit of GPP synthase in peppermint, a member of the *A. thaliana* GGPP synthase gene family may also function as a component of GPP synthase.[55] By providing access to all *A. thaliana* prenyl transferase genes, the *Arabidopsis* genome sequencing project will allow an assessment of the phylogenetic relationships of all three classes of short chain prenyltransferases in a single plant species.

Comparatively little is known about prenyltransferases in conifers. GGPP synthase has been investigated in *Taxus* cell cultures, and the corresponding cDNA has been isolated from Canadian yew (*Taxus canadensis*).[72,73] The deduced amino acid sequence closely resembles those of previously isolated angiosperm GGPP synthases, and the properties of the recombinant enzyme are similar to those of GGPP synthases from other sources. The level of the GGPP synthase cDNA was closely correlated with the corresponding enzyme activity in *T. canadensis* cell suspension cultures that had been induced by treatment with methyl jasmonate a signal molecule frequently released by plants upon herbivore damage.[72] A rise in both transcript level and enzyme activity precede the maximum accumulation of the anti-cancer diterpenoid paclitaxel (taxol).[72]

Terpenoid Synthases and Secondary Transformations

In the third stage of terpenoid biosynthesis, the acyclic prenyl diphosphates, GPP, FPP, and GGPP, undergo a range of intramolecular transformations catalyzed by terpenoid synthases, also known as terpenoid cyclases, based upon variations on the same fundamental reaction mechanism involved in prenyltransferase catalysis. The terpenoid synthases produce the parent skeletons of all of the major classes of terpenes. Thus, GPP gives rise to monoterpenes, FDP to sesquiterpenes, and GGDP to diterpenes.[74-76] A large family of plant terpenoid synthases has been characterized, including a distinct subfamily of conifer terpenoid synthases.[78] Because of their central role in generating terpenoid diversity and because of their regulatory function in constitutive and induced terpenoid defenses in conifers, the terpene synthases are the major focus of this review.

The fourth stage of terpenoid biosynthesis comprises the secondary transformation of the parent skeleta produced by the terpene synthases. These transformations include oxidations, reductions, isomerizations, hydrations, and conjugations that produce the many thousands of different terpenoid metabolites found in nature.[9] Only a few such secondary transformations have been studied in conifers. The biosynthesis of the highly functionalized diterpene paclitaxel in *Taxus* involves more than ten modifications of the parental tricyclic diterpene taxadiene, including a series of cytochrome P450-dependent oxygenations and several side-chain additions.[78-81] In grand fir (*Abies grandis*) and lodgepole pine (*Pinus contorta*), two microsomal cytochrome P450-dependent monooxygenases and a soluble aldehyde dehydrogenase catalyze the step-wise oxidation of the diterpene abietadiene into abietic acid, which is a major resin acid component in this species.[82]

TERPENOID SYNTHASES IN CONIFERS

Terpenoid synthases generate the enormous diversity of terpenoid carbon skeletons found in nature.[74-76] Since most terpenoid skeletons are cyclic, containing one or more ring systems (Fig. 1), terpenoid synthases are sometimes referred to as 'terpenoid cyclases', although examples of synthases that produce only acyclic products are known. All terpenoid synthases are quite similar in basic properties, such as molecular weight, cofactor requirement, and kinetic parameters.[54] They all carry out similar electrophilic cyclizations involving common steps, *i.e.*, the generation, transformation, and stabilization of highly reactive carbocations and their ultimate quenching by deprotonation or nucleophile capture.[74-76] From a regulatory standpoint, terpenoid synthases may be involved in the control of pathway flux since they operate at branch points of isoprenoid metabolism and catalyze the first committed steps leading to monoterpenes, sesquiterpenes, and diterpenes (Fig. 2).[27,77]

Table 2: Conifer terpenoid synthases.

Conifer Species	Enzyme Class	Enzyme	Products	References
Abies grandis	Monoterpene Synthase	Myrcene Synthase	Myrcene	95
		(−)-Limonene Synthase	(−)-Limonene, (−)-α-Pinene, (−)-β-Pinene , (−)-β-Phellandrene	95
		(−)-Pinene Synthase	(−)-β-Pinene, (−)-α-Pinene	95, 99
		(−)-Camphene Synthase	(−)-Camphene, (−)-α-Pinene, (−)-Limonene	96
		(−)-β-Phellandrene Synthase	(−)-β-Phellandrene, (−)-β-Pinene, (−)-α-Pinene, (−)-Limonene	96
		Terpinolene Synthase	Terpinolene, (−)-α-Pinene, (−)-Limonene, (−)-β-Pinene, 4 minor products	96
		(−)-Limonene/(−)-α-Pinene Synthase	(−)-Limonene, (−)-α-Pinene, (−)-β-Phellandrene, (−)-β-Pinene, (−)-Sabinene	96
		Cyclase I (partially purified)	Sabinene, Terpinolene, Limonene, other minor products	98
		Cyclase III (partially purified)	3-Carene, other minor products	98
Abies grandis	Sesquiterpene Synthase	E-(α)-Bisabolene Synthase	*E*-(α)-Bisabolene	100
		δ-Cadinene Synthase	δ-Cadinene	23
		γ-Humulene Synthase	γ-Humulene, Sibirene, Longifolene, total of 52 different products	23
		δ-Selinene Synthase	δ-Selinene, (*E,E*)-Germacrene B, total of 34 different products	23
Abies grandis	Diterpene Synthase	Abietadiene Synthase	(−)-Abieta-7(8),13(14)-diene	101, 102
Pinus taeda	Monoterpene Synthase	Synthase I	(−)-α-Pinene, (−)-β-Pinene	89
		Synthase II	(+)-α-Pinene	89
		Synthase III	(−)-β-Pinene, (−)-α-Pinene	89
Pinus contorta	Monoterpene Synthase	Cylase I	(−)-Sabinene, (−)-α-Pinene	86, 87
		Cyclase II	(−)-β-Phellandrene, α-Pinene, Sabinene, β-Pinene	86, 87
		Cyclase II	(+)-3-Carene, Sabinene, α-Pinene	85, 86, 87
		Cyclase IV	(−)-β-Pinene, 3-Carene, α-Pinene	86, 87
Pinus pinaster	Sesquiterpene Synthase	*trans*-β-farnesene synthase	*trans*-β-farnesene	88
Pseudotsuga menziesii	Monoterpene Synthase	(+)-3-Carene Synthase	(+)-3-Carene	85
Taxus brevifolia *Taxus canadensis*	Diterpene Synthase	Taxadiene Synthase	Taxa-4(5),11(12)-diene	90, 91, 92, 93, 94

Much of our knowledge about terpenoid synthases in conifers is based upon studies with grand fir (*Abies grandis*), which has proved to be a remarkable system for the investigation of three classes of terpenoid synthases, monoterpene synthases, sesquiterpene synthases, and diterpene synthases (Table 2).[14,77,83,84] Terpene synthases have also been characterized from Douglas fir (*Pseudotsuga menziesii*), lodgepole pine (*Pinus contorta*), maritime pine (*P. pinaster*), loblolly pine (*P. taeda*), Pacific yew (*Taxus brevifolia*), and Canadian yew (*Taxus canadensis*).[85-94] Typical reactions catalyzed by these three classes of terpene synthases are illustrated in Fig. 3 – 5.

Grand fir contains at least seven different monoterpene synthases (Table 2) that form acyclic, monocyclic, and bicyclic monoterpene olefins based on a common mechanistic theme (Fig. 3).[95-96] The reaction mechanism of monoterpene synthases is thought to be initiated by the ionization of GPP (to cation **1**) assisted by a divalent metal ion cofactor, followed by the return of the diphosphate moiety and isomerization to form the enzyme-bound intermediate linalyl diphosphate (LPP). A given monoterpene synthase forms specifically either 3*R*-LPP or 3*S*-LPP, depending on the initial folding of the geranyl substrate. The stereochemistry of the final product is, thus, determined at this early stage. After rotation about C2–C3 to form the cisoid conformer, LPP undergoes a second ionization (to cation **2**) by cleavage of the diphosphate group, followed by cyclization to the 4*R*- or 4*S*-α-terpinyl cation, the first cyclic intermediate. Acyclic monoterpenes, such as myrcene, may arise by deprotonation of carbocations **1** or **2**. In contrast, the formation of cyclic monoterpenes requires the preliminary isomerization to LPP because GPP cannot cyclize directly on account of the *E* configuration at the C2-C3 double bond at C2–C3.[74,97] Deprotonation of the α-terpinyl cation gives rise to monocyclic monoterpenes, such as limonene and terpinolene, and, following 1,2- and 1,3-hydride shifts to other monocyclic monoterpenes, such as β-phellandrene, and bicyclic monoterpenes, such as sabinene. Additional electrophilic cyclizations of the α-terpinyl cation, *via* the remaining double bond, afford bicyclic carbocationic intermediates *en route* to bicyclic monoterpenes, such as 3-carene, camphene, α-pinene, and β-pinene. The reaction mechanism of monoterpene synthases helps explain an unusual property of these catalysts: their ability to form multiple products. Thus, the monoterpene synthases of grand fir all produce one or two major products plus a few minor products (Table 2).

Most of the properties of conifer monoterpene synthases (molecular mass of approximately 60 kDa, K_m for GPP of 3 - 10 μM, requirement for a divalent cation cofactor (Mn^{2+} preferred))[85,86,89,95,96,98,99] resemble those of angiosperm monoterpene synthases.[54,74] Conifer monoterpene synthases share an unusual requirement for activation by a monovalent cation (K^+ preferred),[86,89,95,98] and have a pH optimum close to neutrality,[86,89,98] in contrast to most angiosperm monoterpene synthases that exhibit acidic pH optima.[54,74]

At least four different sesquiterpene synthases are expressed in stems of wounded grand fir saplings (Table 2), two of which are the multiple product enzymes, δ-selinene synthase and γ-humulene synthase, that form an enormous array of 34 and 52 different terpenoid olefins, respectively.[23] The product profiles of the other grand fir sesquiterpene synthases are more limited. δ-Cadinene synthase forms δ-cadinene as its major product, and *E*-α-bisabolene synthase forms *E*-α-bisabolene as its only product.[23,100] As with monoterpenes, the formation of cyclohexanoid sesquiterpenes, such as α-bisabolene, requires preliminary isomerization of the *trans*-farnesyl preursor to the C_{15} analog of LPP, *i.e.* 3*R*- or 3*S*-nerolidyl diphosphate, followed by ionization-dependent cyclizations (Fig. 4). The increased size of the farnesyl chain and number of double bonds permit cyclization to 10- and 11-membered macrocycles, such as the γ-humulene and germacrene type sesquiterpenoids.[23] Internal additions to the remaining double bonds of the initially formed cyclic carbocations also occur along with hydride shifts and methyl migrations that permit generation of a broad range of structures,products are also derived from FPP, including *trans*-β-farnesene, which is formed by a sesquiterpene synthase purified from the needles of maritime pine (*Pinus pinaster*).[88] Like other sesquiterpene synthases, those from including δ-cadinene, δ-selinene, and longifolene (Fig. 1).[23] Acyclicconifers have a molecularmass of approximately 50 kDa, and prefer Mg^{2+} to Mn^{2+} as their essential divalent ion cofactor. Sesquiterpene synthases also convert GPP into the simple cyclohexanoid monoterpene limonene *in vitro*, but probably never encounter this substrate *in vivo* due to the separation of monoterpene and sesquiterpene formation in different subcellular compartments.[23,77,100]

Although diterpenoids are widespread in the plant kingdom and are major components of conifer oleoresin, only two diterpene synthases have been characterized from conifers, namely abietadiene synthase from grand fir and taxadiene synthase from *Taxus*.[90,91,93,94,101,102] Abietadiene synthase represents the major wound-inducible diterpene synthase in grand fir, while taxadiene synthase catalyzes the committed step of paclitaxel biosynthesis. These two enzymes represent two fundamentally different mechanisms of cyclization in the diterpene series (Fig. 5). The macrocyclic diterpene taxa-4(5),11(12)-diene is formed by a cyclization analogous to those of the monoterpene and sesquiterpene series, involving initial ionization of the prenyl diphosphate substrate GGPP.[92] However, abietadiene synthase is a bifunctional enzyme, which first generates an enzyme bound copalyl diphosphate (CPP) as an intermediate (Fig. 5). This reaction is initiated by protonation of the terminal double bond of GGPP, followed by two

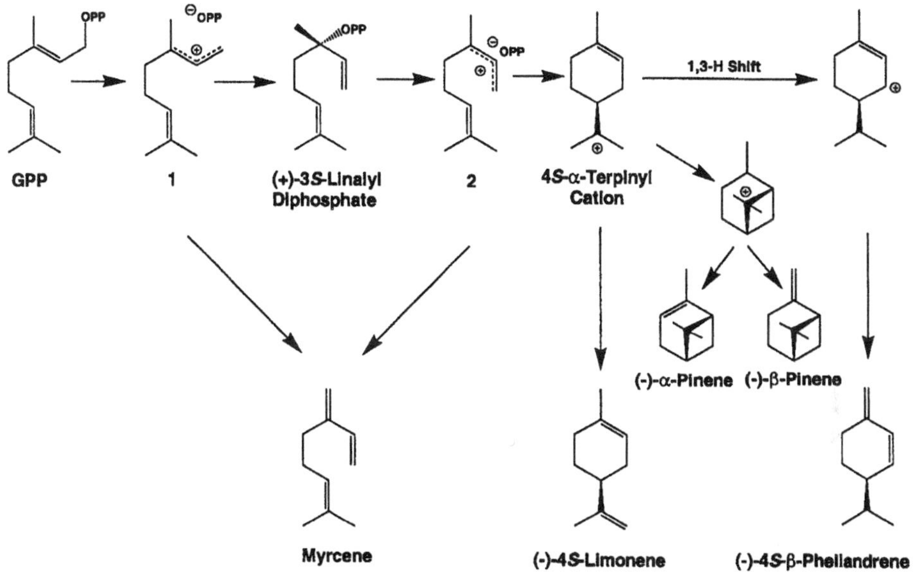

Figure 3. Formation of representative acyclic, monocyclic, and bicylic monoterpenes from geranyl diphosphate.

ring closures and proton elimination to generate (+)-CPP.[101] In the second step, ionization of the diphosphate ester of (+)-CPP initiates a further cyclization *via* pimaradiene as an intermediate *en route* to (−)-abieta-7(8),13(14)-diene.[101] The reaction cascade catalyzed by abietadiene synthase resembles the conversion of GGPP to (−)-CPP, and the subsequent conversion of (−)-CPP to (−)-*ent*-kaurene, the precursor of the gibberellin family of plant hormones.[103] Two mechanistically different terpene cyclases are involved in *ent*-kaurene formation, one catalyzing the protonation-dependent cyclization of GGPP to CPP (copalyl diphosphate synthase), and a second catalyzing the ionization-dependent cyclization of CPP to *ent*-kaurene (*ent*-kaurene synthase). In abietadiene synthase, these two mechanisms are combined in a single bifunctional enzyme.[102] Interestingly, a bifunctional kaurene synthase is also known from the fungus *Phaeospaeria*.[104] It is intriguing to speculate on whether the bifunctional abietadiene synthase evolved from a monofunctional enzyme, participating in the gibberellin pathway, or whether the separate enzymes that cyclize GGPP to *ent*-kaurene are derived from an ancestral bifunctional type.

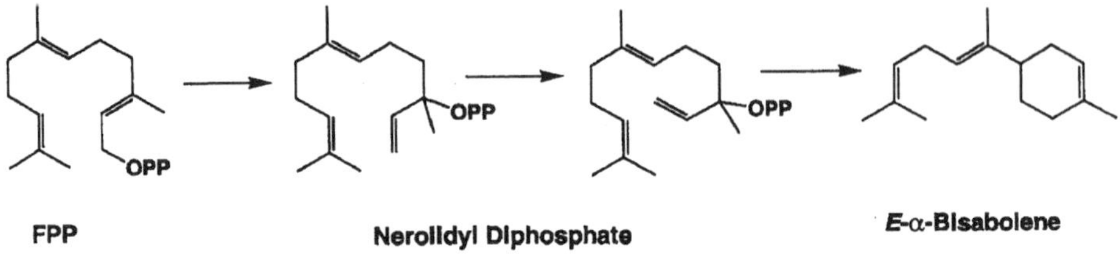

Figure 4. Formation of the sesquiterpenoid (*E*)-α-bisabolene from farnesyl diphosphate.

Molecular Characterization of a Distinct Subfamily of Terpenoid Synthases in Conifers

In recent years, the molecular cloning of terpene synthases has contributed much to our understanding of terpenoid biosynthesis in plants. Cloning strategies and sequence analysis of plant terpenoid synthases have recently been reviewed.[77] Approximately forty plant terpenoid synthases of known function have now been cloned (Table 3), not counting the numerous terpene synthase-like genes of unknown function identified in plant genome sequencing projects. Many of the cDNAs have been expressed in *E. coli* for characterization of the recombinant enzymes. Sequence comparison and phylogenetic reconstruction classify all plant terpenoid synthases into six subfamilies of the large *TPS* gene family (Fig. 6).[77] Terpene synthases from conifers form a distinct group, designated as the *TPSd* subfamily. Members of the *TPSd* group were cloned either in a reverse genetic approach, as in the case of grand fir abietadiene, or by similarity-based cloning with PCR.[102] Similarity-based cloning yielded seven monoterpene synthases and three sesquiterpene synthases from grand fir, as well as a cDNA encoding taxadiene synthase from *Taxus brevifolia*.[23,93,95,96,101] The cloned conifer terpenoid synthase cDNAs encode proteins of 550 to 850 amino acids, in agreement with the observed native molecular masses of 50 to 100 kDa.[77] In general, grand fir monoterpene synthases are between 600 and 650 amino acids in length, and are larger by 50 to 70 amino acids than sesquiterpene synthases. This difference is due largely to the *N*-terminal transit peptides required for plastidial import of the nuclear encoded monoterpene synthases.[95,96] A conserved tandem arginine element has been demonstrated to approximate the putative *N*-terminus of mature monoterpene synthases after cleavage of the transit peptide, and to be involved in the initial diphosphate migration step of the coupled monoterpene isomerization-

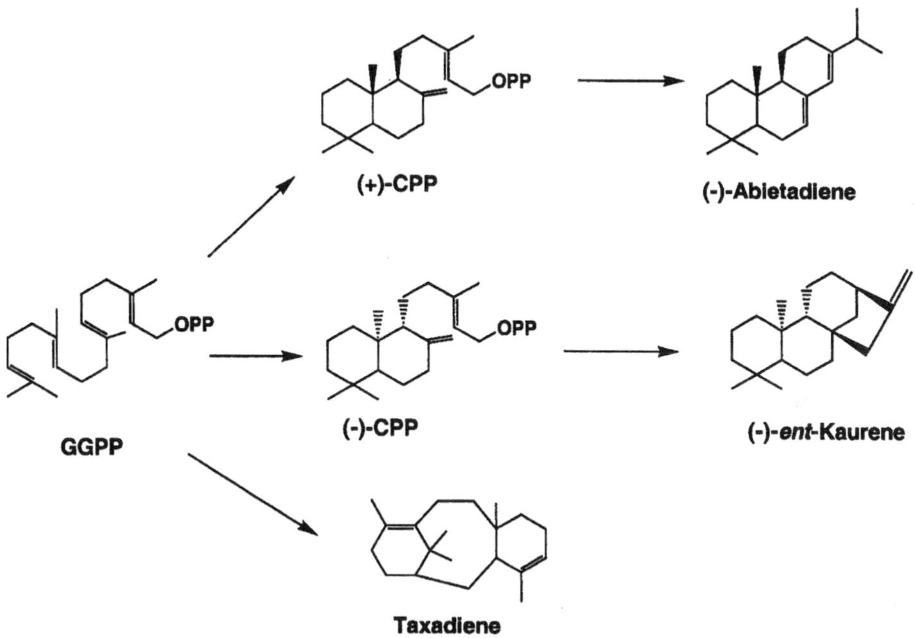

Figure 5. Fromation of diterpenes from geranylgeranyl diphosphate.

cyclization reaction sequence.[105] The plastidial diterpene synthases, abietadiene and taxadiene synthase, are approximately 210 amino acids longer than the monoterpene synthases, due to a large sequence insertion of unknown function and origin.[77,93,102] This insertion is also found in the grand fir sesquiterpene synthase, (*E*)-α-bisabolene synthase, the monoterpene synthase, linalool synthase from *Clarkia breweri* and related enzymes, and the two diterpene cyclases that yield (–)-copalyl diphosphate and (–)-*ent*-kaurene.[77,106]

The *TPSd* subfamily of gymnosperm terpene synthases now represents the catalytically most diverse group of functionally characterized terpene synthases (Fig. 6).[77] Curiously, sequence relatedness has no apparent correlation with the nature of the products formed. Studies with grand fir monoterpene cyclases demonstrated that genes that have as high as 70 to 90 % identity at the amino acid level may encode terpene synthases with different product profiles.[95,96] At the other extreme, synthases that share less than 30% amino acid identity may catalyze the same cyclization reaction, such as (–)-limonene synthase from mint and (–)-limonene synthase from grand fir.[95,107] Therefore, for functional characterization, each cloned terpene synthase must be heterologously expressed as an active rccombinant protein, and the enzymatic reaction products formed with C_{10}, C_{15} and C_{20} prenyl diphosphate substrates must be identified. For monoterpene synthases and diterpene synthases, which are plastid localized, truncation of the *N*-terminal transit peptides may be necessary for satisfactory expression in *E. coli*.[96,105] Heterologous expression of cloned terpene synthases has not only furnished essential information for characterization of these catalysts, but also provided unambiguous proof that many form multiple products in fixed ratios (Table 2). For example, recombinant (–)-pinene synthase from grand fir, in both native and recombinant form, produces (–)-α- and (–)-β-pinene in a ratio of 2:3.[95,99]

Expression of the cDNAs for grand fir monoterpene synthases has demonstrated that the complex turpentine mixture is produced by the action of a closely related family of *TPS* genes.[95,96,108] The cloned monoterpene synthases from grand fir include both single product enzymes, such as myrcene synthase, and enzymes that produce one or more major products and several minor products, such as (–)-limonene synthase, (–)-β-phellandrene synthase, (–)-limonene/(–)-α-pinene synthase, terpinolene synthase, and (–)-camphene synthase (Table 1).[95,96] Differences in the active sites of the grand fir cyclases are required to convert the single substrate GPP to so many different monoterpene products.[74,97] Interestingly, characterization of pinene synthases from loblolly pine (*P. taeda*) revealed three different cyclases, each responsible for the formation of mixtures of (–)-α-pinene, (–)-β-pinene, and (+)-α-pinene (Table 2).[89] Cloning of closely related, yet functionally distinct, enzymes, and elucidation of their three-dimensional structure, combined with site directed mutagenesis, will aid in the identification of the active site elements responsible for the formation of specific monoterpene products. By similar means, it should also be possible to identify the structural requirements that determine the substrate specificity of terpenoid synthases. While plastidial monoterpene synthases and diterpene synthases have evolved strict specificity for their respective prenyl diphosphate substrates, GPP and GGPP, cytosolic sesquiterpene synthases studied *in vitro* catalyze simple cyclizations that use GPP as well as the native substrate FPP[77] it has been suggested that the strict substrate specificity of monoterpene synthases and diterpene synthases might be an evolutionary consequence of the co-compartmentation of these enzymes in plastids to allow for independent regulation of monoterpene and diterpene biosynthesis.[77] Given that GPP and GGPP formation is restricted to the plastids, sesquiterpene synthases, which reside in the cytosol, may not have been selected to discriminate against these alternative substrates.

Table 3: cDNAs encoding plant terpenoid synthases and their classification into subfamilies of the *TPS* gene family.

Terpene Synthase	**Species**	***TPS*-Subfamily**	**Product Class**	**Accession**	**Reference**
Abietadiene synthase ag22	*Abies grandis*	*TPSd*	C_{20}	U50768	102
epi-Aristolochene synthase	*Nicotiana tabacum*	*TPSa*	C_{15}	L04680	148
epi-Aristolochene synthase	*Capsicum annuum*	*TPSa*	C_{15}	AF061285	149
E-α-Bisabolene synthase	*Abies grandis*	*TPSd*	C_{15}	AF006195	100
(+)-Bornyl diphosphate synthase	*Salvia officinalis*	*TPSb*	C_{10}	AF051900	150
(+)-δ-Cadinene synthase XC14	*Gossypium arboreum*	*TPSa*	C_{15}	U23205	151
(+)-δ-Cadinene synthase XC1	*Gossypium arboreum*	*TPSa*	C_{15}	U23206	151
(+)-δ-Cadinene synthase A	*Gossypium arboreum*	*TPSa*	C_{15}	U27535	152
(+)-δ-Cadinene synthase CAD1A	*Gossypium arboreum*	*TPSa*	C_{15}	X96429	152
(+)-δ-Cadinene synthase CAD1C2	*Gossypium arboreum*	*TPSa*	C_{15}	Y16432	153
(+)-δ-Cadinene synthase cdn1	*Gossypium hirsutum*	*TPSa*	C_{15}	U88318	1548
(-)-Camphene synthase	*Abies grandis*	*TPSd*	C_{10}	U87910	96
Casbene synthase	*Ricinus communis*	*TPSa*	C_{20}	L32134	155
epi-cedrol synthase	*Artemisia annua*	*TPSa*	C_{15}	AJ001539	156
epi-cedrol synthase	*Artemisia annua*	*TPSa*	C_{15}	AF157059	157
1,8-Cineole synthase	*Salvia officinalis*	*TPSb*	C_{10}	AF051899	150
(-)-Copalyl diphosphate synthase	*Arabidopsis thaliana*	*TPSc*	C_{20}	U11034	158
(-)-Copalyl diphosphate synthase 1	*Cucurbita maxima*	*TPSc*	C_{20}	AF049905	159
(-)-Copalyl diphosphate synthase 2	*Cucurbita maxima*	*TPSc*	C_{20}	AF049906	159

Table 3: cont.

Copalyl diphosphate synthase	*Lycopersicon esculentum*	*TPSc*	C_{20}	AB015675	Genbank direct submission
(-)-Copalyl diphosphate synthase	*Pisum sativum*	*TPSc*	C_{20}	U63652	160
(-)-Copalyl diphosphate synthase	*Zea mays*	*TPSc*	C_{20}	L37750	161
(-)-Copalyl diphosphate synthase	*Stevia rebaudiana*	*TPSc*	C_{20}	AF034545	162
E-β-Farnesene synthase	*Mentha x piperita*	*TPSa*	C_{15}	AF024615	163
Germacrene C synthase	*Lycopersicon esculentum*	*TPSa*	C_{15}	AF035630	164
γ-Humulene synthase	*Abies grandis*	*TPSd*	C_{15}	U92267	23
(-)-Kaurene synthase	*Cucurbita maxima*	*TPSe*	C_{20}	U43904	165
(-)-Kaurene synthase	*Arabidopsis thalinana*	*TPSe*	C_{20}	AF034774	166
(-)-Kaurene synthase	*Stevia rebaudiana*	*TPSe*	C_{20}	AF097310	162
(-)-Kaurene synthase	*Stevia rebaudiana*	*TPSe*	C_{20}	AF097311	162
(-)-Limonene synthase	*Abies grandis*	*TPSd*	C_{10}	AF006193	95
(-)-Limonene synthase	*Mentha longifolia*	*TPSb*	C_{10}	AF175323	Genbank direct submission
(-)-Limonene synthase	*Mentha spicata*	*TPSb*	C_{10}	L13459	107
(-)-Limonene synthase	*Perilla frutescens*	*TPSb*	C_{10}	D49368	167
(-)-Limonene/(-)-α-Pinene synthase	*Abies grandis*	*TPSd*	C_{10}	AF139207	96
S-Linalool synthase	*Clarkia breweri*	*TPSf*	C_{10}	U58314	168
S-Linalool synthase 2	*Clarkia breweri*	*TPSf*	C_{10}	AF067603	106
(3*R*)-linalool synthase QH1	*Artemisia annua*	*TPSb*	C_{10}	AF154125	169
(3*R*)-linalool synthase QH5	*Artemisia annua*	*TPSb*	C_{10}	AF154124	169
Myrcene synthase	*Abies grandis*	*TPSd*	C_{10}	U87908	95
Myrcene/*E*-β-Ocimene synthase	*Arabidopsis thaliana*	*TPSb*	C_{10}	AF178535	170
(-)-β-Phellandrene synthase	*Abies grandis*	*TPSd*	C_{10}	AF139205	95

Table 3: cont.

(-)-Pinene synthase	*Abies grandis*	*TPSd*	C_{10}	U87909	95
(+)-Sabinene synthase	*Salvia officinalis*	*TPSb*	C_{10}	AF051901	150
δ-Selinene	*Abies grandis*	*TPSd*	C_{15}	U92266	23
Sesquiterpene synthase	*Artemisia annua*	*TPSa*	C_{15}	AJ249561	Genbank direct submission
Sesquiterpene synthase	*Elaeis oleifera*	*TPSa*	C_{15}	AF080245	Genbank direct submission
Taxadiene synthase	*Taxus breivifolia*	*TPSd*	C_{20}	U48796	93
Terpinolene synthase	*Abies grandis*	*TPSd*	C_{10}	AF139206	96
Vetispiradiene synthase 1	*Hyoscyamus muticus*	*TPSa*	C_{15}	HMU20188	171
Vetispiradiene synthase PVS1	*Solnaum tuberosum*	*TPSa*	C_{15}	AB022598	172
Vetispiradiene synthase PVS2	*Solnaum tuberosum*	*TPSa*	C_{15}	AB022719	172
Vetispiradiene synthase PVS4	*Solnaum tuberosum*	*TPSa*	C_{15}	AB023816	Genbank direct submission
Vetispiradiene synthase VS1	*Solnaum tuberosum*	*TPSa*	C_{15}	AF042382	Genbank direct submission
Vetispiradiene synthase 3	*Solnaum tuberosum*	*TPSa*	C_{15}	AF043298	Genbank direct submission
Vetispiradiene synthase 4	*Solnaum tuberosum*	*TPSa*	C_{15}	AF043299	Genbank direct submission
Vetispiradiene synthase 5	*Solnaum tuberosum*	*TPSa*	C_{15}	AF043300	Genbank direct submission

A major advance in the study of these plant enzymes occured with the recent description of the three-dimensional structure of a tobacco sesquiterpene synthase called *epi*-aristolochene synthase.[109] This protein was composed entirely of α-helices with short connecting loops and turns, forming a two-layer α-barrel active site. At the entrance of the active site, the diphosphate of the prenylsubstrate moiety is complexed by two Mg^{2+} ions. The aspartate residues that coordinate one of the Mg^{2+} ions are part of an aspartate-rich motif, DDxxD, absolutely conserved in all plant terpenoid synthases studied to date.[74,77] Several conserved amino acid residues appear to be involved in stabilization of the negative charge of the diphosphate following ionization and in stabilization of the carbocationic intermediates. The three-dimensional structure of *epi*-aristolochene synthase resembles the two known structures of microbial terpene synthases despite the lack of obvious sequence similarity at the level of their primary structures.[110,111] This common three-dimensional framework is also shared in part with prenyl transferases, suggesting that these two large groups of enzymes, which catalyze consecutive steps in terpenoid metabolism by similar electrophilic mechanisms, may have a common phylogenetic origin.[53,112,113]

While all plant terpenoid synthases share a common evolutionary origin, the conifer synthases form a distinct branch of the *TPS* tree (Fig. 6).[77] Sequence comparison and phylogenetic reconstruction reveal that conifer monoterpene, sesquiterpene, and diterpene synthases (represented by genes from grand fir and *Taxus brevifolia*) are more closely related to each other than they are to their counterparts of angiosperm origin.[77] This pattern implies that terpene synthases in conifers and angiosperms diverged after separation of these two lineages. Within the monophyletic conifer *TPSd* group, the seven monoterpene synthases are more closely related to each other than to other conifer terpene synthases.[95,96] Among the sesquiterpene synthases, two enzymes, γ-humulene synthase and δ-selinene synthase, form a separate branch of the *TPSd* group, while a third, (*E*)-α-bisabolene synthase, is more closely related to the gymnosperm diterpene synthases, abietadiene synthase and taxadiene synthase, all of which share an unusual insertion of 210 amino acids. Thus, the specialized single-product sesquiterpene synthases and multiple-product sesquiterpene synthase in grand fir may have traced different evolutionary paths from a common ancestor.[77] Extant terpenoid synthases of the conifer *TPSd* subfamily show an enormous functional diversity, which may have evolved by *TPS* gene amplification and independent structural and functional mutations of amplified genes giving rise to new terpene synthases.

Regulation of Inducible Terpenoid Biosynthesis in Conifers

Conifers have evolved both constitutive and inducible terpenoid-based defense strategies.[12,14,114-116] As previously discussed, resin accumulates in secretory structures, such as blisters and ducts, during the normal development of stems and foliage. In addition, wounding caused by stem boring insects or leaf feeding insects, as well as application of chemical elicitors derived from insect-associated fungi, trigger induced resinosis.[12,116-122] . For experimental purposes, mechanical wounding has frequently been employed to test trees for their capacity to activate inducible defenses, measured by assessing changes in terpene concentration, enzyme activity, anatomy, or gene expression.[121,123-125] A comparative study among ten conifer species belonging to seven genera of the pine family revealed a negative correlation between the capacity to form large quantities of constitutive resin and the capacity to activate *de novo* biosynthesis of terpenoids upon mechanical wounding, as measured by *in vitro* terpene synthase activity.[121]

Species that accumulate large amounts of resin in extensive resin canal systems, such as pines (*Pinus*), rely primarily on constitutive resinosis, and wounding of stems does not significantly increase terpenoid biosynthesis.[89,121] In contrast, mechanical wounding elicits a strong induction of monoterpene synthase activity in fir (*Abies*) and spruce (*Picea*).[121] In grand fir stems, constitutive resin biosynthesis is localized to isolated resin cells and the epithelial cells that line multi-cellular resin blisters.[10,11] In spruce, constitutive resin is formed in axial resin ducts in the bark, but only few resin canals exist in constitutive xylem (wood). However, after insect herbivory, fungal elicitation, or mechanical wounding, grand fir and several spruce species produce secondary resin in newly differentiated resin systems.[11,114,126] In white spruce (*Picea glauca*), cambium cells, which normally form xylem, are reprogrammed to produce axial traumatic resin ducts after attack of the leader shoot by the white pine weevil (*Pissodes strobi*).[125,127] Traumatic resin ducts in Norway spruce (*Picea abies*) develop after bark beetle attack, mechanical wounding, or inoculation with the bark beetle-associated blue stain fungus *Ceratocystis polonica*.[128] Thus, induced terpenoid defenses in conifers rely not only on metabolic changes, but can also be accompanied by anatomical changes. Neither the molecular events that coordinate such induced changes of cell differentiation nor those that lead to induced terpenoid biosynthesis are known. Their identification may provide a valuable means of enhancing natural defense mechanisms in conifers.

The regulation of induced terpenoid biosynthesis has been extensively studied in grand fir.[14,83,84] This species responds to mechanical wounding of sapling stems by forming increased amounts of monoterpenes, sesquiterpenes, and diterpene resin acids. Enzyme activities of all three classes of terpene synthases, monoterpene, sesquiterpene, and diterpene synthases increase in wounded stem tissues, and the levels of monoterpene synthase protein (measured with polyclonal antibodies) also increase.[23,124,129,130] In general, terpene synthases reach their maximum induced levels of activity 10 - 14 days after wounding of trees.[124] Within this period of induction, the kinetics of individual enzymes may vary. Two of the enzymes involved in the conversion of abietadiene to abietic acid, the major wound-induced diterpene resin acid, exhibit similar kinetics of induction.[129] Induced terpenoid biosynthesis in conifers is not restricted to the stems. Simulated herbivory also induced monoterpene synthase enzyme activity in needles of ponderosa pine (*Pinus ponderosa*), lodgepole pine (*P. contorta*), and white fir (*Abies alba*), but not in Engelmann spruce (*Picea engelmannii*).[123]

Cloning of the cDNAs for eleven different terpene synthases from wound-induced grand fir sapling stems provided specific gene probes to perform expression of the three classes of terpene synthases.[124] RNA-blot hybridization showed that the accumulation of monoterpene synthase transcripts increases within a few hours after wounding and reaches a maximum within two to four days. Gene expression for diterpene synthases is also activated, but the time course is slightly delayed. For the sesquiterpene synthases of grand fir, only δ-cadinene synthase and (*E*)-α-bisabolene synthase appear to be inducible.[23,100] In contrast to the rapid accumulation of mRNAs for monoterpene and diterpene synthases, (*E*)-α-bisabolene synthase mRNA reaches its maximum level of accumulation two weeks after wounding.[100] (*E*)-α-Bisabolene is the direct precursor to todomatuic acid and the juvabione-type insect juvenile hormone analogs that may interfere with insect reproduction.[100] These differences in the timing of induced gene expression may reflect the different defensive roles of the various terpenoids.

Intraspecific Variation in Conifer Terpenoid Biosynthesis

Many conifers show extensive intraspecific variation in oleoresin terpene composition that appears to be genetically controlled.[5,131] Crossing studies have revealed that the inheritance of individual monoterpenes is sometimes under the control of a single gene having a pair of dominant/recessive alleles.[51,131] However, more frequently the inheritance of different monoterpenes is linked in complex ways, probably as a result of the ability of individual terpene synthases to produce multiple products, and the presence in a single species of distinct synthases.[132]

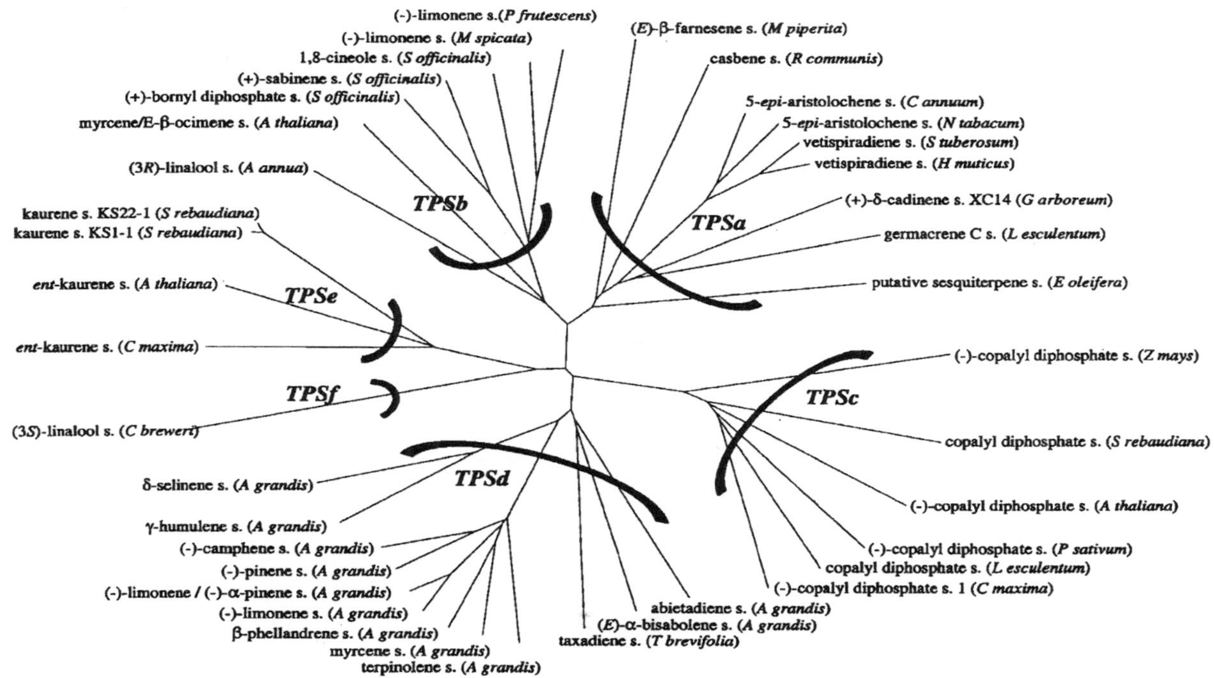

Figure 6. Phylogenetic tree of representative plant terpenoid synthases. *TPSa* through *TPSf* designate subfamilies of the *TPS* gene family. Members were cloned as cDNAs and were functionally characterized. Two subfamilies represent typical monoterpenoid synthases (*TPSb*) and sesquiterpenoid synthases (*TPSa*) of angiosperms. The *TPSa* group also contains a diterpene synthase of angiosperm origin.[155] The *TPSd* subfamily of conifers contains members of all three classes of monoterpenoid synthases, sesquiterpenoid synthases, and diterpenoid synthases. Angiosperm diterpenoid synthases involved in the biosynthesis of gibberellic acid phytohormones are classified in two subfamilies, *TPSc* containing (-)-copalyl diphosphate synthases and *TPSe* containing *ent*-kaurene synthases. Notably, a member of the *TPSe* sub-family from *Stevia rebaudiana* has aquired a function in secondary metabolism of steviol biosynthesis.[162] The *TPSf* branch represents the unusual monoterpene synthase linalool synthase.

In grand fir, our current knowledge of the *TPSd* gene family explains much of the diversity in monoterpene composition and provides a foundation for the investigation of genetic controls. To explore the incidence of genetic variation in this species, 46 saplings were started from seed collected from a single wild population and were examined individually to determine the products of constitutive and wound-induced monoterpene synthase activities.[133] Striking differences were observed among individuals, which allowed the identification of seven distinct monoterpene biosynthetic chemotypes. The existence of such chemotypes can be partially explained by independent expression of a family of monoterpene synthase genes.[95,96] When present in a natural conifer population, such variation may provide defense against herbivores adapted to tolerate the average terpene chemistry of the population.[21,134] The fact that terpenoids typically occur in complex mixtures is important in this regard as mixtures increase the opportunities for an individual plant to be chemically distinct from others in the same population. Adaptation of herbivores to overcome multi-gene and multi-component chemical defense is likely to be more difficult compared to a situation in which tree defense builds upon a single component or on a constant and invariable mixture of monoterpenoids. The two multiproduct sesquiterpene synthases of grand fir, δ-selinene synthase and γ-humulene synthase, which form 34 and 52 different products, respectively, provide a major contribution to terpenoid complexity in this species through the expression of only two genes. Although the multiproduct sesquiterpene synthases are not induced by wounding in grand fir, differences in the level of constitutive expression have been observed among individual trees leading to further variation in terpene composition.[23]

PERSPECTIVE

The pathway of terpenoid biosynthesis in conifers and other plants has been shaped by millions of years of evolution. As discussed above, there are two distinct routes for the construction of the basic C_5 unit: the cytosolic mevalonic acid pathway and the plastid-localized DXP pathway, which may have originated with the acquisition of an endosymbiotic cyanobacterium by an ancestor to green plants. Among the prenyltransferases, the enzymes catalyzing the union of the C_5 units to form acyclic prenyl diphosphates, GPP, and GGPP synthases may also be ascribed to an endosymbiotic origin. Once formed, the prenyl diphosphates are converted to an enormous variety of basic carbon skeletons by the terpene synthases. These enzymes have radiated to form a large gene family in conifers that has evolved independently after phylogenetic separation from the angiosperm lineage (Fig. 6).

The acquisition of new terpene biosynthetic genes by endosymbiotic capture or independent evolution can now be simulated on a much shorter time scale by molecular breeding efforts directed towards improving the properties of economically-important plant species. In contrast to agricultural crop plants, the "domestication" of forest trees has only been practiced for the last 100 years and is naturally a slow process. A common approach is the selection of individuals with certain desirable traits for clonal propagation. However, limiting the gene pool of replanted forests may significantly reduce the terpene diversity as compared to a natural stand, thus increasing the susceptibility of trees to herbivores and pathogens. Reducing genetic diversity in forest environments is even more likely to result in losses to herbivores and pathogens than in agricultural systems because of the large difference in generation times between trees and their pests that permits their adaptation to tree resistance traits during its long lifetime. Knowledge of the genetic and biochemical regulation of terpenoid chemical defenses in conifers should allow forest geneticists to maintain some of the terpenoid diversity that has been selected over more than 300 million years of evolution, and to further create new terpenoid mixtures that might prove benefical in conifer defenses.

REFERENCES

1. PILGER, R. 1926. Coniferae. pp. 121-407 *in* Die Natürlichen Pflanzenfamilien, Vol. 13 (A. Engler, K. Prantl eds.), W. Engelmann, Leipzig.
2. STEFANOVIC, S., JAGER, M., DEUTSCH, J., BROUTIN, J., MASSLOT, M. 1998. Phylogenetic relationship of conifers inferred from partial 28S rRNA gene sequences. Am. J. Bot. 85: 688-697.
3. ROWE, J.W. (ed.) 1989. Natural Products of Woody Plants. Vol. I and II. Springer-Verlag, Berlin.
4. GERSHENZON, J., CROTEAU, R. 1991. Terpenoids. pp. 165-219 *in* Herbivores: Their Interactions with Secondary Plant Metabolites, Vol. 1, 2nd ed. (G. A. Rosenthal, M. R. Berenbaum, eds.) Academic Press, San Diego.
5. VON RUDLOFF, E. 1975. Volatile leaf oil analysis in chemosystematic studies of North American conifers. Biochem. System. Ecol. 2: 131-167.
6. JÜTTNER, F. 1988. Changes of monoterpene concentrations in needles of pollution-injured *Picea abies* exhibiting montane yellowing. Physiol. Plant. 72: 48-56.
7. MERK, L., KLOOS, M., SCHÖNWITZ, R., ZIEGLER, H. 1988. Influence of various factors on quantitative composition of leaf monoterpenes of *Picea abies* (L.) Karst. Trees 2: 45-51.

8. SCHÖNWITZ, R., LOHWASSER, K., KLOOS, M., ZIEGLER, H. 1990. Seasonal variation in the monoterpenes of *Picea abies* (L.) Karst. Trees 4: 34-40.
9. CONOLLY, J.D., HILL, R.A. 1991. Dictionary of Terpenoids. Chapman and Hall, London.
10. FAHN, A. 1979. Secretory Tissues in Plants. Academic Press, New York, 218 p.
11. BANAN, M.W. 1936. Vertical resin ducts in the secondary wood of the Abietineae. New Phytol. 35: 11-46.
12. JOHNSON, M.A., CROTEAU, R. 1987. Biochemistry of conifer resistance to bark beetles and their fungal symbionts. ACS Symp. Ser. 325: 67-91.
13. GIJZEN, M., LEWINSOHN, E., SAVAGE, T.J., CROTEAU, R. 1993. Conifer Monoterpenes. ACS Symp. Ser. 525: 8-22.
14. PHILLIPS, M.A., CROTEAU, R. 1999. Resin-based defenses in conifers. Trends Plant Sci. 4: 184-190.
15. WILKINSON, R.C. 1980. Relationship between cortical monoterpenes and susceptibility of eastern white pine to white-pine weevil attack. Forest Sci. 26: 581-589.
16. HARRIS, L.J., BORDEN, J.H., PIERCE, H.D., OEHLSCHLAGER, A.C. 1983. Cortical resin monoterpenes in Sitka spruce and resistance to the white pine weevil, *Pissodes strobi* (Coleoptera: Curculionidae). Can. J. For. Res. 13: 350-352.
17. BROOKS, J.E., BORDEN, J.H., PIERCE, H.D. 1987. Foliar and cortical monoterpenes in Sitka spruce: Potential indicators of resistance to the white pine weevil, *Pissodes strobi* Peck (Coleoptera: Curculionidae). Can. J. For. Res. 17: 740-745.
18. BROOKS, J.E., BORDEN, J.H., PIERCE, H.D., LISTER, G.R. 1987. Seasonal variation in foliar and bud monoterpenes in Sitka spruce. Can. J. Bot. 65: 1249-1252.
19. TOMLIN, E.S., BORDEN, J.H., PIERCE, H.D. 1996. Relationship between cortical resin acids and resistance of Sitka spruce to the white pine weevil. Can. J. Bot. 74: 599-606.
20. TOMLIN, E.S., BORDEN, J.H., PIERCE, H.D. 1997 Relationship between volatile foliar terpenes and resistance of Sitka spruce to the white pine weevil. Forest Science 43: 501-508.
21. LANGENHEIM, J.H. 1994. Higher plant terpenoids: A phytocentric overview of their ecological roles. J. Chem. Ecol. 20: 1223-1280.
22. WOOD, D.L. 1982. The role of pheromones, kairomones and allomones in the host selection and colonization behavior of bark beetles. Annu. Rev. Entomol.

27: 411-446.

23. STEELE, C.L., CROCK, J., BOHLMANN, J., CROTEAU, R. 1998a. Sesquiterpene synthases from grand fir (*Abies grandis*): Comparison of constitutive and wound-induced activities, and cDNA isolation, characterization, and bacterial expression of δ-selinene synthase and γ-humulene synthase. J. Biol. Chem. 273: 2078-2089.
24. BOWERS, W.S. 1991. Insect hormones and anti-hormones. pp. 431-456 *in* Herbivores: Their Interaction with Secondary Plant Metabolites, Vol. 1 (G.A. Rosenthal and M.R. Berenbaum, eds.), Academic Press, San Diego.
25. XIE, Y., ISMAN, M.B., FENG, Y., WONG, A. 1993. Diterpene resin acids: Major active principles in tall oil against variegated cutworm, *Peridroma saucia* (Lepidoptera: Noctuidae). J. Chem. Ecol. 19: 1075-1084.
26. BJÖRKMAN, C., LARSSON, S. 1991. Pine sawfly defence and variation in host plant resin acids: A trade-off with growth. Ecol. Entomol. 16: 283-289.
27. GERSHENZON, J., CROTEAU, R. 1993. Terpenoid Biosynthesis: the basic pathway and formation of monoterpenes, sesquiterpenes, and diterpenes. pp. 339-388 *in* Lipid Metabolism in Plants. (T.S Moore Jr., ed.), CRC Press, Boca Raton.
28. GERSHENZON, J., KREIS, W. 1999. Biochemistry of terpenoids: Monoterpenes, sesquiterpenes, diterpenes, sterols, cardiac glycosides and steroid saponins. Annu. Plant Rev. 4: 222-299.
29. BACH, T.J., BORONAT, A., CAMPOS, N., FERRER, A., VOLLACK, K.-U. 1999. Mevalonate biosynthesis in plants. Crit. Rev. Biochem. Mol. Biol. 34: 107-122.
30. NEWMAN, J.D., CHAPPELL, J. 1999. Isoprenoid biosynthesis in plants: Carbon partitioning within the cytoplasmic pathway. Crit. Rev. Biochem. Mol. Biol. 34: 95-106.
31. BOLL, M., KARDINAL, A. 1990. 3-Hydroxy-3-methylglutaryl coenzyme A reductase from spruce (*Picea abies*). Properties and regulation. Z. Naturforsch. 45c: 973-979.
32. BOLL, M., KARDINAL, A., BERNDT, J. 1987. Properties and regulation of 3-hydroxy-3-methylglutaryl coenzyme A reductase from spruce (*Picea abies*). Biol. Chem. Hoppe-Seyler 368: 1024.
33. EISENREICH, W., SCHWARZ, M., CARTAYRADE, A., ARIGONI, D., ZENK, M.H., BACHER, A. 1998. The deoxyxylulose phosphate pathway of terpenoid biosynthesis in plants and microorganisms. Chemistry and Biology 5: R221-R233.
34. ROHMER, M. 1999. The discovery of a mevalonate-independent pathway for isoprenoid biosynthesis in bacteria, algae and higher plants. Nat. Prod. Rep.16:

565-574.

35. LICHTENTHALER, H.K. 1999. The 1-deoxy-D-xylulose 5-phosphate pathway of isoprenoid biosynthesis in plants. Annu. Rev. Plant Physiol. Plant Mol. Biol. 50: 47-65.
36. SPRENGER, G.A., SCHÖRKEN, U., WIEGERT, T., GROLLE, S., DE GRAAF, A.A., TAYLOR, S.V., BEGLEY, T.P., BRINGER-MEYER, S., SAHM, H. 1997. Identification of a thiamin-dependent synthase in *Escherichia coli* required for the formation of 1-deoxy-D-xylulose 5-phosphate precursor to isoprenoids, thiamin, and pyridoxal. Proc. Natl. Acad. Sci. USA 94: 12857-12862.
37. LANGE, B.M., WILDUNG, M.R., MCCASKILL, D., CROTEAU, R. 1998. A family of transketolases that directs isoprenoid biosynthesis via a mevalonate-independent pathway. Proc. Natl. Acad. Sci. USA 95: 2100-2104.
38. LOIS, L.M., CAMPOS, N., PUTRA, S.R., DANIELSEN, K., ROHMER, M., BORONAT, A. 1998. Cloning and characterization of a gene from *Escherichia coli* encoding a transketolase-like enzyme that catalyzes the synthesis of D-1-deoxyxylulose 5-phosphate, a common precursor for isoprenoid, thiamin, and pyridoxal biosynthesis. Proc. Natl. Acad. Sci. USA 95: 2105-2110.
39. BOUVIER, F., D'HARLINGUE, A., SUIRE, C., BACKHAUS, R.A., CAMARA, B. 1998. Dedicated roles of plastid transketolases during the early onset of isoprenoid biogenesis in pepper fruits. Plant Physiol. 117: 1423-1431.
40. TAKAHASHI, S., KUZUYAMA, T., WATANABE, H., SETO, H. 1998. A 1-deoxy-D-xylulose 5-phosphate reductoisomerase catalyzing the formation of 2-C-methyl-D-erythritol 4-phosphate in an alternative nonmevalonate pathway for terpenoid biosynthesis. Proc. Natl. Acad. Sci. USA 95: 9879-9884.
41. LANGE, B.M., CROTEAU, R. 1999a. Isoprenoid biosynthesis via a mevalonate-independent pathway in plants: Cloning and heterologous expression of 1-deoxyxylulose-5-phosphate reductoisomerase from peppermint. Arch. Biochem. Biophys. 365: 170-174.
42. ARIGONI, D., GINER, J.-L., SAGNER, S., WUNGSINTAWEEKUL, J., ZENK, M.H., KIS, K., BACHER, A., EISENREICH, W. 1999. Stereochemical course of the reduction step in the formation of 2-c-methylerythritol from the terpene precursor 1-deoxyxylulose in higher plants. Chem. Commun. 1999: 1127-1128.
43. ROHDICH, F., WUNGSINTAWEEKUL, J., FELLERMEIER, M., SAGNER, S., HERZ, S., KIS, K., EISENREICH, W., BACHER, A., ZENK, M.H. 1999. Cytidine 5'-triphosphate-dependent biosynthesis of isoprenoids: YgbP protein of *Escherichia coli* catalyzes the formation of 4-phosphocytidyl-2-*C*-methylerythritol. Proc. Natl. Acad. Sci. USA 96: 11758-11763.

44. CHARON, L., PALE-GROSDEMANGE, C., ROHMER, M. 1999. On the reduction steps in the mevalonate independent 2-C-methyl-D-erythrithol 4-phophate (MEP) pathway for isoprenoid biosynthesis in the bacterium *Zymomonas mobilis*. Tetrahed. Lett. 40: 7231-7234.
45. LANGE, B.M., CROTEAU, R. 1999b. Isopentenyl diphosphate biosynthesis via a mevalonate-independent pathway: Isopentenyl monophosphate kinase catalyzes the terminal enzymatic step. Proc. Natl. Acad. Sci. USA 96: 13714-13719.
46. ADAM, K.-P., THIEL, R., ZAPP, J. 1999. Incorporation of 1-[1-^{13}C]-deoxy-D-xylulose in Chamomile sesquiterpenes. Arch. Biochem. Biophys. 369: 127-132.
47. DISCH, A., ROHMER, M. 1998. On the absence of the glyceraldehyde 3-phosphate/pyruvate pathway for isoprenoid biosynthesis in fungi and yeasts. FEMS Microbiol. Lett. 168: 201-208.
48. JOOMA, H., WIESNER, J., SANDERBRAND, S., ALTINCICEK, B., WEIDEMEYER, C., HINTZ, M., TURBACHOVA, I., EBERL, M., ZEIDLER, J., LICHTENTHALER, H.K., SOLDATI, D., BECK, E. 1999. Inhibitors of the nonmevalonic acid pathway of isoprenoid biosynthesis as antimalarial drugs. Science 285: 1573-1576.
49. MARTIN, W., HERRMANN, R.G. 1998. Gene transfer from organelles to the nucleus: How much, what happens, and why? Plant Physiol. 118: 9-17.
50. EISENREICH, W., MENHARD, B., HYLANDS, P.J., ZENK, M.H., BACHER, A. 1996. Studies on the biosynthesis of taxol: The taxane carbon skeleton is not of mevalonoid origin. Proc. Natl. Acad. Sci. USA 93: 6431-6436.
51. BERNARD-DAGAN, C., PAULY, G., MARPEAU, A., GLEIZES, M., CARDE, J.-P., BARADAT, P. 1982. Control and compartmentation of terpene biosynthesis in leaves of *Pinus pinaster*. Physiol. Veg. 20: 775-795.
52. KOYAMA, T., OGURA, K. 1999. Isopentenyl diphosphate isomerase and prenyltransferase. pp. 69-96 *in* Comprehensive Natural Products Chemistry Vol 2: Isoprenoids Including Carotenoids and Steroids (D.E. Cane, ed.), Pergamon Press, Oxford.
53. WANG, K., OHNUMA, S.-I. 1999. Chain-length determination mechanism of isoprenyl diphosphate synthases and implications for molecular evolution. Trends Biochem. Sci. 24: 445-451.
54. ALONSO, W.R., CROTEAU, R. 1993. Prenyltransferases and Cyclases. Meth. Plant Biochem. 9: 239-260.
55. BURKE, C.C., WILDUNG, M.R., CROTEAU, R. 1999. Geranyl diphosphate synthase: Cloning, expression, and characterization of the prenyltransferase as

a heterodimer. Proc. Natl. Acad. Sci. USA 96: 13062-13067.

56. TARSHIS, L.C., YAN, M., POULTER, C.D., SACCHETTINI, J.C. 1994. Crystal structure of recombinant farnesyl diphosphate synthase at 2.6-Å resolution. Biochemistry. 33: 10871-10877.
57. TARSHIS, L.C., PROTEAU, P.J., KELLOGG, B.A., SACCHETTINI, J.C., POULTER, C.D. 1996. Regulation of chain length by isoprenyl diphosphate synthases. Proc. Natl. Acad. Sci. USA 93: 15018-15023.
58. OHNUMA, S.-I., NAKAZAWA, T., HEMMI, H., HALLBERG, A.-M., KOYAMA, T., OGURA, K., NISHINO, T. 1996 Conversion from farnesyl diphosphate synthase to geranylgeranyl diphosphate synthase by random chemical mutagenesis. J. Biol. Chem. 271: 10087-10095.
59. OHNUMA, S.-I., HIROOKA, K., HEMMI, H., ISHIDA, C., OHTO, C., NISHINO, T. 1996. Conversion of archaebacterial geranylgeranyl-diphosphate synthase. J. Biol. Chem. 271: 18831-18837.
60. OHNUMA, S.-I., NARITA, K., NAKAZAWA, T., ISHIDA, C., TAKEUCHI, Y., OHTO, C., NISHINO, T. 1996. A role of the amino acid residue located on the fifth position before the first aspartate-rich motif of farnesyl diphosphate synthase on determination of the final product. J. Biol. Chem. 271: 30748-30754.
61. OHNUMA, S.-I., HIROOKA, K., OHTO, C., NISHINO, T. 1997. Conversion from archaeal geranylgeranyl diphosphate synthase to farnesyl diphosphate synthase. J. Biol. Chem. 272: 5192-5198.
62. NARITA, K., OHNUMA, S.-I., NISHINO, T. 1999. Protein design of geranyl diphosphate synthase. Structural features that define the product specificities of prenyltransferases. J. Biochem. 126: 566-571.
63. OHTO, C., ISHIDA, C., NAKANE, H., MURAMATSU, M., NISHINO, T., OBATA, S. 1999. A thermophilic cyanobacterium *Synechococcus elongatus* has three different Class I prenyltransferase genes. Plant Mol. Biol. 40: 307-321.
64. ATTUCCI, S., AITKEN, S.M., GULICK, P.J., IBRAHIM, R. 1995. Farnesyl pyrophosphate synthase from white lupin:molecular cloning, expression, and purification of the expressed protein. Arch. Biochem. Biophys. 321: 493-500.
65. PAN, Z., HERICKHOFF, L., BACKHAUS, R.A. 1996. Cloning, characterization, and heterologous expression of cDNAs for farnesyl diphosphate synthase from the guayule rubber plant reveals that this prenyltransferase occurs in rubber particles. Arch. Biochem. Biophys. 332: 196-204.
66. CUNILLERA, N., ARRO, M., DELOURME, D., KARST, F., BORONAT, A., FERRER, A. 1996. *Arabidopsis thaliana* contains tow differentially expressed

farnesyl-diphosphate synthase genes. J. Biol. Chem. 271: 7774-7780.
67. CUNILLERA, N., BORONAT, A., FERRER, A. 1997. The *Arabidopsis thaliana FPS1* gene generates a novel mRNA that encodes a mitochondrial farnesyl-diphosphate synthase isoform. J. Biol. Chem. 272: 15381-15388.
68. SCOLNIK, P.A., BARTLEY, G.E. 1994. Nucleotide sequence of an *Arabidopsis* cDNA for geranylgeranyl pyrophosphate synthase. Plant Physiol. 104: 1469-1470.
69. SCOLNIK, P.A., BARTLEY, G.E. 1995. Nucleotide sequence of a putative geranylgeranyl pyrophosphate synthase from *Arabidopsis*. Plant Physiol. 108: 1343.
70. ZHU, X.F., SUZUKI, K., OKADA, K., TANAKA, K., NAKAGAWA, T., KAWAMUKAI, M., MATSUDA, H. 1997. Cloning and functional expression of a novel geranylgeranyl pyrophosphate synthase gene from *Arabidopsis thaliana* in *Escherichia coli*. Plant Cell Physiol. 38: 357-361.
71. ZHU, X.F., SUZUKI, K., SAITO, T., OKADA, K., TANAKA, K., NAKAGAWA, T., MATSUDA, H., KAWAMUKAI, M. 1997. Geranylgeranyl pyrophosphate synthase encoded by the newly isolated gene *GGPS6* from *Arabidopsis thaliana* is localized in mitochondria. Plant Mol. Biol. 35: 331-341.
72. HEFNER, J., KETCHUN, R.E.B., CROTEAU, R. 1998. Cloning and functional expression of a cDNA encoding geranylgeranyl diphosphate synthase from *Taxus canadensis* and assessment of the role of this prenyltransferase in cells induced for taxol production. Arch. Biochem. Biophys. 360: 62-74.
73. LASKARIS, G., DE JONG, C.F., JAZIRI, M., VAN DER HEIJDEN, R., THEODORIDIS, G., VERPOORTE, R. 1999. Geranylgeranyl diphosphate synthase activity and taxane production in *Taxus baccata* cells. Phytochemistry 50: 939-946.
74. WISE, M.L., CROTEAU, R. 1999. Monoterpene biosynthesis. pp 97-154 *in* Comprehensive Natural Products Chemistry Vol 2: Isoprenoids Including Carotenoids and Steroids (D.E. Cane, ed.), Pergamon Press, Oxford.
75. CANE, D.E. 1999. Sesquiterpene biosynthesis: cyclization mechanisms. pp. 154-200 *in* Comprehensive Natural Products Chemistry Vol 2: Isoprenoids Including Carotenoids and Steroids (D.E. Cane, ed.), Pergamon Press, Oxford.
76. MACMILLAN, J., BEALE, M.H. 1999. Diterpene biosynthesis. pp. 217-244 *in* Comprehensive Natural Products Chemistry Vol 2: Isoprenoids Including Carotenoids and Steroids (D.E. Cane, ed.), Pergamon Press, Oxford.
77. BOHLMANN, J., MEYER-GAUEN, G., CROTEAU, R. 1998. Plant terpenoid synthases: molecular biology and phylogenetic analysis. Proc. Natl. Acad. Sci.

USA 95: 4126-4133.

78. HEFNER, J., RUBENSTEIN, S.M., KETCHUN, R.E.B., GIBSON, D.M., WILLIAMS, R.M., CROTEAU, R. 1996. Cytochrome P450-catalyzed hydroxylation of taxa-4(5), 11(12)-diene to taxa-4(20),11(12)-dien-5α-ol: The first oxygenation step in taxol biosynthesis. Chemistry and Biology 3:479-489.
79. HEZARI, M., CROTEAU, R. 1997. Taxol biosynthesis: an update. Planta Med. 63: 291-295.
80. WALKER, K., KETCHUN, R.E.B., HEZARI, M., GATFIELD, D., GOLENIOWSKI, M., BARTHOL, A., CROTEAU, R. 1999. Partial purification and characterization of acetyl coenzyme A: Taxa-4(20),11(12)-dien-5α-ol O-acetyl transferase that catalyzes the first acylation step of taxol biosynthesis. Arch. Biochem. Biophys. 364: 273-279.
81. MENHARD, B., ZENK, M.H. 1999. Purification and characterization of acetyl coenzyme A: 10-Hydroxytaxane O-acetyltransferase from cell suspension cultures of *Taxus chinensis*. Phytochemistry. 50: 763-774.
82. FUNK, C., CROTEAU, R. 1994. Diterpenoid resin acid biosynthesis in conifers: Characterization of two cytochrome P450-dependent monooxygenases and an aldehyde dehydrogenase involved in abietic acid biosynthesis. Arch. Biochem. Biophys. 308: 258-266.
83. STEELE, C.L., LEWINSOHN, E., CROTEAU, R. 1995. Induced oleoresin biosynthesis in grand fir as a defense against bark beetles. Proc. Natl. Acad. Sci USA. 92: 4164-4168.
84. BOHLMANN, J., CROTEAU, R. 1999. Diversity and variability of terpenoid defences in conifers: Molecular genetics, biochemistry and evolution of the terpene synthase gene family in grand fir (*Abies grandis*). Novartis Found. Symp. Ser. 223: 132-149.
85. SAVAGE, T.J., CROTEAU, R. 1993. Biosynthesis of monoterpenes: regio- and stereochemistry of (+)-3-carene biosynthesis. Arch. Biochem. Biophys. 305: 581-587.
86. SAVAGE, T.J., HATCH, M.W., CROTEAU, R. 1994. Monoterpene synthases of *Pinus contorta* and related conifers: A new class of terpenoid cyclase. J. Biol. Chem. 269: 4012-4020.
87. SAVAGE, T.J., ICHII, H., HUME, S.D., LITTLE, D.B., CROTEAU, R. 1995. Monoterpene synthases from gymnosperms and angiosperms: Stereospecificity and inactivation by cysteinyl- and arginyl-directed modifying reagents. Arch. Biochem. Biophys. 320: 257-265.
88. SALIN, F., PAULY, G., CHARON, J., GLEIZES, M. 1995. Purification and characterization of trans-β-farnesene synthase from maritime pine (*Pinus pinaster* Ait.) needles. J. Plant Physiol. 146: 203-209.

89. PHILLIPS, M.A., SAVAGE, T.J., CROTEAU, R. 1999. Monoterpene synthases of loblolly pine (*Pinus taeda*) produce pinene isomers and enantiomers. Arch. Biochem. Biophys. 372: 197-204.
90. KOEPP, A.E., HEZARI, M., ZAJICEK, J., STOFER VOGEL, B., LAFEVER, R.E., LEWIS, N.G., CROTEAU, R. 1995. Cyclization of geranylgeranyl diphosphate to taxa-4(5),11(12)-diene is the committed step of taxol biosynthesis in Pacific yew. J. Biol. Chem. 270: 8686-8690.
91. HEZARI, M., LEWIS, N.G., CROTEAU, R. 1995. Purification and characterization of taxa-4(5),11(12)-diene synthase from pacific yew (*Taxus brevifolia*) that catalyzes the first committed step of taxol biosynthesis. Arch. Biochem. Biophys. 322: 437-444.
92. LIN, X., HEZARI, M., KOEPP, A.E., FLOSS, H.G., CROTEAU, R. 1996. Mechanism of taxadiene synthase, a diterpene cyclase that catalyzes the first step of taxol biosynthesis in Pacific yew. Biochemistry. 35: 2968-2977.
93. WILDUNG, M.R., CROTEAU, R. 1996. A cDNA clone for taxadiene synthase, the diterpene cyclase that catalyzes the committed step of taxol biosynthesis. J. Biol. Chem. 271: 9201-9204.
94. HEZARI, M., KETCHUN, R.E.B., GIBSON, D.M., CROTEAU, R. 1997. Taxol production and taxadiene synthase activity in *Taxus canadensis* cell suspension cultures. Arch. Biochem. Biophys. 337: 185-190.
95. BOHLMANN, J., STEELE, C.L., CROTEAU, R. 1997. Monoterpene synthases from grand fir (*Abies grandis*): cDNA isolation, characterization and functional expression of myrcene synthase, (-)-(4*S*)-limonene synthase, and (-)-(1*S*,5*S*)-pinene synthase. J. Biol. Chem. 272: 21784-21792.
96. BOHLMANN, J., PHILLIPS, M., RAMACHANDIRAN, V., KATOH, S., CROTEAU, R. 1999. cDNA cloning, characterization and functional expression of four new members of the *Tpsd* gene family from grand fir (*Abies grandis*). Arch. Biochem. Biophys. 368: 232-243.
97. CROTEAU, R. 1987. Biosynthesis and catabolism of monoterpenoids. Chem. Rev. 87: 929-954.
98. GIJZEN, M., LEWINSOHN, E., CROTEAU, R. 1991. Characterization of the constitutive and wound-inducible monoterpene cyclases of grand fir (*Abies grandis*). Arch. Biochem. Biophys. 289: 267-273.
99. LEWINSOHN, E., GIJZEN, M., CROTEAU, R. 1992. Wound-inducible pinene cyclase from grand fir: Purification, characterization, and renaturation after SDS-PAGE. Arch. Biochem. Biophys. 293: 167-173.
100. BOHLMANN, J., CROCK, J., JETTER, R., CROTEAU, R. 1998. Terpenoid-based defenses in conifers: cDNA cloning, characterization, and functional expression of wound-inducible (*E*)-α-bisabolene synthase from grand fir

(*Abies grandis*). Proc. Natl. Acad. Sci. USA 95: 6756-6761.

101.LAFEVER, R.F., STOFER VOGEL, B., CROTEAU, R. 1994. Diterpenoid resin acid biosynthesis in conifers: Enzymatic cyclization of geranylgeranyl pyrophosphate to abietadiene, the precursor of abietic acid. Arch. Biochem. Biophys. 313: 130-149.

102.STOFER VOGEL, B., WILDUNG, M.R., VOGEL, G., CROTEAU, R. 1996. Abietadiene synthase from grand fir (*Abies grandis*): cDNA isolation, characterization, and bacterial expression of a bifunctional diterpene cyclase involved in resin acid biosynthesis. J. Biol. Chem. 271: 23262-23268.

103.HEDDEN, P., KAMIYA, Y. 1997. Gibberellin biosynthesis: enzymes, genes and their regulation. Annu. Rev. Plant Physiol. Plant Mol. Biol. 48:431-460.

104.KAWAIDE, H., IMAI, R., SASSA, T., KAMIYA, Y. 1997. *ent*-Kaurene synthase from the fungus *Phaeosphaeria* sp. L487. J. Biol. Chem. 272: 21706-21712.

105.WILLIAMS, D.C., MCGARVEY, D.J., KATAHIRA, E.J., CROTEAU, R. 1998. Truncation of limonene synthase preprotein provides a fully active pseudomature form of this monoterpene cyclase and reveals the function of the amino-terminal arginine pair. Biochemistry. 37: 12213-12220.

106.CSEKE, L., DUDAREVA, N., PICHERSKY, E. 1998. Structure and evolution of linalool synthase. Mol. Biol. Evol. 15: 1491-1498.

107.COLBY, S.M., ALONSO, W.R., KATAHIRA, E., MCGARVEY, D.J., CROTEAU, R. 1993. 4*S*-limonene synthase from the oil glands of spearmint (*Mentha spicata*): cDNA isolation, characterization, and bacterial expression of the catalytically active monoterpene cyclase. J. Biol. Chem. 268: 23016-23024.

108.LEWINSOHN, E., SAVAGE, T.J., GIJZEN, M., CROTEAU, R. 1993. Simultaneous analysis of monoterpenes and diterpenoids of conifer oleoresin. Phytochem. Anal. 4: 220-225.

109.STARKS, C.M., BACK, K., CHAPPELL, J., NOEL, J.P. 1997 Structural basis for cyclic terpene biosynthesis by tobacco 5-*epi*-aristolochene synthase. Science 277: 1815-1820.

110.WENDT, K.U., PORALLA, K., SCHULZ, G.E. 1997. Structure and function of squalene cyclase. Science 277: 1811-1815.

111.LESBURG, C.A., ZHAI, G., CANE, D.E., CHRISTIANSON, D.W. 1997. Crystal structure of pentalenene synthase: Mechansitic insights on terpenoid cyclization reactions in biology. Science 277: 1820-1824.

112.HOLM, L., SANDER, C. 1996. Mapping the protein universe. Science 273: 595-602.

113.LESBURG, C.A., CARUTHERS, J.M., PASCHALL, C.M., CHRISTIANSON, D.W. 1998. Managing and manipulating carbocations in biology: terpenoid cyclase structure and mechanism. Curr. Opinion Struct. Biol. 8: 695-703.
114.BERRYMAN, A.A. 1972. Resistance of conifers to invasion by bark beetle fungus associations. BioScience 22: 598-602.
115.RAFFA, K.F. 1991. Induced defensive reactions in conifer bark-beetle systems. pp. 245-276 *in* Phytochemical Induction by Herbivores (D.W. Tallamy, M.J. Raupp, eds.), Academic Press, New York.
116.RAFFA, K.F., SMALLEY, E.B. 1995. Interaction of pre-attack and induced monoterpene concentrations in host conifer defense against bark beetle-fungal complexes. Oecologia 102: 285-295.
117.RAFFA K.F., BERRYMAN, A.A. 1982. Accumulation of monoterpenes and associated volatiles following inoculation of grand fir with a fungus transmitted by the fir engraver *Scolytus ventralis* (Coleoptera: Scolytidae). Can. Entomol. 114: 797-810.
118.MILLER, R.H., BERRYMAN, A.A., RYAN, C.A. 1986. Biotic elicitors of defense reactions in lodgepole pine. Phytochemistry. 25: 611-612.
119.CROTEAU, R., GURKEWITZ, S., JOHNSON, M.A., FISK, H.J. 1987. Biochemistry of oleoresinosis. Plant Physiol. 85: 1123-1128.
120.CHENICLET, C. 1987. Effects of wounding and fungus inoculation on terpene producing systems of maritime pine. J. Exp. Bot. 38: 1557-1572.
121.LEWINSOHN, E., GIJZEN, M., CROTEAU, R. 1991. Defense mechanisms of conifers: differences in constitutive and wound-induced monoterpene biosynthesis among species. Plant Physiol. 96: 44-49.
122.KLEPZIG, K.D., KRUGER, E.L., SMALLEY, E.B., RAFFA, K.F. 1995. Effects of biotic and abiotic stress on induced accumulation of terpenes and phenolics in red pines inoculated with bark beetle-vectores fungus. J. Chem. Ecol. 21: 601-627.
123.LITVAK, M.E., MONSON, R.K. 1998. Patterns of induced and constitutive monoterpene production in conifer needles in relation to in sect herbivory. Oecologia 114: 531-540.
124.STEELE, C.L., KATOH, S., BOHLMANN, J., CROTEAU, R. 1998b. Regulation of oleoresinosis in grand fir (*Abies grandis*): Differential transcriptional control of monoterpene, sesquiterpene, and diterpene synthase genes in response to wounding. Plant Physiol. 116: 1497-1504.
125.TOMLIN, E.S., ALFARO, R.I., BORDEN, J.H., HE, F. 1998. Histological response of resistant and susceptible white spruce to simulated white pine weevil damage. Tree Physiol. 18: 21-28.

126.THOMPSON, R.B., SIFTON, H.B. 1926. Resin canals in the Canadian spruce (*Picea canadensis* (Mill.) B.S.P.) – an anatomical study, especially in relation to traumatic effects and their bearing on phylogeny. Philos. Trans. R. Soc. Lond. B. 214: 111.

127.ALFARO, R.I. 1995. An induced defense reaction in white spruce to attack by the white pine weevil, *Pissodes strobi.* Can. J. For. Res. 25: 1725-1730.

128.NAGY, N.E., FRANCESCHI, V.R., SOLHEIM, H., KREKLING, T., CHRISTIANSEN, E. 2000. Wound-induced traumatic resin duct formation in stems of Norway spruce (Pinaceae); anatomy and cytochemical traits. Am. J. Bot. 87 in press.

129.FUNK, C., LEWINSOHN, E., STOFER VOGEL, B., STEELE, C.L., CROTEAU, R. 1994. Regulation of oleroresinosis in grand fir (*Abies grandis*). Plant Physiol. 106: 999-1005.

130.GIJZEN, M., LEWINSOHN, E., CROTEAU, R. 1992. Antigenic cross-reactivity among monoterpene cyclases from grand fir and induction of these enzymes upon stem wounding. Arch. Biochem. Biophys. 294: 670-674.

131.SQUILLACE, A.E., SWINDEL, B.F. 1986. Linkage among genes controlling monoterpene constituent levels in loblolly pine. For. Sci. 32: 97-112.

132.ESTEBAN, I., BERGMANN, F., GREGORIUS, H.-R., HUHTINEN, O. 1976. Composition and genetics of monoterpenes from cortical oleoresin of Norway spruce and their significance for clone identification. Silv. Genet. 25: 59-66.

133.KATOH, S., CROTEAU, R. 1998. Individual variation in constitutive and induced monoterpene biosynthesis in grand fir. Phytochemistry. 47: 577-582.

134.CATES, R.G., REDAK, R.A., HENDERSON, C.B. 1983. Patterns in defensive natural product chemistry: Douglas fir and western spruce budworm interactions. pp. 3-19 *in* Plant Resistance to Insects. (P. A. Hedin, ed.) American Chemical Society, Washington, D. C..

135.DELOURME, D., LACROUTE, F., KARST, F. 1994. Cloning of an *Arabidopsis thaliana* cDNA coding for farnesyl diphosphate synthase by functional complementation in yeast. Plant Mol. Biol. 26: 1867-1873.

136.MATSUSHITA, Y., KANG, W.K., CHARLWOOD, B.V. 1996. Cloning and analysis of a cDNA encoding farnesyl diphosphate synthase from *Artemisia annua.* Gene 172: 207-209.

137.HUGUENEY, P., BOUVIER, F., BADILLO, A., QUENNEMET, J., HARLINGUE, A., CAMARA, B. 1996. Developmental and stress regulation of gene expression for plastid and cytosolic isoprenoid pathways in pepper fruits. Plant Physiol. 111: 619-629.

138.LIU, C.-J., HEINSTEIN, P., CHEN, X.-Y. 1999. Expression pattern of genes encoding farnesyl diphosphate synthase and sesquiterpene cyclase in cotton suspension-cultures cells treated with fungal elicitor. Mol. Plant-Microbe Interact. 12: 1095-1104.

139.ADWILAGA, K., KUSH, A. 1996. Cloning and characterization of cDNA encoding farnesyl diphosphate synthase from rubber tree (*Hevea brasiliensis*). Plant Mol. Biol. 30: 935-946.

140.SANMIYA, K., IWASAKI, T., MATSUOKA, M., MIYAO, M., YAMAMOTO, N. 1997. Cloning of a cDNA that encodes farnesyl diphosphate synthase and the blue-light-induced expression of the corresponding gene in the leaves of rice plants. Biochim. Biophys. Acta: Gene Structure Expression 1350: 240-246.

141.LI, C.P., LARKINS, B.A. 1996. Identification of a maize endosperm-specific cDNA encoding farnesyl pyrophosphate synthetase. Gene 171: 193-196.

142.SCOLNIK, P.A., BARTLEY, G.E. 1996. Two more members on an *Arabidopsis* geranylgeranyl pyrophosphate synthase gene family. Plant Physiol. 110: 1435-1436.

143.KUNZ, M., RÖMER, S., SUIRE, C., HUGUENEY, P., WEIL, J.H., SCHANTZ, R., CAMARA, B. 1992. Identification of a cDNA for the plastid-located geranylgeranyl pyrophosphate synthase from *Capsicum annuum*: correlative increase in enzyme activity and transcript level during fruit ripening. Plant J. 2: 25-34.

144.BADILLO, A., STEPPUHN, J., DERUERE, J., CAMARA, B., KUNTZ, M. 1995. Structure of a functional geranylgeranyl pyrophosphate synthase gene from *Capsicum annuum*. Plant Mol. Biol. 27: 425-428.

145.BANTIGNIES, B., LIBOZ, T., AMBID, C. 1996. Nucleotide sequence of a *Catharanthus roseus* geranylgeranyl diphosphate synthase gene. Plant Physiol. 110: 336.

146.AITKEN, S.M., ATTUCCI, S., IBRAHIM, R.K., GULICK, P.J. 1995. cDNA encoding geranylgeranyl pyrophosphate synthase from white lupin. Plant Physiol. 108: 837-838.

147.BONK, M., HOFFMAN, B., VON LINTIG, J., SCHLEDZ, M., AL-BABILI, S., HOBEIKA, E., KLEINIG, H., BEYER, P. 1997. Chloroplast import of four carotenoid biosynthetic enzymes *in vitro* reveals differential fates prior to membrane binding and oligomeric assembly. Eur. J. Biochem. 247: 942-950.

148.FACCHINI, P.J., CHAPPELL, J. 1992 Gene family for an elicitor-induced sesquiterpene cyclase in tobacco. Proc. Natl. Acad. Sci. USA 89: 11088-11092.

149.BACK, K., HE, S., KIM, K.U., SHIN, D.H. 1998. Cloning and bacterial expression of sesquiterpene cyclase, a key branch point enzyme for the synthesis of sesquiterpenoid phytoalexin capsidiol in UV-challenged leaves of Capsicum annuum. Plant Cell Physiol. 39: 899-904.

150.WISE, M.L., SAVAGE, T.J., KATAHIRA, E., CROTEAU, R. 1998. Monoterpene synthases from common sage (*Salvia officinalis*). cDNA isolation, characterization, and functional expression of (+)-sabinene synthase, 1,8-cineole synthase, and (+)-bornyl diphosphate synthase. J. Biol. Chem. 273: 14891-14899.

151.CHEN, X.Y., CHEN, Y., HEINSTEIN, P. DAVISSON, V.J. 1995. Cloning, expression, and characterization of (+)-delta-cadinene synthase: a catalyst for cotton phytoalexin biosynthesis. Arch. Biochem. Biophys. 324: 255-266.

152.CHEN, X.Y., WANG, M., CHEN, Y., DAVISSON, V.J., HEINSTEIN, P. 1996 Cloning and heterologous expression of a second (+)-delta-cadinene synthase from *Gossypium arboreum*. J. Nat. Prod. 59: 944-951.

153.MENG, Y.L., JIA, J.W., LIU, C.J., LIANG, W.Q., HEINSTEIN, P., CHEN, X.Y. 1999. Coordinated accumulation of (+)-delta-cadinene synthase mRNAs and gossypol in developing seeds of *Gossypium hirsutum* and a new member of the cad1 family from *G. arboreum*. J. Nat. Prod. 6: 248-252.

154.DAVIS, E.M., CHEN, Y.-S., ESSENBERG, M., PIERCE, M.L. 1998. cDNA sequence of a (+)-delta-cadinene synthase gene (Accession No.U88318) induced in *Gossypium hirsutum* L. by bacterial infection. Plant Physiol. 116: 1192.

155.MAU, C.J.D., WEST, C.A. 1994. Cloning of casbene synthase cDNA: evidence for conserved structural features among terpenoid cyclases in plants. Proc. Natl. Acad. Sci. USA 91: 8497-8501.

156.MERCKE, P., CROCK, J., CROTEAU, R., BRODELIUS, P.E. 1999. Cloning, expression and characterization of epi-cedrol synthase, a sesquiterpene cyclase from *Artemisia annua* L. Arch. Biochem. Biophys. 369: 213-222.

157.HUA, L., MATSUDA S.P.T. 1999. The molecular cloning of 8-epicedrol synthase from *Artemisia annua*. Arch. Biochem. Biophys. 369: 208-212.

158.SUN, T.-P., KAMIYA, Y. 1994. The Arabidopsis GA1 locus encodes the cyclase ent-kaurene synthetase A of gibberellin biosynthesis. Plant Cell 6: 1509-1518.

159.SMITH, M.W., YAMAGUCHI, S., AIT-ALI, T., KAMIYA, Y. 1998. The first step of gibberellin biosynthesis in pumpkin is catalyzed by at least two copalyl diphosphate synthases encoded by differentially regulated genes. Plant Physiol. 118: 1411-1419.

160.AIT-ALI, T., SWAIN, S.M., REID, J.B., SUN, T.-P., KAMIYA, Y. 1997. The LS locus of pea encodes the gibberellin biosynthesis enzyme ent-kaurene synthase A. Plant J. 11: 443-454.

161.BENSEN, R.J., JOHAL, G.S., CRANE, V.C., TOSSBERG, J.T., SCHNABLE, P.S., MEELEY, R.B., BRIGGS, S.P. 1995. Cloning and characterization of the maize An1 gene. Plant Cell 7: 75-84.

162.RICHMAN, A.S., GIJZEN, M., STARRATT, A.N., YANG, Z., BRANDLE J. 1999. Diterpene synthesis in *Stevia rebaudiana*: recruitment and up-regulation of key enzymes from the gibberellin biosynthetic pathway. Plant J. 19: 1-11.

163.CROCK, J., WILDUNG, M.R., CROTEAU, R. 1997. Isolation and bacterial expression of a sesquiterpene synthase cDNA clone from peppermint (*Mentha xpiperita*, L.) that produces the aphid alarm pheromone (*E*)-β-farnesene. Proc. Natl. Acad. Sci. USA 94: 12833-12838.

164.COLBY, S.M., CROCK, J., DOWDLE-RIZZO, B., LEMAUX, P.G., CROTEAU, R. 1998 Germacrene C synthase from *Lycopersicon esculentum* cv VFNT cherry tomato: cDNA isolation, characterization, and bacterial expression of the multiple product sesquiterpene cyclase. Proc. Natl. Acad. Sci. USA 95: 2216-2221.

165.YAMAGUCHI, S., SAITO, T., ABE, H., YAMANE, H., MUROFUSHI, N., KAMIYA,Y. 1996. Molecular cloning and characterization of a cDNA encoding the gibberellin biosynthetic enzyme ent-kaurene synthase B from pumpkin (*Cucurbita maxima* L.). Plant J. 10: 203-213.

166.YAMAGUCHI,S., SUN,T.P., KAWAIDE,H., KAMIYA,Y. 1998. The GA2 locus of *Arabidopsis thaliana* encodes ent-kaurene synthase of gibberellin biosynthesis. Plant Physiol. 116: 1271-1278.

167.YUBA, A., YAZAKI, K., TABATA, M., HONDA, G., CROTEAU, R. 1996. cDNA cloning, characterization, and functional expression of 4*S*-(-)-limonene synthase from *Perilla frutescens*. Arch. Biochem. Biophys. 332: 280-287.

168.DUDAREVA, N., CSEKE, L., BLANC, V.M., PICHERSKY, E. 1996. Evolution of floral scent in *Clarkia*: Novel patterns of *S*-linalool synthase gene expression in the *C. breweri* flower. Plant Cell 8: 1137-1148.

169.JIA, J.W., CROCK, J., LU, S., CROTEAU, R., CHEN, X.Y. 1999. (3*R*)-Linalool synthase from *Artemisia annua* L.: cDNA isolation, characterization, and wound induction. Arch. Biochem. Biophys. 372: 143-149.

170.BOHLMANN, J., MARTIN, D., OLDHAM, N.J., GERSHENZON, J. 2000. Terpenoid secondary metabolism in *Arabidopsis thaliana*: cDNA cloning, characterization, and functional expression of a myrcene/(*E*)-β-ocimene synthase. Arch. Biochem. Biophys. 375, in press.

171.BACK, K., CHAPPELL, J. 1995. Cloning and bacterial expression of a sesquiterpene cyclase from *Hyoscyamus muticus* and its molecular comparison to related terpene cyclases. J. Biol. Chem. 270: 7375-7381

172.YOSHIOKA, H., YAMADA, N., DOKE, N. 1999. cDNA cloning of sesquiterpene cyclase and squalene synthase, and expression of the genes in potato tuber infected with *Phytophtora infestans*. Plant Cell Physiol. 40: 993-998.

Chapter Six

FUNCTION AND EVOLUTION OF PLANT CYTOCHROME P450

Rachel A. Kahn and Francis Durst

CNRO-Institut de Biologie Moléculaire des Plantes
Département d'Enzymologie Celluaire et Moléculaire
28 rue Goethe
F-67083 Strasbourg, France

INTRODUCTION

Cytochrome P450s (P450s) form an extraordinary group of catalysts outranking any other family of enzymes by the number of substrates recognized, the number of reactions catalyzed, and the number of inducers. Reaction types are so so varied and substrates so diverse and numerous that it has been proposed to rename these enzymes "diversozymes".[1] The name has not caught on, but the definition by Coon is an excellent summary of what P450s are and why they exert such an attraction and foster thousands of papers per year since their discovery in the late 50's: "Cytochrome P450, the most versatile biological catalyst known, was originally named as a pigment having a carbon monoxide spectrum at about 450 nm and no known function. Recent progress in many laboratories has revealed that the P450 superfamily has immense diversity in functions, with hundreds of isoforms in many species catalyzing many types of chemical reactions. We believe it safe to predict that each mammalian species may be found to have up to a hundred P450 isoforms that respond *in toto* to a thousand or more inducers and that, along with P450s from other sources, metabolize a million or more potential substrates."[1]

In this chapter, we shall give a short introduction on P450 protein organization and catalytic mechanism, describe the P450 superfamily of genes, and summarize the present view of the function and evolution of CYP genes in plants. Additional reading: an excellent overview of P450 mechanisms and genetics can be found in several recent reviews, some of which have addressed specifically different aspects of P450 in plants.[2-8]

THE P450 ENZYMATIC COMPLEX AND CATALYTIC MECHANISM

Cytochrome P450s are heme-thiolate proteins, *i.e.* they bear, in a hydrophobic pocket and well shielded from the outside of the molecule, a protoporphyrin IX linked to the apoprotein by a unique bond between the heme iron center and the sulfur of a conserved cysteinyl residue. They have been found in all parts of plants and in all cellular types that have been investigated. Except for the deviant P450s in family 74 (see below), all those characterized to date in plants are localized in the microsomal fraction, and most detailed studies have found them tightly bound to the ER.

When P450 functions as a monooxygenase or an oxidase, which is the case in most reactions, molecular oxygen binds to the heme iron and is activated by a complex suite of redox reactions that culminates with the formation of a reactive oxygen species that is then transferred to the substrate (Fig. 1). Carbon monoxide may bind, instead of oxygen, to reduced P450, producing a characteristic CO-binding difference spectrum (reduced P450 + CO *versus* reduced P450) with an absorbance maximum at 450 nm. This P450-CO complex, which has given the name to P450 (= pigment absorbing at 450 nm) is inactive. Inhibition by CO and reversion of this inhibition by 450 nm light is characteristic of most P450 reactions. Oxygen activation requires an additional redox partner, the flavoprotein NADPH-P450 reductase. This enzyme, which is also ER-bound, transfers in a sequential manner 2 electrons from NADPH to the heme iron and ultimately to oxygen. While mammalian P450 enzymes all rely on a single P450-reductase, several plants have been shown to possess 2 or 3 such flavoproteins.[9-12] It has been suggested that one reductase isoform might be specifically induced under stress conditions.[13]

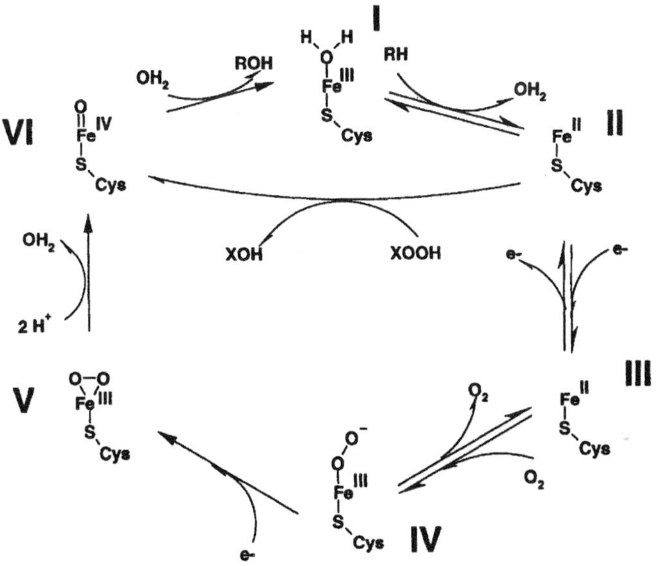

Figure 1. Catalytic cycle of a P450 monooxygenase
In the resting state, water is bound to the heme center, blocking out oxygen and protecting against auto-oxidation and generation of harmful oxygen species. Upon binding of an oxidizable substrate RH (I), water is displaced (II), triggering electron transfer (III) from the reductase, and oxygen binding (IV). Oxygen is activated (V) and transfered to RH forming ROH, the reaction product (VI).

The stoichiometry of a typical monooxygenase reaction is as follows: S + O_2 + $NADPH_2$ -> SO + H_2O + $NADP^+$, where S is an oxidizable substrate. The exceptions to this generalization occur with members of the CYP74 family, the allene oxide synthase (AOS), first identified in flaxseed, and the fatty acid hydroperoxide lyase (HPL).[14,15] Enzymes in the CYP74 family do not require molecular oxygen and NADPH, but act upon fatty acid hydroperoxides and catalyze the extra or intra molecular transfer of one oxygen atom of the hydroperoxide that is reduced to an alcohol. These enzymes do not show the typical ER insertion signal found at the N-terminus of most other eukaryotic P450s. In fact, AOS and HPL activities have been measured recently on the inner chloroplast membrane (Blée and Joyard, personal communication).

DIVERSITY AND EVOLUTION OF PLANT P450 GENES

CYPs Form a Vast and Divergent Superfamily of Genes.

The P450s from all organisms share a common ancestor. In fact this is tautological since the nomenclature committee accepts as P450 only those proteins that can be satisfactorily aligned with, and show discernible homology to, other P450 sequences. For example, the nitric oxide synthase, which is a hemoprotein oxygenase that exhibits the same carbon monoxide difference spectrum as P450s has been excluded because it does not satisfy these phylogenetic criteria.

P450s form a vast superfamily of genes. They are categorized in families (sequences sharing ≥40% aminoacid positional identity) and subfamilies (≥55% identity). For example, CYP71A1 is the first member of subfamily A of family 71. Generally, the gene is noted *CYP71A1,* while CYP71A1 stands for the protein. The CYP classification is purely phylogenetic. There is no consideration of function. In fact, the function of most of the 1400 genes cloned to date remains unknown. The P450 superfamily is not only vast, it is also highly divergent. Even among plant P450s, identity may be as low as 16% with CYP51, or even 12% with the deviant CYP74 (see below). Furthermore, only 5 residues, including the cysteine in the heme-binding domain, are strictly conserved among all cytochrome P450s. How can genes with such low identity be assigned to a particular CYP family or be considered new P450s? This is because P450s share some conserved domains that are typical: 1) the so-called P450 signature motif that contains the cysteine that links the heme iron to the apoprotein. Generally, the form of this heme binding domain is FXXGXXXCXG. Noted exceptions are CYP74A, CXG, and CYP74B, where only **C** is conserved; 2) two charged residues in helix K forming the EXXR motif; 3) the 'meander' region, rich in aromatic residues and prolines, between helix K and the heme-binding region; 4) a motif of the general form AGX[E/D]T located in Helix I, which is part of the oxygen binding pocket.[16] This motif is not conserved in P450s of the CYP74 family, where it is the hydroperoxy function of the substrate and not O_2 that interacts with the heme iron.

How far back can one trace a P450 ancestor, and what its function was, is an open question. Present day P450s, including those from archeabacteria, are highly evolved proteins. Their structure and function, when known, offer only partial clues to the origin of the P450. It is generally believed that oxidases have evolved through modification of preexisting electron carrier proteins.[17] It has been proposed that P450s derive from such proteins and have existed since before the appearance of oxygen to carry out reductive reactions (Fig. 2). Indeed, several present day P450s, when operating under hypoxic conditions and with polyhalogenated compounds or azodyes as substrates, are excellent reductases. One fungal P450, CYP55, is an effective NO reductase.[18] When oxygen appeared in trace levels about 3300 MYA (million years ago), a possible function of P450 was to detoxify it by catalyzing the concerted 4 electron reduction of O_2 to H_2O, thus avoiding the generation of harmful activated oxygen species. When O_2 concentration rose and the atmosphere shifted from reducing to oxidizing, about 2000 MYA, the function of P450 may have similarly shifted from oxygen reduction to oxygen activation and transfer to endogenous and then xenobiotic substrates. This concept of P450 evolution has been summarized by Nebert and Feyereisen.[19] Since sterols and fatty acids are so intimately associated with membranes and all aspects of cell life, it is believed that oxidation of sterols and fatty acids was the first oxygenase type of P450 function. At present, CYP51, the sterol 14-demethylase, is the only P450 with orthologous forms in animals, plants, yeast, and bacteria.

It has been argued recently that all plant P450s originate from a CYP51 ancestor.[20] However, our phylogenetic studies suggest that the unicellular ancestor of plants had at least 3 distinct P450 branches. A first branch shared common ancestors with CYP1 and CYP2 that are mainly involved in the metabolism of polycyclic aromatics and drugs in animals. This branch gave rise to the Group A of plant P450s,[4] a group comprising the enzymes catalyzing typical "plant reactions", such as synthesis of lignin, alkaloids, flavonoids, etc. A second branch shared a common ancestor with animal CYP4s and fungal CYP52s that catalyze the ω and (ω-1)-hydroxylation of fatty acids. This branch gave rise to the plant fatty acid hydroxylases of families CYP86 and CYP94 and further comprises families CYP96 and CYP97 that have not yet been functionally characterized. A third branch shared a common ancestor with animal, fungal, and bacterial sterol oxidases. This branch gave rise to the plant obtusifoliol 14-demethylase (CYP51) and brassinolide hydroxylases (CYP85 and CYP90 families). The two latter branches were originally defined as Group non-A P450s by Durst and Nelson.[4] One difficult and, at present unresolved, problem is the placement of the CYP74 in the general tree of cytochrome P450s. In all phylogenies that we have considered, whichever method (distance, parsimony, maximum likelihood) was used, family 74 branches out and appears as the most divergent of all P450 families. This suggests that CYP74 is a present day representative of an ancestral P450 form. This is supported by : 1) the fact that CYPs 74 neither require NADPH nor oxygen but catalyze internal oxygen transfer in polyunsaturated fatty acid hydroperoxide substrates, a type of reaction likely to have occurred in early photosynthetic organisms when partial O_2 pressure was still low on Earth; 2) recent biochemical data that localize these enzymes on the inner envelope of the chloroplast (see above). A simplified tree of plant P450s, showing the main branches having evolved in Group A and Group non-A, is presented in Figure 3.

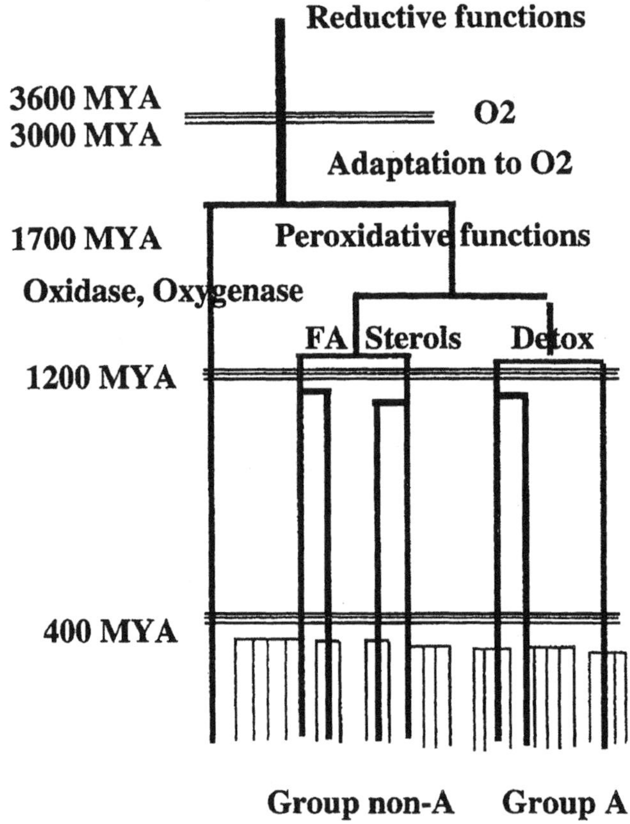

Figure 2 . Putative evolution of cytochrome P450 enzymes (see text).

All phylogenetic studies (see examples by David Nelson at http://drnelson.utmem.edu/) point to an explosive development of the P450 superfamily starting 400 MYA. Since this corresponds to the period of land colonization by plants, it has been proposed that one of the driving forces at work behind this sudden diversification of *CYP* genes is the so-called chemical war between plants and animals. To deter predators and pathogens, plants have evolved increasingly more sophisticated chemicals with poisonous, anti-feedant, or behavior-modifying properties. This has required new P450s able to diversify the structure of terpenes, alkaloids, phenolics, and phytoalexins. Conversely, rapid evolution of P450s, essentially in the CYP1, CYP2, and CYP3 families, has conferred on animals the capacity to detoxify these compounds.

Lessons from the Systematic Sequencing of the *Arabidopsis thaliana* Genome.

Much of the material discussed here can be found at http://drnelson.utmem.edu/ and http://ag.arizona.edu/p450/plantp450.html. Extrapolating from the 235 distinct P450s found in 81% of the *Arabidopsis* genome, the total number may be estimated at 290. This is a remarkable number, even considering that 20-25 may prove to be pseudogenes. The characterization of the catalytic and physiological functions of all these genes is a challenging task, the first difficulty being, of course, to envisage all the different substrates. It is interesting to note, in this context, that half of the P450s identified at present have no corresponding EST within the 36500 entries of the *A. thaliana* EST database. This suggests that a number of the P450s are not constitutive and may be expressed only at defined developmental stages or under particular conditions, such as biotic and abiotic stress. Only 5 of the 42 CYP families presently defined in plants are not represented in *Arabidopsis*, one of which, CYP80, contains enzymes involved in alkaloid synthesis (see below). The four others have not been characterized. Some families contain a large number of genes, such as *CYP*71 with 45 or more genes, *CYP*705 with 25 or more, and *CYP*81 with 16 or more. This, and the fact that some of the *CYPs* are highly clustered (for example 14 *CYP*71B in a cluster on chromosome III, four *CYP*71B clustered on chromosome I, and four more on chromosome V), indicates that gene conversion, crossing over, and intense gene duplication are responsible for the high number of *CYP*s in plants.

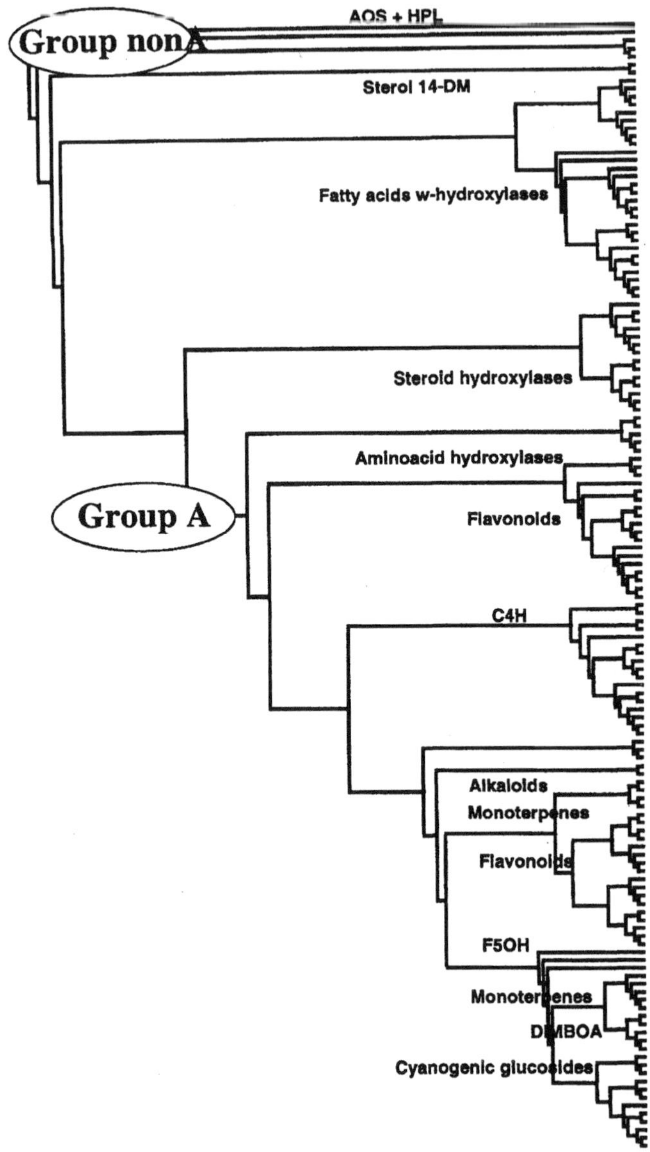

Figure 3. A phylogenetic tree of plant P450s.
The tree indicates genes with identified physiological functions. C4H and F5H: cinnamate and ferulate hydroxylases, DIMBOA: P450s involved in DIMBOA synthesis, AOS and HPL: allene oxide synthase and fatty acid hydroperoxide lyase. See text for details.

ISOLATION AND CLONING OF PLANT P450s

The first cDNAs encoding plant P450s were cloned following an initial purification of the enzyme.[5] Relatively few sequences were used to predict a certain consensus, and this led to the isolation of many more P450s by a molecular approach. At present, the major task is to link a physiological function to the large number of plant P450 sequences. This involves biochemical work, often combined with genetics of relevant mutants.

Biochemical Approach

The most direct way to obtain the gene encoding a P450 of interest is to purify the enzyme, obtain peptide-sequences and/or antibodies, and screen an appropriate cDNA library. This method is time consuming, and the risk of failure is high. The enzymes to be purified are present in small amounts, possibly only under certain developmental conditions or in specialized tissues. They are unstable in crude plant extracts, and the detergents that are essential for their isolation create additional problems with instability. For these reasons, only relatively few plant P450s have been purified to homogeneity,[21-24] and an unknown number have resisted isolation. A molecular approach, therefore, is warranted in cases where purification fails or cannot be envisioned.

Molecular Approaches

For any molecular approach to be successful, it is important to select carefully the optimal tissue. Furthermore, functional expression in a heterologous host and/or complementation is necessary to demonstrate function.

The polymerase chain reaction (PCR). PCR has often been applied to clone novel plant P450s. If cloning of just any P450 is the option, success is assured in advance. However, if the option is to isolate a plant P450 with a distinct physiological function, the matter is more complicated.

Two cDNAs encoding F3'5'H involved in anthocyanin biosynthesis in petunia were isolated by using a PCR strategy, with degenerate primers corresponding to the conserved heme binding domain. Restriction fragment length polymorphism (RFLP) mapping, complementation of mutant petunia lines, and heterologous expression in yeast were used to demonstrate the function of the isolated genes. The clones corresponding to the *Hf1* and *Hf2* genetic loci were designated CYP75A1 and CYP75A3.[25] The petunia CYP75A1 cDNA sequence was used as a probe in a low stringency screen of a lisianthus (*Eustoma grandiflorum*) petal cDNA library, leading to the cloning of the lisianthus 3'5'-hydroxylase, CYP75A5.[26]

In a search for P450s involved in the biosynthesis of indole alkaloids, 16 cDNA fragments were isolated from a *Catharanthus roseus* root cDNA library. PCR was performed by using degenerate primers corresponding to the heme binding region (FLPFGxGxR) against either the poly(A)tail or a vector sequence near the poly(A)tail.[27]

Three A-type P450s, CYP71E1, CYP98A1, and CYP99A1, and a fragment corresponding to the sorghum cinnamic acid hydroxylase (C4H) were cloned from a sorghum seedling cDNA library. Three consensus regions in A-type cytochrome P450s - (V/I)KEx(L/F)R, FxPERF, and PFGxGRRxCxG were used for primer design. CYP71E1 was identified by functional expression in *E. coli* as the P450 that catalyzes the conversion of an aldoxime to a mandelonitrile in the biosynthesis of the cyanogenic glucoside dhurrin. No function could be attributed to CYP98A1 or CYP99A1.[28]

Multiple *Arabidopsis thaliana* P450s were isolated from a seedling cDNA library by using primers based on sequence conservation between CYP71A1 and CYP73A2. This led to the isolation of 11 P450s from *A. thaliana*.[29] One represents the *A. thaliana* C4H and was identical to the CYP73A5 previously isolated from *A. thaliana*. Several represent CYP71B subfamily members. Some defined new subfamilies: CYP76C1, CYP81D1, CYP81D8, CYP81F1, and CYP83B1.[30,31] Four of the P450 genes were mapped by RFLP, and each showed unique expression patterns.[29] The physiological function was attributed to one of the isolated P450s, namely the cinnamic acid hydroxylase. This demonstrates the relative ease of obtaining P450 clones, which is followed by the difficult task of elucidating their physiological roles.

Reverse transcriptase-polymerase chain reaction (RT-PCR). An RT-PCR strategy was used to identify wound-induced P450s in pea. mRNA was prepared from wounded pea stem sections, reverse transcribed, and PCR amplified by using a 3' oligo(dT) primer and a degenerate 5'PCR primer encoding EEFxPERF. Two full-length cDNA clones were isolated. One, designated CYP73A9, is the C4H from pea, and the other, CYP82A1, represents a P450 of unknown function.[32]

Pauli and Kutchan[33] isolated P450s involved in alkaloid biosynthesis by RT-PCR. RNA from methyl jasmonate induced cell cultures of California poppy was used as a template, and 5'PCR primers were based on conserved sequences in the heme-binding domain.[33] Three full-length cDNAs were isolated, two of which represent alleles of *CYP80B1*, and the third represents *CYP82B1*. No function was ascribed to CYP82B1, however, heterologous expression in yeast and insect cells followed by activity assays demonstrated that *CYP80B1* encodes the (*S*)-*N*-methylcoclaurine 3'-hydroxylase involved in benzylisoquinoline alkaloid biosynthesis.

Differential display of mRNA (DD-RT-PCR) allows the identification and molecular cloning of genes differentially expressed at two states in a given tissue.[34] Elicitor induced soybean cDNAs encoding P450s that are potentially involved in the biosynthesis of the phytoalexin glyceollin were isolated by DD-RT-PCR. Upstream primers were based on the conserved PFG motif in the heme-binding region. Sixty differential bands were identified, leading to the cloning of eight full-length cDNAs encoding P450s that are transcriptionally activated under glyceollin inducing conditions.[35] Four CYP families are represented (CYP71, CYP73, CYP82, and CYP93), and, so far, functional expression in yeast has confirmed that at least two of the clones encode P450s of the glyceollin pathway. *CYP73A11* encodes soybean C4H, and *CYP93A1* encodes the dihydroxypterocarpan 6a-hydroxylase[36] purified by Kochs and Grisebach in 1989.[21] The cloning of CYP93A1 from soybean by differential display was first reported by Suzuki et al., however, no function was attributed to the gene at the time.[37]

Database cloning. ESTs: The expressed sequence tag (EST) databases have proven to be valuable resources for the isolation of P450s.[38] The following are examples of EST based cloning, but these are by no means exhaustive.

A C4H *A. thaliana* cDNA clone was identified in the *A. thaliana* EST database and used to isolate a corresponding genomic clone designated CYP73A5.[31] Sorghum CYP79A1 catalyzes the conversion of tyrosine to an aldoxime in cyanogenic glucoside biosynthesis, and the deduced amino acid sequence of CYP79A1 was used to screen the *A. thaliana* ESTs. An EST with significant sequence identity to CYP79A1 was identified and was subsequently used to isolate CYP79B1 from *Sinapis alba*. CYP79B1 presumably represents the P450 that catalyzes the conversion of tyrosine to aldoxime in the biosynthesis of glucosinolates.[39] The unusual hydroperoxide lyase (HPL) was purified from green pepper, cloned, and designated CYP74B1.[40] An *A. thaliana* EST with a high degree of homology to the green pepper sequence was identified and used to clone a full length cDNA encoding an *A. thaliana* HPL. The activity was confirmed by functional expression in the *E. coli* pGEX system.[41]

The *A. thaliana* EST database was searched with consensus sequences derived from the mammalian CYP4 and yeast CYP52 families. A candidate EST fragment was identified, and the full-length cDNA cloned and designated CYP86A1. Functional expression in yeast demonstrated that CYP86A1 catalyzes the omega-hydroxylation of fatty acids, and, indeed, is functionally similar to the mammalian and yeast enzymes used as a basis for the cloning strategy.[42]

Mutants. The availability of mutants that are blocked in P450 catalyzed reactions provides an alternative method for the cloning of plant P450 genes. Either natural or induced mutants can be used, and the approach requires a well-defined pathway to correlate phenotype with gene function. Mutant complementation and heterologous expression is subsequently used to demonstrate gene function.

T-DNA tagged mutants: The *cpd* (constitutive photomorphogenesis and dwarfism) mutant was identified in an *A. thaliana* T-DNA tagged collection. T-DNA-plant insert junctions were cloned from a genomic library constructed from the *cpd* mutant, and used as probes for further screening of wild-type cDNA. A *CPD* cDNA, was isolated that encodes a P450 designated CYP90A1. Genetic complementation of the *cpd* mutation by *CYP90A1* was demonstrated, and feeding experiments with brassinolide restored the wild-type phenotype, indicating that CYP90A1 is involved in the biosynthesis of active brassinosteroids.[43] *Dwarf4* mutants of *A. thaliana* were identified in a collection of 14000 T-DNA tagged *A. thaliana*. The *DWF4* gene was cloned by plasmid rescue, and encodes a P450 designated CYP90B1 with significant homology to the CPD enzyme. CYP90B1 is involved in a key regulatory step in brassinosteroid biosynthesis where it catalyzes steroid 22alpha-hydroxylations.[44] The *rotundifolia3* (*rot3*) mutant of *A. thaliana* has a defect in the polar elongation of leaf cells. The mutation was observed in *A. thaliana* populations submitted to either fast neutron irradiation of seeds, ethyl methanesulfonate (EMS) mutagenesis, or T-DNA tagging. The *ROT3* gene was cloned by plasmid rescue, by using the T-DNA tagged line. *ROT3* encodes CYP90C1, which may be involved in steroid biosynthesis, however, probably not in brassinosteroid biosynthesis as application of exogenous brassinosteroids does not restore the wild-type.[45] Recently, it was shown that the *ROT3* gene seems to play an important role in the polar elongation of leafy organs.[46]

The gene encoding ferulate 5-hydroxylase in *A. thaliana* was cloned by a T-DNA tagging approach. The *fah1* mutant was identified in a visual screen of a collection of tagged *A. thaliana* lines, and the tagged line was used to clone the gene encoding F5H, designated *CYP84A1*. Complementation of the *fah1* mutant with *CYP84A1* restored the wild-type phenotype.[47]

EMS mutants: Following EMS mutagenesis of *A. thaliana*, the *ga3* mutant was identified as a non-germinating, GA-responsive dwarf. The *GA3* gene was mapped to a single BAC, and a single putative P450 gene located on this BAC was identified and designated CYP701A3. Mutants were complemented with the wild-type P450 gene leading to a restoration of the wild-type phenotype. It is likely that *GA3* encodes *ent*-kaurene oxidase, the first P450 mediated step in GA biosynthesis.[48]

Transposon tagged mutants: Strategies for gene cloning that use transposon tagging have been reviewed by Walbot.[49] Transposon tagging by using the *Mutator* (*Mu*) family transposons was applied to clone the maize *Dwarf3* (*D3*) gene that encodes an early step in gibberellin (GA) biosynthesis. The *D3* gene was identified and was shown to encode a P450 designated CYP88A1. Biochemical analysis indicates that the *d3* mutation blocks an early step in the pathway of GA biosynthesis, and it is predicted that the *D3* gene encodes an early 13-hydroxylase activity.[50] The tomato *Dwarf* gene (*D*) was isolated by heterologous transposon tagging by using the maize *Activator* transposable element, and it was classified as CYP85.[51] CYP85 is involved in brassinosteroid biosynthesis, which was confirmed recently by complementation of the mutant, and by functional expression in yeast. CYP85 catalyzes the C-6 oxidation of 6-deoxycastasterone to castasterone, a late step in brassinosteroid production.[52]

The entire secondary metabolic pathway for the biosynthesis of DIBOA in maize was established *via* molecular genetics. The cyclic hydroxamic acids DIBOA and DIMBOA are gramineae specific defense compounds. The genes *Bx1* through *Bx5* are all clustered on chromosome four and encode proteins that catalyze the conversion of indole-3-glycerol phosphate to DIBOA. They were isolated by using several molecular approaches, including subtractive cDNA cloning from high *versus* low DIMBOA lines, and by the use of the *Mu* transposon tagging system. Four of the genes, *Bx2* through *Bx5*, share similar gene structures and were identified as members of the CYP71C subfamily.[53,54] The four P450s were heterologously expressed in yeast, and their proposed activities and high substrate specificity were confirmed.[55] It remains to be seen whether clustering of genes involved in the same pathway is a common phenomenon.

P450s INVOLVED IN THE BIOSYNTHESIS OF NATURAL COMPOUNDS

Alkaloids

Alkaloids are organic nitrogenous bases found mainly in plants, but also to a lesser extent in micro-organisms and animals. Within alkaloid biosynthesis one finds an intricate interplay between soluble and membrane-associated enzymes, and the P450s seem to represent a point of regulation (for a review see

Chou and Kutchan[56]). Some of the most fascinating chemistries, difficult to mimic in organic chemistry, are catalyzed by P450s found in alkaloid biosynthesis. These enzymes are highly substrate specific, and appear to have evolved solely for a role in secondary metabolism.[56]

Hydroxylases. A single, specific hydroxylation at C-6 of protopine leads to the spontaneous rearrangement of this molecule to yield dihydrosanguinarine. The protopine-6-hydroxylase reaction was studied in cell cultures of *Eschscholzia californica*, and the hydroxylase was shown to be a P450 specifically present in plant species that produce benzo(*c*)phenanthridine alkaloids.[57,58]

Nineteen enzymatic reactions are required for the conversion of tyrosine to macarpine, as demonstrated by using elicited cell cultures of *Thalictrum bulgaricum*. Seven of the reactions are catalyzed by highly substrate specific P450s, which are all activated from three- to ten-fold by elicitation, whereas the cytosolic methyltransferases involved are not. This interplay of cytosolic and membrane-associated enzymes is thought to have a regulatory role.[59] The methyl jasmonate-induced CYP80B1 of *E. californica* was recently cloned by a direct, genetic approach, and it was demonstrated to be the (S)-*N*-methylcoclaurine 3'-hydroxylase involved in an early step of macarpine biosynthesis (Fig. 4). CYP80B1 shows a high degree of stereo-and regio-selectivity.[33]

P450s play an important role in the biosynthesis of monoterpenoid indole alkaloids. These alkaloids are found in *e.g. Catharanthus roseus,* from which the powerful cytotoxic drugs vinblastine and vincristine are isolated.[60] Vinblastine and vincristine are dimers formed *in vivo* by condensation of catharanthine and vindoline. The biosynthesis of vindoline from tabersonine has been suggested to include three enzymatic hydroxylations. The enzyme that converts tabersonine to 16-hydroxytabersonine was shown to be a P450 monooxygenase.[61] It was recently cloned and identified as CYP71D12.[62] The second to the last step in vindoline biosynthesis is catalyzed by a 2-oxoglutarate-dependent dioxygenase.[63] The third oxygenation is a yet uncharacterized hydration. Within one pathway a requirement for both membrane-bound and soluble enzymes has evolved, several classes of oxygenases are implicated, and only the first and the last two steps are regulated by light.[61]

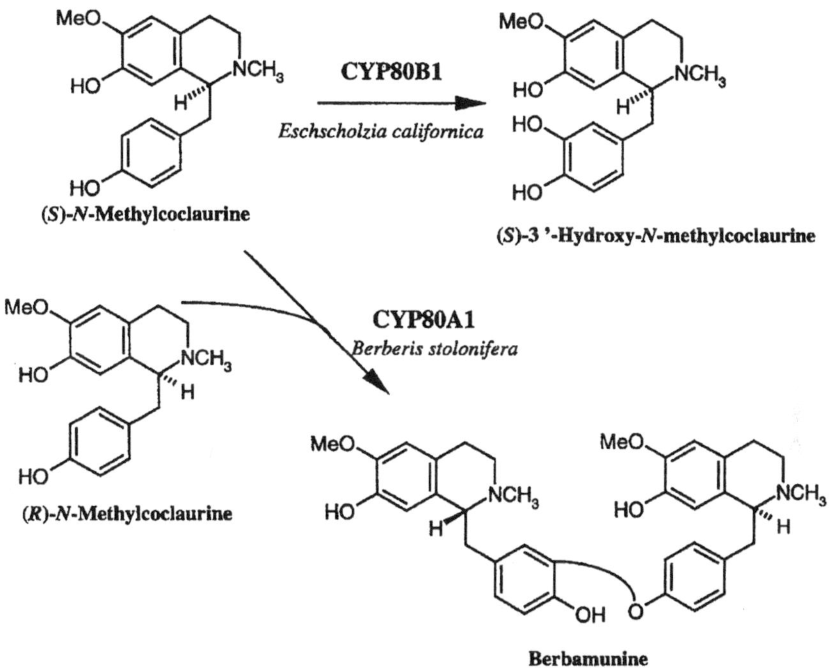

Figure 4. Reactions catalyzed by CYP80.
CYP80B1 is the *(S)-N*-methylcoclaurine 3'-hydroxylase involved in the biosynthesis of macarpine in *Eschscholzia californica.* CYP80A1 from *Berberis stolonifera* catalyzes the C-O-coupling of two enantiomeric *N*-methylcoclaurines leading to the formation of the bisbenzylisoquinoline alkaloid berbamunine.

Methylenedioxy bridge enzymes. (S)-Stylopine is a central intermediate in the biosynthesis of benzylisoquinoline alkaloids in higher plants. Stylopine contains two methylenedioxy groups in rings A and D. Formation of a methylenedioxy group is initiated *via* hydroxylation of an *ortho*-methoxyphenol, resulting in the production of a hemiformal that *via* either a radical or cationic mechanism eliminates water during or prior to ring closure (Bauer and Zenk, and references therein[64]). The formation of the two methylenedioxy bridges is catalyzed by two independent P450s, the (S)-cheilanthifoline synthase, and the (S)-stylopine synthase.[64,65] Similarly, it was demonstrated that the methylenedioxy bridge formation of berberine is catalyzed by a specific P450 designated (S)-canadine synthase.[66]

C-O coupling and C-C coupling. Several phenol coupling reactions in alkaloid biosynthesis have been shown to involve highly specific P450s. The oxidation of phenols by one-electron transfer involves phenolic radicals, which by radical pairing form new C-C or C-O bonds, either intra- or inter-molecularly. The catalyst behind the mechanism remained obscure for a long period.[67]

In microsomes prepared from immature capsules of *Papaver somniferum*, a P450 was demonstrated to catalyze the C-C coupling that transforms (R)-reticuline into salutaridine.[67] Salutaridine is a precursor of thebaine, codeine, and morphine. Similarly, a P450 catalyzes the C-C coupling in the formation of isoandrocymbine from autumnaline in the biosynthesis of colchicine in *Colchicum autumnale*.[68] A second P450 in this pathway was shown to be involved in a peculiar ring-expansion reaction of *O*-methylandrocymbine to demecolcine.[69]

The C-O coupling of two enantiomeric *N*-methylcoclaurines leading to the formation of the bisbenzylisoquinoline alkaloid berbamunine is catalyzed by a P450 berbamunine synthase (Fig. 4).[67] Berbamunine synthase was purified to homogeneity from microsomes of *Berberis stolonifera* cell cultures, and the oxidative coupling reaction was shown to proceed in a highly regio- and stereo-specific manner.[24] The cDNA encoding berbamunine synthase was isolated (*CYP80A1*), and heterologously expressed.[12,70]

Phenylpropanoids

The phenylpropanoids are derived from the shikimate pathway, present in plants and micro-organisms but not in animals. The phenylpropanoid pathway leads to the synthesis of a large variety of compounds based on C6-C3 units. Cinnamic acids, lignin monomers, coumarins, flavonoids, isoflavonoids, and pterocarpans are all compounds derived from this pathway, and they seem to have evolved as protection against biotic and abiotic factors, such as plant pests and UV light (for reviews see Werck-Reichhart, and Dixon and Paiva[71,72]). The first step in the phenylpropanoid pathway is the conversion of phenylalanine to cinnamic acid, catalyzed by phenylalanine ammonia-lyase (PAL). This is followed by the hydroxylation of *trans*-cinnamic acid to *p*-coumaric acid, catalyzed by the ubiquitous cinnamic acid hydroxylase (C4H), designated CYP73. *p*-Coumaric acid is subsequently activated to its CoA thioester 4-coumaroyl CoA by the 4-coumarate:CoA ligase (4CL).

C4H. The cinnamic acid hydroxylase catalyzes a vital step in plant secondary metabolism (Fig. 5) and is present throughout the plant kingdom. Recent data suggest the existence of two different classes of C4H in monocots (maize) and in dicots (*Phaseolus vulgaris*).[73] Genes encoding both classes within one species so far are only available in maize.[74]

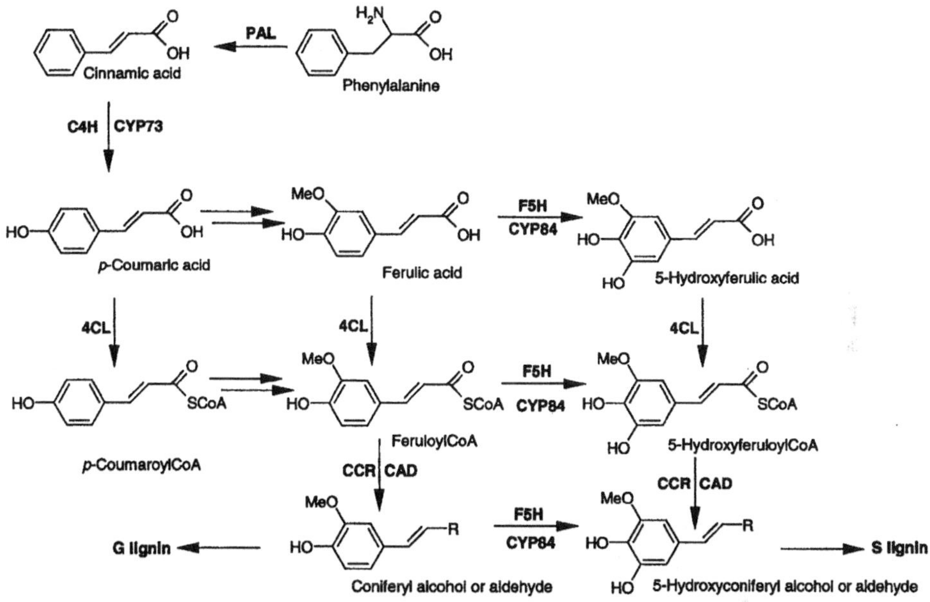

Figure 5. Phenylpropanoid metabolism.
Phenylalanine is converted to cinnamic acid by the phenylalanine ammonia-lyase (PAL). Cinnamic acid is subsequently hydroxylated by the cinnamic acid 4-hydroxylase (C4H, CYP73) yielding *p*-coumaric acid. *p*-Coumaric is activated to its CoA thioester *p*-coumaroyl CoA by the 4-coumarate:CoA ligase (4CL). CYP84, the ferulate 5-hydroxylase (F5H), is involved in the biosynthesis of both guaiacyl and sinapyl lignin monomers.

Both the substrate and the product of C4H have been implicated in the regulation of upstream and downstream enzymes of the phenylpropanoid pathway.[71,72] CYP73A1, the C4H of *Helianthus tuberosus*, was shown to be induced in response to wounding and chemical treatment, and the induction was regulated at the transcriptional level.[75] *A. thaliana* C4H was shown to be expressed coordinately with other enzymes of the phenylpropanoid pathway, PAL1 and 4CL. Analysis of the 5'promoter region of the *C4H* gene revealed the presence of three putative regulatory motifs also found in *PAL* and *4CL* genes.[30,31] The expression of parsley C4H (CYP73A10) is strictly co-regulated with PAL at both the mRNA and protein level. Distribution patterns of mRNA and protein were identical for C4H and PAL, as shown in cross-sections from different parsley tissues. It has been proposed that the genes encoding the core reactions of phenylpropanoid metabolism form a tight regulatory unit. C4H could serve in a multienzyme complex as a structural and metabolic anchor of the cytosolic activities to the ER.[76]

The substrate specificity of *H. tuberosus* C4H was investigated, and certain small planar molecules with dipolar character were metabolized by CYP73A1, however, with low efficiency.[77] 2-Naphthoic acid (2-NA) and its analogues were shown to be excellent substrates with high turnovers, however, no alternative physiological substrates were identified.[78] Piperonylic acid (PA) is a natural compound, which was identified as an efficient mechanism based inhibitor of CYP73A1, *in vitro* as well as *in vivo*. PA is a useful tool for investigation of phenylpropanoid metabolism and can be used to block the synthesis of *p*-coumaric acid. This leads to an accumulation of cinnamic acid in the plant, which, in turn, may cause a shift in the precursor flow leading to the formation of C6-C1 compounds, such as benzoic and salicylic acids.[79]

F5H. In the lignin biosynthetic branch of the phenylpropanoid pathway (Fig. 5), *p*-coumaric acid is converted to caffeic acid, and subsequently to ferulic acid. Ferulate-5-hydroxylase (F5H, CYP84) catalyzes the hydroxylation of ferulate to 5-hydroxyferulate, and, only recently, alternative substrates of F5H were demonstrated. F5H was expressed in yeast, and coniferaldehyde and coniferyl alcohol were shown to have substantially lower Km´s than ferulate. This result implies that our whole understanding of lignin biosynthesis needs revision, and that ferulate may not be relevant as a substrate *in vivo*.[80] Regulation of F5H has been studied in wild-type as well as in transgenic *A. thaliana* expressing F5H under the control of either the 35S or the C4H promoter. The 35S promoter failed to promote high levels of *F5H* gene expression, whereas the C4H promoter was far more effective. A lignin composed almost entirely of syringyl units was synthesized throughout the plant when F5H was expressed under the control of the C4H promoter. This implied that F5H catalyzes a rate-limiting step in syringyl lignin biosynthesis.[81] Transgenic tobacco over-expressing either PAL (from bean) or C4H (from alfalfa) showed no change in lignin composition, and it was concluded that the control points in the lignin biosynthetic pathway must be downstream of PAL and C4H.[82]

The recent results on F5H substrate choice indicate that instead of involving sinapic acid as an intermediate, the biosynthesis of syringyl lignin monomers most likely proceeds by hydroxylation or methylation of coniferaldehyde and /or coniferyl alcohol.[80]

Flavonoids. A wide range of enzymes catalyzing steps of flavonoid biosynthesis have been identified, including cytochrome P450s (Fig. 6). Molecular approaches have led to the cloning of some of these P450s, notably the flavonoid 3',5'-hydroxylase from petunia (CYP75A1);[25] the licodione synthase (CYP93B1),[83] the isoflavone 2′ hydroxylase (CYP81E1),[84] and the isoflavone synthase (CYP93C2),[85] all from licorice; the pterocarpan 6-hydroxylase (CYP93A1),[35-37] and the isoflavone synthase (CYP93C1), both from soybean.[86]

Licodione synthase (LS) is involved in the conversion of the flavanone (2S)-liquiritigenin to licodione. Induced cell cultures of the licorice plant (*Glycyrrhiza echinata*) were used for microsome preparation, and it was demonstrated that a P450 catalyzes the conversion of (2S)-liquiritigenin to licodione. Licodione synthase, thus, hydroxylates the C-2 of liquiritigenin yielding 2-hydroxyliquiritigenin, which spontaneously undergoes a hemiacetal opening to give licodione. When the same microsomes were tested with (2S)-naringenin as substrate, the flavanone was converted to 2-hydroxynaringenin, which after a subsequent dehydration yields a flavone. This process is catalyzed by an enzyme designated flavone synthase II (FNSII). In both cases, the flavanone skeleton is hydroxylated at C-2.[87] Several P450 cDNAs were isolated from an elicited licorice cell culture library by using a PCR approach based on conserved P450 sequences.[88] One of these, designated CYP93B1, was expressed in insect cells and yeast, and shown to catalyze the formation of licodione from (2S)-liquiritigenin, and of 2-hydroxynaringenine from (2S)-naringenin.[89] *CYP93B1,* thus, encodes both (2S)-flavanone 2-hydroxylase, designated licodione synthase (LS), and flavone synthase II (FNSII), depending on the substrates employed. Furthermore, CYP93B1 was shown to metabolize three different hydroxylated flavanones, namely liquiritigenin, naringenin, and eriodictyol.[83]

The isoflavonoids form a distinct subclass of flavonoid compounds and are almost entirely restricted to the Leguminosae. They represent structural flavonoid variants in which the shikimate-derived aromatic ring has migrated to the adjacent carbon of the heterocycle. The rearrangement process is P450 catalyzed and is quite rare in nature. The mechanism behind this aryl migration is still not well understood.[60]

The rearrangement of flavanone to isoflavone in the conversion of (2S)-naringenin to genistein was investigated in elicitor treated soybean cell suspension cultures. It was proposed that the conversion proceeds in two steps involving the formation of a 2-hydroxyisoflavanone, which is subsequently dehydrated to genistein. The rearrangement process was shown to be catalyzed by a P450 designated isoflavone synthase (IFS).[90] Recently, the cloning of two P450s encoding IFS has been reported. One is the IFS from soybean (CYP93C1),[86] the other is the IFS from licorice.[85] Since IFS and LS basically share the same substrates and seem to be encoded by members of the same family, this opens up the possibility of interesting structure/function comparisons. Furthermore, the mechanism behind the aryl migration will be easier to study once a heterologous expression system is in place.

Terpenoids

The terpenoids form a large and structurally diverse family of natural products derived from C_5 isoprene units. This family contains compounds that play an important role in the interaction of plants with their surroundings, such as hormones (gibberellic acids, abscisic acid), volatile oils, phytosterols, carotenoids, and phytoalexins.[91] Many natural products contain terpenoid elements in their molecules (*e.g.* indole alkaloids) adding to the diversity. P450s play an essential role in the metabolism of terpenoids. Unfortunately, we still have little information concerning the genes encoding these P450s.

Monoterpenes. The geraniol/nerol 10-hydroxylase of *Catharanthus roseus* was among the first plant P450s to be characterized.[92,93] The conversion of geraniol to its 10-hydroxy derivative represents an initial step in monoterpenoid indole alkaloid biosynthesis. The geraniol 10-hydroxylase was purified from seedlings and cell cultures, and the activity reconstituted.[93,94] The first plant P450 to be cloned was the CYP71A1 of avocado.[95] It was initially believed that CYP71A1 was a geraniol/nerol 10-hydroxylase.[96] This was rectified when it was shown that CYP71A1 metabolizes nerol to two epoxides (2,3- and 6,7-epoxide).[97] The physiological role of avocado CYP71A1 remains unknown. Nerol 10-hydroxylase activities have been measured in catmint (*Nepeta racemosa*),[97] from which CYP71A5 and CYP71A6 were cloned. The expression pattern of CYP71A5 was consistent with that expected for a gene encoding geraniol 10-hydroxylase, but heterologous expression is required to elucidate the function of CYP71A5.[98] Recently, geraniol 10-hydroxylase was purified, cloned, and functionally characterized as CYP76B6.[99] Monoterpene hydroxylases of several mint species have been thoroughly characterized (for review see Lupien et al.[100]). The hydroxylation of (-)-limonene was investigated in microsomal preparations of glandular trichomes of three species: *Mentha piperita*, *Mentha spicata*, and *Perilla frutescens*. The hydroxylation of (-)-limonene led to the formation of the

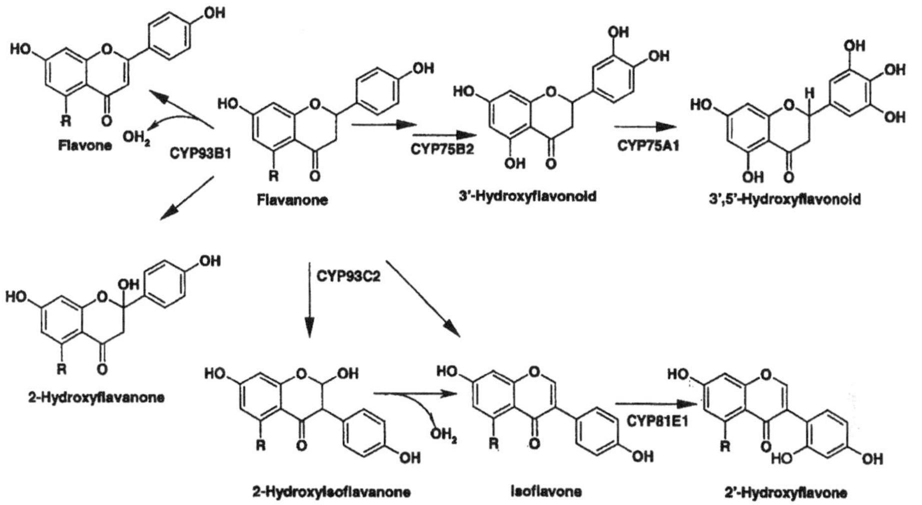

Figure 6. P450s involved in flavonoid and isoflavonoid biosynthesis.
CYP93B1 from licorice hydroxylates the flavonone skeleton at the C-2 position. Depending on the flavonone, the hydroxylation is followed either by a dehydration yielding a flavone, or by a hemiacetal opening of the 2-hydroxyflavonone. CYP93C1 from soybean and CYP93C2 from licorice catalyze the rearrangement of flavanone to isoflavone. CYP81E1 is the isoflavone 2'-hydroxylase from licorice. F3'H (CYP75B2) and F3'5'H (CYP75A1) from petunia are involved in the production of colored anthocyanins.

corresponding C-3, C-6, and C-7 alcohols, respectively. Only a single product was generated from (-)-limonene in each species, and the reactions were demonstrated to be catalyzed by highly substrate-specific and regio-selective P450s. A large family of similar, but clearly distinct, P450s is expected to participate in monoterpene metabolism.[101] The limonene-6-hydroxylase of *M. spicata* has been cloned and designated CYP71D18, together with the limonene-3-hydroxylase of *M. piperita*, designated CYP71D13 (Fig. 7).[102]

The availability of these sequences should help in the understanding of how such intricate regio-selectivities have evolved. The biosynthesis of carvone in the fruit of caraway (*Carum carvi*) involves the hydroxylation of (+)-limonene at the C-6 position, yielding (+)-*trans*-carveol. This reaction is catalyzed by a P450 designated (+)-limonene-6-hydroxylase, and it seems that the hydroxylation of limonene to *trans*-carveol represents a rate-limiting step in carvone biosynthesis.[103]

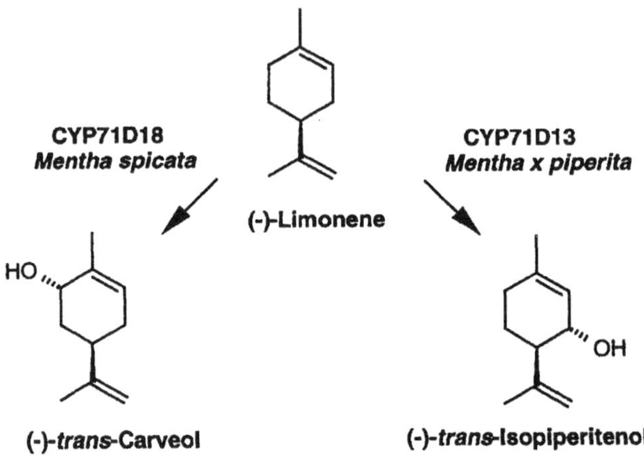

Figure 7. The regiospecific limonene hydroxylases from mint.
CYP71D13 is the (-)-limonene-3-hydroxylase from peppermint (*Mentha x piperita*), and CYP71D18 is the (-)-limonene-6-hydroxylase from spearmint (*Mentha spicata*).

Terpenoid derived phytohormones. The gibberellins (GAs) are kaurene derived phytohormones required for shoot elongation in higher plants. The overall transformation of *ent*-kaurene through *ent*-kaurenoic acid to the gibberellins is highly oxidative in nature. The first step in the metabolism of *ent*-kaurene is the hydroxylation of the C-19 methyl to give *ent*-kaurenol. Next follows the conversion of *ent*-kaurenol to *ent*-kaurenal, which is hydroxylated to yield *ent*-kaurenoic acid. Murphy and West demonstrated the involvement of P450s in the first and the last of these steps, and proposed the participation of a P450 in the conversion of *ent*-kaurenol to *ent*-kaurenal.[104] Recent data suggest that the entire conversion from *ent*-kaurene to *ent*-kaurenoic acid could be catalyzed by a single P450. The pea *lh* mutants are blocked in the oxidation of *ent*-kaurene, *ent*-kaurenol, and *ent*-kaurenal, suggesting that *ent*-kaurene oxidase is multifunctional and catalyzes three steps in the GA biosynthetic pathway.[105] A gene encoding *ent*-kaurene oxidase was isolated by the use of the *ga3* mutant of *A. thaliana* and designated CYP701A3.[48] A P450 involved in an early step in the GA proper part of the pathway has been isolated. As described above, CYP88A1 was isolated by using the *d3* mutant of maize. *CYP88A1* is thought to encode an early 13-hydroxylase activity, namely the conversion of GA_{12} to GA_{53}.[50]

Brassinosteroids (BRs) are a group of naturally occurring polyhydroxysteroids involved in different developmental events throughout the life cycle of plants. BRs frequently induce responses similar to those induced by the main classes of growth-promoting hormones, but are to be regarded as a distinct group of plant hormones.[106-108] The precursors of BRs are plant sterols, which are synthesized from cycloartenol through a series of reactions, including the 14alpha-demethylation of obtusifoliol catalyzed by a P450. This P450 was purified from etiolated sorghum seedlings, cloned, and designated CYP51, and shown to be orthologous to the CYP51s of mammals and fungi.[23,109]

Brassinolide (BL) is the most active BR. The proposed biosynthetic pathway predicts that at least 20 genes are involved in the conversion of squalene to brassinolide. The BR-specific part of the pathway diverges into the early and the late C-6 oxidation pathways.[110] Within the BR pathway, several reactions involve P450s (Fig. 8).

The mutants *cpd* (CYP90A1), *dwf4* (CYP90B1), and *dwarf* (CYP85), mentioned earlier in this chapter, correspond to mutations in three BR specific P450s. Using feeding and complementation experiments, CYP90B1 and CYP90A1 were predicted to catalyze early steps in BL biosynthesis, namely 22alpha-hydroxylation of 6-oxo campestanol to cathasterone, and 23alpha-hydroxylation of cathasterone to teasterone, respectively.[43,44] Transcription of the *CPD* gene is negatively controlled by brassinosteroids, and brassinolide, as well as several side-chain hydroxylated BRs, is efficient in negative feedback of the *CPD* gene.[111]

The *dwarf* mutation (CYP85) represents a late step in BL biosynthesis. CYP85 catalyzes the C-6 oxidation of 6-deoxocastasterone to castasterone in the late C-6 oxidation pathway, as confirmed by functional expression in yeast.[51,52] Thorough analysis of the endogenous levels of BR intermediates in tomato suggested that the late C-6 oxidation pathway may be the major pathway leading to BR production in tomato.[52] This is unlike *A. thaliana* that has both the early and late C-6 oxidation pathways. In all cases, the P450s seem to represent key regulatory steps in BR biosynthesis.

Fatty Acids

Fatty acid hydroxylases. Cytochrome P450-dependent monooxygenases from plants catalyze in-chain- and omega-hydroxylation, as well as epoxidation of medium- and long-chain fatty acids. Recent research efforts have clarified that there are multiple forms of cytochrome P450 involved in these reactions, each of which possesses distinguishable substrate specificity (for a review see Salaün and Helvig[112]). The biological roles of these distinct P450 forms are still poorly understood. However, evidence suggests that some may play an important role in the biosynthesis of plant cuticles[113,114] and fatty acid derived signals.[115,116] The P450-dependent fatty acid hydroxylases could also be considered as a means of preventing the accumulation of significantly high, possibly toxic, concentrations of free FAs in the plant cells that are liberated by the action of phospholipases in the early response to different kinds of stress. The ω–hydroxylated long-chain fatty acids generated could be converted into dicarboxylic acid derivatives that then can be shortened by the peroxisomal and the mitochondrial ß-oxidation pathways.

Early work in this laboratory has indicated that plants possess distinct P450s that hydroxylate fatty acids either at the methyl terminus (ω position), at (ω-1), or at more in-chain carbon positions.[112] The in-chain hydroxylase, producing 8-, 9- and 10-hydroxyfatty acids from capric (C10), lauric (C12), and myristic (C14)acids, was first described in *Helianthus tuberosus* and subsequently identified in microsomes from many plant species.[117] This enzyme has been cloned and designated CYP81B1. Functional expression in yeast confirmed that all these reactions are supported by this single P450 protein.[118] Interestingly, P450s from the CYP81E subfamily were identified as isoflavone hydroxylases.[84] This, and the fact that short chain fatty acids are minor components in most plants, suggests that despite the very low Km measured for these compounds,[118] fatty acids may not be the physiological substrate of this enzyme. A predominant (ω-1)-hydroxylase activity, capable of hydroxylating C12 to C18 saturated and unsaturated fatty acids as well as herbicides (see below), was characterized in wheat microsomes.[119] Because of their suggested involvement in cutin and suberin synthesis and their possible role in the formation of fatty acid derived signals, the ω-hydroxylases have been investigated in detail.[120] Early on it was recognized that these enzymes hydroxylate saturated and unsatured fatty acids ranging from C10 to C18 and that they are highly inducible by compounds collectively designated as peroxisomal proliferators.[121]

The first ω-hydroxylases have been cloned, founding two new families of plant P450s, CYP86 and CYP94.[42,122] The number of these P450s in a single plant is surprisingly large. At least 8 more CYP86s, the functions of which remain to be determined, are present in *A. thaliana*. We further cloned three members from the CYP94 family in *Vicia sativa* and 5 from *Nicotiana tabacum*. Functional expression and regulation studies show that these fatty acid hydroxylases have distinct substrate specificity and regulation patterns. For example, CYP94A1 that metabolizes substrates ranging from C10 to C18:3 catalyzes predominantly (>95%) hydroxylation at the ω position. It is rapidly induced by clofibrate and jasmonate.[122,123] In contrast, CYP94A2, from the same plant, is clearly a myristate hydroxylase and is barely induced by jasmonate or peroxisomal proliferators.[124]

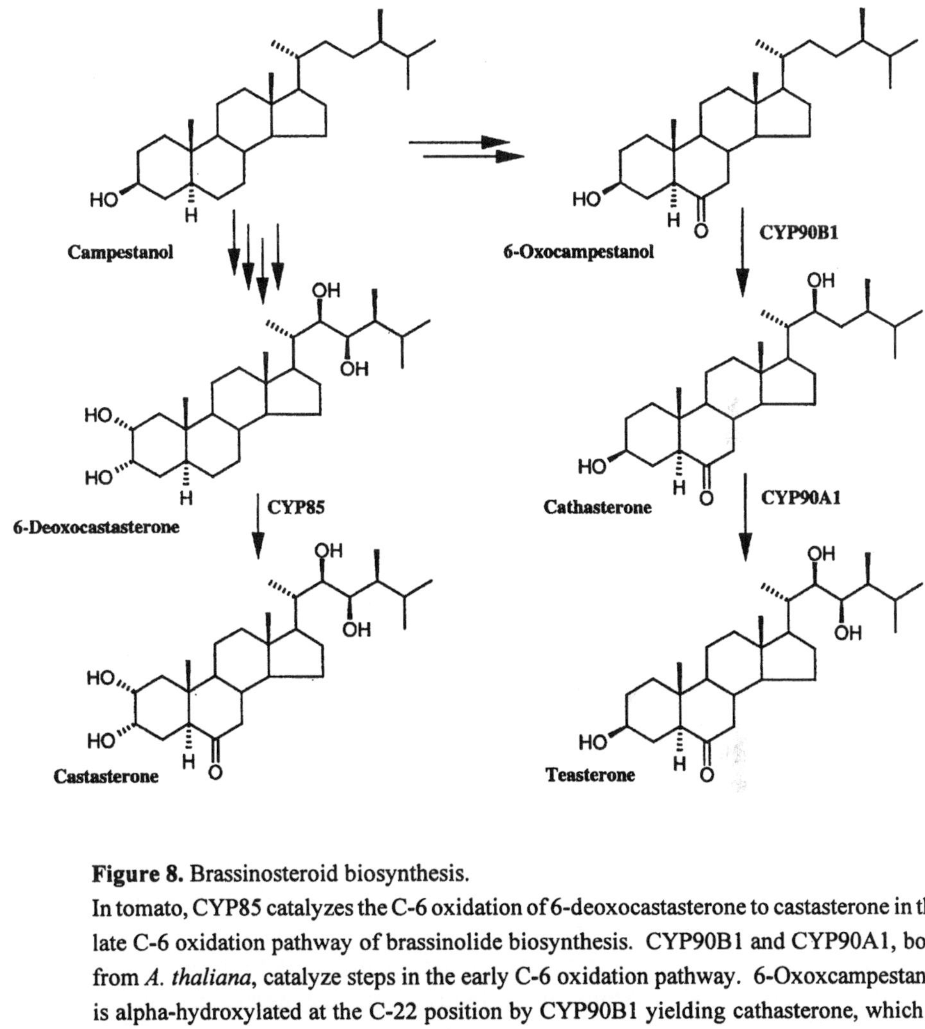

Figure 8. Brassinosteroid biosynthesis.
In tomato, CYP85 catalyzes the C-6 oxidation of 6-deoxocastasterone to castasterone in the late C-6 oxidation pathway of brassinolide biosynthesis. CYP90B1 and CYP90A1, both from *A. thaliana*, catalyze steps in the early C-6 oxidation pathway. 6-Oxoxcampestanol is alpha-hydroxylated at the C-22 position by CYP90B1 yielding cathasterone, which is subsequently hydroxylated at the C-23 position by CYP90A1 to yield teasterone. Multiple arrows indicate more than one enzymatic step.

Fatty acid hydroperoxide lyase and dehydratase. As already described, fatty acid hydroperoxides are substrates of P450s from family CYP74. The allene oxide synthase (CYP74A) was first cloned from flaxseeds and guayule.[125,126] This P450 catalyzes the first committed step in the biosynthesis of jasmonic acid, 12-oxo-PDA, and short-chain aldehydes. CYP74B, encoding the fatty acid hydroperoxide lyase, was isolated and cloned from bell pepper.[40] Both enzymes were subsequently identified in *Arabidopsis* and play key roles in the oxylipin cascade involved in plant defense reactions (for a recent review see Blée[127]).

Other Natural Compounds

In the biosynthesis of cyanogenic glucosides, two multifunctional P450s, CYP79A1 and CYP71E1, are involved in the conversion of amino acids to cyanogenic glucoside.[128,28] This pathway was elucidated by using *Sorghum bicolor* seedlings. The chapter by Möller in the present volume is devoted to the biosynthesis of cyanogenic glucosides and glucosinolates.

METABOLISM OF XENOBIOTICS

In many cases, the first committed step in the metabolism of a xenobiotic molecule is an oxidation reaction. Although other enzymes like peroxidases or phenolases appear capable of oxidizing chemicals, in the vast majority of cases where the nature of the enzyme has been studied, the involvement of a P450 enzyme has been demonstrated. Cytochrome P450 catalyzed reactions are both irreversible and slow. Typical turn-over is under 100 mole product/mole P450 per minute, with many xenobiotic oxidations well below this level. As a consequence, the P450-catalyzed step is likely to determine the overall pharmacokinetics of an exogenous compound. Most studies have been concerned with herbicide metabolism in crop plants or in weeds. Alkyl and aryl hydroxylations, *N*- and *O*-dealkylations of foreign molecules increase their polarity, allow their further conjugation and storage in the cell-wall or in the vacuole, and, in the case of herbicides, limit their activity at the target sites. It is now well- recognized that P450s constitute a major factor of resistance of crops to several classes of herbicides,[129] and also play a central role in the occurrence of multiple- or cross-resistant weeds.[130] As other P450 activities in plants, xenobiotic metabolism is highly inducible by environmental factors and by agrochemicals, including safeners.[131] Safeners are chemicals used to increase the tolerance of crop plants towards herbicides. It is probable that metabolism of xenobiotics is catalyzed by "physiological" P450s that happen per chance to accommodate one or several xenobiotics in their active site and are able to transfer oxygen to them. Despite the numerous reports of xenobiotic metabolism by microsomes of diverse plant origin, and clear demonstration of P450-dependence, there are few cases when such an activity has been linked to a specific enzyme with a known physiological function. In wheat, it has been shown that the fatty acid ($\omega-1$)–hydroxylase catalyzes diclofop hydroxylation[132] and constitutes the main factor of resistance of this plant to this herbicide,[133] but not to phenylureas, thus demonstrating that phenylureas and aryloxyphenoxypropionates are metabolized by distinct P450 isoforms in wheat.[134]

At the molecular level, CYP76B1, a P450 induced by wounding and xenobiotics in tuber tissues from *Helianthus tuberosus*, catalyzed actively the N-demethylation of phenylurea herbicides.[135] Interestingly, CYP76B6, a closely related P450 cloned from *Catharanthus roseus*, catalyzes the 10-hydroxylation of geraniol and nerol, two monoterpenols without structural relationship to phenylureas, suggesting that it will be difficult to predict xenobiotic metabolizing capacity from knowing the physiological function of a P450, and *vice versa*. Another interesting finding is that CYP71A10, a P450 from soybean, which is less than 40% identical to CYP76B1, catalyzes the same set of reactions on this class of herbicides, but probably has a different physiological function from that of CYP76B1.[136]

CONCLUSION

The number of distinct P450s in plants is exceptionally large. In *A. thaliana*, the estimated number is 285-290. Some plants, with larger or amphiploid genomes, may have substantially more. In sharp contrast, *Caenorhabditis elegans* has 81 P450s. The estimated number in humans is 49 + 15 pseudogenes. Yeast has only three. The remarkable diversification of this gene superfamily in higher plants appears to be linked to the crucial roles of P450s at the interface between plants and their environment.

ACKNOWLEDGMENTS

Rachel Alice Kahn acknowledges a postdoctoral fellowship from the Carlsberg Foundation, Denmark.

REFERENCES

1. COON, M.J., VAZ, A.D.N., BESTERVELT, L.L., 1996. Cytochrome P450. 2. Peroxidative reactions of diversozymes. FASEB. J. 10:428-434.
2. PETERSON, J. A., GRAHAM-LORENCE, S. E. 1995. Bacterial P450s. Structural similarities and functional differences. pp. 151-180 *in* Cytochrome P450. Structure, Mechanism, and Biochemistry, 2nd Edition (Ortiz de Montellano, P.R. ed.), Plenum Press, New York.
3. BOLWELL, G.P., BOZAK, K., ZIMMERLIN, A., 1994. Plant cytochrome P450. Phytochemistry 37:1491-1506.

4. DURST, F., NELSON, D.R., 1995. Diversity and evolution of plant P450 and P450 reductases. Drug Metab. Drug Interact. 12:189-206.
5. DURST, F., O'KEEFE, D.P., 1995. Plant cytochromes P450: An overview. Drug Metab. Drug Interact. 12:171-188.
6. SCHULER, M.A., 1996. Plant cytochrome P450 monooxygenases. Crit. Rev. Plant. Sci. 15:235-284.
7. CHAPPLE, C. 1998. Molecular-genetic analysis of plant cytochrome P450-dependent monooxygenases. Annu. Rev. Plant Physiol. 49:311-343.
8. HALKIER, B.A., 1996. Catalytic reactivities and structure/function relationships of cytochrome P450 enzymes. Phytochemistry. 43:1-21.
9. BENVENISTE, I., LESOT, A., HASENFRATZ, M.-P., KOCHS, G., DURST, F., 1991. Multiple forms of NADPH-cytochrome P450 reductase in higher plants. Biochem. Biophys. Res. Commun. 177:105-112.
10. SHET, M.S., SATHASIVAN, K., ARLOTTO, M.A., MEHDY, M.C., ESTABROOK, R.W., 1993. Purification, characterization, and cDNA cloning of an NADPH-cytochrome-P450 reductase from mung bean. Proc. Natl. Acad. Sci. USA. 90:2890-2894.
11. LOUERAT-ORIOU, B., PERRET, A., POMPON, D., 1998. Differential redox and electron-transfer properties of purified yeast, plant and human NADPH-cytochrome P-450 reductases highly modulate cytochrome-p-450 activities. Eur. J. Biochem. 258:1040-1049.
12. ROSCO, A., PAULI, H.H., PRIESNER, W., KUTCHAN, T.M. 1997. Cloning and heterologous expression of NADPH-cytochrome P450 reductases from the papaveraceae. Arch. Biochem. Biophys. 348:369-377.
13. KOOPMANN, E., HAHLBROCK, K., 1997. Differentially regulated NADPH:cytochrome P450 oxidoreductases in parsley. Proc. Natl. Acad. Sci. USA. 94:14954-14959.
14. SONG, W.C., BRASH, A.R., 1991. Purification of an allene oxide synthase and identification of the enzyme as a cytochrome-P450. Science 253:781-784.
15. MATSUI, K., WILKINSON, J., HIATT, B., KNAUF, V., KAJIWARA, T., 1999. Molecular-cloning and expression of *Arabidopsis* fatty-acid hydroperoxide lyase. Plant Cell. Physiol. 40:477-481.
16. HASEMANN, C.A., KURUMBAIL, R.G., BODDUPALLI, S.S., PETERSON, J.A., DEISENHOFER, J., 1995. Structure and function of cytochromes-P450 - A comparative-analysis of 3 crystal-structures. Structure 3:41-62.
17. MASON, H.S. 1968. Homology among oxidases. pp. 69-91 *in* Homologous Enzymes and Biochemical Evolution (N.V. Thoai and J. Roche, eds.), Gordon & Breach, New York.

18. OBAYASHI, E., TAKAHASHI, S., SHIRO, Y., 1998. Electronic-structure of reaction intermediate of cytochrome P450Nor in its nitric-oxide reduction. J. Amer. Chem. Soc. 120:12964-12965.
19. NEBERT, D.W., FEYEREISEN, R. 1994. Evolutionary argument for a connection between drug metabolism and signal transduction. pp. 3-13 *in* Cytochrome P450. Biochemistry, Biophysics and Molecular Biology (M.C. Lechner, ed.), John Libbey Eurotext, Paris.
20. NELSON, D.R., 1999. Cytochrome P450 and the individuality of species. Arch Biochem Biophys. 369: 1-10.
21. KOCHS, G., GRISEBACH, H. 1989. Phytoalexin synthesis in soybean: Purification and reconstitution of cytochrome P450 3,9-dihydroxypterocarpan 6α-hydroxylase and separation from cytochrome P450 cinnamate 4-hydroxylase. Arch. Biochem. Biophys. 273: 543-553.
22. GABRIAC, B., WERCK-REICHHART, D., TEUTSCH, H., DURST, F. 1991. Purification and immunocharacterisation of a plant cytochrome P450: the cinnamic acid 4-hydroxylase. Arch. Biochem. Biophys. 288: 302-309.
23. KAHN, R. A., BAK, S., OLSEN, C. E., SVENDSEN, I., MØLLER, B. L., 1996. Isolation and reconstitution of the heme-thiolate protein obtusifoliol 14 alpha-demethylase from *Sorghum bicolor* (L.) Moench. J. Biol. Chem. 271: 32944-32950.
24. STADLER, R., ZENK, M. H. 1993. The purification and characterization of a unique cytochrome P-450 enzyme from *Berberis stolonifera* plant cell cultures. J. Biol. Chem. 268: 823-831.
25. HOLTON, T., BRUGLIERA, F., LESTER, D. R., TANAKA, Y., HYLAND, C. C., MENTING, J. G. T., LU., C-H., FARCY, E., STEVENSON, T. W., CORNISH, E. C. 1993. Cloning and expression of cytochrome P450 genes controlling flower colour. Nature 366: 276-279.
26. NIELSEN, K. M., PODIVINSKY, E. 1997. cDNA cloning and endogenous expression of a flavonoid 3'5'-hydroxylase from petals of lisianthus (*Eustoma grandiflorum*). Plant Science 129: 167-174.
27. MEIJER, A. H., SOUER, E., VERPOORTE, R., HOGE, J. H. C. 1993. Isolation of cytochrome P-450 cDNA clones from the higher plant *Catharanthus roseus* by a PCR strategy. Plant Mol. Biol. 22: 379-383.
28. BAK, S., KAHN, R. A., NIELSEN, H. L., MØLLER, B. L., HALKIER, B. A. 1998. Cloning of three A-type cytochromes P450, CYP71E1, CYP98, and CYP99 from *Sorghum bicolor* (L.) Moench by a PCR approach and identification by expression in *Escherichia coli* of CYP71E1 as a multifunctional cytochrome P450 in the biosynthesis of the cyanogenic glucoside dhurrin. Plant Mol. Biol. 36: 393-405.

29. MIZUTANI, M., WARD, E., OHTA, D. 1998. Cytochrome P450 superfamily in *Arabidopsis thaliana*: Isolation of cDNAs, differential expression, and RFLP mapping of multiple cytochromes P450. Plant Mol. Biol. 37: 39-52.
30. MIZUTANI, M., OHTA, D., SATO, R. 1997. Isolation of a cDNA and a genomic clone encoding cinnamate 4-hydroxylase from *Arabidopsis* and its expression manner in planta. Plant Physiol. 113: 755-763.
31. BELL-LELONG, D. A., CUSUMANO, J. C., MEYER, K., CHAPPLE, C. 1997. Cinnamate-4-hydroxylase expression in *Arabidopsis*. Plant Physiol. 113: 729-738.
32. FRANK, M. R., DEYNEKA, J. M., SCHULER, M. A. 1996. Cloning of wound-induced cytochrome P450 monooxygenases expressed in pea. Plant Physiol. 110: 1035-1046.
33. PAULI, H. H., KUTCHAN, T. M. 1998. Molecular cloning and functional heterologous expression of two alleles encoding *(S)-N*-methylcoclaurine 3'-hydroxylase (CYP80B1), a new methyl jasmonate-inducible cytochrome P-450-dependent mono-oxygenase of benzylisoquinoline alkaloid biosynthesis. Plant J. 13: 793-801.
34. LIANG, P., PARDEE, A. B. 1992. Differential display of eukaryotic messenger RNA by means of the polymerase chain reaction. Science 257: 967-971.
35. SCHOPFER, C., R., EBEL, J. 1998. Identification of elicitor-inducible cytochrome P450s of soybean (*Glycine max* L.) using differential display of mRNA. Mol. Gen. Genet. 258: 315-322.
36. SCHOPFER, C. R., KOCHS, G., LOTTSPEICH, F., EBEL, J. 1998. Molecular characterization and functional expression of dihydroxypterocarpan 6α-hydroxylase, an enzyme specific for pterocarpanoid phytoalexin biosynthesis in soybean (*Glycine max* L.). FEBS Lett. 432: 182-186.
37. SUZUKI, G., OHTA, H., KATO, T., IGARASHI, T., SAKAI, F., SHIBATA, D.. TAKANO, A., MASUDA, T., SHIOI, Y., TAKAMIYA, K-I. 1996. Induction of a novel cytochrome P450 (CYP93 family) by methyl jasmonate in soybean suspension-cultured cells. FEBS Lett. 383: 83-86.
38. NEWMAN, T., DE BRUIJN, F. J., GREEN, P., KEEGSTRA, K., KENDE, H., MCINTOSH, L., OHLROGGE, J., RAIKHEL, N., SOMERVILLE, S., THOMASHOW, M. 1994. Genes galore: A summary of methods for accessing results from large-scale partial sequencing of anonymous *Arabidopsis* cDNA clones. Plant Physiol. 106:1241-1255.
39. BAK, S., NIELSEN, H. L., HALKIER, B. A. 1998. The presence of CYP79 homologues in glucosinolate-producing plants shows evolutionary conservation of the enzymes in the conversion of amino acid to aldoxime in

the biosynthesis of cyanogenic glucosides and glucosinolates. Plant Mol. Biol. 38: 725-734.

40. MATSUI, K., SHIBUTANI, M., HASE, T., KAJIWARA, T. 1996. Bell pepper fruit fatty acid hydroperoxide lyase is a cytochrome P450 (CYP74B). FEBS Lett. 394: 21-24.
41. BATE, N., J., SIVASANKAR, S., MOXON, C., RILEY, J. M.C., THOMPSON, J. E., ROTHSTEIN, S. J. 1998. Molecular characterization of an *Arabidopsis* gene encoding hydroperoxide lyase, a cytochrome P-450 that is wound inducible. Plant Physiol. 117: 1393-1400.
42. BENVENISTE, I., TIJET, N., ADAS, F., PHILIPPS, G., SALAUN, J.P., DURST, F. 1998. CYP86A1 from *Arabidopsis thaliana* encodes a cytochrome P450-dependent fatty acid omega-hydroxylase. Biochem. Biophys. Res. Commun. 243: 688-693.
43. SZEKERES, M., NEMETH, K., KONCZ-KALMAN, Z., MATHUR, J., KAUSCHMANN, A., ALTMANN, T., REDEI, G. P., NAGY, F., SCHELL, J., KONCZ, C. 1996. Brassinosteroids rescue the deficiency of CYP90, a cytochrome P450, controlling cell elongation and de-etiolation in *Arabidopsis*. Cell 85: 171-182.
44. CHOE, S., DILKES, B. P., FUJIOKA, S., TAKATSUTO, S., SAKURAI, A., FELDMANN, K.A. 1998. The *DWF4* gene of *Arabidopsis* encodes a cytochrome P450 that mediates multiple 22alpha-hydroxylation steps in brassinosteroid biosynthesis. The Plant Cell 10: 231-243.
45. KIM, G-T., TSUKAYA, H., UCHIMIYA, H. 1998. The *ROTUNDIFOLIA3* gene of *Arabidopsis thaliana* encodes a new member of the cytochrome P-450 family that is required for the regulated polar elongation of leaf cells. Genes & Development 12: 2381-2391.
46. KIM, G-T., TSUKAYA, H., SAITO, Y., UCHIMIYA, H. 1999. Changes in the shapes of leaves and flowers upon overexpression of cytochrome P450 in *Arabidopsis*. Proc. Natl. Acad. Sci. USA. 96: 9433-9437.
47. MEYER, K., CUSUMANO, J., C., SOMERVILLE, C., CHAPPLE, C. C. S. 1996. Ferulate-5-hydroxylase from *Arabidopsis thaliana* defines a new family of cytochrome P450-dependent monooxygenases. Proc. Natl. Acad. Sci. USA 93: 6869-6874.
48. HELLIWELL, C., A., SHELDON, C. C., OLIVE, M. R., WALKER, A.R., ZEEVART, J. A. D., PEACOCK, W. J., DENNIS, E. S. 1998. Cloning of the *Arabidopsis ent*-kaurene oxidase gene *GA3*. Proc. Natl. Acad. Sci. USA 95: 9019-9024.
49. WALBOT, V. 1992. Strategies for mutagenesis and gene cloning using transposon tagging and T-DNA insertional mutagenesis. Annu. Rev. Plant

Physiol. Plant Mol. Biol. 43: 49-82.
50. WINKLER, R. G., HELENTJARIS, T. 1995. The maize *Dwarf3* gene encodes a cytochrome P450-mediated early step in gibberellin biosynthesis. The Plant Cell 7:1307-1317.
51. BISHOP, G. J., HARRISON, K., JONES, J. D. G. 1996. The tomato *Dwarf* gene isolated by heterologous transposon tagging encodes the first member of a new cytochrome P450 family. The Plant Cell 8: 959-969.
52. BISHOP, G. J., NOMURA, T., YOKOTA, T. HARRISON, K., NOGUCHI, T., FUJIOKA, S., TAKATSUTO, S., JONES, J. D. G., KAMIYA, Y., 1999. The tomato DWARF enzyme catalyzes C-6 oxidation in brassinosteroid biosynthesis. Proc. Natl. Acad. Sci. USA 96: 1761-1766.
53. FREY, M., CHOMET, P., GLAWISCHNIG, E., STETTNER, C., GRUN, S., WINKLMAIR, A., EISENREICH, W., BACHER, A., MEELEY, R. B., BRIGGS, S. P., SIMCOX, K., GIERL, A. 1997. Analysis of a chemical plant defence mechanism in grasses. Science 277: 696-699.
54. FREY, M., KLIEM, R., SAEDLER, H., GIERL, A. 1995. Expression of a cytochrome P450 gene family in maize. Mol. Gen. Genet. 246: 100-109.
55. GLAWISCHNIG, E., GRUN, S., FREY, M., GIERL, A. 1999. Cytochrome P450 monooxygenases of DIBOA biosynthesis: specificity and conservation among grasses. Phytochemistry 50: 925-930.
56. CHOU, W-M., KUTCHAN, T. M. 1998. Enzymatic oxidations in the biosynthesis of complex alkaloids. Plant J. 15: 289-300.
57. TANAHASHI, T., ZENK, M. H. 1988. One step enzymatic synthesis of dihydrosanguinarine from protopine. Tetrahedron Lett. 29: 5625-5628.
58. TANAHASHI, T., ZENK, M. H. 1990. Elicitor induction and characterization of microsomal protopine-6-hydroxylase, the central enzyme in benzophenanthridine alkaloid biosynthesis. Phytochemistry 29: 1113-1122.
59. KAMMERER, L., DE-EKNAMKUL, W., ZENK, M. H. 1994. Enzymic 12-hydroxylation and 12-*O*-methylation of dihydrochelirubine in dihydromacarpine formation by *Thalictrum bulgaricum*. Phytochemistry 36: 1409-1416.
60. DEWICK, P. M. 1997. Medicinal Natural Products: A Biosynthetic Approach. John Wiley, 466 p.
61. ST-PIERRE, B., DE LUCA, V. 1995. A cytochrome P-450 monooxygenase catalyzes the first step in the conversion of tabersonine to vindoline in *Catharanthus roseus*. Plant Physiol. 109: 131-139.
62. SCHROEDER, G., UNTERBUSCH, E., KALTENBACH, M., SCHMIDT, J., STRACK, D., DE LUCA, V., SCHROEDER, J. 1999. Light-induced cytochrome P450-dependent enzyme in indole alkaloid biosynthesis:

tabersonine 16-hydroxylase(1). FEBS Lett. 458: 97-102.
63. DE CAROLIS, M., CHAN, F., BALSEVICH, J., DE LUCA, V. 1990. Isolation and characterization of a 2-oxoglutarate dependent dioxygenase involved in the second-to-last step in vindoline biosynthesis. Plant Physiol. 94: 1323-1329.
64. BAUER, W., ZENK, M. H. 1991. Two methylenedioxy bridge forming cytochrome P-450 dependent enzymes are involved in (S)-stylopine biosynthesis. Phytochemistry 30: 2953-2961.
65. BAUER, W., ZENK, M. H. 1989. Formation of both methylenedioxy groups in the alkaloid (S)-stylopine is catalyzed by cytochrome P-450 enzymes. Tetrahedron Lett. 39: 5257-5260.
66. RUEFFER, M., ZENK, M. H. 1994. Canadine synthase from *Thalictrum tuberosum* cell cultures catalyzes the formation of the methylenedioxy bridge in berberine synthesis. Phytochemistry 36: 1219-1223.
67. ZENK, M.H., GERARDY, R., STADLER, R. 1989. Phenol oxidative coupling of benzylisoquinoline alkaloids is catalysed by regio- and stereo-selective cytochrome P-450 linked plant enzymes: Salutaridine and berbamunine. J. Chem. Soc. Chem. Commun. 1725 -1727.
68. NASREEN, A., RUEFFER, M., ZENK, M. H. 1996. Cytochrome P-450-dependent formation of isoandrocymbine from autumnaline in colchicine biosynthesis. Tetrahedron Lett. 37: 8161-8164.
69. RUEFFER, M., ZENK, M. H. 1998. Microsome-mediated transformation of *O*-methylandrocymbine to demecolcine and colchicine. FEBS Lett. 438: 111-113.
70. KRAUS, P. F. X., KUTCHAN, T. M. 1995. Molecular cloning and heterologous expression of a cDNA encoding berbamunine synthase, a C-O phenol-coupling cytochrome P450 from the higher plant *Berberis stolonifera*. Proc. Natl. Acad. Sci. USA. 92: 2071-2075.
71. WERCK-REICHHART, D. 1995. Cytochromes P450 in phenylpropanoid metabolism. Drug Metab. Drug Interact. 12: 221-243.
72. DIXON, R. A., PAIVA, N. L. 1995. Stress-induced phenylpropanoid metabolism. The Plant Cell 7: 1085-1097.
73. NEDELKINA, S., JUPE, S. C., BLEE, K. A., SCHALK, M., WERCK-REICHHART, D., BOLWELL, G. P. 1999. Novel characteristics and regulation of a divergent cinnamate 4-hydroxylase (CYP73A15) from French bean: Engineering expression in yeast. Plant Mol. Biol. 39: 1079-1090.
74. POTTER, S., MORELAND, D. E., KREUZ, K., WARD, E. 1995. Induction of cytochrome P450 genes by ethanol in maize. Drug Metab. Drug Interact. 12: 317-327.

75. BATARD, Y., SCHALK, M., PIERREL, M-A., ZIMMERLIN, A., DURST, F., WERCK-REICHHART, D. 1997. Regulation of the cinnamate 4-hydroxylase (CYP73A1) in Jerusalem artichoke tubers in response to wounding and chemical treatments. Plant Physiol. 113: 951-959.
76. KOOPMANN, E., LOGEMANN, E., HAHLBROCK, K. 1999. Regulation and functional expression of cinnamate 4-hydroxylase from parsley. Plant Physiol. 119: 49-55.
77. PIERREL, M. A., BATARD, Y., KAZMAIER, M., MIGNOTTE-VIEUX, C., DURST, F., WERCK-REICHHART, D. 1994. Catalytic properties of the plant cytochrome P450 CYP73 expressed in yeast. Substrate specificity of a cinnamate hydroxylase. Eur. J. Biochem. 224: 835-844.
78. SCHALK, M., BATARD, Y., SEYER, A., NEDELKINA, S., DURST, F., WERCK-REICHHART, D. 1997. Design of fluorescent substrates and potent inhibitors of CYP73As, P450s that catalyze 4-hydroxylation of cinnamic acid in higher plants. Biochemistry 36: 15253-15261.
79. SCHALK, M., CABELLO-HURTADO, F., PIERREL, M-A., ATANOSSOVA, R., SAINDRENAN, P., WERCK-REICHHART, D. 1998. Piperonylic acid, a selective, mechanism-based inactivator of the *trans*-cinnamate 4-hydroxylase: A new tool to control the flux of metabolites in the phenylpropanoid pathway. Plant Physiol. 118: 209-218.
80. HUMPHREYS, J. M., HEMM, M. R., CHAPPLE, C. 1999. New routes for lignin biosynthesis defined by biochemical characterization of recombinant ferulate 5-hydroxylase, a multifunctional cytochrome P450-dependent monooxygenase. Proc. Natl. Acad. Sci. USA 96: 10045-10050.
81. MEYER, K., SHIRLEY, A. M., CUSUMANO, J. C., BELL-LELONG, D. A., CHAPPLE, C. 1998. Lignin monomer composition is determined by the expression of a cytochrome P450-dependent monoxygenase in *Arabidopsis*. Proc. Natl. Acad. Sci. USA 95: 6619-6623.
82. SEWALT, V. J. H., NI, W., BLOUNT, J. W., JUNG, H. G., MASOUD, S. A., HOWLES, P. A., LAMB, C., DIXON, R. A. 1997. Reduced lignin content and altered lignin composition in transgenic tobacco down-regulated in expression of L-phenylalanine ammonia-lyase or cinnamate 4-hydroxylase. Plant Physiol. 115: 41-50.
83. AKASHI, T., AOKI, T., AYABE, S. 1998. Identification of a cytochrome P450 cDNA encoding (2*S*)-flavanone 2-hydroxylase of licorice (*Glycyrrhiza echinata* L.;Fabaceae) which represents licodione synthase and flavone synthase II. FEBS Lett. 431: 287-290.
84. AKASHI, T., AOKI, T., AYABE, S. 1998. CYP81E1, a cytochrome P450 cDNA of licorice (*Glycyrrhiza echinata* L.), encodes isoflavone 2'-

hydroxylase. Biochem. Biophys. Res. Commun. 251: 67-70.
85. AKASHI, T., SHIMADA, N., AOKI, T., AYABE, S. 1999. Molecular characterization of cytochrome P450s involved in flavonoid biosynthesis active in interactions of leguminous plants and environmental organisms. Proceedings, 11th International Conference on Cytochrome P450, Sendai, Japan.
86. STEELE, C. L., GIJZEN, M., QUTOB, D., DIXON, R., A. 1999. Molecular characterization of the enzyme catalyzing the aryl migration reaction of isoflavonoid biosynthesis in soybean. Arch. Biochem. Biophys. 367: 146-150.
87. OTANI, K., TAKAHASHI, T., FURUYA, T., AYABE, S. 1994. Licodione synthase, a cytochrome P450 monooxygenase catalyzing 2-hydroxylation of 5-deoxyflavanone, in cultured *Glycyrrhiza echinata* L. cells. Plant Physiol. 105: 1427-1432.
88. AKASHI, T., AOKI, T., TAKAHASHI, T., KAMEYA, N., NAKAMURA, I., AYABE, S. 1997. Cloning of cytochrome P450 cDNAs from cultured *Glycyrrhiza echinata* L. cells and their transcriptional activation by elicitor-treatment. Plant Sci. 126: 39-47.
89. AKASHI, T., AOKI, T., TAKAHASHI, T., KAMEYA, N., NAKAMURA, I., AYABE, S. 1997. Two new cytochrome P450 cDNAs (Accession Nos. AB001379 and AB001380) from elicitor-induced licorice (*Glycyrrhiza echinata* L.) cells. Plant Physiol. 115: 1288.
90. KOCHS, G., GRISEBACH, H. 1986. Enzymic synthesis of isoflavones. Eur. J. Biochem. 155: 311-318.
91. MCGARVEY, D., CROTEAU, R. 1995. Terpenoid metabolism. The Plant Cell 7: 1015-1026.
92. MEEHAN, T. D., COSCIA, C. J. 1973. Hydroxylation of geraniol and nerol by a monoxygenase from *Vinca rosea*. Biochem. Biophys. Res. Commun. 53: 1043-1048.
93. MADYASTHA, K. M., MEEHAN, T. D., COSCIA, C. J. 1976. Characterization of a cytochrome P-450 dependent monoterpene hydroxylase from the higher plant *Vinca rosea*. Biochemistry 15: 1097-1102.
94. MEIER, A., H., DE WAAL, A., VERPOORTE, R. 1993. Purification of the cytochrome P-450 enzyme geraniol 10-hydroxylase from cell cultures of *Catharanthus roseus*. J. Chromatogr. 635: 237-249.
95. BOZAK, K. R., YU, H., SIREVAG, R., CHRISTOFFERSEN, R. E. 1990. Sequence analysis of ripening-related cytochrome P-450 cDNAs from avocado fruit. Proc. Natl. Acad. Sci. USA 87: 3904-3908.
96. HALLAHAN, D. L., NUGENT, J. H. A., HALLAHAN, B. J., DAWSON, G. W., SMILEY, D. W., WEST, J. M., WALLSGROVE, R. M. 1992. Interactions

of avocado (*Persea americana*) cytochrome P-450 with monoterpenoids. Plant Physiol. 98: 1290-1297.

97. HALLAHAN, D. L., LAU, S-M. C., HARDER, P. A., SMILEY, D. W. M., DAWSON, G. W., PICKETT, J. A., CHRISTOFFERSEN, R. E., O'KEEFE, D. P. 1994. Cytochrome P-450-catalysed monoterpenoid oxidation in catmint (*Nepeta racemosa*) and avocado (*Persea americana*); evidence for related enzymes with different activities. Biochim. Biophys. Acta 1201: 94-100.
98. CLARK, I. M., FORDE, B. G., HALLAHAN, D. L. 1997. Spatially distinct expression of two new cytochrome P450s in leaves of *Nepeta racemosa*: Identification of a trichome-specific isoform. Plant Mol. Biol. 33: 875-885.
99. COLLU, G. 1999. Geraniol 10-hydroxylase. A cytochrome P450 enzyme involved in the biosynthesis of terpenoid indole alkaloids in higher plants: An enzymological and molecular study. PhD thesis, University of Leiden, NL.
100. LUPIEN, S., KARP, F., PONNAMPERUMA, K., WILDUNG, M., CROTEAU, R. 1995. Cytochrome P450 limonene hydroxylases of *Mentha* species. Drug Metab. Drug Interact. 12: 245-260.
101. KARP, F., MIHALIAK, C. A., HARRIS, J. L., CROTEAU, R. 1990. Monoterpene biosynthesis: Specificity of the hydroxylations of (-)-limonene by enzyme preparations from peppermint (*Mentha piperita*), spearmint (*Mentha spicata*), and Perilla (*Perilla frutescens*) leaves. Arch. Biochem. Biophys. 276: 219-226.
102. LUPIEN, S., KARP, F., WILDUNG, M., CROTEAU, R. 1999. Regiospecific cytochrome P450 limonene hydroxylases from mint (*Mentha*) species: cDNA isolation, characterization, and functional expression of (-)-4S-limonene-3-hydroxylase and (-)-4S-limonene-6-hydroxylase. Arch. Biochem. Biophys. 368:181-192.
103. BOUWMEESTER, H. J., GERSHENZON, J., KONINGS, M. C. J. M., CROTEAU, R. 1998. Biosynthesis of the monoterpenes limonene and carvone in the fruit of caraway. Plant Physiol. 117: 901-912.
104. MURPHY, P. J., WEST, C.A., 1969. The role of mixed function oxidases in kaurene metabolism in *Echinocystis macrocarpa* Greene endosperm. Arch. Biochem. Biophys. 133: 395-407.
105. SWAIN, S. M., REID, J. B., KAMIYA, Y. 1997. Gibberellins are required for embryo growth and seed development in pea. Plant J. 12: 1329-1338.
106. MANDAVA, N. B., 1988. Plant growth-promoting brassinosteroids. Annu. Rev. Plant Physiol. Plant Mol. Biol. 39: 23-52.
107. NAGPAL, P., VITART, V., MCMORRIS, T. C., CHORY, J., 1996. A role for brassinosteroids in light-dependent development of *Arabidopsis*. Science 272: 398-401.

108. YOKOTA, T. 1997. The structure, biosynthesis and function of brassinosteroids. Trends in Plant Science 2: 137-143.

109. BAK, S., KAHN, R. A., OLSEN, C. E., HALKIER, B. A., 1997. Cloning and expression in *Escherichia coli* of the obtusifoliol 14 alpha-demethylase of *Sorghum bicolor* (L.) Moench, a cytochrome P450 orthologous to the sterol 14alpha-demethylases (CYP51) from fungi and mammals. Plant J. 11: 191-201.

110. CHOE, S., NOGUCHI, T., FUJIOKA, S., TAKATSUTO, S., TISSIER, C.P, GREGORY, B. D., ROSS, A.S., YOSHIDA, S., TAX, F.E., FELDMANN, K.A. 1999. The *Arabidopsis dwf7/ste1* mutant is defective in the delta^{7}sterol C-5 desaturation step leading to brassinosteroid biosynthesis. The Plant Cell 11: 207-221.

111. MATHUR, J., MOLNAR, G., FUJIOKA, S., TAKATSUTO, S., SAKURAI, A., YOKOTA, T., ADAM, G., VOIGT, B., NAGY, F., SCHELL, J., KONCZ, C., SZEKERES, M., 1998. Transcription of the *Arabidopsis* CPD gene, encoding a steroidogenic cytochrome P450, is negatively controlled by brassinosteroids. Plant J. 14: 593-602.

112. SALAUN, J.-P., HELVIG, C., 1995. Cytochrome P450-dependent oxidation of fatty acids. Drug Metab. Drug Interact. 12:262-284.

113. KOLATTUKUDY, P.E., 1980. Biopolyester membranes of plants: cutin and suberin. Science. 208: 990-1000.

114. PINOT, F., BENVENISTE, I., SALAUN, J.P., LOREAU, O., NOEL, J.P., SCHREIBER, L., DURST,F., 1999. Production *in vitro* by the cytochrome P450 CYP94A1 of major C18 cutin monomers and potential messengers in plant-pathogen interactions: enantioselectivity studies. Biochem. J. 342: 27-32.

115. SCHWEIZER, P., JEANGUENAT, A., WHITACRE, D., METRAUX, J.P., MOSINGER, E., 1996. Induction of resistance in barley against Erysiphe graminis f sp hordei by free cutin monomers. Physiol. Mol. Plant. Pathol. 49: 103-120.

116. FAUTH, M., SCHWEIZER, P., BUCHALA, A., MARKSTADTER, C., RIEDERER, M., KATO, T., KAUSS, H., 1998. Cutin monomers and Surface Wax Constituents Elicit H_2O_2 in Conditioned Cucumber Hypocotyl Segments and Enhance the Activity of Other H_2O_2 Elicitors. Plant Physiol. 117: 1373-1380.

117. SALAUN, J.-P., BENVENISTE, I., REICHHART, D., DURST, F., 1981. Induction and specificity of a cytochrome P-450 dependent laurate in-chain hydroxylase from higher plant microsomes. Eur. J. Biochem. 119: 651-655.

118.CABELLO-HURTADO, F., BATARD, Y., SALAUN, J.P., DURST, F., PINOT, F., WERCK-REICHHART, D., 1998. Cloning, expression in yeast, and functional characterization of CYP81B1, a plant cytochrome P450 that catalyzes in-chain hydroxylation of fatty acids. J. Biol. Chem. 273: 7260-7267.
119.PINOT, F., ALAYRAC, C., MIOSKOWSKI, C., DURST, F., SALAUN, J.P., 1994. New cytochrome P450-dependent reactions from wheat: Terminal and sub-terminal hydroxylation of oleic acid by microsomes from naphthalic acid anhydride and phenobarbital induced wheat seedlings. Biochem. Biophys. Res. Commun. 198: 795-803.
120.BENVENISTE, I., SALAUN, J.-P., SIMON, A., REICHHART, D., DURST, F., 1982. Cytochrome P-450 dependent w-hydroxylation of lauric acid by microsomes from pea seedlings. Plant Physiol. 70: 122-126.
121.SALAUN, J.-P., SIMON, A., DURST, F., 1986. Specific induction of lauric acid ω-hydroxylase by clofibrate, diethylhexyl-phtalate and 2,4-dichlorophenoxyacetic acid in higher plants. Lipids 21: 776-779.
122.TIJET, N., HELVIG, C., PINOT, F., LEBOUQUIN, R., LESOT, A., DURST, F., SALAUN, J.P., BENVENISTE, I. 1998. Functional expression in yeast and characterization of a clofibrate-inducible plant cytochrome-P450 (CYP94A1) involved in cutin monomers synthesis. Biochem. J. 332:583-589.
123.PINOT, F., BENVENISTE, I., SALAUN, J.P., DURST, F., 1998. Methyl jasmonate induces lauric acid omega-hydroxylase activity and accumulation of CYP94A1 transcripts but does not affect epoxide hydrolase activities in *Vicia sativa* seedlings. Plant Physiol. 118: 1481-1486.
124.LE BOUQUIN, R., PINOT, F., BENVENISTE, I., SALAUN, J.P., DURST, F., 1999. Cloning and functional characterization of CYP94A2, a medium chain fatty acid hydroxylase from *Vicia sativa*. Biochem. Biophys. Res. Commun. 261: 156-162.
125.SONG, W.C., FUNK, C.D., BRASH, A.R., 1993. Molecular Cloning of an Allene Oxide Synthase - A cytochrome-P450 specialized for the metabolism of fatty acid hydroperoxides. Proc. Natl. Acad. Sci. USA. 90: 8519-8523.
126.PAN, Z.Q., DURST, F., WERCK-REICHHART, D., GARDNER, H.W., CAMARA, B., CORNISH, K., BACKHAUS, R.A., 1995. The major protein of guayule rubber particles is a cytochrome-P450 - Characterization based on cdna cloning and spectroscopic analysis of the solubilized enzyme and its reaction-products. J. Biol. Chem. 270: 8487-8494.
127.BLEE, E., 1998. Phytooxylipins and plant defense reactions. Prog. Lipid Res. 37: 33-72.
128.KOCH, B. M., SIBBESEN, O., HALKIER, B. A., SVENDSEN, I., MØLLER, B. L. 1995. The primary sequence of cytochrome P450tyr, the multifunctional

N-hydroxylase catalyzing the conversion of L-tyrosine to p-hydroxyphenylacetaldehyde oxime in the biosynthesis of the cyanogenic glucoside dhurrin in Sorghum bicolor (L.) Moench. Arch. Biochem. Biophys. 323: 177-186.

129. BARRETT, M., POLGE, N., BAERG, R., BRADSHAW, L., PONELEIT, C., 1997. Role of cytochrome P450s in herbicide metabolism and selectivity and multiple herbicide metabolizing cytochromes P450 activities in Maize. pp. 35-50 *in* Regulation of Enzymatic Systems Detoxifying Xenobiotics In Plants (K.K. Hatzios, ed.), Kluwer, Dordrecht/Boston/London.

130. HOLT, J.S., POWLES, S.B., HOLTUM, J.A.M., 1993. Mechanisms and agronomic aspects of herbicide reistance. Annu. Rev. Plant Physiol. Plant Mol. Biol. 44: 203-230.

131. DURST, F., BENVENISTE, I., LESOT, A., SALAUN, J.-P., WERCK-REICHHART, D., 1997. Induction of plant cytochrome P450. pp. 19-34 *in* Regulation of Enzymatic Systems Detoxifying Xenobiotics In Plants (K.K. Hatzios, ed.), Kluwer, Dordrecht/Boston/London.

132. ZIMMERLIN, A., DURST, F., 1992. Aryl hydroxylation of the herbicide diclofop by a wheat cytochrome-P-450 monooxygenase - substrate specificity and physiological activity. Plant Physiol. 100: 874-881.

133. ZIMMERLIN, A., SALAUN, J.-P., DURST, F., MIOSKOWSKI, C., 1992. Cytochrome P-450 dependent hydroxylation of lauric acid at the subterminal position and oxidation of unsaturated analogs in wheat microsomes. Plant Physiol. 100: 863-873.

134. HELVIG, C., TARDIF, F.J., SEYER, A., POWLES, S.B., MIOSKOWSKI, C., DURST, F., SALAUN, J.P., 1996. Selective-inhibition of a cytochrome-P450 enzyme in wheat that oxidizes both the natural substrate lauric acid and the synthetic herbicide diclofop. Pestic. Biochem. Physiol. 54: 161-171.

135. ROBINEAU, T., BATARD, Y., NEDELKINA, S., CABELLOHURTADO, F., LERET, M., SOROKINE, O., DIDIERJEAN, L., WERCK-REICHHART, D., 1998. The chemically inducible plant cytochrome-P450 CYP76B1 actively metabolizes phenylureas and other xenobiotics. Plant Physiol. 118: 1049-1056.

136. SIMINSZKY, B., CORBIN, F.T., WARD, E.R., FLEISCHMANN, T.J., DEWEY, R.E., 1999. Expression of a soybean cytochrome P450 monooxygenase cDNA in yeast and tobacco enhances the metabolism of phenylurea herbicides. Proc. Natl. Acad. Sci. U S A. 96: 1750-1755.

Chapter Seven

THE BIOSYNTHESIS, DEGRADATION, TRANSPORT AND POSSIBLE FUNCTION OF CYANOGENIC GLUCOSIDES

Patrik R. Jones,[1,2,3] Mette Dahl Andersen,[1,2] John Strikart Nielsen,[1,2] Peter Bordier Høj,[3] and Birger Lindberg Møller[1,2]

[1]Plant Biochemistry Laboratory
Department of Plant Biology
Royal Veterinary and Agricultural University
40 Thorvaldsensvej, DK-1871 Frederiksberg C
Copenhagen, Denmark

[2]Center for Molecular Plant Physiology (PlaCe)
Royal Veterinary and Agricultural University
40 Thorvaldsensvej, DK-1871 Frederiksberg C
Copenhagen, Denmark

[3]Department of Horticulture, Viticulture and Oenology
University of Adelaide, Waite Campus
Glen Osmond, South Australia, Australia.

Evolution of Metabolic Pathways – J.T. Romeo et al. (Editors)

INTRODUCTION

The ability of living organisms to produce hydrogen cyanide from constitutive compounds is characterized as cyanogenesis. It is a common trait among plant species, including ferns and gymnosperms as well as monocotyledonous and dicotyledonous angiosperms.[1-7] Most precursors have been found to be cyanogenic glucosides, although other forms, *e.g.* cyanogenic lipids, are known to exist.[2,8] All known cyanogenic glucosides are *O*-β-glycosides of α-hydroxynitriles (Fig. 1), and are with a few exceptions synthesized from six different parent amino acids. Several agriculturally important crop plants are known to be cyanogenic, *e.g.* cassava, sorghum, barley, bamboo, lima beans, white clover, elder, flax, rubber tree, cherry, plum, peach, apricot, almond, and macadamia.[3,6,7,9,10] Upon tissue disruption, cyanogenic glucosides are rapidly degraded to the corresponding aldehyde or ketone, cyanide, and sugar (Fig. 2). *In vivo*, the glucosides are compartmentalized away from the specific catabolic

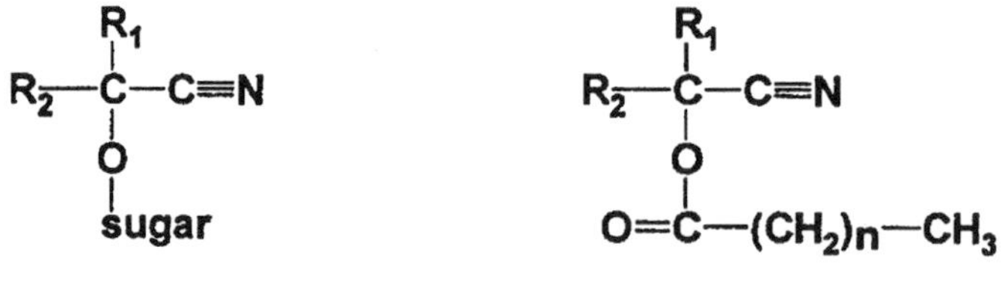

Figure 1. General structures of cyanogenic glucosides and cyanolipids.

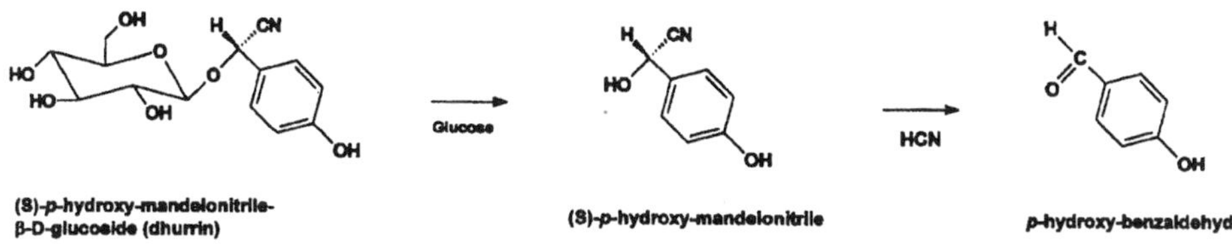

Figure 2. Catabolism of the cyanogenic glucoside dhurrin in *Sorghum bicolor* is catalyzed by a specific α-glucosidase[1] and an α-hydroxynitrilase[2]

enzymes, the β-glucosidases and α-hydroxynitrilases, thereby repressing cyanide release within the intact plant.[5,11-16]

The ability of cyanogenic glucosides to release poisonous cyanide upon tissue disruption renders the presence of large quantities of them in a crop plant a nutritional problem for man and animals eating or feeding on such plants. This problem is of special importance in cassava, which is a major staple crop grown in most tropical regions of Africa, Latin American, and Asia. The level of cyanide released from cassava tubers from different cultivars ranges between 15 and 200 mg per kg tuber, as calculated on a fresh weight basis of the tuber pith. The peel has a much higher cyanide potential. Over three-fold differences in cyanide potential of roots from the same cassava plant are not uncommon.[17] In the Democratic Republic of Congo, the average annual per capita consumption of cassava is 333 kg (FAO homepages, FAOSTAT nutrition data, food supply statistics, http://apps.fao.org). Use of cassava as a staple food, thus, requires careful processing to remove the content of cyanogenic glucosides and their degradation products. Populations that obtain most of their dietary calories from cassava-derived products generally have a poor nutritional status, which in turn aggravates the effect of residual amounts of cyanide.[18-21]

The biological functions of cyanogenic glucosides in plants have been difficult to assess.[22-26] A major goal within cyanogenic glucoside research, therefore, is to find out why a specific plant produces this type of natural product constitutively, or at certain developmental stages, or as a response to abiotic or biotic stress. Based on such knowledge, it will be possible to assess whether it will be feasible to down-regulate the production of these compounds in crop plants without diminishing the physiological fitness of the plant. Hopefully, the elimination of a highly abundant cyanogenic glucoside, in addition to increasing food safety, would be accompanied by other positive traits such as increased deposition of storage protein due to increased nitrogen availability. If complete elimination of a cyanogenic glucoside results in sensitive plants with other undesirable characteristics, then the cyanogenic glucosides should be eliminated or specifically reduced only in those parts of the plants used for human consumption. The techniques of molecular biology enable the production of such plants.

For efficient regulation and optimization of the accumulation of cyanogenic glucosides, it is necessary to study the processes responsible for their biosynthesis, mobilization, and degradation. The entire biosynthetic pathway has been characterized as being catalyzed by only three enzymes (Fig. 3). The first two are multifunctional cytochrome P450s,[27-32] and the third is an α-hydroxynitrile glucosyltransferase.[33] In addition, it has been proposed that cyanogenic glucosides are transported in the form of diglucosides.[23] Cyanogenic glucoside glucosyltransferases, the enzymes that catalyze the production of cyanogenic diglucosides, are yet to be described and isolated. Degradation of cyanogenic glucosides requires β-glucosidases[34,35] and α-hydroxynitrilases,[36-40] both of which have been isolated and cloned. This review summarizes our current knowledge about biosynthesis, mobilization, and degradation of cyanogenic glucosides.

COMMON FEATURES OF NATURAL PRODUCT SYNTHESIS AND XENOBIOTIC METABOLISM

Depending on the plant species, production of cyanogenic glucosides may proceed in all parts of a plant or in a single or few specific tissues. Seedlings, inflorescences, and immature fruits are among the most active tissues with respect to cyanogenic glucoside synthesis.[41-43] Because cyanogenic glucosides are constitutively produced in healthy plant tissue, they belong to the class of natural products referred to as "phytoanticipins".[44] Natural products are defined as compounds with no participation in primary physiological processes,[45] although this view has been challenged.[46] The biosynthesis and compartmentation of different groups of natural products share many common features. All groups of compounds are derived from primary precursors, are commonly found as conjugates, and often accumulate in vacuoles. It is no surprise, therefore, that the enzymes involved in their biosynthesis and compartmentation share similarities. Typically, the biosynthesis of natural products involves cytochrome P-450 monooxygenases[47,48] and glucosyltransferases.[46,49,50] Evidence has been presented for the existence of independent active vacuolar transporters of malonyl-,[51,52] glycosyl-,[53,54] and glutathione-S-conjugated secondary metabolites.[55,56] The transfer of glycosyl-, malonyl-, and glutathione-S-groups to natural products may, therefore, be a prerequisite to rapid vacuolar transport.

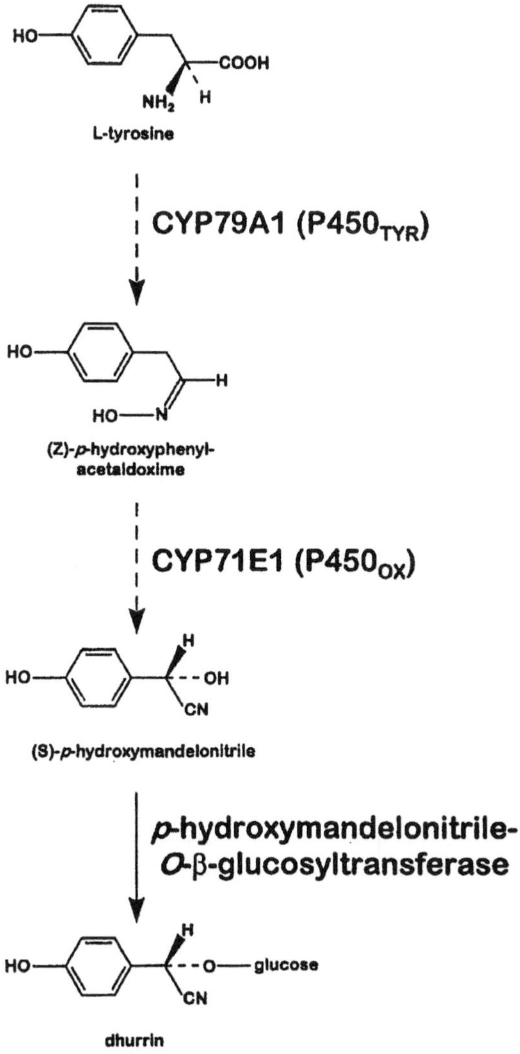

Figure 3. The enzymes catalyzing the biosynthesis of the cyanogenic glucoside dhurrin in *S. bicolor*

The general steps involved in xenobiotic metabolism have been outlined in several reviews.[57-60] These consist of phase I (activation), phase II (conjugation), and phase III (compartmentation) processes. Cytochrome P-450s commonly introduce a hydroxyl-group through oxidation (phase I), while one or more forms of conjugations (phase II) are necessary prior to compartmentation (phase III) in the vacuole or in the cell wall.[59] Novel xenobiotics, such as pesticides and other environmental contaminants, are metabolized by these systems.[61,62] Accordingly, plant xenobiotic metabolism is based on enzyme and transportation systems similar to those involved in the biosynthesis and sequestration of plant natural products. The question, therefore, arises whether there is a general xenobiotic metabolism system in plants that can handle novel xenobiotics, or whether the observed metabolism represents "chance" activities from broad-substrate specificity secondary metabolite systems. The possibility that highly specific proteins have evolved together with general, broader substrate-specificity proteins can not be excluded.

The group of cyanogenic glucosides precisely follows this general synthetic scheme by incorporating membrane-bound cytochrome P-450 monooxygenases, followed by an apparently soluble glucosyltransferase in their biosynthesis (Fig. 3).[27,29-33,63] The glucosides are deposited in vacuoles by an unknown mechanism, although a malonyl-glucoside has been identified in tissues of *Merremia dissecta*.[64] In addition, evidence has been presented to suggest that cyanogenic glucosides are highly mobile in certain plants.[65,66] Cyanogenesis, therefore, represents a good opportunity to further the study of many aspects of natural product biosynthesis in plants.

BIOLOGICAL IMPLICATIONS OF CYANOGENESIS

Cyanide is volatile and has a high affinity for metal functional groups,[67] which gives cyanide its toxic properties. These are particularly attributed to the inhibition of metallo-proteins, such as cytochrome oxidase A_3, a key-enzyme of the mitochondrial respiratory pathway of most organisms, and reactions with several other non-metallo-proteins.[68] These nutritionally negative side effects become pronounced when plant material containing large amounts of cyanogenic glucosides or their corresponding α-hydroxynitriles (*e.g.* insufficiently processed cassava products) are consumed by animals or man. Cyanogenic glucosides that enter the mammalian digestive system intact may be catabolized by the resident microflora, which results in the release and absorption of cyanide by the consumer.[21] The symptoms of acute cyanide intoxications are nausea and vomiting.[18,20,21] Chronic cyanide intoxication may cause ataxic neuropathy and optic atrophy.[69] In combination with severe malnutrition and deficient sulfur intake, high dietary cyanide exposure may cause konzo, a distinct form of tropical myelopathy characterized by abrupt onset of spastic paraparesis (partial paralysis affecting the lower limbs).[19,70] The rapidity of cyanide release from cyanogenic glucosides is increased greatly by simultaneous consumption of fresh plant material rich in β-glucosidase and α-hydroxynitrilases,[71] and is dependent on the bacterial flora of the gut.[72] Feeding from highly cyanogenic crops can, therefore, result in chronic and acute intoxication.[73] Cyanogenic glucosides are also known to be precursors to ethyl carbamate in fermented foods such as bread, wine, beer, and distilled spirits.[74]

The presence of cyanogenic glucosides may improve pest or pathogen resistance of plants.[26,75] Cyanogenic glucosides are present in tissues and intracellular compartments different from their corresponding catabolic enzymes. Upon tissue disruption, the two components are combined, resulting in the release of hydrogen cyanide.[5,11-16,76,77] Given the toxicity of the end-products generated by catabolism of cyanogenic glucosides and the method of release, it is possible that plant cyanogenesis has evolved in response to certain herbivore and pathogen pressures. Numerous ecological studies have favored this argument.[78] However, often the evidence has been circumstantial.[79-81] Several reports indicate that the presence of cyanoglucosides has no deterrent effect on particular pathogens,[82] or herbivores.[26] Woodhead and Bernays[83,84] showed that neither dhurrin nor *p*-hydroxybenzaldehyde, a catabolite of dhurrin (Fig. 3), had a deterrent effect on the lesion nematode *Locusta migratoria* feeding on young *S. bicolor*, while HCN did have a significant effect. However, later work by Woodhead,[85,86] in contrast to previous results, showed that *p*-hydroxybenazaldehyde indeed was active as a deterrent against *Locusta migratoria* in mature *Sorghum* plants. Similarly, Dreyer *et al.*[87] found *p*-hydroxybenzaldehyde to be a more effective feeding deterrent than dhurrin against greenbug (*Schizapis graminum*) when tested in a synthetic diet. Interestingly, Woodhead *et al.*[88] found that up to 30% of epicuticular *Sorghum* wax was made up of *p*-hydroxybenzaldehyde at the young seedling stage. It was speculated that the presence of high amounts of *p*-hydroxybenazaldehyde may be related to the considerable turnover of dhurrin that occurs in young seedlings.[89,90]

Cyanogenesis is widespread in nature, and co-evolution of cyanogenic plants and their pests has provided ample time for development of pests able to metabolize cyanide into non-toxic constitutents. It has been documented that the cyanogenic rubber tree (*Hevea brasiliensis*) has reduced pathogen defense abilities towards the fungus *Microcyclus ulei* due to the presence of cyanogenic glucosides.[24] Upon infection, the cyanide released inhibits *de novo* biosynthesis of the major phytoalexin scopoletin.[25] When the level of endogenously released hydrogen cyanide from highly cyanogenic clones of *H. brasiliensis* was lowered by flushing with moistened air, the biosynthesis of the phytoalexin scopoletin was equivalent to the level observed in weakly cyanogenic clones.[24] Similarly, a positive correlation between cyanogenic glucoside content in barley (*Hordeum vulgare*) and the sporulation efficiency of powdery mildew has been reported.[9,10]

The effects of particular natural products will most likely be different on different organisms. In addition, studies by Hain *et al.*[91] have conclusively shown that the speed of biosynthesis of induced defense systems is of major importance for successful defense against fungi such as *Botrytis cinerea*. Similarly, Selmar *et al.*[92] showed by determining the rate of HCN liberation that α-hydroxynitrile lyase activity of *Hevea brasiliensis* was pivotal for effective cyanogenesis. Ecological studies would, therefore, benefit from considering both the quantities and the nature of potential defense compounds, as well as the speed of their formation. These aspects have been partly addressed by Woodhead and Bernays.[83,84] Alternative "roles" for cyanoglucosides also have been proposed. Weston *et al.*[93] provided evidence for the allelopathic potential of cyanogenic *Sorghum* hybrids towards several weed and vegetable species. The observed effect was attributed to dhurrin and *p*-hydroxybenzaldehyde. However, Forney and Foy[94] suggested that the most phytotoxic compounds isolated in *Sorghum* rhizospheres were anthocyanins. It has also been suggested that cyanogenic glucosides may constitute mobilizable nitrogen-storage forms.[23,65]

BIOSYNTHESIS OF CYANOGENIC GLUCOSIDES

Classical Studies

Compared with other groups of natural products, such as terpenes, alkaloids, and flavonoids, the number of known cyanogenic glucosides is limited.[6,7] The structural diversity of these compounds is limited by the narrow substrate specificity of the enzymes catalyzing the initial reactions in their biosynthesis. Only the protein amino acids, tyrosine, phenylalanine, valine, isoleucine, and leucine and the non-protein amino acid, 2-(2-cyclopentenyl)glycine, are used as precursors[1,2,5,7] An additional compound, acalyphin, appears to be derived from nicotinic acid.[95] Subsequent glycosylation and acylation reactions increase the structural diversity. In contrast to other groups of plant natural products, structural modifications due to secondary hydroxylation and methylation reactions are of rare occurrence.

Much early biosynthetic work was carried out with flax (*Linum usitatissimum*), cherry laurel (*Prunus laurocerasus*), and sorghum (*Sorghum bicolor*).[96-102] By double-labeling experiments, it was demonstrated that the C-N bond is not broken during the biosynthetic process.[103] Accordingly, all intermediates in the pathway contain nitrogen. Likewise, double labeling experiments showed that the covalent bond linking the α and β carbon atoms of the parent amino acid was not severed during biosynthesis.[104] Major intermediates were thought to include an *N*-hydroxyamino acid, an aldoxime, and an α-hydroxynitrile. Labeled amino acid precursors introduced into these plants resulted in extremely high introduction of label (up to 49%), but, nevertheless, it was not possible to isolate the intermediates thought to be involved in the pathway from these biosynthetically active tissues.[105]

In Vitro Biosynthetic Studies

The pathway for the different cyanogenic glucosides is thought to follow a common biosynthetic scheme (Fig. 4).[106,107] A major breakthrough in the efforts to identify the intermediates involved was the isolation of biosynthetically active microsomal fractions. This was first accomplished from etiolated seedlings of sorghum (*S. bicolor*).[108,109] Sorghum contains the tyrosine-derived cyanogenic glucoside dhurrin.[41,110] The microsomal fraction was active when supplemented with a substrate and NADPH in the presence of molecular oxygen. The microsomal fraction from sorghum was found to catalyze the *in vitro* conversion of tyrosine to *p*-hydroxymandelonitrile, the aglycone of dhurrin, when prepared in the presence of a reductant such as β-mercaptoethanol or dithiothreitol. When the microsomal system was prepared in the absence of thiol reagents, the last part of the pathway was inactivated, and (Z)-*p*-hydroxyphenylacetaldoxime was obtained as the end product.[108,111,112] Subsequently, active microsomal preparations have been obtained from several other cyanogenic plants including flax (*Linum usitatissimum*),[113] white clover (*Trifolium repens*),[114] *Triglochin maritima*,[43,115,116] California poppy (*Eschscholtzia californica*),[117] and cassava (*Manihot esculenta*).[118] Although the studies carried out by using these additional *in vitro* systems have not been as thorough as those carried out in sorghum, the data obtained lend support to the assumption that the cyanogenic glucosides present in each of these plant species are synthesized by the basic scheme first outlined for sorghum (Fig. 4). These studies also demonstrate that all steps in the biosynthesis, except the final glycosylation step, are catalyzed by membrane-bound enzymes. A common feature of the different microsomal systems used was that only small amounts of intermediates accumulated. Secondly, the kinetic parameters

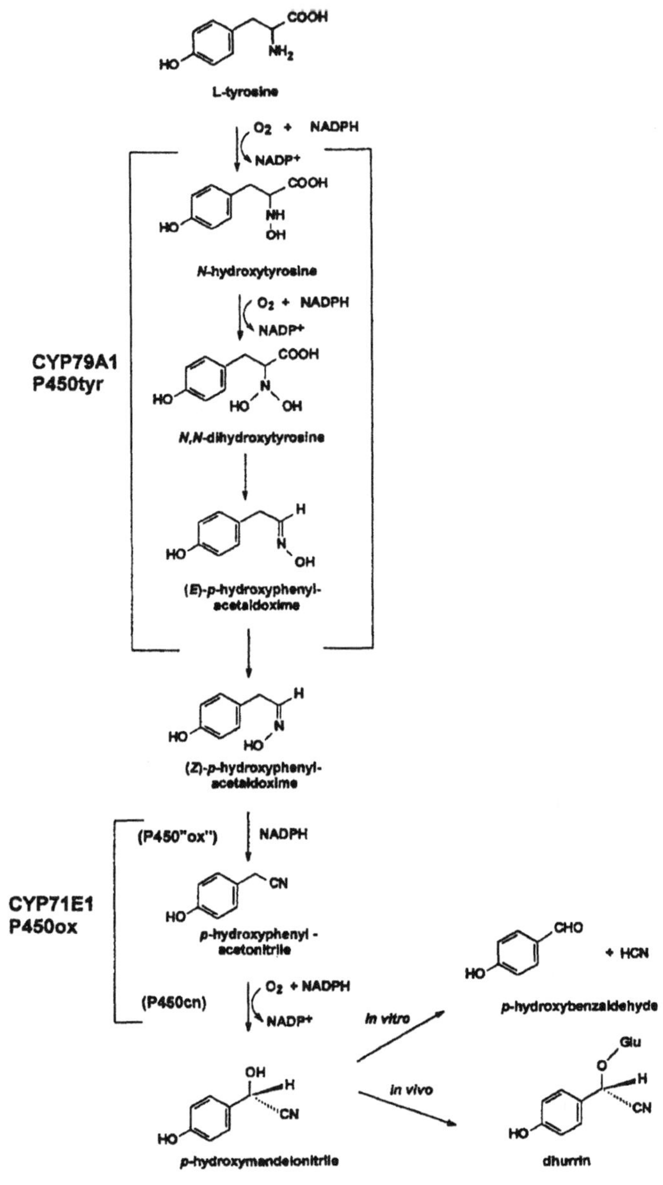

Figure 4. Intermediates in the biosynthetic pathway for the cyanogenic glucoside dhurrin in *S. bicolor*

measured for the individual reaction steps were not consistent with those of the overall pathway.[119] Studies that used microsomes prepared from sorghum[111] and subsequently from *Triglochin maritima*,[116] demonstrated that the pathway is highly channelled. In sorghum, the only intermediate that equilibrated freely with externally added intermediates was the Z-oxime (Z-*p*-hydroxyphenylacetaldoxime).[120] In *T. maritima*, the nitrile (*p*-hydroxymandelonitrile) was the only non-channelled intermediate.[116] These observations explain the general lack of accumulation of intermediates by using the *in vitro* systems, as well as the results of earlier studies in which radiolabeled precursors were administered to excised plant parts, and no intermediates were detectable in spite of efficient incorporation rates into the cyanogenic glucoside.[105] The channelling mechanism was thought to protect labile intermediates from wasteful and deleterious side reactions by ensuring a rapid and efficient flow of carbon and nitrogen through the pathway.

A common characteristic for the microsomal systems appears to be that the microsomal preparations are inactivated if the plant material used is not completely free of seed coats.[5,43,113,118] The nature of the powerful inhibitors in the seed coats remains unknown. Preliminary studies in cassava show that the inhibitory effect is the result of the action of multiple components (Andersen and Møller, unpublished results).

Identification of Intermediates by Using the Sorghum Microsomal System

Biosynthetic studies with radioactively labeled precursors and trapping experiments in which unlabeled putative intermediates were included in the microsomal reaction mixtures identified *N*-hydroxyamino acids, *N,N*-dihydroxyamino acids, (*E*)-aldoximes, (*Z*)-aldoximes, nitriles, and cyanohydrins as key intermediates in the biosynthetic pathway (Fig. 4). All these compounds, except the *N,N*-hydroxyamino acid, have been chemically synthesized and are metabolized by the microsomal system.[5,109,112,119,121-125] The extreme lability of the *N,N*-dihydroxyamino acid makes it impossible to investigate directly, because it can be neither chemically synthesized nor isolated.

It is not understood why (*E*)- as well as (*Z*)-aldoximes are involved as intermediates, but trapping experiments clearly indicate that the (*E*)-isomer (*E*-*p*-hydroxyphenylacetaldoxime) is produced first and converted to the nitrile (*p*-hydroxymandelonitrile) with the (*Z*)-isomer as an obligatory intermediate.[120]

Microsomal activity is dependent on the presence of oxygen and NADPH. Stoichiometric measurements of oxygen consumption demonstrated that conversion of the parent amino acid tyrosine to the corresponding cyanohydrin *p*-

hydroxymandelonitrile proceeds with the consumption of three molecules of oxygen, indicating the involvement of three hydroxylation reactions. Two molecules of oxygen are consumed in the conversion of the amino acid to the aldoxime, whereas a single oxygen molecule is consumed in the conversion of the aldoxime to the cyanohydrin.[126] Quantitative measurements of NADPH consumption were not possible because of the formation of an $NADP^+$-CN adduct that interferes with absorption measurements at 340 nm.[127,128] Biosynthetic experiments that use stable isotopes have also helped elucidate the nature of the intermediates involved.[120,125] Thus, use of an amino acid, in which the hydrogen atom at the α-carbon atom was labeled with deuterium, disclosed that the hydrogen atom at this position is completely conserved in the aldoxime[120] This excludes compounds, such as ketoximes, that do not retain this α-hydrogen atom as intermediates in the biosynthetic pathway (Fig. 4).[120] Biosynthetic experiments with $^{18}O_2$ also have been of central importance for elucidation of the pathway.[125] With the amino acid as substrate and incubation in an $^{18}O_2$ atmosphere, the aldoxime produced was demonstrated to be entirely labeled with ^{18}O in the hydroxylamine function, as expected. However, 100% incorporation of ^{18}O in the hydroxylamine function of the aldoxime was also observed when the *N*-hydroxyamino acid was provided as substrate and incubation in an $^{18}O_2$ atmosphere, although free rotation around the C-N bond would be expected to cause a 50% loss of ^{18}O-labeling. The complete retention af ^{18}O-labeling in the *N*-hydroxy group of the oxime function in the latter experiment demonstrated that the oxygen atom recovered in the aldoxime is different from the one initially introduced by *N*-hydroxylation of the amino acid. Similarly, these studies demonstrate that the *N,N*-dihydroxyamino acid is an intermediate, although the extreme lability of this compound, as described earlier, prevented direct demonstration of this fact (Fig. 4).[125]

Biosynthetic studies with microsomes confirm the conclusions based on more classic biosynthetic experiments that the C-N bond is not broken during the synthesis of cyanogenic glucosides,[103] and that the covalent bond linking the α and β carbon atoms of the parent amino acid is not severed during biosynthesis.[104]

Two Multifunctional P450s Catalyze All the Membrane-Bound Steps in Cyanogenic Glucoside Synthesis

Two steps in the biosynthesis of dhurrin were initially shown to be catalyzed by cytochrome P450 reactions in sorghum. These are the formation of *N*-hydroxytyrosine from tyrosine and the formation of *p*-hydroxymandelonitrile from *p*-hydroxyphenylacetonitrile.[126] Subsequently, two multifunctional P450s in sorghum were found to catalyze all the membrane-associated steps in dhurrin biosynthesis (*i.e.*, the conversion of tyrosine to *p*-hydroxymandelonitrile).[27,30,31] The first P450, designated P450tyr, catalyzes the conversion of tyrosine to Z-*p*-hydroxyphenylacetaldoxime,[27,29,30] whereas the second, designated P450ox, catalyzes the conversion of Z-*p*-hydroxyphenylacetaldoxime to *p*-hydroxymandelonitrile (Fig. 5).[1,32]

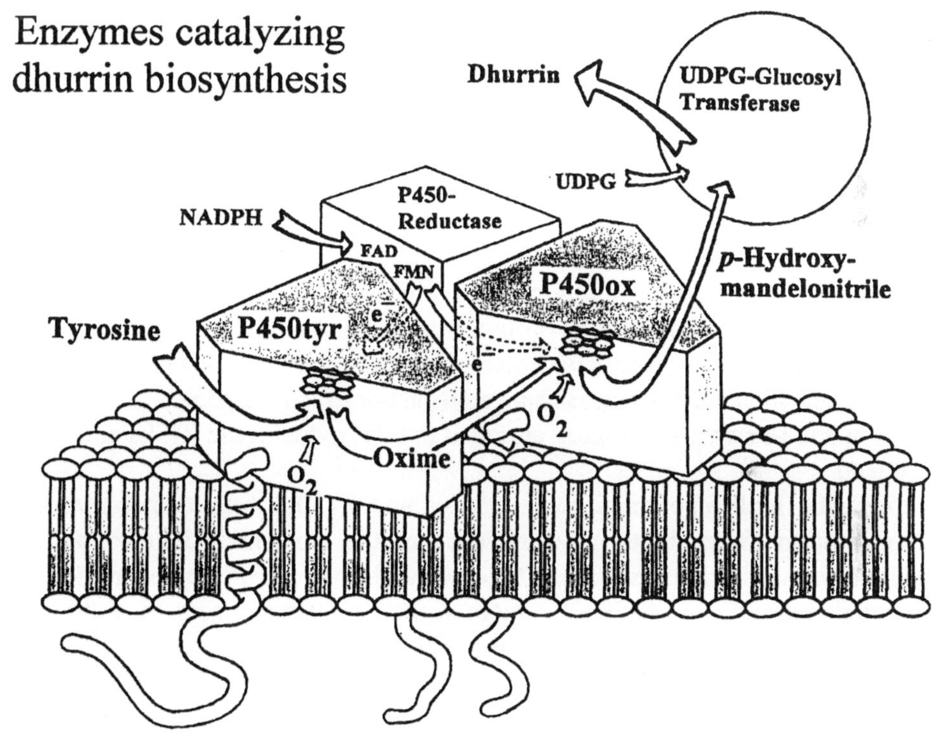

Figure 5. Model depicting the membrane bound and soluble enzymes involved in dhurrin biosynthesis adapted after a model prepared by Nelson and Strobel.[131]

Plant cytochrome P450s are membrane-bound enzymes.[48,129-131] Generally, the isolation of plant P450s has proved difficult because they are present in minute amounts and only at specific developmental stages, or under specific environmental conditions.[132-136] Microsomal preparations obtained from etiolated sorghum seedlings were used as starting material for the isolation of cytochrome P450tyr and P450ox. Although total cytochrome P450s constitute approximately

10% of the protein in rat liver microsomes, the total cytochrome P450s in microsomes isolated from sorghum seedlings constitute less than 1%.[27] Thus, new approaches had to be developed for the isolation of P450tyr and P450ox. The microsomal membranes were solubilized by using a combination of nonionic detergents, and the solubilized membrane proteins were applied to an anion-exchange column. Yellow pigments and cytochrome P450ox do not bind to this column,[31] whereas P450tyr does and is recovered free of pigments by EDTA elution.[27] Temperature-induced Triton X-114 phase partitioning,[137] in the presence of 30% glycerol, permitted separation of P450ox from the pigments. The two P450s were subsequently isolated by dye column chromatography.[27,31] A double Triton X-114 phase partitioning procedure that allows separation of cytochrome P450s from green pigments and provides an extract highly enriched in total cytochrome P450s was subsequently developed.[138] Upon phase partitioning in phosphate buffer by using Triton X-114, plant P450s partition in the pigmented detergent rich phase. It was discovered that the partitioning of the cytochrome P450s could be shifted to a pigment-free Triton X-114 poor phase by changing the buffer component to borate. The protein extract containing the cytochrome P450s but devoid of green pigments was then subjected to a second phase partitioning step before which the buffer was changed from borate to phosphate. This provided a Triton X-114 rich phase highly enriched in total cytochrome P450s and free of pigments.

The multifunctionality of P450tyr and P450ox was demonstrated by reconstitution experiments in which each of the isolated P450 enzymes was inserted into artificial membranes together with the NADPH-cytochrome P450 oxidoreductase.[30,31] Artificial membranes prepared from L-a-dilauroyl phosphatidylcholine were the best, as measured by the turnover number of the reconstituted cytochrome P450s.[30] By combined insertion of P450tyr, P450ox, and the reductase, artificial membranes were obtained that catalyzed the complete conversion of tyrosine to *p*-hydroxymandelonitrile. After addition of a soluble extract containing the UDPG-glucosyltransferase, the entire dhurrin biosynthetic pathway from tyrosine to dhurrin was reconstituted.[31]

Recently, the UDPG-p-hydroxymandelonitrile glucosyltransferase has been isolated, cloned, and functionally expressed.[33] This has made it possible to reconstitute the complete pathway by using pure UDPG-p-hydroxy-mandelonitrile glucosyltransferase instead of the crude soluble extract (Fig. 6).[33]

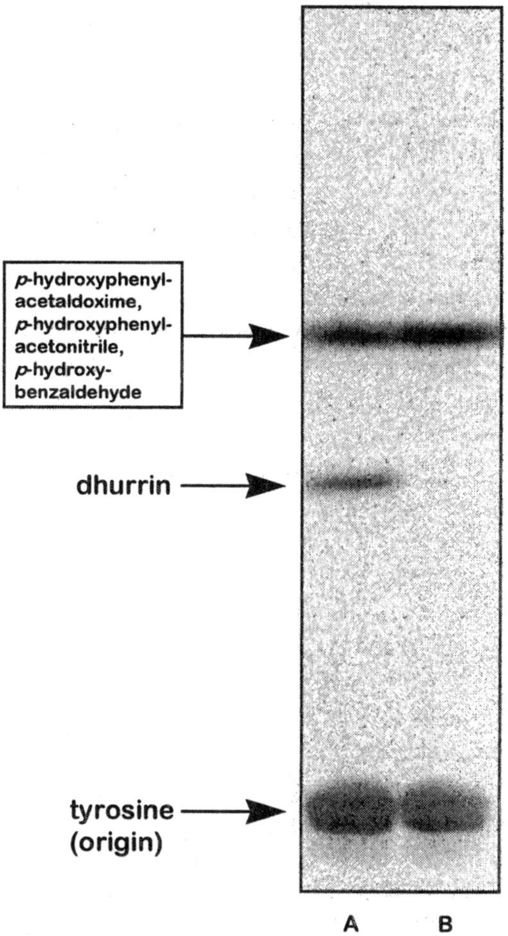

Figure 6. *In vitro* reconstitution of dhurrin synthesis by using isolated plant enzymes (NADPH-cytochrome P450 oxidoreductase) and recombinant enzymes (CYP79A1, CYP79E1 and *p*-hydroxymandelonitrile). Lane A: CYP79A1, CYP71E1 and NADPH-reductase. Lane B: As lane A but without NADPH-reductase. ^{14}C-Tyrosine was used as substrate and samples subjected to TLC and subsequent exposure to phosphorimager screens.

In the purification of P450tyr and P450ox, the formation of substrate-binding spectra on addition of tyrosine or *p*-hydroxyphenylacetaldehyde oxime was used to monitor the progress and efficiency of the purification procedures tested.[139,140] This worked well for isolation of P450tyr, which forms a type I spectrum in the presence of tyrosine as well as with *N*-hydroxytyrosine, in agreement with the reconstitution experiments, demonstrating the multifunctionality of P450tyr.[30] A similar type I spectrum was observed on administration of *p*-hydroxyphenylacetaldoxime to crude microsomal extracts. However, the P450 isolated by monitoring the type I spectrum turned out to be obtusifoliol 14a-demethylase.[141,142] Later, when the cytochrome P450ox protein was isolated and when a cDNA clone encoding P450ox was isolated and expressed in *Escherichia coli*, the protein turned out to produce a reversed type I spectrum with *p*-hydroxyphenylacetaldoxime.[32,143] In contrast, for unknown reasons, the isolated obtusifoliol 14a-demethylase did produce a clear Type I spectrum with the aldoxime. This serves to illustrate that substrate binding is not always a reliable assay for identification of a specific cytochrome P450 enzyme.

Cytochrome P450tyr was cloned by using a monospecific polyclonal antibody and oligonucleotide probes designed on the basis of amino acid sequences of tryptic fragments derived from the isolated protein.[29] The highest positional identity (30.8%) is with the 3′5′-flavonoid hydroxylase of petunia,[144] and with a cytochrome P450 sequence (CYP71A1) of unknown function from avocado. This last cytochrome is most likely involved in terpenoid hydroxylation.[132,145] According to the International P450 Nomenclature Committee, a P450 protein sequence from one family is usually defined as having less than 40% amino acid identity with a P450 protein from any other family.[146,147] Accordingly, P450tyr constitutes the first member of a new cytochrome family, and was assigned the name CYP79A1. The heme-binding cysteine residue of cytochrome P450tyr is recognizable at position 493, but this region deviated from the consensus sequence by having an unusual alanine residue at position 495. The central region of helix I contains three residues, Ala-352, Asn-355, and Pro-356, deviating from the consensus sequence. CYP56 is the only other known cytochrome P450 that uses tyrosine as a substrate, and this cytochrome contains the same Asn-Pro substitution in helix I, suggesting an importance of these residues in defining the substrate specificity.[148] Subsequently, CYP79D1 and CYP79D2 from cassava (*Mannihot esculenta*) were shown to carry the same Asn-Pro substitution, although these cytochrome P450s utilize valine and isoleucine as substrates.[149] Cytochrome P450tyr is the first functionally characterized P450 from a mono-cotyledonous plant and is the first membrane-bound *N*-hydroxylase found to have high substrate specifity. A final and unambiguous proof of the multifunctionality of P450tyr was obtained by expression of the protein in *E. coli*.[150,151] The recombinant P450tyr showed the same multifunctional catalytic properties as the P450tyr isolated from sorghum (Fig. 6).[152]

The *E. coli* expression system is highly efficient. The amount of P450tyr that can be isolated from a single liter of *E. coli* culture equals the amount obtained from approximately 1000 kg of sorghum seedlings. *E. coli* does not contain endogenous P450s or an NADPH-cytochrome P450 oxidoreductase. Nevertheless, *E. coli* is able to support catalytic activity of heterologously expressed cytochrome P450 because of the presence of two flavoproteins - flavodoxin and NADPH-flavodoxin oxidoreductase - that are able to donate the reducing equivalents required for the P450 to carry out its catalytic function.

In $^{18}O_2$ experiments performed with microsomal preparations, it was not initially possible to understand how the enzyme system could discriminate between the two oxygen atoms introduced by the two subsequent *N*-hydroxylations of tyrosine because the *N,N*-dihydroxytyrosine intermediate would be expected to show free rotation around the C-N single bond.[125] Given the multifunctionality of P450tyr and that both *N*-hydroxylations take place in the same active site, it appears that the decarboxylation and dehydration reactions proceed at a stage where hydrogen abstraction from *N*-hydroxytyrosine has taken place and while the nitrogen atom is still bound to the active iron-oxo complex. This would result in complete loss of the oxygen atom incorporated by *N*-hydroxylation of tyrosine as experimentally observed.[125]

1-*aci*-Nitro-2-(*p*-hydroxyphenyl)ethane has previously been reported to be produced in significant amounts from *N*-hydroxytyrosine and to be metabolized into *p*-hydroxymandelonitrile by the microsomal system.[112] These results, in combination with the demonstrated consumption of two molecules of oxygen in the conversion of tyrosine to the oxime, were interpreted to indicate that 2-*aci*-nitro-3-*p*-hydroxyphenylethane is an intermediate between *N*-hydroxytyrosine and the oxime.[112] With our present knowledge of the multifunctionality of P450tyr, 1-*aci*-nitro-2-(*p*-hydroxyphenyl)ethane must be considered to represent a side product of the pathway. 1-*aci*-Nitro-2-(*p*-hydroxyphenyl)ethane may arise when *N,N*-dihydroxytyrosine dissociates so slowly from the active site of P450tyr that a third *N*-hydroxylation reaction takes place. Alternatively, the *N,N*-dihydroxytyrosine undergoes a dehydration reaction, resulting in the formation of 2-nitroso-3-*p*-hydroxyphenylpropionate, which then becomes hydroxylated to form 2-nitro-3-*p*-hydroxyphenylpropionic acid. The latter compound would then decarboxylate spontaneously into 1-*aci*-nitro-2-(*p*-hydroxyphenyl)-ethane.

Cytochrome P450ox was cloned by polymerase chain reaction (PCR) approaches.[32] Plant P450s have been divided into the A type and non-A type,[153] and more recently into four main clans.[147] The A-type catalyzes reactions in plant secondary metabolism, whereas those involved in general household reactions are assigned as belonging to the non-A type. Computer-based alignment of all available A-type sequences identified three consensus sequences: FXPERF, PFGXGRRXCXG, and (V/I)KEX(L/F)R.[32] Based on highly degenerated inosine-containing primers derived from the consensus sequences, three novel cytochrome P450s were cloned in addition to cinnamic acid 4-hydroxylase (CYP73) from *Sorghum bicolor*.[32] Database sequence comparisons identified two of the novel P450s as the first representatives of two new P450 families, CYP98 and CYP99, whereas the third clustered with the CYP71 family and belongs to the first member of the CYP71E1 subfamilily. All four P450s are grouped into the 71 clan.[147] Originally, members of the CYP71 family were thought to be involved in terpenoid synthesis,[132,145] but subsequent analyses have shown that the CYP71C subfamily is involved in the biosynthesis of the plant defense compounds DIBOA and DIMBOA found in grasses.[154,155] As for P450tyr, expression studies were carried out in *E. coli*.[150,151] These studies identified the CYP71E1 clone as the one that encodes a P450 enzyme able to use *p*-hydroxyphenylacetaldoxime as a substrate and convert it into *p*-hydroxymandelonitrile. P450ox is labile compared with the P450s previously isolated from sorghum.[31,143] With the microsomal *in vitro* system, the conversion of *p*-hydroxyphenylacetaldehyde oxime to *p*-hydroxymandelonitrile had earlier been shown to constitute the most labile part of the system.[108,109] This may be because cytochrome P450ox catalyzes a dehydration reaction in addition to a *C*-hydroxylation reaction.[31,32,143] The mechanism allowing P450ox to catalyze a dehydration reaction is not understood. The P450-mediated dehydration of aldoximes to nitriles has previously been reported with liver microsomes.[156-158] A major difference between the liver microsomal system and P450ox is that the former requires strict anaerobic conditions, whereas the latter proceeds aerobically and catalyzes a subsequent *C*-hydroxylation reaction. The liver P450 that catalyzes aldoxime dehydration has not yet been isolated.

The availability of recombinant CYP79A1 and CYP71E1 permitted a more detailed analysis of the substrate specificity of the isolated enzymes after reconstitution with NADPH-cytochrome P450 reductase.[143] All intermediates in dhurrin synthesis, as well as phenylalanine and cinnamic acid, produced Type I substrate binding spectra with CYP79A1, although the amplitudes observed were smaller compared to the Type I spectra obtained with the genuine substrates tyrosine and N-hydroxytyrosine. As observed with the microsomal system, tyrosine was the only substrate found to be metabolized. CYP71E1 produced a reverse Type I spectrum with its substrate, *p*-hydroxyphenylacetaldoxime. This is surprising because CYP71E1 is isolated in the low spin state.[143] The peculiar spectral properties of CYP71E1 were attributed to the fact that this particular cytochrome catalyzes an unusual dehydration reaction of a Z-oxime to a nitrile that may represent the survival of a traditional cytochrome P450 catalyzed reaction from the preoxygenic era. In the subsequent *C*-hydroxylation reaction, CYP71E1 catalyzes the conversion of *p*-hydroxyphenylacetonitrile to the cyanohydrin *p*-hydroxymandelonitrile. Recombinant CYP71E1 was found to produce a Type I binding spectrum with *p*-hydroxyphenylacetonitrile and with several other nitriles when analyzed at 4°C. Again, the amplitude was the largest when measured with the nitrile serving as the genuine substrate. The substrate specificity of CYP71E1 is not as absolute as that of CYP79A1.[143]

In P450tyr and P450ox, the NH_2-terminal segment in front of the proline-rich region contains 55 and 68 amino acids, respectively.[29,32] Generally, the length of this segment in A-type cytochrome P450s is about 35 amino acids, including the hydrophobic segment that forms the membrane anchor. The substrates for P450tyr and P450ox are hydrophilic compared with most other P450 substrates and, therefore, will be found as solutes in the cytoplasm and not trapped within the lipid matrix of membranes. The unusually long NH_2-terminal segment may serve to facilitate the access of substrates to the respective active sites.

The identification of P450tyr and P450ox as multifunctional enzymes explains why all intermediates in dhurrin biosynthesis except the Z-oxime are channelled.[111] It also raises the question of the true nature of the intermediates in the biosynthetic pathway. Because all of these intermediates except the Z-oxime are kept within the catalytic site of one or the other of the two P450s and do not freely exchange with the exogenously added compounds, possibly they should not be considered genuine intermediates. More likely, the intermediates are enzyme-bound, short-lived, reactive complexes, or more-or-less stable transition states. A similar situation has appeared in studies with the multifunctional $P450_{scc}$ from adrenal cortex that catalyzes the channeled conversion of cholesterol to pregnenolone.[159,160]

Studies with Microsomal Systems from Other Plant Sources

Cassava. A microsomal system that catalyzes *in vitro* synthesis of the aglucones of the two cyanogenic glucosides linamarin and lotaustralin has been isolated from young seedlings of cassava (*Manihot esculenta*; Euphorbiaceae) (Fig. 7). Valine and isoleucine were converted in the presence of NADPH and oxygen to linamarin and lotaustralin in a process involving P450 enzymes.[118] 2-(2-Cyclopentenyl)glycine was also converted to cyanogenic glucosides with this enzyme system.[118] The enzyme system is located in the cotyledons and their petioles. After synthesis, linamarin and lotaustralin are transported rapidly to other parts of the growing seedling.[118] Because 2-(2-cyclopentenyl)glycine served as a precursor for cyanohydrin biosynthesis when administered to cassava microsomes, the question arose as to whether cyclopentenoid derived cyanogens might occur in cassava. Subsequent studies demonstrated that such compounds are not present, or present in amounts less than 0.03% of the linamarin content.[161] This indicates that cassava *in vivo* is not able to synthesize 2-(2-cyclopentenyl)glycine.

After synthesis in the cotyledons and petioles, linamarin and lotaustralin are transported rapidly to the root of the cassava plant.[118] Transport of cyanogenic glucosides has previously been proposed to occur as diglycosides, special transport forms that are not cleaved by β-glucosidases.[23,162,163] This so-called linustatin pathway was first demonstrated by Selmar in studies with the rubber tree (*Hevea brasiliensis*).[23] In agreement with Selmar's transport hypothesis, minute quantities of the two diglycosides linustatin and neolinustatin, derived from linamarin and lotaustralin, respectively, were detected in the cassava seedlings.[164] In older cassava plants, however, these transport compounds could not be detected.[165] This may be because *de novo* synthesis of cyanogenic glucosides also takes place in the cassava roots,[166] and that transport of cyanogenic glucosides from leaves to roots may not take place during all developmental stages. Alternatively, the sensitivity of the assays used was not high enough.

The presence of cyanogenic glucosides constitutes a major obstacle for food quality improvement in cassava.[167] Attempts to eliminate the production of cyanogenic glucosides through traditional breeding have not been successful. An alternative approach would be to block the initial steps in linamarin and lotaustralin biosynthesis by using anti-sense technology. Towards this goal, two genes belonging to the CYP79 family, CYP79D1 and D2, that control the rate limiting steps in linamarin and lotaustralin biosynthesis have recently been isolated from cassava.[149]

Triglochin maritima. The biosynthesis of taxiphyllin has been studied with a microsomal preparation from etiolated seedlings of *T. maritima* (Fig. 8).[115,116] In the presence of NADPH and molecular oxygen, this microsomal system catalyzes the conversion of tyrosine into *p*-hydroxymandelonitrile. Initial biosynthetic experiments showed that *p*-hydroxyphenylacetonitrile was the only detectable intermediate.[115] As previously observed in sorghum, the biosynthetic pathway from the amino acid to the cyanohydrin was also channelled in *T. maritima*.[116] In sorghum, the (Z)-aldoxime was the only intermediate that equilibrated freely with exogenously added intermediates.[111] In *T. maritima*, the nitrile was the only intermediate that exchanged freely with exogenously added intermediates, although small amounts of the oxime could be detected in trapping experiments.[116] The most obvious interpretation of this observation would be that in *T. maritima* the conversion of the amino acid to the cyanohydrin is not necessarily catalyzed

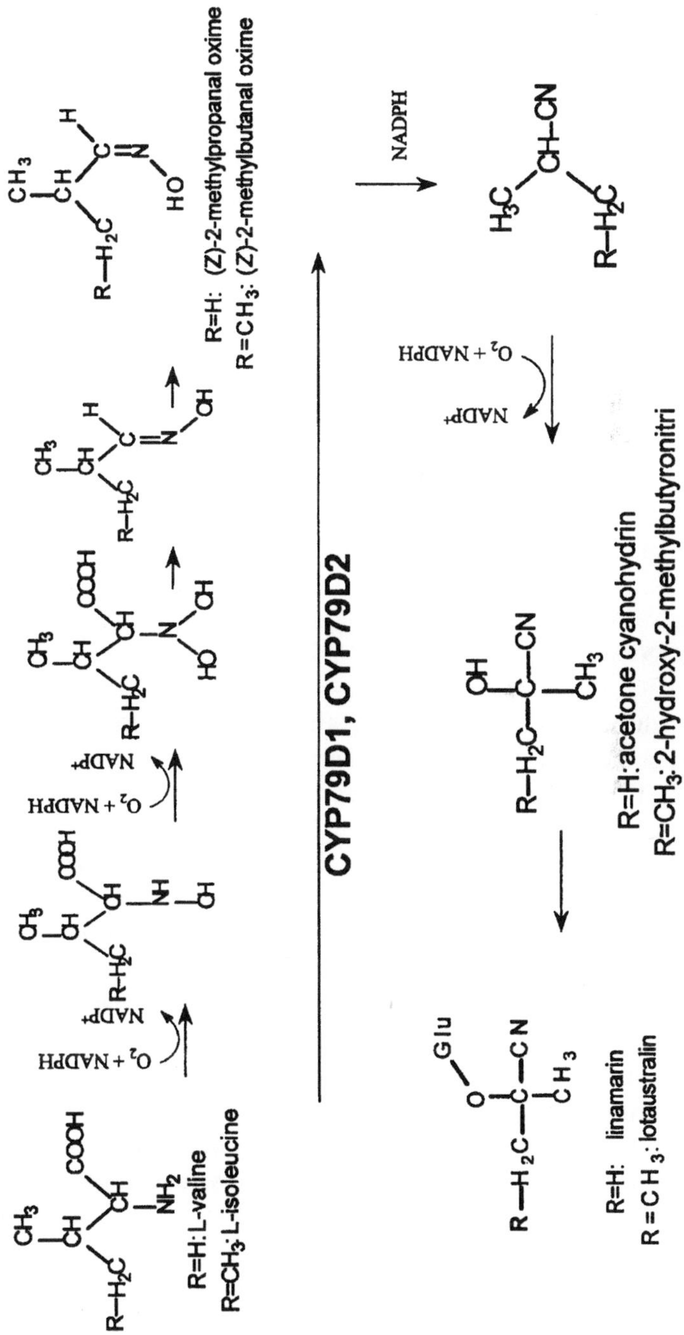

Figure 7. Intermediates in the biosynthesis of the cyanogenic glucosides linamarin and lotaustralin in *Manihot esculenta*

by two multifunctional P450s in the same manner as in sorghum. To resolve this enigma, cyanogenic glucoside synthesis in *T. maritima* was reinvestigated.[13] In this study, the oxime as well as the nitrile was found to accumulate in the microsomal reaction mixtures. Using radiolabelled tyrosine as substrate, the radiolabel was easily trapped in the oxime as well as the nitrile when these were added as unlabelled compounds. It was hypothesized that cyanogenic glucoside synthesis in *T. maritima* is catalyzed by multifunctional cytochrome P450 enzymes similar to CYP79A1 and CYP71E1 in sorghum, except that the homolog to CYP71E1 in *T. maritima* exhibits a less tight binding of the nitrile, thus permitting the release of this intermediate and its conversion into triglochinin. According to this scheme, the nitrile would constitute the branch point between the biosynthetic pathways for taxiphyllin and triglochinin. Recently, two cDNA clones encoding homologs (CYP79E1 and CYP79E2) to CYP79A1 from sorghum have been isolated and characterized from *T. maritima*.[168] Each of these homologs catalyzes the conversion of tyrosine to *p*-hydroxyphenylacetaldoxime, as documented by heterologous expression in *E. coli*.[168] No information is available on the enzyme system responsible for the ring opening reaction resulting in triglochinin formation.

EVOLUTION

Cytochrome P450s are found in all kingdoms of life. Phylogenetic trees have been made that include many different origins (see:http:/drnelson.utmem.edu/P450trees.htlm). The fungal cytochrome P450 sequences do not all cluster together, nor do the plant and animal sequences. However, there is a distinct assortment of these into kingdom-specific clusters.[147] According to the approved nomenclature system for cytochrome P450s, each of these clusters includes multiple families,[146,147] and these are now referred to as clans.[146] It is thought that the cytochrome P450s within each clan have diverged from a single common ancestor. Plant cytochrome P450s are divided into four main clans designated the 71, 72, 85, and 86 clans.[147] The Clan 71 is the largest, and the two multifunctional sorghum cytochrome P450s, CYP79A1 and CYP71E1, both belong to this clan.[147]

In all cases studied, the conversion of amino acids to Z-oximes in the biosynthesis of cyanogenic glucosides has been shown to involve cytochrome P450 enzymes belonging to the CYP79 family: CYP79A1 in *S. bicolor*, CYP79D1 and CYP79D1 in *M. esculenta*, and CYP79E1 and CYP79E2 in *T. maritima*. In *S. bicolor* and *T. maritima*,[168] tyrosine is the amino acid substrate; in *M. esculenta*,[149]

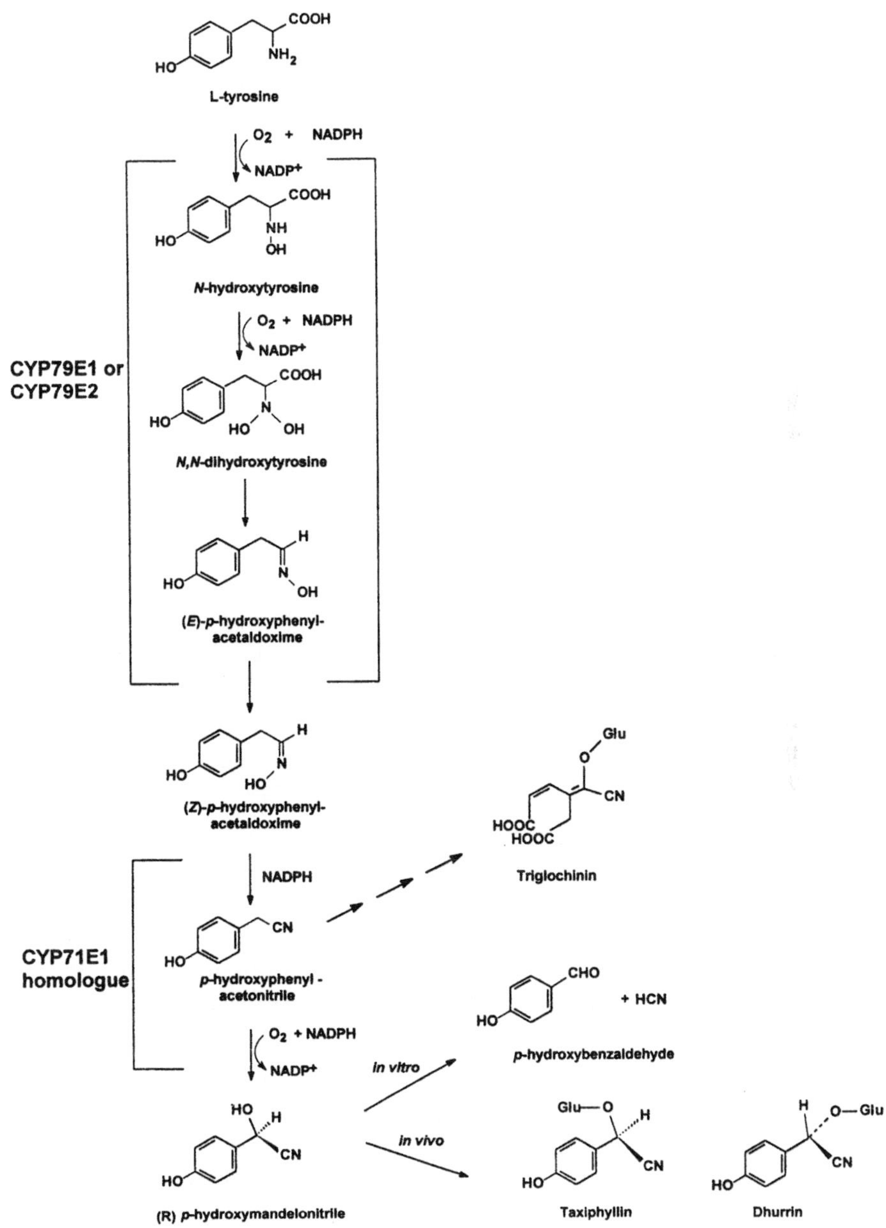

Figure 8. Intermediates in the biosynthesis of the cyanogenic glucosides taxiphyllin and triglochinin in *Triglochin maritima*

the substrates are valine and isoleucine. Also, the biosynthetic pathways of at least some glucosinolates involve cytochrome P450 enzymes in the conversion of the parent amino acid to the corresponding aldoxime.[169] CYP79 homologs have been identified in the two glucosinolate producing plants *S. alba* (CYP79B1) and *A. thaliana* (CYP79B2), but so far no successful expression of the isolated clones has been achieved.[170] Both plants are members of the Brassicaceae family but contain glucosinolates derived from different substrates.[169,171] Despite this difference, the two full length clones showed 89% identity at the amino acid level (Table I). Alignment of the CYP79 sequences obtained has identified some highly conserved CYP79 specific regions covering part of the I-helix, β6-1, and β1-4.[172,173] PCR approaches based on these sequences should provide easy access to CYP79 homologues in other species. Conserved regions within groups and families of cytochrome P450s have previously successfully been used to design degenerate primers for amplification of specific cytochrome P450 clones.[134,144,170,174] CYP79E1 and CYP79E2 show 94% identity, whereas CYP79D1 and CYP79D2 show 85% identity (Table I). CYP79E1 showed 45-49% identity to other members of the CYP79 family. In sequence alignment to other members of the CYP79 family, CYP79E1 showed highest identity and similarity to CYP79A1, the only other sequence from a monocotyledoneous plant currently available. In contrast, CYP79A1 showed higher identity/similarity to CYP79B1, CYP79B2, and CYP79D1, all from dicotyledonous plants. This is surprising considering that CYP79B1 and CYP79B2 are putatively involved in the biosynthesis of glucosinolates, whereas CYP79D1 is involved in cyanogenic glucoside synthesis. Although isolated from two different species, CYP79B1 and CYP79B2 show 89% identity. A similar high identity is not seen between the three CYP79 enzymes isolated from different species producing cyanogenic glucosides.

Cytochrome P450s are thought to possess the same tertiary structure.[172] Nevertheless, they only exhibit one conserved sequence FXXGXRXCXG (X being any amino acid), harboring the heme-binding cysteine residue.[146,170] When restricted groups or families of cytochrome P450s are aligned, additional conserved amino acid residues are found, such as the proposed highly conserved A-group heme binding consesus sequence - PFGXGRRXCXG,[153,170] and the CYP79 specific sequence - (S/T)F(T/S)TG(R/K)RGCX(G/A).[149,153,168] The CYP79 family diverges from the A-group heme binding consensus sequence. A notable variation in the CYP79 specific sequence is that the generally highly conserved glycine residue that is positioned two amino acids downstream from the heme-binding cysteine residue (amino acid 479 in CYP79E1), in most CYP79s is replaced by alanine. This glycine residue is positioned closely to the heme plane and allows a sharp turn from the cysteine-pocket into the L-helix.[172] In the highly conserved PERF region within microsomal cytochrome P450s (448-451 in CYP79E1), the phenylalanine residue has been replaced by histidine in all currently known CYP79s.[149,168,170] The differences, both within the CYP79 family, and in relation to residues that are more or less conserved within the A-group or microsomal P450s in general, illustrate how difficult it is, based on sequence alignments, to predict amino acids important for different functions.

All eukaryotic cytochrome P450 enzymes found so far, except CYP55A,[175] are membrane-bound. Some are localized in the mitochondrial membranes, while most, including plant cytochrome P450s, are localized in the endoplasmic reticulum.[176] The N-terminal of the microsomal cytochrome P450s functions as a hydrophobic signal sequence directing the proteins to the endoplasmic reticulum. The CYP79Es from *T. maritima* contain the four structurally conserved domains in the N-terminal.[152] CYP79Es have a short stretch of four uncharged amino acids between the positively charges region and the proline-rich region,[168] as opposed to CYP79A1 that has a long stretch of 18 uncharged amino acids at this position.[29] It has previously been suggested that a long stretch of uncharged residues serves to facilitate the access of the rather hydrophilic substrate tyrosine to the active site.[29] The fact that the cyanogenic glucosides in *T. maritima* are derived from tyrosine contradicts this suggestion.

TABLE I

Amino acid sequence identity and similarity (%) between six members of the CYP79 family

Similarity / Identity	CYP79E1	CYP79E2	CYP79A1	CYP79B1	CYP79B2	CYP79D1
CYP79E1		95.2	61.7	58.1	58.9	60.0
CYP79E2	94.1		61.5	57.6	58.5	59.2
CYP79A1	48.8	48.8		65.5	67.1	65.8
CYP79B1	44.9	44.9	51.3		92.3	65.1
CYP79B2	44.5	44.6	52.6	89.3		67.3
CYP79D1	46.4	46.5	51.5	49.1	50.7	

GLYCOSYLTRANSFERASES IN PLANT NATURAL PRODUCT METABOLISM

Enzyme Characteristics

The last step of cyanogenic glucoside biosynthesis employs an apparently soluble UDP-glucose:glucosyltransferase, which catalyzes the formation of a cyanogenic glucoside from the corresponding α-hydroxynitrile and UDP-glucose (Fig. 9).[33,63,111,115,177-179] Generally, glycosyltransferases catalyze the transfer of sugars from nucleotide-diphosphate sugars onto a wide range of substrates, predominantly at oxygen atoms of carboxyl and hydroxyl groups.[180] The reaction can be summarized as follows:

XDP-sugar + R-OH -----> XDP + R-O-sugar (R=aglycone, X=nucleotide)

Glucosyltransferases involved in the synthesis of plant natural products only transfer glucose moieties from UDP-glucose, resulting in the formation of glucosides. Phenolic compounds such as simple phenols and flavonoids,[180] terpenoids and other isoprenoids,[181] plant growth regulators,[182] and xenobiotics[58] are commonly found as glycoconjugates in plants. The sugar moieties are more polar than most aglycones, and glycosylation, therefore, results in products with increased hydrophilicity.[183] Glycoconjugates have also been found to be less chemically reactive than their respective aglycones,[184,185] while glycosylation in other cases has been associated with a change in biological activity.[186-188] It would seem likely that glycosylation, therefore, would be beneficial prior to compartmentation in aqueous and acidic vacuoles. Natural products as well as xenobiotic metabolites would thereby become more chemically inert and stable, ensuring that the vacuolar structure and chemical complement remain intact.[189] Modulation of biological activity by glycosylation and de-glycosylation, as with certain plant growth regulators,[182] would seem far more efficient than catabolism and *de novo* synthesis, both in terms of velocity and energy consumption. In some cases, glycoconjugates may also serve as transport forms,[23,163,190-192] or glycosylation may constitute a requirement for transport across membranes.[53,54,193] An interesting question, therefore, is whether the presence of a particular glycosyltransferase is neccesary for the accumulation of the resultant glycoside? Given the chemical lability of aglycones, such as α-hydroxynitriles[92,194] and anthocyanidins,[184] this may well be the case for certain plant natural products, in particular when the aglycone form has not been observed to accumulate *in vivo*. Given that at least 50 different glycosidic forms have been identified in single plant species,[195,196] it is interesting to ask whether these conjugation reactions are catalyzed by few broad-specificity glycosyltransferases or whether numerous high-specificity enzymes are present (Fig. 10)? Several partially purified protein preparations have shown activities towards a wide range of substrates,[61,62,197-199] although only closely related substances usually are assayed. Glycosyltransferases purified to homogeneity have shown high[200] as well as broad[49] substrate-specificity, and it, therefore, seems most likely that both classes exist within plants. An important concept highlighted by Vogt and Taylor[193] is that of substrate accessibility *in vivo*. While certain purified glycosyltransferases can show broader substrate specificity *in vitro* than what is observed *in vivo*, differences between *in vivo* and *in vitro* specificity can be shadowed by factors such as compartmentation,[193] channelling,[193] and selective endogenous inhibitors.[201]

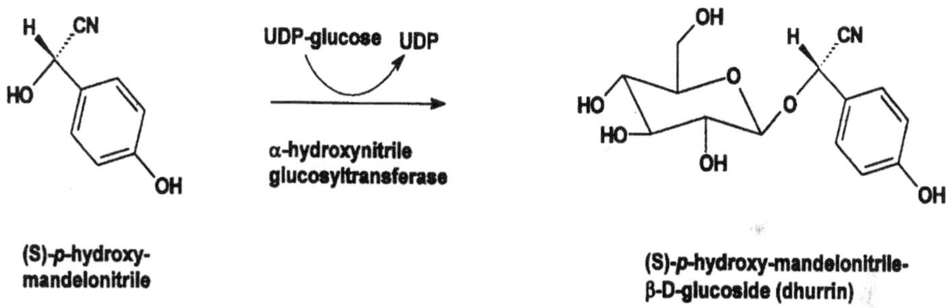

Figure 9. The last step in the biosynthesis of the cyanogenic glucoside dhurrin is catalyzed by a UDPG-*p*-hydroxymandelonitrile glucosyltransferase.

From more than 50 reports of partially purified glycosyltransferases, few aspire to have achieved purification to homogeneity.[33,49,200,202-204] The apparent lack of success in their purification may be attributed to the lability of certain glycosyltransferases[203] and the low level of protein present in cells.[49,200] For example, in young fruits of Pummelo (*Citrus*), the flavanone-glucoside naringin constitutes up to 75% dry weight, while the glucosyltransferase protein only comprised approximately 0.08% of the total soluble protein.[49]

The apparent molecular masses of most plant glycosyltransferases partially purified to date have ranged between 43 and 61 kDa.[49,62,177-179,199,201,202,205-214] A number of purified and partially-purified glycosyltransferases in the 45 to 59 kDa range have upon denaturing gel electrophoresis (SDS-PAGE) appeared to be composed of two equivalent sized sub-units.[199,202,214] However, all putative and

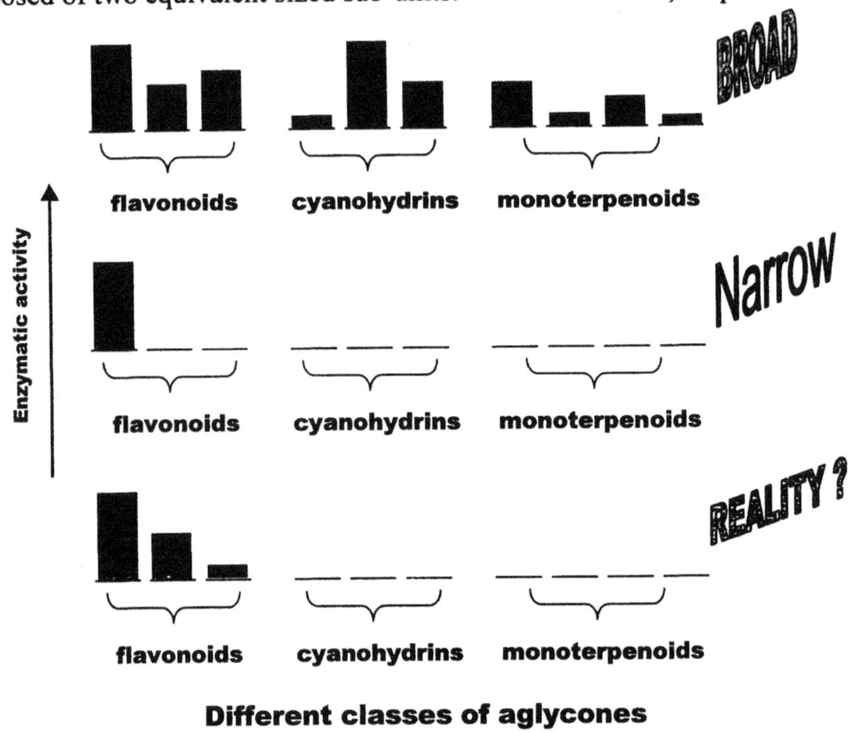

Figure 10. Do UDPG-glucosyltransferases exhibit broad, narrow or very high substrate specificity? A UDPG-Glucosyltransferase with broad substrate specificity would be able to glucosylate a large number of aglycons belonging to different classes of compounds in primary as well as secondary metabolism. A very high substrate specificity would require the existence of an immense number of UDPG-glucosyltransferase genes. As documented with the cyanohydrine glucosyltransferase, the reality most likely is that UDPG-glucosyltransferases preferentially glucosylates structurally similar compounds within a specific class of compounds but in addition exhibits activity towards a few selected compounds belonging to completely different classes.

known glycosyltransferase-encoding cDNAs encode proteins larger than 50 kDa in size (see below). The greater majority of plant glycosyltransferases display maximum activity at a slightly alkaline pH of 7.5-8.0, while activity maxima around pH 5.0-5.8 also have been reported.[62,193,199,214] Metal cations have been found generally to inhibit activity to some degree,[209] while Cu^{2+}, in all cases but one,[62] inhibits glycosyltransferase activity strongly. This may, however, be due to the chemical lability of the substrate or end-product,[206] as Cu^{2+} is a strong catalyst of oxidation.

Molecular Biology of Glycosyltransferases

cDNAs and genes that putatively encode for glucosyltransferases have been isolated from mammals, plants, viruses, and bacteria. A maize UDP-glucose:flavonoid-glucosyltransferase was the first to be cloned and sequenced from plants, through the use of transposon tagging.[215] Subsequently, other putative flavonoid-glucosyltransferase-encoding cDNAs have been obtained by library screening from plants such as *Anthirrinium majus*,[216] *Vitis vinifera*,[217] and *Hordeum vulgare*,[218] and through differential screening from tomato,[219] and *Petunia hybrida*.[220] Only three plant glucosyltransferase-encoding cDNAs have to date been identified conclusively with their corresponding proteins. These utilize anthocyanidins,[221] indole-3-acetic acid,[222] and the alkaloid solanidine[50] as accepting substrates.

All putative glucosyltransferase-encoding cDNAs encode for proteins of similar mass, namely between 50-60 kDa. Moreover, they all show an overall positional amino acid identity, between 20 and 30%, with varous sections of markedly higher homology. One particular section towards the 3′-end of all sequences has attracted much attention, due to the high homology between organisms as diverse as mammals and microbes.[218,219,223,224] The common denominator of all of these sequences is the belief that they encode enzymes that catalyze the transfer of UDP-sugars to a suitable acceptor. As suggested by Hundle *et al.*,[225] and Szerszen *et al.*,[222] a common UDP-binding site seems the most likely reason behind this high degree of homology.

Further compelling evidence for the identity of the homologous section and the general structure of mammalian glycosyltransferases was provided by Mackenzie,[226] and by Ritter *et al.*[227] They showed that unique human mRNA sequences, each having two distinct domains and encoding six complete UDP-glucoronosyltransferases with different acceptor substrate specificity, are formed by combination of a single mRNA exon (representing one domain) with one of six different mRNA exons (representing the second domain) *in vivo*. The first mRNA exon, which encodes for the UDP-glucoronosyl binding domain, was shown to be identical for all glycosyltransferases in this family, while the second mRNA exon conferred specificity for the acceptor substrate (*e.g.* phenol and bilirubin). Although the same biology may not function in plants, it points to a possible common evolution of these enzymes, with differing domains conveying different substrate binding affinities and functions. The two partially purified polypeptide subunits, allegedly found in some plant species, may represent two such differing domains.[199,202,214]

CYANOHYDRIN UDPG-GLYCOSYLTRANSFERASES

β-Glycosyltransferases have been partly purified from a few cyanogenic species.[63,97,115,178,179,228] They have pH optima in the range of 6.5-9.0 and usually lack a requirement for metal ions or cofactors. Although they have an absolute specificity for UDP-glucose, they are less specific toward cyanohydrins.

Hahlbrock and Conn[97] purified an enzyme system from flax that catalyzed the glucosylation of a series of 2-hydroxynitriles. This enzyme system exhibited maximum rates with the cyanohydrins of acetone and methyl ethyl ketone, the aglycones of linamarin and lotaustralin. The enzyme preparation was inactive towards aromatic cyanohydrins. The enzyme glucosylated both the (*R*) and (*S*)-forms of the cyanohydrin of 2-butanone. Usually, racemic mixtures of cyanogenic glucosides are not encountered in plants. This suggests that, in general, only one stereoisomer of the cyanohydrin is glucosylated by the enzyme. In contrast to cyanohydrins derived from aromatic amino acids, the cyanohydrin derived from isoleucine carries two substituents (methyl and ethyl groups) that structurally are similar. This may be the reason why the flax enzyme system glucosylates the *(R)*- as well as the *(S)*-form of the cyanohydrin of 2-butanone, thereby producing two different glucosides *in vitro*.

Studies in *T. maritima* demonstrated that the microsomal system supplemented with UDP-glucose and a soluble fraction catalyzed the synthesis of taxiphyllin. Incubation of the glucosyltransferase fraction with *p*-hydroxy-(*R,S*)-mandelonitrile and UDP-glucose produced taxiphyllin, but not dhurrin, demonstrating that the UDPG-glucosyltransferase specifically glucosylates only one of the stereoisomeric cyanohydrins.[177,229]

Studies that use cell-free extracts from *Prunus serotina* have demonstrated that racemic mandelonitrile is specifically converted into prunasin. These preparations did not convert prunasin to the diglycoside amygdalin.[16,228]

The homology of the putative UDP-binding domain of glycosyltransferases was utilized by Hughes and co-workers to isolate six different putative glucosyltransferase cDNAs.[223] A cassava cDNA library was screened at low stringency, *i.e.* under conditions that allow some degree of mismatch between the probe and target sequence, by using a flavonoid-glucosyltransferase-encoding sequence fragment from *Anthirrhinum majus* as a probe. No reports of attempts to identify these sequences through transformation studies or functional expression in *E. coli* have appeared. However, it is possible that one of these cDNAs encodes an acetone cyanohydrin glucosyltransferase, as a tissue actively synthesizing cyanogenic glucosides was selected for cDNA library construction.

In *S. bicolor*, the last step in dhurrin biosynthesis is the glucosylation of *p*-hydroxymandelonitrile by a UDP-glucose:glucosyltransferase.[63,111,177] This enzyme has recently been isolated from etiolated seedlings by employing Reactive Yellow 3 chromatography with UDP-glucose elution as the critical step.[33] Amino acid sequencing allowed the cloning of a full-length cDNA encoding the glucosyltransferase. Among the few characterized glucosyltransferases, the deduced translation product showed highest overall sequence identity to *Zea mays* flavonoid-glucosyltransferase (*Bz-Mc-2 allele*).[230] The substrate specificity of the enzyme was established by using isolated recombinant protein produced in *E. coli*. Compared to endogenous *p*-hydroxymandelonitrile, mandelonitrile, benzylalcohol, and benzoic acid were utilized at maximum rates of 87%, 15%, and 6%, respectively.[33] The picture that is emerging regarding plant glucosyltransferase substrate specificity is one of limited but extended plasticity towards metabolites of related structure. This in turn ensures that a relatively high, but finite, number of glucosyltransferases can give rise to the large number of glucosides found in plants.[33]

SUBCELLULAR LOCALIZATION OF ENZYMES AND INTERMEDIATES

Biosynthetically active *in vitro* systems catalyzing the conversion of a parent amino acid to the corresponding oxime have been isolated from a number of plant species.[43,108,111-118] In all cases, these are obtained as microsomal preparations. Further fractionation of the microsomal membranes has shown that the biosynthetic activity co-purifies with the membranes of the endoplasmic reticulum.[231] In agreement with this, the multifunctional cytochrome P450 enzymes catalyzing these reactions contain motifs in their N-terminal region pointing to such a location.[29,149] However, it can not be excluded that vesicles enriched in these membrane-bound enzymes are subsequently derived from the endoplasmatic reticulum. In contrast to the conversion of amino acids to oximes, the last step in cyanogenic glucoside biosynthesis is catalyzed by an apparently soluble UDP-glucose glucosyltransferase. Given that the optimum activity of the glucosyltransferase, where reported, has been observed at pH 8-9, it would seem more likely that the glucosyltransferases are cytosolic rather than vacuolar. In a more detailed study by Wurtele *et al.*,[12] it was found that a significant proportion of the glucosyltransferase activity was associated with a plastid fraction, after separation by sucrose density gradient centrifugation. In *Sorghum*, dhurrin accumulates in vacuoles.[76,77] *p*-Hydroxy-mandelonitrile, the α-hydroxynitrile aglucone, is labile to dissociation at the physiological pH of plant cytosols,[92,194] while dhurrin is relatively stable.[232] Some of the synthesized *p*-hydroxymandelonitrile would, therefore, be expected to dissociate into *p*-hydroxybenzaldehyde and HCN prior to glucosylation, if *p*-hydroxymandelonitrile is not channelled *in vivo* or in any other way avoids release into the cytosolic space. This is particularly so in species such as *Sorghum bicolor* and *Linum usitatissimum* where cytosolic α-hydroxynitrile lyases are present in tissue where biosynthesis is known to occur.[233] Given the toxicity of both of these catabolites and the apparent futility of catabolism subsequent to biosynthesis, it has been suggested that the α-hydroxynitrile glucosyltransferase is co-located with the two P450s.[111] However, experiments, where a bifunctional reagent was included in the protein extraction buffer in order to bind closely linked proteins, failed to show any such association.[111] Recently, McMahon *et al.*[234] claimed that vacuoles isolated from cassava were able to synthesise acetone cyanohydrin (the α-hydroxynitrile) from exogenous valine (amino acid precursor), suggesting a vacuolar location of the corresponding glucosyltransferase. Given that the activity optima of two acetone cyanohydrin glucosyltransferase preparations from cassava leaves were between pH 7 and 9,[179] this would seem unlikely. Likewise, tonoplast preparations from sorghum do not possess the ability to metabolize tyrosine into the corresponding oxime or cyanohydrin (Møller, unpublished data). Any "accidental" release of HCN during biosynthesis may be rapidly taken care of by incorporation of the cyanide into β-cyanoalanine by cytosolic β-cyanoalanine synthase.[235] In a complex paper by Adewusi,[90] it was concluded that sorghum seedlings had a metabolic turnover of dhurrin equivalent to 27% of total biosynthesis. It is not clear whether the experiments precluded the possibility of dissociation prior to dhurrin biosynthesis or mobilization. Hence, some degree of α-hydroxynitrile breakdown prior to glucosylation may occur. Nevertheless, the subcellular localization of α-hydroxynitrile glucosyltransferases and the two cytochrome-P450s are yet to be determined conclusively.

ACCUMULATION

In *S. bicolor*, the accumulation of cyanogenic glucosides is highly tissue -and development-specific. For example, while the seeds contain very low amounts of cyanogenic glucosides,[236] dhurrin may constitute up to 30% dry weight of the shoot apex in young seedlings.[110] Despite reports that no cyanoglucosides are present nor synthesized in roots of young sorghum seedlings,[110] other researchers have found evidence for cyanogenic glucoside accumulation in root tissue.[90,237,238] In seedlings and roots of mature cassava plants, the cyanogenic glucosides linamarin and lotaustralin are present in all tissues.[118,194] In almonds and flax, the monoglucosides have different tissue and development specificities to their corresponding diglucosides,[239] whereas plants such as sorghum and cassava have been observed to contain diglucoside forms in their apoplastic fluids.[163,164,192] A common trend among most cyanogenic plant species is that young plants contain the highest concentrations of cyanoglucosides.[16,110,239]

CATABOLISM

Some cyanogenic glucosides such as dhurrin[232] and most α-hydroxynitriles[92,194] are vulnerable to spontaneous breakdown in macerated plant tissue. In addition, certain plants may contain non-specific β-glucosidases with activity towards cyanogenic glucosides.[240] However, the efficacy of cyanogenesis for defense may be dependent upon how rapid the toxic catabolic by-products are formed.[92] The hydrolysis of the glucosyl moiety is facilitated by specific β-glucosidases,[13,35,241] while the dissociation of the α-hydroxynitrile is catalyzed by specific α-hydroxynitrile lyases.[13,38,40,233,242] Cyanogenic glucosides as well as the catabolizing enzymes are constitutive,[44] necessitating some form of differential compartmentation between substrates and enzymes *in vivo* in order to avoid large-scale release of HCN prior to maceration.

Two studies have investigated whether this compartmentation is at a cellular or a subcellular level,[242,243] while further studies have aimed at investigating the location of the catabolizing enzymes alone.[11,15,233,240,241,249] Early studies on *Sorghum* concluded that dhurrin was present in epidermal leaf tissue, while both catabolizing enzymes were present exclusively in mesophyll leaf cells.[242] However, subsequent studies by Wajant *et al.*[233] that used immunocytological techniques found α-hydroxynitrile lyases present in the cytoplasm of all types of leaf tissue. Therefore, the results by Kojima *et al.*[242] may warrant reconsideration. Thayer and Conn[11] argued that β-glucosidase activity was associated with chloroplast fractions in sorghum leaf mesophyll tissue, as evidenced by sucrose density gradients and activity assays. As dhurrin is located in vacuoles of *S. bicolor*,[76,77] compartmentation in sorghum would certainly seem to occur at a subcellular level, although it cannot be ruled out to occur at the cellular level as well. In other species investigated, β-glucosidase activity has been exclusively, or at least partially, associated with the extracellular space of cyanogenic glucoside-containing cells.[66,240,243-245] In cassava, the non-isotopic digoxigenin labeling system has been used to visualize the presence of the mRNA for linamarase (β-glucosidase) in cells of young leaves.[14] These data demonstrate localization of linamarase gene expression in the lactifers (latex vessels).

Likewise, mRNA encoding linamarase was present in the latex exudate. High levels of linamarase activity were also demonstrated in the latex of leaf petioles dependent on the presence of attached leaflets.[14] In contrast, the α-hydroxynitrilase was only present in minute levels in the latex.[14]

Subcellular localization studies in a number of different plants should be repeated by using more modern techniques, such as immunolocalization, in particular, since clones and antibodies recently have been isolated for the biosynthetic and catabolic enzymes in a number of species.[15,27,246]

MOBILITY, TURNOVER, AND TRANSPORT

Clegg *et al.*[65] monitored the tissue distribution of the cyanogenic glucoside linamarin in wild lima beans (*Phaseolus lunatus*) for 26 days after germination. The total level remained constant throughout the experimental period, while the cyanogenic glucosides disappeared from the cotyledons, only to seemingly reappear in other tissues of the plant. As the break-down products of linamarin are not readily converted to valine, the precursor of linamarin, it was suggested that linamarin was transported intact within the plant.[65] Later studies by Koch *et al.*[118] that used cassava (*M. esculenta*) arrived at a similar conclusion, as no biosynthetic activity could be detected in root tissue where cyanoglucosides are known to accumulate. However, later studies that used antibodies for *S. bicolor* P-450$_{TYR}$ did find evidence for at least some biosynthesis in cassava roots.[165]

Further compelling evidence for cyanoglucoside mobility has been provided by Selmar and co-workers. In *H. brasiliensis*, the mono-glucoside linamarin is di-glucosylated to form linustatin in developing endosperm tissue (Fig. 11).[23,247] It has been hypothesized that linustatin subsequently is transported from the endosperm to the developing leaves where the di-glucoside moiety is cleaved, the α-hydroxynitrile catabolized, and the resultant cyanide is metabolized to asparagine *via* β-cyanoalanine.[23,66,190] Evidence has been given for (1) formation of linustatin from linamarin,[247] (2) exudation of linustatin from endosperm at a particular developmental stage, (3) simultaneous loss of cyanoglucoside content in the endosperm, (4) simultaneous increase in specific di-glucosidase and β-cyanoalanine activity in the developing seedling,[23] (5) movement of linustatin from endosperm to cotyledon tissue and subsequent metabolism,[190] and (6) presence of apoplastic β-glucosidases with specific monoglucosidase activity in the absence of apoplastic diglucosidase activity in leaf tissue.[66]

As cyanogenic glucosides ultimately are derived from a limited number of amino acids, cyanogenic glucoside biosynthesis, mobilization, and metabolism to asparagine may, therefore, constitute a transport system for nitrogen in plants such as *H. brasiliensis*. Where cyanogenic di-glucosides are transported and cleaved to form the corresponding mono-glucoside, this system may play a role in high-efficiency mobilization of constitutive defense compounds. Although no evidence has so far been presented for the mobilization of cyanogenic glucosides in sorghum, the occurrence of a di-glucoside of dhurrin has recently been reported (Fig. 11).[192] Interestingly, significant amounts of the di-glucoside were present in apoplasmic exudate, in support of the hypothesis brought forward by Selmar and co-workers. Minute amounts of the di-glucosides corresponding to linamarin and lotaustralin have also been detected in cassava.[164]

FUTURE PROSPECTS

The pathways for cyanogenic glucoside biosynthesis and degradation have been elucidated, and the enzymes and structural genes involved have now been identified and isolated from a number of plants. This provides important new tools for studies at the plant, tissue, cellular, and subcellular level. Future challenges are to resolve the biology of cyanogenic glucoside function. What are the biochemical mechanisms in the cell that ensure efficient conversion of the labile α-hydroxynitrile (the aglycone) into the more stable glucoside? Does the glucosyltransferase form a metabolon (tight complex) multienzyme with the two multifunctional cytochrome P450 enzymes so that the previously reported channeling of the pathway[111] also includes the glucosylation step?[248] Do such complexes help to exclude the accessibility of other substrates to the active site of the glucosyltransferase, thereby changing otherwise promiscuous glucosyltransferases into enzymes, the substrate specificity of which are defined by their site of binding. In this manner, a limited number of glucosyltransferases (*e.g.* the 65 glucosyltransferases estimated to be present in *A. thaliana*) could be enough to mediate specific glucosylation of a much larger number of aglycones. What are the controlling steps in cyanogenic glucoside synthesis? What are the transport proteins mediating transport of cyanogenic glucosides across membranes? How are these compounds transported between cells and between tissues? What are the turn-over rates? Do cyanogenic glucosides serve a function as easily stored and mobilizable nitrogen? Is their main function rather to serve as phytoanticipins protecting the plant from microorganisms and herbivory? Are cyanogenic glucosides "old fashioned nuclear weapons" that plants might as well get rid of, thus saving the energy cost related to their formation and storage? Does

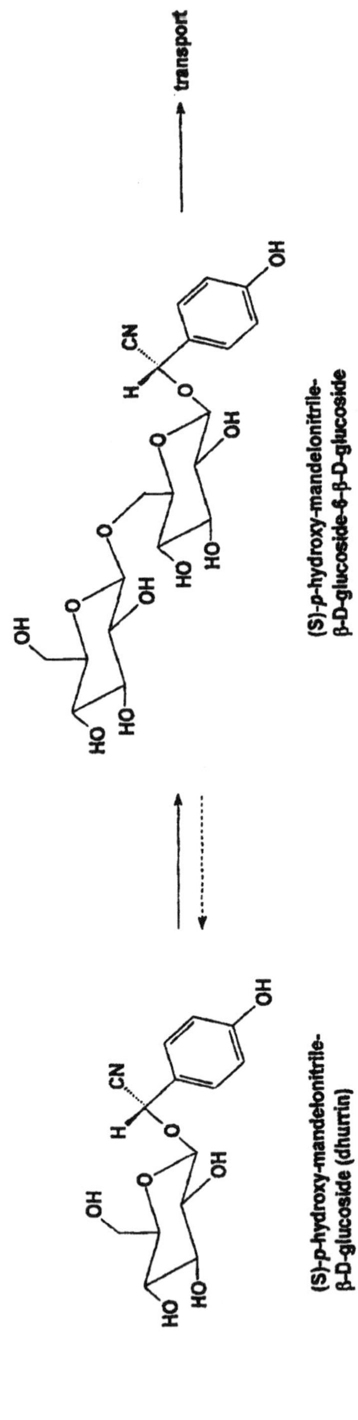

Figure 11. Diglucosides as putative intercellular transport forms of cyanogenic glucosides with dhurrin outlined as an example.

the widespread occurrence of cyanogenic glucosides in the plant kingdom indicate that co-evolution has resulted in different functions of cyanogenic glucosides in different plants in different environments? The membrane bound steps in cyanogenic glucoside and glucosinolate synthesis are the same, and metabolic cross-talk between these two pathways resulting in the formation of "unnatural" natural products may now be engineered. As a first result, transgenic *Arabidopsis* plants that produce large amounts of *p*-hydroxybenzylglucosinolate have been obtained.[249] Likewise, cyanogenic tobacco and *Arabidopsis* plants have been produced (Bak *et al*, unpublished data). It will be interesting to study the changes in pest resistance and herbivory caused by the introduction of cyanogenesis into a previously acyanogenic species. By using anti-sense technology, it may also be possible to eliminate or reduce the cyanogenic glucoside content of important crop plants like cassava to improve food safety. If necessary to retain resistance to pests and herbivores, the reduction of cyanogenic glucoside content may also be achieved in a tissue specific manner to increase food safety. Combined with parallel work on other groups of natural products, the next decade is likely to provide answers to several of the questions raised above.

ACKOWLEDGMENTS

The careful technical support provided by principal Karina Peitersen is gratefully acknowledged. The basic and applied work on cyanogenic glucosides carried out at the Plant Biochemistry Laboratory at the Royal Veterinary and Agricultural University in Copenhagen has been financially supported by grants from the EU Science and Technology for Development Programme II and III, the Danish Governmental Biotechnology Programme, the Danish National Research Foundation, DANIDA, the Danish Agricultural and Veterinary Research Council, the Danish Natural Science Research Council, the Carlsberg Foundation, Director Ib Henriksens Foundation, the Australian Research Council and the Rockefeller Foundation.

REFERENCES

1. CONN, E.E. 1969. Cyanogenic glucosides. J. Agr. Food Chem. 7: 519-526.
2. SEIGLER, D.S. 1975. Isolation and characterization of naturally occurring cyanogenic compounds. Phytochemistry 14: 9-29.
3. CONN, E.E. 1981. Cyanogenic glucosides. pp. 479-500 *in* Secondary Plant Products. The Biochemistry of Plants (P.K. Stumpf and E.E. Conn, eds). Academic Press, New York.

4. POULTON, J.E. 1990. Cyanogenesis in plants. Plant Physiol. 94: 401-405.
5. MØLLER, B.L., POULTON, J. 1993. Cyanogenic glucosides. pp. 183-207 *in* Methods of Plant Biochemistry (P.J. Lea, ed.) Vol. 9. Academic Press, New York.
6. LECHTENBERG, M., NAHRSTEDT, A. 1999. Cyanogenic glucosides. pp. 147-191 *in* Naturally Occurring Glycosides (R. Ikan, ed.), John Wiley and Sons, Toronto.
7. MØLLER, B.L., SEIGLER, D.S. 1999. Biosynthesis of cyanogenic glucosides, cyanolipids, and related compunds, pp. 563-609 *in* Plant Amino Acids. (B.K. Singh, ed.) Biochemistry and Biotechnology.
8. SEIGLER, D.S., BRINKER, A.M. 1993. Characterisation of cyanogenic glycosides, cyanolipids, nitroglycosides, organic nitro compounds and nitrile glucosides from plants. pp. 51-131 *in* Alkaloids and Sulphur Compounds (P.G. Waterman, ed.), Meth. Plant Biochem. (P.M. Dey and J.B. Harborne, eds.). Volume 8. Academic Press, New York.
9. POURMOHSENI, H., IBENTHAL, W.D. 1991. Novel b-cyanoglucosides in the epidermal tissue of barley and their possible role in the barley-powdery mildew interaction. Angew. Bot. 65: 341-350.
10. POURMOHSENI, H., IBENTHAL, W.D., MACHINEK, R., REMBERG, G., WRAY, V. 1993. Cyanoglucosides in the epidermis of *Hordeum vulgare*. Phytochemistry 33: 295-297.
11. THAYER, S.S., CONN, E.E. 1981. Subcellular localization of dhurrin β-glucosidase and hydroxynitrile lyase in the mesophyll cells of *Sorghum* leaf blades. Plant Physiol. 67: 617-622.
12. WURTELE, E.S., THAYER, S.S., CONN, E.E. 1982. Subcellular localization of a UDP-glucose:aldehyde cyanohydrin β-glucosyl transferase in epidermal plastids of *Sorghum* leaf blades. Plant Physiol. 70: 1732-1737.
13. POULTON, J.E. 1988. Localization and catabolism of cyanogenic glucosides. pp. 67-91 *in* Cyanide Compounds in Biology. Ciba Foundation Symposium 140. (D. Evered, S. Harnett, eds.) John Wiley & Sons, Chichester, UK.
14. PANCORO, A., HUGHES, M.A. 1992. *In-situ* localization of cyanogenic β-glucosidase (linamarase) gene expression in leaves of cassava (*Manihot esculenta* Crantz) using non-isotopic riboprobes. Plant J. 2: 821-827.
15. SWAIN, E., LI, C.P., POULTON, J.E. 1992. Tissue and subcellular localization of enzymes catabolizing (*R*)-amygdalin in mature *Prunus serotina* seeds. Plant Physiol. 100: 291-300.
16. SWAIN, E., POULTON, J.E. 1994. Utilization of amygdalin during seedling development of *Prunus serotina*. Plant Physiol. 106: 437-445.

17. BOKANGA, M. 1994. Distribution of cyanogenic potential in cassava germplasm. Acta Hort. 375:117-123.
18. ROSLING, H. 1987. Cassava toxicity and food security. UNICEF. pp. 1-40, ISBN 91-971 029-0-3.
19. HOWLETT,P., BRUBAKER, G.R., MLINGI, N., ROSLING, H. 1990. Konzo, an epidemic upper motor neuron disease studied in Tanzania. Brain 113: 223-235.
20. MLINGI, N. 1992. An outbreak of acute intoxications from consumption of insufficiently processed cassava in Tanzania. Nutrition Res. 12: 677-687.
21. ROSLING, H. 1994. Measuring effects in humans of dietary cyanide exposure from cassava. Acta Hort. 375: 271-283.
22. JONES, D.A. 1988. Cyanogenesis in animal-plant interactions. pp. 151-176 *in* Cyanide Compounds in Biology. Ciba Foundation Symposium 140 (D. Evered and S. Harnett, eds.). John Wiley & Sons, Chichester, UK.
23. SELMAR, D., LIEBEREI, R., BIEHL, B. 1988. Mobilization and utilization of cyanogenic glucosides. The linustatin pathway. Plant Physiol. 86: 711-716.
24. LIEBEREI, R., BIEHL, B., GIESMANN, A., JUNQUEIRA, N.T.V. 1989. Cyanogenesis inhibits active defense reactions in plants. Plant Physiol. 90: 33-36.
25. LIEBEREI, R., FOCK, H.P., BIEHL, B. 1996. Cyanogenesis inhibits active pathogen defence in plants: Inhibition by gaseous HCN of photosynthetic CO_2 fixation and respiration in intact leaves. Angew. Bot. 70: 230-238.
26. JONES, D. 1998. Why are so many food plants cyanogenic? Phytochemistry 47: 155-162.
27. SIBBESEN, O., KOCH, B., HALKIER, B., MØLLER, B.L. 1994. Isolation of a heme-thiolate enzyme cytochrome $P450_{TYR}$, which catalyzes the committed step in the biosynthesis of the cyanogenic glucoside dhurrin in *Sorghum bicolor* (L.) Moench. Proc. Natl. Acad. Sci. USA 91: 9740-9744.
28. HALKIER, B.A., SIBBESEN, O., KOCH, B., MØLLER, B.L. 1995. Characterization of cytochrome $P450_{TYR}$, a multi-functional haem-thiolate *N*-hydroxylase involved in the biosynthesis of the cyanogenic glucoside dhurrin. Drug Metab. Drug Interact. 12: 285-297.
29. KOCH, B., SIBBESEN, O., HALKIER, B.A., SVENDSEN, I., MØLLER, B.L. 1995. The primary sequence of cytochrome P450tyr, the multifunctional *N*-hydroxylase catalyzing conversion of L-tyrosine to *p*-hydroxyphenylacetaldehyde oxime in the biosynthesis of the cyanogenic glucoside dhurrin in *Sorghum bicolor* (L.) Moench. Arch. Biochem. Biophys. 323: 177-186.

30. SIBBESEN, O., KOCH, B., HALKIER, B.A., MØLLER, B.L. 1995. Cytochrome P-450$_{TYR}$ is a multifunctional heme-thiolate enzyme catalyzing the conversion of L-tyrosine to *p*-hydroxyphenylacetaldehyde oxime in the biosynthesis of the cyanogenic glucoside dhurrin in *Sorghum bicolor* (L.) Moench. J. Biol. Chem. 270: 3506-3511.
31. KAHN, R., BAK, S., SVENDSEN, I., HALKIER, B.A., MØLLER, B.L. 1997. Isolation and reconstitution of cytochrome P450ox and *in vitro* reconstitution of the entire biosynthetic pathway of the cyanogenic glucoside dhurrin from sorghum. Plant Physiol. 115: 1661-1670.
32. BAK, S., KAHN, R.A., NIELSEN, H.L., MØLLER, B.L, HALKIER, B.A. 1998. Cloning of three novel A-type cytochromes P450, CYP71E1, CYP98 and CYP99 from *Sorghum bicolor* (L.) Moench using a PCR approach and identification by expression in *Escherichia coli* of CYP71E1 as a multifunctional cytochrome P450 in the biosynthesis of the cyanogenic glucoside dhurrin. Plant Mol. Biol. 36: 393-405.
33. JONES, P.R., MØLLER, B.L., HØJ, P.B. 2000. The UDP-glucose-*p*-hydroxymandelonitrile-*O*-glucosyltransferase that catalyzes the last step in synthesis of the cyanogenic glucoside dhurrin in *Sorghum*. J. Biol. Chem. In press.
34. HUGHES, M.A. 1993. The molecular genetics of plant cyanogenic β-glucosidases. pp. 153-169 *in* β-Glucosidases (A. Esen, ed.) ACS Symposium Series 533, Washington.
35. BARRETT, T., SURESH, C.G., TOLLEY, S.P., DODSON, E.J., HUGHES, M.A. 1995. The crystal structure of a cyanogenic β-glucosidase from white clover, a family 1 glycosyl hydrolase. Structure 3: 951-960.
36. HUGHES, J., DE C. CARVALHO, F.J.P., HUGHES, M.A. 1994. Purification, characterization, and cloning of α-hydroxynitrile lyase from cassava (*Manihot esculenta* Crantz). Arch. Biochem. Biophys. 311: 496-502.
37. WAJANT, H., FÖRSTER, S., SELMAR, D., EFFENBERGER, F., PFIZENMAIER, K. 1995. Purification and characterization of a novel *(R)*-mandelonitrile lyase from the fern *Phlebodium aureum*. Plant Physiol. 109: 1231-1238.
38. HU, Z., POULTON, J.E. 1997. Sequencing, genomic organization, and preliminary promotor analysis of a black cherry (*R*)-(+)-mandelonitrile lyase gene. Plant Physiol. 115: 1359-1369.
39. TRUMMLER, K., WAJANT, H. 1997. Molecular cloning of acetone cyanhydrin lyase from flax (*Linum usitatissimum*. Definition of a novel class of hydroxynitrile lyases. J. Biol. Chem. 272: 4770-4774.

40. HU, Z., POULTON, J.E. 1999. Molecular analysis of (*R*)-(+)-Mandelonitrile lyase microheterogeneity in black cherry. Plant Physiol. 119: 1535-1546.
41. AKAZAWA, T., MILJANICH, P., CONN, E.E. 1960. Studies on cyanogenic glycoside of *Sorghum vulgare*. Plant Physiol. 35: 535-538.
42. SWAIN, E., LI, C.P., POULTON, J.E. 1992. Development of the potential for cyanogenesis in maturing black cherry (*Prunus serotina* Ehrh.) fruits. Plant Physiol. 98: 1423-1428.
43. NIELSEN, J.S., MØLLER, B.L. 1999. Biosynthesis of cyanogenic glucosides in *Triglochin maritima* and the involvement of cytochrome P450 enzymes. Arch. Biochem. Biophys. 368: 121-130.
44. OSBOURN, A.E. 1996. Preformed antimicrobial compounds and plant defense against fungal attack. Plant Cell 8: 1821-1831.
45. BERENBAUM, M.R.1995. The chemistry of defense: Theory and practice. Proc. Natl. Acad. Sci. USA 92: 2-8.
46. RHODES, M.J.C. 1994. Physiological roles for secondary metabolites in plants: Some progress, many outstanding problems. Plant Mol. Biol. 24: 1-20.
47. BOLWELL, G.P., BOZAK, K., ZIMMERLIN, A. 1994. Plant cytochrome P450. Phytochemistry 37: 1491-1506.
48. SCHULER, M.A. 1996. Plant cytochrome P450 monooxygenases. Crit. Rev. Plant Sci. 15: 235-284.
49. BAR-PELED, M., LEWINSOHN, E., FLUHR, R., GRESSEL, J. 1991. UDP-rhamnose: flavanone-7-*O*-glucoside-2"-*O*-rhamnosyltransferase. Purification and characterization of an enzyme catalyzing the production of bitter compounds in citrus. J. Biol. Chem. 266: 20953-20959.
50. MOEHS, C.P., ALLEN, P.V., FRIEDMAN, M., BELKNAP, W.R. 1997. Cloning and expression of solanidine UDP-glucose glucosyltransferase from potato. Plant J. 11: 227-236.
51. MATERN, U., REICHENBACH, C., HELLER, W. 1986. Efficient uptake of flavonoids into parsley (*Petroselinum hortense*) vacuoles requires acylated glycosides. Planta 167: 183-189.
52. BOUZAYEN, M., LATCHE, A., PECH, J-C., MARIGO, G. 1989. Carrier-mediated uptake of 1-(malonylamino)-cyclopropane 1-carboxylic acid in vacuoles isolated from *Catharanthus roseus* cells. Plant Physiol. 91: 1317-1322.
53. GAILLARD, C., DUFAUD, A., TOMMASINI, R., KREUZ, K., AMRHEIN, N., MARTINOIA, E. 1994. A herbicide antidote (safener) induces the activity of both the herbicide detoxifying enzyme and of a vacuolar transporter for the detoxified herbicide. FEBS Lett. 352: 219-221.

54. KLEIN, M., WEISSENBÖCK, G., DUFAUD, A., GAILLARD, C., KREUZ, K., MARTINOIA, E. 1996. Different energization mechanisms drive the vacuolar uptake of a flavonoid glucoside and a herbicide glucoside. J. Biol. Chem. 271: 29666-29671.
55. MARRS, K.A., ALFENITO, M.R., LLOYD, A.M., WALBOT, V. 1995. A glutathione S-transferase involved in vacuolar transfer encoded by the maize gene *Bronze-2*. Nature 375: 397-400.
56. LI, Z-S., ALFENITO, M., REA, P.A., WALBOT, V., DIXON, R.A. 1997. Vacuolar uptake of the phytoalexin medicarpin by the glutathione conjugate pump. Phytochemistry 45: 689-693.
57. FARAGO, S., BRUNOLD, C., KREUZ, K. 1994. Herbicide safeners and glutathione metabolism. Physiol. Plant. 91: 537-542.
58. SANDERMANN, H. 1994. Higher plant metabolism of xenobiotics: the "green liver" concept. Pharmacogenetics 4: 225-241.
59. KREUZ, K., TOMMASINI, R., MARTINOIA, E. 1996. Old enzymes for a new job. Plant Physiol. 111: 349-353.
60. COLEMAN, J.O.D., BLAKE-KALFF, M.M.A., DAVIES, E.T.G. 1997. Detoxification of xenobiotics by plants: Chemical modification and vacuolar compartmentation. Trends Plant Sci. 2: 144-151.
61. WETZEL, A., SANDERMANN, H. 1994. Plant biochemistry of xenobiotics: Isolation and characterization of a soybean *O*-glucosyltransferase of DDT metabolism. Arch. Biochem. Biophys. 314: 323-328.
62. GALLANDT, E.R., BALKE, N.E. 1995. Xenobiotic glucosyltransferase activity from suspension-cultured *Glycine max* cells. Pesticide Sci. 43: 31-40.
63. REAY, P.F., CONN, E.E. 1974. The purification and properties of a uridine diphosphate glucose:aldehyde cyanohydrin β-glucosyltransferase from sorghum seedlings. J. Biol. Chem. 249: 5826-5830.
64. NAHRSTEDT, A., SKJOTTGAARD JENSEN, P., WRAY, V. 1989. Prunasin-6′-malonate, a cyanogenic glucoside from *Merremia dissecta*. Phytochemistry 28: 623-624.
65. CLEGG, D.O., CONN, E.E., JANZEN, D.H. 1979. Developmental fate of the cyanogenic glucoside linamarin in Costa Rican wild lima bean seeds. Nature 278: 343-344.
66. GRUHNERT, C., BIEHL, B., SELMAR, D. 1994. Compartmentation of cyanogenic glucosides and their degrading enzymes. Planta 195: 36-42.
67. HALL, A.H., RUMACK, B.H. 1986. Clinical toxicology of cyanide. Ann. Emerg. Med. 15: 1067-1074.

68. SOLOMONSON, L.P. 1981. Cyanide as a metabolic inhibitor. pp. 11-28 *in* Cyanide in Biology (B. Vennesland, E.E. Conn, C.J. Knowles, J. Westley, and F. Wissing, eds.). Academic Press, New York.
69. OSUNTUKUN, B.O. 1968. An ataxic neuropathy in Nigeria. Brain 91:215-245.
70. TYLLESKÄR, T., BANEA, M., BIKANGI, N., COOKE, R., POULTER, N.H., ROSLING, H. 1992. Cassava cyanogens and konzo, an upper motoneuron disease found in Africa. Lancet 339: 208-211.
71. SCHMIDT, E.S., NEWTON, G.W., SANDERS, S.M., LEWIS, J.P., CONN, E.E. 1978. Laetrile toxicity studies in dogs. JAMA 239: 943-947.
72. NEWTON, G.W., SCHMIDT, E.S., LEWIS, J.P., CONN, E.E., LAWRENCE, R. 1981. Amygdalin toxicity studies in rats predict cronic cyanide poisoning in humans. Western J. Med. 134: 97-103.
73. NARTEY, F. 1981. Cyanogenesis in tropical feeds and foodstuffs. pp. 115-132 *in* Cyanide in Biology (B. Vennesland, E.E. Conn, C.J. Knowles, J. Westley, and F. Wissing, eds.). Academic Press, New York.
74. COOK, R., MCCAIG, N., MCMILLAN, J.M.B., LUMSDEN, W.B. 1990. Ethyl carbamate formation in grain based spirits. Part III. The primary source. J. Inst. Brew. 96: 233-244.
75. NICOLLIER, G.F., POPE, D.F., THOMPSON, A.C. 1983. Biological activity of dhurrin and other compounds from Johnson Grass (*Sorghum halepense*). J. Agr. Food Chem. 31: 744-748.
76. SAUNDERS, J.A., CONN, E.E., LIN, C.H., STOCKING, R. 1977. Subcellular localization of the cyanogenic glucoside of sorghum by autoradiography. Plant Physiol. 59: 647-652.
77. SAUNDERS, J.A., CONN, E.E. 1978. Presence of the cyanogenic glucoside dhurrin in isolated vacuoles from *Sorghum*. Plant Physiol. 61: 154-157.
78. FERREIRA, C., PARRA, J.R.P., TERRA, W.R. 1997. The effect of dietary plant glycosides on larval midgut β-glucosidases from *Spodoptera frugiperda* and *Diatraea saccharalis*. Insect Biochem. Mol. Biol. 27: 55-59.
79. WOODHEAD, S., PADGHAM, D.E., BERNAYS, E.A. 1980. Insect feeding on different sorghum cultivars in relation to cyanide and phenolic acid content. Ann. Appl. Biol. 95: 151-157.
80. NAHRSTEDT, A. 1985. Cyanogenic compounds as protecting agents for organisms. Plant System. Evol. 150: 35-47.
81. HRUSKA, A.J. 1988. Cyanogenic glucosides as defense compounds. A review of the evidence. J. Chem. Ecol. 14: 2213-2217.

82. STARR, J.L., NEWTON, R.J., MILLER, F.R. 1984. Presence of dhurrin in sorghum root tissue and the effect of pathogenesis on hydrogen cyanide potential. Crop Sci. 24: 739-742.
83. WOODHEAD, S., BERNAYS, E. 1977. Changes in release rates of cyanide in relation to palatability of *Sorghum* to insects. Nature 270: 235-236.
84. WOODHEAD, S., BERNAYS, E.A. 1978. The chemical basis of resistance of *Sorghum bicolor* to attack by *Locusta migratoria*. Ent. Exp. Appl. 24: 123-142.
85. WOODHEAD, S. 1983. Surface chemistry of *Sorghum bicolor* and its importance in feeding by *Locusta migratoria*. Physiol. Entomol. 8: 345-352.
86. WOODHEAD, S. 1982. *p*-Hydroxybenzaldehyde in the surface wax of sorghum: Its importance in seedling resistance to acridids. Ent. Exp. Appl. 31: 296-302.
87. DREYER. D.L., REESE, J.C., JONES, K.C. 1981. Aphid feeding deterrents in sorghum. Bioassay, isolation and characterization. J. Chem. Ecol. 7: 273-284.
88. WOODHEAD, S., GALEFFI, C., BETTOLO, G.B.M. 1982. *p*-Hydroxybenzaldehyde as a major constituent of the epicuticular wax of seedling *Sorghum bicolor*. Phytochemistry 21: 455-456.
89. BOUGH, W.A., GANDER, J.E. 1971. Exogenous L-tyrosine metabolism and dhurrin turnover in sorghum seedlings. Phytochemistry 10: 67-77.
90. ADEWUSI, S.R. 1990. Turnover of dhurrin in green sorghum seedlings. Plant Physiol. 94: 1219-1224.
91. HAIN, R., REIF, H-J., KRAUSE, E., LANGEBARTELS, R., KINDL, H., VORNAM, B., WIESE, W., SCHMELZER, E., SCHREIER, P.H., STÖCKER, R.H., STENZEL, K. 1993. Disease resistance results from foreign phytoalexin expression in a novel plant. Nature 361: 153-156.
92. SELMAR, D., LIEBEREI, R., BÖHLE, B., CONN, E.E. 1989. a-Hydroxynitrile lyase in *Hevea brasiliensis* and its significance for rapid cyanogenesis. Physiol. Plant. 75: 97-101.
93. WESTON, L.A., HARMON, R., MUELLER, S. 1989. Allelopathic potential of sorghum-sudangrass hybrid (SUDEX). J. Chem. Ecol. 15: 1855-1865.
94. FORNEY, D.R., FOY, C.L. 1985. Phytotoxicity of products from rhizospheres of a Sorghum-Sudangrass hybrid (*Sorghum bicolor x Sorghum sudanese*). Weed Sci. 33: 597-604.
95. NAHRSTEDT, A., KANT, J.D., WRAY, V. 1982. Acalyphin, a cyanogenic glucoside from *Acalypha indica.* Phytochemistry 21: 101-105.

96. HAHLBROCK, K.H., TAPPER, B.A., BUTLER, G.W., CONN, E.E. 1968. Conversion of nitriles and α-hydroxynitriles to cyanogenic glucosides in flax seedlings and cherry laurel leaves. Arch. Biochem. Biophys. 125: 1013-1016.
97. HAHLBROCK, K., CONN, E.E. 1970. The biosynthesis of cyanogenic glycosides in higher plants. Purification and properties of a uridine diphosphate-glucose-ketone cyanohydrin β-gluco-syltransferase from *Linum usitatissimum* L. J. Biol. Chem. 245: 917-922.
98. TAPPER, B.A., BUTLER, G.W. 1971. Oximes, nitriles and 2-hydroxynitriles as precursors in the biosynthesis of cyanogenic glucosides. Biochem. J. 124: 935-941.
99. TAPPER, B.A., BUTLER, G.W. 1972. Intermediates in the biosynthesis of linamarin. Phytochemistry 11: 1041-1046.
100. FARNDEN, K.J.G., ROSEN, M.A., LILJEGREN, D.R. 1973. Aldoximes and nitriles as intermediates in the biosynthesis of cyanogenic glucosides. Phytochemistry 12: 2673-2677.
101. ZILG, H., CONN, E.E. 1974. Stereochemical aspects of lotaustralin biosynthesis. J. Biol. Chem. 249: 3112-3115.
102. ROSEN, M.A., FARNDEN, K.J.F., CONN, E.E 1975. Stereochemical aspects of the biosynthesis of the epimeric cyanogenic glucosides dhurrin and taxiphyllin J. Biol. Chem. 250: 8302-8308.
103. URIBE, E.G., CONN, E.E. 1966. The metabolism of aromatic compounds in higher plants. VII. The origin of the nitrile nitrogen atom of dhurrin (β-D-glucopyranosyloxy-L-*p*-hydroxymandelonitrile). J. Biol. Chem. 241: 92-94.
104. KOUKOL, J., MILJANICH, P., CONN, E.E. 1962. The metabolism of aromatic compounds in higher plants. VI: Studies on the biosynthesis of dhurrin, the cyanogenic glucoside of *Sorghum vulgare*. J. Biol. Chem. 237: 3223-3228.
105. REAY, P.F., CONN, E.E. 1970. Dhurrin synthesis in excised shoots and roots of sorghum seedlings. Phytochemistry 9: 1825-1827.
106. CONN, E.E. 1973. Biosynthesis of cyanogenic glucosides. Biochem. Soc. Symp. 38: 277-302.
107. CONN, E.E. 1980. Cyanogenic glucosides. Annu. Rev. Plant Physiol. 31: 433-451.
108. MCFARLANE, I.J, LEES, E.M., CONN, E.E. 1975. The *in vitro* biosynthesis of dhurrin, the cyanogenic glucoside of *Sorghum bicolor*. J. Biol. Chem. 250:4708-4713.
109. MØLLER, B.L., CONN, E.E. 1979. The biosynthesis of cyanogenic glucosides in higher plants. *N*-Hydroxytyrosine as an intermediate in the

biosynthesis of dhurrin by *Sorghum bicolor* (Linn) Moench. J. Biol. Chem. 254: 8575-8583.

110.HALKIER, B.A., MØLLER, B.L 1989. Biosynthesis of the cyanogenic glucoside dhurrin in seedlings of *Sorghum bicolor* (L.) Moench and partial purification of the enzyme system involved. Plant Physiol. 90: 1552-1559.

111.MØLLER, B.L., CONN, E.E. 1980. The biosynthesis of cyanogenic glucosides in higher plants. Channeling of intermediates in dhurrin biosynthesis by a microsomal system from *Sorghum bicolor* (Linn) Moench. J. Biol. Chem. 255: 3049-3056.

112.HALKIER, B.A., MØLLER, B.L. 1990. The biosynthesis of cyanogenic glucosides in higher plants. Identification of three hydroxylation steps in the biosynthesis of dhurrin in *Sorghum bicolor* (L.) Moench and the involvement of 1-*aci*-nitro-2-(*p*-hydroxyphenyl)ethane as an intermediate. J. Biol. Chem. 265: 21114-21121.

113.CUTLER, A.J., STERNBERG, M., CONN, E.E. 1985. Properties of a microsomal enzyme system from *Linum usitatissimum* (linen flax) which oxidizes valine to acetone cyanohydrin and isoleucine to 2-methylbutanone cyanohydrin. Arch. Biochem. Biophys. 238: 272-279.

114.COLLINGE, D.B., HUGHES, M.A. 1982. *In vitro* characterization of the *Ac* locus in white clover (*Trifoleum repens* L.) Arch. Biochem. Biophys. 218: 38-45.

115.HÖSEL, W., NAHRSTEDT, A. 1980. *In vitro* biosynthesis of the cyanogenic glucoside taxiphyllin in *Triglochin maritima*. Arch. Biochem. Biophys. 203: 753-757.

116.CUTLER, A.J., HÖSEL, W., STERNBERG, M., CONN, E.E. 1981. The *in vitro* biosynthesis of taxiphylin and the channeling of intermediates in *Triglochin maritima*. J. Biol. Chem. 256: 4253-4258.

117.HÖSEL, W., BERLIN, J., HANZLIK, T.N., CONN, E.E. 1985. *In-vitro* biosynthesis of 1-(4′-hydroxyphenyl)-2-nitroethane and production of cyanogenic compounds in osmotically stressed cell suspension cultures of *Eschscholtzia californica* Cham., Planta 166: 176-181.

118.KOCH, B., NIELSEN, V.S., HALKIER, B.A., OLSEN, C.E., MØLLER, B.L. 1992. The biosynthesis of cyanogenic glucosides in seedlings of cassava (*Manihot esculenta* Crantz). Arch. Biochem. Biophys. 292: 141-150.

119.SHIMADA, M., CONN, E.E. 1977. The enzymatic conversion of *p*-hydroxyphenylacetaldoxime to *p*-hydroxyphenylacetonitrile. Arch. Biochem. Biophys. 180: 199-207.

120.HALKIER, B.A., OLSEN, C.E., MØLLER, B.L. 1989. The biosynthesis of cyanogenic glucosides in higher plants. The *(E)*- and *(Z)*-isomers of *p*-

hydroxyphenylacetaldehyde oxime as intermediates in the biosynthesis of dhurrin in *Sorghum bicolor* (L.) Moench. J. Biol. Chem. 264: 19487-19494.

121.MØLLER, B.L., MCFARLANE, I.J., CONN, E.E. 1977. Chemical synthesis and disproportionation of *N*-hydroxytyrosine. Acta Chem. Scand. B31: 343-344.

122.MØLLER, B.L. 1978. Intermediates in the biosynthesis of cyanogenic glucosides determined by use of a gas chromatograph coupled with a gas proportional counter. Anal. Biochem. 81: 292-304.

123.MØLLER, B.L., CONN, E.E., SWEELEY, C.C. 1979. Mass spectrometric identification of intermediates in the biosynthesis of cyanogenic glucosides. Carlsberg Res. Commun. 44: 367-379.

124.MØLLER, B.L. 1981. The involvement of *N*-hydroxyamino acids as intermediates in metabolic transformations. pp. 197-215 *in* Cyanide in Biology (B. Vennesland, E.E.Conn, C.J.Knowles, J. Westley and F.Wissing, eds.) Academic Press, London.

125.HALKIER, B.A., LYKKESFELDT, J., MØLLER, B.L. 1991. 2-Nitro-3-(*p*-hydroxy-phenyl)propionate and *aci*-1-nitro-2-(*p*-hydroxyphenyl)ethane, two intermediates in the biosynthesis of the cyanogenic glucoside dhurrin in *Sorghum bicolor* (L.) Moench. Proc. Natl. Acad. Sci. USA 88: 487-491.

126.HALKIER, B.A., MØLLER, B.L. 1990. Involvement of cytochrome P-450 in the biosynthesis of dhurrin in *Sorghum bicolor* (L.) Moench. Plant Physiol. 96: 10-17.

127.COLOWICK, S.P., KAPLAN, N.O., CIOTTI, M. 1951. The reaction of pyridine nucleotide with cyanide and its analytical use. J. Biol. Chem. 191: 447-459.

128.HALKIER, B.A., SCHELLER, H.V., MØLLER, B.L. 1988. Cyanogenic glucosides: the biosynthetic pathway and the enzyme system involved. pp. 49-66 *in* Cyanide Compounds in Biology (D. Evered and S. Harnett, eds.) Ciba Found. Symp. 140, J. Wiley, Chichester.

129.DONALDSON, R.P., LUSTER, D.G. 1991. Multiple forms of plant cytochromes P-450. Plant Physiol. 96: 669-674.

130.HALKIER, B.A. 1996. Catalytic reactivities and structure/function relationships of cytochrome P450 enzymes. Phytochemistry 43: 1-21.

131.NELSON, D.R., STROBEL, H.W. 1988. On the membrane topology of vertebrate cytochrome P-450 proteins. J. Biol. Chem. 263: 6038-6050.

132.HALLAHAN, D.L., NUGENT, J.H.A., HALLAHAN, B.J., DAWSON, G.W., SMILEY, D.W., WEST, J.M., WALLSGROVE, R.M. 1992. Interactions of avocado (*Persea americana*) cytochrome P-450 with monoterpenoids. Plant Physiol. 98: 1290-1297.

133.WERCK-REICHART, D. 1995. Cytochromes P450 in the phenylpropanoid metabolism. Drug Metab. Drug Interact. 12:221-243.
134.FRANK, M.R., DEYNEKA, J.M., SCHULER, M.A. 1996. Cloning of wound-induced cytochrome P450 monooxygenases expressed in pea. Plant Physiol. 110: 1035-1046.
135.BATARD, Y., SCHALK, M., PIERREL, M.A., ZIMMERLIN, A., DURST, F., WERCK-REICHART, D. 1997. Regulation of the cinnamate 4-hydroxylase (CYP73A1) in Jerusalem artichoke tubers in response to wounding and chemical treatments. Plant Physiol. 113: 931-959.
136.ROBINEAU, T., BATARD, Y., NEDELKINA, S., CABELLO-HURTADO, F., LERET, M., SOROKINE, O., DIDERJEAN, L., WERCK-REICHART, D. 1998. The chemically inducible plant cytochrome P450 CYP76B1 actively metabolizes phenylureas and other xenobiotics. Plant Physiol. 118: 1049-1056.
137.WERCK-REICHART, D., BENVENISTE, I., TEUTSCH, H., DURST, F., GABRIAC, B. 1991. Glycerol allows low-temperature phase separation of membrane proteins solibilized in Triton X-114: Application to the purification of plant cytochromes P-450 and b_5. Anal. Biochem. 197: 125-131.
138.ANDERSEN, M.D., MØLLER, B.L. 1998. Double Triton X-114 phase partitioning for the purification of plant cytochromes P450 and removal of green pigments. Prot. Express. Purific. 13: 366-372.
139.JEFCOTE, C.R. 1978. Measurement of substrate and inhibitor binding to microsomal cytochrome P-450 by optical difference spectroscopy. Methods Enzymol. 27: 258-279.
140.LEWIS, D.F.V. 1996. Introduction. pp. 1-54 *in* Cytochromes P450. Structure, Function and Mechanism. Taylor and Francis, London.
141.BAK, S., KAHN, R.A., OLSEN, C.E., HALKIER, B.A. 1997. Cloning and expression in *Escherichia coli* of the obtusifoliol 14α-demethylase of *Sorghum bicolor* (L.) Moench, a cytochrome P450 orthologous of the lanosterol 14α-demethylases (CYP51) from fungi and mammals. Plant J. 11: 191-201.
142.KAHN, R.A., BAK, S., OLSEN, C.E., SVENDSEN, I., MØLLER, B.L. 1996. Isolation and reconstitution of the heme-thiolate protein obtusifoliol 14α-demethylase from *Sorghum bicolor* (L.) Moench. J. Biol. Chem. 271: 32944-32950.
143.KAHN, R.A., MØLLER, B.L. 1999. Substrate specificity of the cytochrome P450 enzymes CYP79A1 and CYP71E1 involved in the biosynthesis of the cyanogenic glucoside dhurrin in *Sorghum bicolor* (L.) Moench. Arch. Biochem. Biophys. 363: 9-18.

144.HOLTON, T.A., BRUGLIERE, F., LESTER, D.R., TANAKA, Y., HYLAND, C.D., MENTING, J.G.T., LU, C.Y., FARCY, E., STEVENSON, T.W., CORNISH, E.C. 1993. Cloning and expression of cytochrome P450 genes controlling flower colour. Nature 366: 276-279.

145.CHRISTOFFERSON, R.E., PERCIVAL, F.W., BOZAK, K. 1995. Functional and DNA sequence divergence of the CYP71 gene family in higher plants. Drug. Metab. Drug Interact. 12: 207-221.

146.NELSON, D.R., KOYMANS, L., KAMATAKI, T., STEGEMAN, J.J., FEIEREISEN, R., WAXMAN, D.J., WATERMAN, M.R., GOTOH, O., COON, M.J., ESTABROOK, R.W., GUNSALUS, I.C., NEBERT, D.W. 1996. P450 superfamily: Update on new sequences, gene mapping, accession numbers and nomenclature. Pharmacogenetics 6: 1-42.

147.NELSON, D.R. 1999. Cytochrome P450 and the individuality of species. Arch. Biochem. Biophys. 369: 1-10.

148.BRIZA, P., BREITENBACH, M., ELLINGER, A., SEGALL, J. 1990. Isolation of two developmentally regulated genes involved in spore wall maturation in *Saccharomyces cerevisiae*. Gen. Dev. 4: 1775-1789.

149.ANDERSEN, M.D., BUSK, P.K., SVENDSEN, I., MØLLER, B.L. 2000. Cytochromes P450 from cassava (*Manihot esculenta* Crantz) catalyzing the first steps in the biosynthesis of the cyanogenic glucosides linamarin and lotaustralin. Cloning, functional expression in *Pichia pastoris* and substrate specificity of the isolated recombinant enzymes. J. Biol. Chem. In press.

150.BARNES, H.J., ARLOTTO, M.P., WATERMAN, M.R. 1991. Expression and enzymatic activity of recombinant cytochrome P450 17α-hydroxylase in *Escherichia coli.* Proc. Natl. Acad. Sci. USA 88: 5597-5601.

151.BARNES, H.J. 1996. Maximizing expression of eucaryotic cytochrome P450s in *Escherichia coli*. Methods Enzymol. 272: 3-14.

152.HALKIER, B.A., NIELSEN, H.L., KOCH, B., MØLLER, B.L. 1995. Purification and characterization of recombinant cytochrome $P450_{TYR}$ expressed at high levels in *Escherichia coli.* Arch. Biochem. Biophys. 322: 369-377.

153.DURST, F., NELSON, D.R. 1995. Diversity and evolution of plant P450 and P450-reductases. Drug Metab. Drug Interact. 12: 198-206.

154.FREY, M., CHOMET, P., GLAWISCHNIG, E., STETTNER, C., GRÜN, S., WINKLMAIR, A., EISENREICH, W., BACHER, A., MEELEY, R.B., BRIGGS, S.P., SIMCOX, K., GIERL, A. 1997. Analysis of a chemical plant defense mechanism in grasses. Science 277: 696-699.

155.FREY, M., KLIEM, R., SAEDLER, H., GIERL, A. 1995. Expression of a cytochrome P450 gene family in maize. Mol. Gen. Genet. 246: 100-109.

156.DEMASTER, E.G., SHIROTA, F.N., NAGASAWA, H.T. 1992. A Beckmann-type dehydration of *n*-butyraldoxime catalyzed by cytochrome P-450. J. Org. Chem. 57: 5074-5075.
157.BOUCHER, J.L., DELAFORGE, M., MANSUY, D. 1994. Dehydration of alkyl- and arylaldoximes as a new cytochrome P450-catalyzed reaction: Mechanism and stereochemical characteristics. Biochemistry 33: 7811-7818.
158.MANSUY, D. 1994. Cytochromes P450 and model systems: Great diversity of catalyzed reactions. Pure Appl. Chem. 66: 737-744.
159.HUME, R., KELLY, R.W., TAYLOR, P.L., BOYD, G.S. 1984. The catalytic cycle of cytochrome P-450_{SCC} and intermediates in the conversion of cholesterol to pregnenolone. Eur. J. Biochem. 140: 583-591.
160.HEYL, B.L., TYRRELL, D.J., LAMBETH, J.D. 1986. Cytochrome P-450_{scc}-substrate interactions. Role of the 3β- and side chain hydroxyls in binding to oxidized and reduced forms of the enzyme. J. Biol. Chem. 261: 2743-2749.
161.LYKKESFELDT, J., MØLLER, B.L. 1995. On the absence of 2-(2′-cyclopentenyl)glycine-derived cyanogenic glucosides in cassava, *Manihot esculenta* Crantz. Acta Chem. Scand. 49: 540-542.
162.SELMAR, D., LIEBEREI, R., JUNQUEIRA, N., BIEHL, B. 1991. Changes in cyanogenic glucoside content in seeds and seedlings of *Hevea* species. Phytochemistry 30: 2135-2140.
163.SELMAR, D. 1994. Translocation of cyanogenic glucosides in cassava. Acta. Hort. 375: 61-67.
164.LYKKESFELDT, J., MØLLER, B.L. 1994. Cyanogenic glucosides in cassava, *Manihot esculenta* Crantz. Acta Chem. Scand. 48: 178-180.
165.WHITE, W.L.B., MCMAHON, J.M., SAYRE, R.T. 1994. Regulation of cyanogenesis in cassava. Acta Hort. 375: 69-77.
166.DU, L., BOKANGA, M., MØLLER, B.L, HALKIER, B.A. 1995. The biosynthesis of cyanogenic glucosides in roots of cassava. Phytochemistry 39: 323-326.
167.KOCH, B.M., SIBBESEN, O., SWAIN, E., KAHN, R.A., DU, L., BAK, S., HALKIER, B.A., MØLLER, B.L. 1994. Possible use of a biotechnological approach to optimize and regulate the content and distribution of cyanogenic glucosides in cassava to increase food safety. Acta Hort. 375: 45-60.
168.NIELSEN, J.S., MØLLER, B.L. 2000. Cloning and expression of cytochrome P450 catalyzing the conversion of tyrosine to *p*-hydroxyphenylacetaldoxime in the biosynthesis of cyanogenic glucosides in *Triglochin maritima*. Plant Physiol. in press.
169.DU, L., LYKKESFELDT, J., OLSEN, C.E., HALKIER, B.A. 1995. Involvement of a cytochrome P450 in glucosinolate biosynthesis as

demonstrated by an *in vitro* microsomal enzyme system isolated from jasmonic acid-induced seedlings of *Sinapis alba* L. Proc. Natl. Acad. Sci. USA 92: 12505-12509.

170.BAK, S., NIELSEN, H.L., HALKIER, B.A. 1998. The presence of CYP79 homologous in glucosinolate-producing plants shows evolutionary conservation of the enzymes in the conversion of amino acid to aldoxime in the biosynthesis of cyanogenic glucosides and glucosinolates. Plant Mol. Biol. 38: 724-734.

171.HOGGE, R.L., REED, D.W., UNDERHILL, E.W. 1988. HPLC separation of glucosinolates from seeds and leaves of *Arabidopsis thaliana* and their identification using thermospray liquid chromatography-mass spectrometry. J. Chromatogr. Sci. 26: 551-560.

172.HASEMANN, C.A., KURUMBAIL, R.G., BODDUPALLI, S.S., PETERSON, J.A., DEISENHOFER, J. 1995. Structure and function of cytochromes P450: A comparative analysis of three crystal structures. Structure 3: 41-62.

173.GOTOH, O. 1992. Substrate recognition sites in cytochrome P450 family 2 (CYP2) proteins inferred from comparative analyses of amino acid and coding nucleotide sequences. J. Biol. Chem. 267: 83-90.

174.MEIJER, A.H., SOUER, E., VERPOORTE, R., HOGE, J.H.C. 1993. Isolation of cytochrome P-450 cDNA clones from the higher plant *Catharanthus roseus* by a PCR strategy. Plant Mol. Biol. 22: 379-383.

175.PARK, S.Y., SHIMIZU, H., ADACHI, S.I., NAKAGAWA, A., TANAKA, I., NAKAHARA, K., SHOUN, H., OBAYASHI, E., NAKAMURA, H., LIZUKA, T., SHIRO, Y. 1997. Crystal structure of nitric oxide reductase from denitrifying fungus *Fusarium oxysporum*. Nature Struct. Biol. 4: 827-832.

176.SAKAGUCHI, M., MIHARA, K., SATO, R. 1987. A short amino-terminal segment of microsomal cytochrome P-450 functions as an insertion signal and as a stop-transfer sequence. EMBO J. 6: 2425-2431.

177.HÖSEL, W., SCHIEL, O. 1984. Biosynthesis of cyanogenic glucosides: *In vitro* analysis of the glucosylation step. Arch. Biochem. Biophys. 229: 177-186.

178.MEDERACKE, H., SELMAR, D., BIEHL, B. 1995. Glucosyltransferases of the cyanogenic plant cassava (*Manihot esculenta*, Crantz). Angew. Bot. 69: 119-124.

179.MEDERACKE, H., BIEHL, B., SELMAR, D. 1996. Characterization of two cyano glucosyltransferases from cassava leaves. Phytochemistry 42: 1517-1522.

180.HÖSEL, W. 1981. Glycosylation and glycosidases. pp. 725-753 *in* The Biochemistry of Plants, Vol. 7. (P.K. Stumpf, and E.E. Conn, eds.). Academic Press, New York.

181.WILLIAMS, P.J., ALLEN, M.S. 1996. The analysis of flavouring compounds in grapes. pp. 37-57 *in* Modern Methods of Plant Analysis, Vol. 18. Fruit Analysis (H.F. Linskens and J.F. Jackson, eds.). Springer-Verlag, Berlin & Heidelberg.

182.SEMBDNER, G., ATZORN, R., SCHNEIDER, G., 1994. Plant hormone conjugation. Plant Mol. Biol. 26: 1459-1481.

183.HELLER, W., FORKMANN, G. 1994. Biosynthesis of flavonoids. pp. 499-535 *in* The Flavonoids: Advances in Research Since 1986 (J.B. Harborne, ed.). Chapman & Hall, London.

184.BROUILLARD, R. 1982. Chemical structure of anthocyanins. pp. XX *in* Anthocyanins as Food Colours (P. Markakis, ed.) Academic Press, New York.

185.YAMAMOTO I., MUTO, N., MURAKAMI, K., SUGA, S., YAMAGUCHI, H. 1990. L-Ascorbic acid α-glucoside formed by regioselective transglucosylation with rat intestinal and rice seed α-glucosidases: Its improved stability and structure determination. Chem. Pharm. Bull. 38: 3020-3023.

186.HARTWIG, U.A., PHILLIPS, D.A., 1991. Release and modification of *nod*-gene-inducing flavonoids from alfalfa seeds. Plant Physiol. 95: 804-807.

187.DAKORA, F.D., JOSEPH, C.M., PHILLIPS, D.A., 1993. Alfalfa (*Medicago sativa* L.) root exudates contain isoflavonoids in the presence of *Rhizobium meliloti*. Plant Physiol. 101: 819-824.

188.BANDURSKI, R.S., COHEN, J.D., SLOVIN, J.P., REINECKE, D.M., 1995. Hormone biosynthesis and metabolism:Auxin biosynthesis and metabolism. pp. 39-65 *in* Plant Hormones (P.J. Davies, ed.) Kluwer Academic Publishers, Netherlands.

189.STAHL-BISKUP, E., INTERT, F., HOLTHUIJZEN, J., STENGELE, M., SCHULZ, G., 1993. Glycosidically bound volatiles - A review 1986-1991. Flav. Frag. J. 8: 61-80.

190.SELMAR, D. 1993. Transport of cyanogenic glucosides: linustatin uptake by *Hevea* cotyledons. Planta 191: 191-199.

191.DAVIES, P.J. 1995. The plant hormone concept: Concentration, sensitivity and transport. pp. 13-38 *in* Plant Hormones (P.J. Davies, ed.). Kluwer Academic Publishers, Netherlands.

192.SELMAR, D., IRANDOOST, Z., WRAY, V. 1996. Dhurrin-6′-glucoside, a cyanogenic diglucoside from *Sorghum bicolor*. Phytochemistry 43: 569-572.

193. VOGT, T., TAYLOR, L.P. 1995. Flavonol 3-*O*-glycosyltransferase associated with petunia pollen produce gametophyte-specific flavonol diglycosides. Plant Physiol. 108: 903-911.

194. COOKE, R.D. 1978. An enzymatic assay for the total cyanide content of cassava (*Manihot esculenta* Crantz). J. Sci. Food Agric. 29: 345-352.

195. SEFTON, M.A., FRANCIS, I.L., WILLIAMS, P.J., 1993. The volatile composition of Chardonnay juices: A study by flavour precursor analysis. Amer. J. Enol. Viticult. 44: 359-370.

196. SEFTON, M.A., FRANCIS, I.L., WILLIAMS, P.J., 1994. Free and bound volatile secondary metabolites of *V. vinifera* grape cv. Sauvignon Blanc. J. Food Sci. 59: 142-147.

197. SALEH, A.M., POULTON, J.E., GRISEBACH, H., 1976. UDP-glucose: cyanidin-3-*O*-glucosyltransferase from red cabbage seedlings. Phytochemistry 15: 1865-1868.

198. JOURDAN, P.S., MANSELL, R.L. 1982. Isolation and partial characterization of three glucosyltransferases involved in the biosynthesis of flavonol triglucosides. Arch. Biochem. Biophys. 213: 434-443.

199. HRAZDINA, G., 1988. Purification and properties of a UDPglucose:flavonoid-3-*O*-glucosyltransferase from *Hippeastrum* petals. Biochim. Biophys. Acta 955: 301-309.

200. REED, D.W., DAVIN, L., JAIN, J.C., DELUCA, V., NELSON, L., UNDERHILL, E.W. 1993. Purification and properties of UDP-glucose: thiohydroximate glucosyltransferase from *Brassica napus* L. seedlings. Arch. Biochem. Biophys. 305: 526-532.

201. PARRY, A.D., TILLER, S.A., EDWARDS, R. 1994. The effects of heavy metals and root immersion on isoflavonoid metabolism in alfalfa (*Medicago sativa* L.). Plant Physiol. 106: 195-202.

202. KEIL, U., SCHREIER, P. 1989. Purification and partial characterization of UDP-glucose phenol-β-D-glucosyltransferase from papaya fruit. Phytochemistry 28: 2281-2284.

203. KOWALCZYK, S., BANDURSKI, R.S. 1991. Enzymic synthesis of 1-O-(indole-3-ylacetyl)-β-D-glucose. Biochem J. 279: 509-514.

204. VOGT, T., GRIMM, R., STRACK, D. 1999. Cloning and expression of a cDNA encoding betanidin 5-*O*-glucosyltransferase, a betanidin- and flavonoid-specific enzyme with high homology to inducible glucosyltransferases from the Solanaceae. Plant J. 19: 509-519.

205. KÖSTER, J., BARZ, W. 1981. UDP-glucose:isoflavone 7-*O*-glucosyltransferase from roots of chick pea (*Cicer arietinum* L.). Arch. Biochem. Biophys. 212: 98-104.

206.JONSSON, L.M.V., AARSMAN, M.E.G., VAN DIEPEN, P., DE VLAMING, N., SMIT, N., SCHRAM, A.W. 1984. Properties and genetic control of anthocyanin 5-*O*-glucosyltransferase in flowers of *Petunia hybrida*. Planta 160: 341-347.
207.ISHIKURA, N., YAMAMOTO, M. 1990. Detection and characterization of UDP-glucose:flavonoid O-glucosyltransferases in the leaves of *Prunus x yedoensis* Matsum. Plant Cell Physiol. 31: 1109-1115.
208.SANDERMANN, H., SCHMITT, R., ECKEY, H., BAUKNECHT, T. 1991. Plant biochemistry of xenobiotics: Isolation and properties of Soybean *O*- and *N*-glucosyl and *O*- and *N*-malonyltransferases for chlorinated phenols and anilines. Arch. Biochem. Biophys. 287: 341-350.
209.ISHIKURA, N., MATO, M., 1993. Partial purification and some properties of flavonol 3-O-glycosyltransferase from seedlings of *Vigna mungo*, with special reference to the formation of kaempferol 3-O-galactoside and 3-*O*-glucoside. Plant Cell Physiol. 34: 329-335.
210.CHENG, G.W., MALENCIK, D.A., BREEN, P.J. 1994. UDP-glucose:flavonoid *O*-glucosyltransferase from strawberry fruit. Phytochemistry 35: 1435-1439.
211.ISHIKURA, N., YANG, Z.-Q. 1994. Multiple forms of flavonol O-glucosyltransferases in young leaves of *Euonymus alatus* F. *Ciliato-Dentatus*. Phytochemistry 36: 1139-1145.
212.DO, C., CORMIER, F., NICOLAS, Y. 1995. Isolation and characterization of a UDP-glucose:cyanidin 3-*O*-glucosyltransferase from grape cell suspension cultures (*V. vinifera* L.). Plant Sci. 112: 43-51.
213.ROSE, A., GLABGEN, W.E., HOPP, W., SEITZ, H.U. 1996. Purification and characterization of glycosyltransferases involved in anthocyanin biosynthesis in cell-suspension cultures of *Daucus carota* L. Planta 198: 397-403.
214.SUN, Y., HRAZDINA, G., 1991. Isolation and characterization of a UDP-glucose:flavonol *O*-3-glucosyltransferase from illuminated red cabbage (*Brassica oleracea* cv red danish) seedlings. Plant Physiol. 95: 570-576.
215.FEDOROFF, N.V., FURTEK, D.B., NELSON, O.E., 1984. Cloning of the *Bronze* locus in maize by a simple and generalizable procedure using the transposable controlling element *Activator* (*Ac*). Proc. Natl. Acad. Sci. USA 81: 2825-3829.
216.MARTIN, C., PRESCOTT, A., MACKAY, S., BARTLETT, J., VRIJLANDT, E. 1991. Control of anthocyanin biosynthesis in flowers of *Anthirrhinum majus*. Plant J. 1: 37-49.
217.SPARVOLI, F., MARTIN, C., SCIENZA, A., GAVAZZI, G., TONELLI, C. 1994. Cloning and molecular analysis of structural genes involved in

flavonoid and stilbene biosynthesis in grape (*V. vinifera* L.). Plant Mol. Biol. 24: 743-755.

218. WISE, R.P., ROHDE, W., SALAMINI, F. 1990. Nucleotide sequence of the *Bronze-1* homologous gene from *Hordeum vulgare*. Plant Mol. Biol. 14: 277-279.

219. PICTON, S., GRAY, J., BARTON, S., ABUBAKAR, U., LOWE, A., GRIERSON, D. 1993. cDNA cloning and characterisation of novel ripening-related mRNAs with altered patterns of accumulation in the *ripening inhibitor (rin)* tomato ripening mutant. Plant Mol. Biol. 23: 193-207.

220. BRUGLIERA, P., HOLTON, T.A., STEVENSON, T.W., FARCY, E., LU, C.-Y., CORNISH, E.C. 1994. Isolation and characterization of a cDNA clone corresponding to the *Rt* locus of *Petunia hybrida*. Plant J. 5: 81-92.

221. MARKHAM, K.R. 1996. Novel anthocyanins produced in petals of genetically transformed *Lisianthus*. Phytochemistry 42: 1035-1038.

222. SZERSZEN, J., SZCZYGLOWSKI, K., BANDURSKI, R. 1994. *iaglu*, a gene from *Zea Mays* involved in conjugation of growth hormone indole-3-acetic acid. Science 265: 1699-1701.

223. HUGHES, J., HUGHES, M.A. 1994. Multiple secondary plant product UDP-glucose glucosyltransferase genes expressed in cassava (*Manihot esculenta Crantz*) cotyledons. DNA Sequence 5: 41-49.

224. O'REILLY, D.R., 1995. Baculovirus-encoded ecdysteroid UDP-glucosyltransferases. Insect Biochem. Molec. Biol. 25: 541-550.

225. HUNDLE, B.S., O'BRIEN, D.A., ALBERTI, M., BEYER, P., HEARST, J.E. 1992. Functional expression of a zeaxanthin glucosyltransferase from *Erwinia herbicola* and a proposed uridine diphosphate binding site. Proc. Natl. Acad. Sci. USA 89: 9321-9325.

226. MACKENZIE, P.I. 1990. Expression of chimeric cDNAs in cell culture defines a region of UDP-glucuronosyltransferase involved in substrate selection. J. Biol. Chem. 265: 3432-3435.

227. RITTER, J.K., CHEN, F., SHEEN, Y.Y., TRAN, H.M., KIMURA, S., YEATMAN, M.T., OWENS, I.S. 1992. A novel complex locus UG1 encodes human bilirubin, phenol, and other UDP-glucoronosyltransferase isozymes with identical carboxyl termini. J. Biol. Chem. 267: 3257-3261.

228. POULTON, J.E., SHIN, S.-I. 1983. Prunasin biosynthesis by cell-free extracts from black cherry (*Prunus serotina* Ehrh.) fruits and leaves. Z. Naturforsch. 38c: 369-374.

229. DEWICK 1984. The biosynthesis of cyanogenic glucosides and glucosinolates. Natural Prod. Rep. 1: 545-549.

230.FURTEK, D., SCHIELFBEIN, J.W., JOHNSTON, F., NELSON, O.E. 1988. Sequence comparisons of three wild-type *Bronze-1* alleles from *Zea mays*. Plant Mol. Biol. 11: 473-481.

231.SAUNDERS, J.A., CONN, E.E., LIN, C.H., SHIMADA, M. 1977. Localization of cinnamic acid 4-monooxygenase and the membrane-bound enzyme system for dhurrin biosynthesis in *Sorghum* seedlings. Plant Physiol. 60: 629-634.

232.MAO, C.H., ANDERSON, L. 1965. Cyanogenesis in *Sorghum vulgare*. II. Mechanism of the alkaline hydrolysis of dhurrin (*p*-hydroxymandelonitrile glucoside). J. Org. Chem. 30: 603-607.

233.WAJANT, H., RIEDEL, D., BENZ, S., MUNDRY, K.W. 1994. Immunocytological localization of hydroxynitrile lyases from *Sorghum bicolor* L. and *Linum usitatissimum* L. Plant Sci. 103: 145-154.

234.MCMAHON, J.M., WHITE, W.L.B., SAYRE, R.T. 1995. Cyanogenesis in cassava (*Manihot esculenta* Crantz). J. Exptl. Bot. 46: 731-741.

235.BLUMENTHAL-GOLDSCHMIDT, S., BUTLER, G.W., CONN, E.E. 1963. Incorporation of hydrocyanic acid labelled with carbon-14 into asparagine in seedlings. Nature 197: 718-719.

236.ERB, N., ZINSMEISTER, H.D., NAHRSTEDT, A. 1981. Die cyanogenen Glykoside von Triticum, Secale und Sorghum. Planta Med. 41: 84-89.

237.LOYD, R.C., GRAY, E. 1970. Amount and distribution of hydrocyanic acid potential during the life cycle of plants of three sorghum cultivars. Agr. J. 62: 394-397.

238.ARIAS, J.A., DUNKLE, L.D., BRACKER, C.E. 1983. Ultrastructural and developmental alterations induced by *Periconia circinata* toxin in the root tip of sorghum. Can. J. Bot. 61: 1491-1505.

239.FREHNER, M., SCALET, M., CONN, E.E. 1990. Pattern of the cyanide-potential in developing fruits. Plant Physiol. 94: 28-34.

240.SELMAR, D., FREHNER, M., CONN, E.E. 1989. Purification and properties of endosperm protoplasts of *Hevea brasiliensis* L. J. Plant Physiol. 135: 105-109.

241.HÖSEL, W., TOBER, I., EKLUND, S.H., CONN, E.E. 1987. Characterization of β-glucosidases with high specificity for the cyanogenic glucoside dhurrin in *Sorghum bicolor* (L.) Moench seedlings. Arch. Biochem. Biophys. 252: 152-162.

242.KOJIMA, M., POULTON, J.E., THAYER, S.S., CONN, E.E. 1979. Tissue distributions of dhurrin and of enzymes involved in its metabolism in leaves of *Sorghum bicolor*. Plant Physiol. 63: 1022-1028.

243.FREHNER, M., CONN, E.E. 1987. The linamarin α-glucosidase in Costa Rican wild lima beans (*Phaseolus lunatus* L.) is apoplastic. Plant Physiol. 84: 1296-1300.

244.KAKES, P. 1985. Linamarase and other β-glucosidases are present in the cell walls of *Trifolium repens* L. leaves. Planta 166: 156-160.

245.KURZHALS, C.H., GRÜTZMACHER, H., SELMAR, D., BIEHL, B. 1989. Linustatin, the linamarin-glucoside protected against cleavage by apoplastic linamarase. Planta Medica 55: 673.

246.KERESZTESSY, Z., HUGHES, J., KISS, L., HUGHES, M.A. 1996. Co-purification from *E. coli* of a plant beta-glucosidase-glutathione S-transferase fusion protein and the bacterial chaperonin GroEL. Biochem. J. 314: 41-47.

247.SELMAR, D., LIEBEREI, R., BIEHL, B., NAHRSTEDT, A., SCHMIDTMANN, V., WRAY, V. 1987. Occurrence of the cyanogen linustatin in *Hevea brasiliensis*. Phytochemistry 26: 2400-2401.

248.WINKEL-SHIRLEY, B. 1999. Evidence for enzyme complexes in the phenylpropanoid and flavonoid pathways. Physiol. Plant. 107: 142-149.

249.BAK, S., OLSEN, C.E., MØLLER, B.L., HALKIER, B.A. 2000. Metabolic engineering of *p*-hydroxybenzylglucosinolate in *Arabidopsis* by expression of the cyanogenic CYP79A1 from *Sorghum bicolor*. Plant J. in press.

Chapter Eight

TWO-OXOACID-DEPENDENT DIOXYGENASES: INEFFICIENT ENZYMES OR EVOLUTIONARY DRIVING FORCE?

Andy G. Prescott

Department of Applied Genetics
John Innes Centre
Norwich Research Park
Colney Lane, Norwich, UK.

INTRODUCTION

Two-oxoacid-dependent dioxygenases (2-ODDs) are widespread in nature, being found in micro-organisms, plants, and animals. The 2-ODDs are an enzyme super-family that presents an unusual problem in terms of defining the common denominators that delineate the group. The super-family can be sub-divided into two groups: (a) those whose membership is determined on the basis of biochemical characteristics, and (b) those that show sequence similarity to one or more members of the first sub-group.[1]

The members of the first sub-group utilize 2-oxoglutarate as a co-substrate and require ferrous iron and ascorbate as co-factors for maximal activity *in vitro*; these enzymes are commonly called 2-oxoglutarate-dependent dioxygenases. Enzymes found within this group: may show no sequence similarity to any other member of the group *e.g.* clavaminate synthase (CAS);[2] may show limited sequence similarity to other members of the group, *e.g.* prolyl 4-hydroxylase (P4H);[3] or may show a high degree of sequence homology to other members of the group, *e.g.* flavanone 3 -hydroxylase (F3H).[4] The group defined on the basis of sequence similarity to one or more members of the first sub-group contains enzymes that are technically oxidases, *e.g.* 1-aminocyclopropane-1-carboxylic acid oxidase (ACCO),[5] and those for which the biochemical reaction(s) catalyzed are unknown.

Unfortunately, even this convoluted definition of 2-ODDs is inadequate for the definition of some enzymes. The ability to catalyze sequential reactions is a common characteristic of enzymes in this family.[6] Thus, it is possible for a single enzyme to be classified as both a dioxygenase and an oxidase depending on which reaction is being studied, *e.g.* deacetoxycephalosporin C synthase (DAOCS).[7] 2-ODDs can catalyze a range of different reaction types including hydroxylations, desaturations, ring closures, ring expansions, epoxidations, and successive reactions oxidizing a methyl group to an alcohol, aldehyde, and carboxylic acid. Examples of these reaction types are shown in the figures of this chapter.

2-ODDs are found in a wide range of pathways and in both primary and secondary metabolism. It is common to find more than one 2-ODD involved in a biosynthetic pathway, *e.g.* in gibberellin biosynthesis and flavonoid biosynthesis.[8,9] 2-ODDs frequently occur in complex biosynthetic pathways that also involve P450 monooxygenases. In these cases, the mono-oxygenases catalyze early reactions in the pathway.

The fact that the super-family of 2-ODDs contains a number of enzymes that show no sequence homology to any other 2-ODD strongly indicates that there have been several incidences of independent evolution of the ability of non-heme, ferrous iron-dependent dioxygenases to catalyze reactions involving the concomitant two-electron oxidation of 2-oxoglutarate to succinate.[1] This is reinforced by the identification of a common structural motif, a 2-His-1-carboxylate facial triad, for the iron active site in a range of mononuclear non-heme iron (II) enzymes including lipoxygenases, extradiol-cleaving dioxygenases, many 2-ODDs, iron superoxide dismutase, and a pterin-dependent hydroxylase.[10]

MICROBIAL 2-ODDS

The 2-ODDs found in micro-organisms represent the best characterized group in terms of structure and mechanism, as well as illustrating many of the points made in the introduction. The microbial 2-ODDs include both oxidases and dioxygenases, enzymes that show strong sequence homology to each other, *e.g.* the three enzymes involved in the cephalosporin/penicillin biosynthetic pathway(s),[4] as well as those that appear to be structurally "unique" enzymes, *e.g.* CAS and 2,4-dichlorophenoxyacetic acid dioxygenase (*tfd*A).[2,11] Crystal structures are available for two microbial 2-ODDs (IPNS, DAOCS), and these form the basis for several mechanistic hypotheses that have implications for all members of the 2-ODD super-family.[4,12,13]

Isopenicillin N Synthase (IPNS)

IPNS catalyzes the four-electron oxidative ring closure of L-δ-(α-aminoadipoyl)-L-cysteinyl-D-valine (ACV) to isopenicillin N (IPN), a common step in the synthesis of both penicillins and cephalosporins (Fig. 1).[7] The reaction catalyzed by IPNS is unusual in that oxygen is not incorporated into the product, IPN, and there is no 2-oxoacid co-substrate. Nevertheless, IPNS shows substantial sequence similarity to other 2-ODDs.

The crystal structures of an IPNS-Mn(II) complex made under aerobic conditions[4] and an anaerobic IPNS-Fe(II)-ACV complex have been determined.[12] The structure of the IPNS-Mn(II) complex revealed that the active site is situated at the end of a distorted, incomplete "jelly-roll" motif, with the metal ion octahedrally coordinated by His214, Asp216, His270, Gln330, and two water molecules. Mutagenesis studies have confirmed the roles of the two His residues and the Asp residue.[14-16] The conserved glutamine residue is not essential for activity *in vitro* but may be involved in maintaining the selectivity of the active site.[17-19]

Comparison of the IPNS-Mn(II) and IPNS-Fe(II)-ACV complexes shows that the basic "jelly roll" structure is maintained on substrate binding.[12] The metal coordination geometry changes from octahedral to square pyramidal on substrate binding, with one of the water molecules displaced and the amide side chain of Gln330 replaced by the thiol of ACV. This displacement of the penultimate residue of the IPNS primary sequence (Gln330) results in an extension of the C-terminal α helix that helps to enclose the substrate in the active site and to protect the reactive Fe center. Analysis of IPNS-Fe(II)-ACV crystals exposed to the dioxygen analog, nitric oxide (NO), shows that the NO binds within the vacant coordination site *trans* to Asp216, suggesting that dioxygen would bind in a similar position.[12]

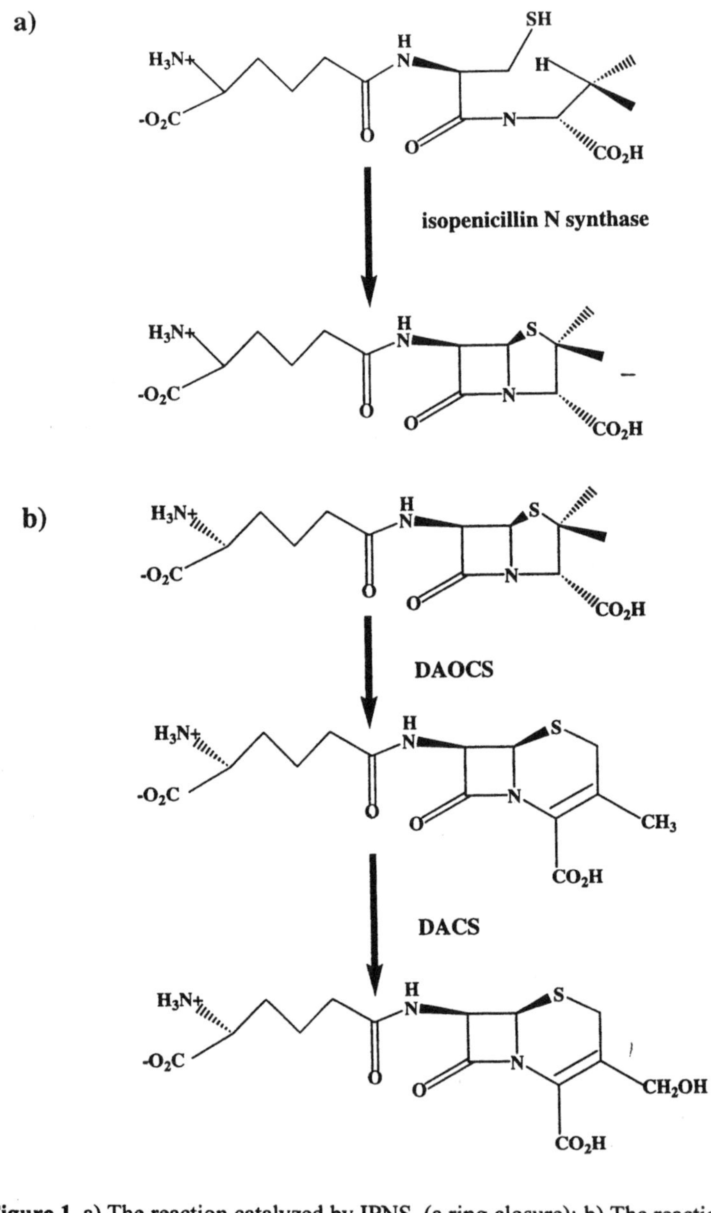

Figure 1. a) The reaction catalyzed by IPNS, (a ring closure); b) The reactions catalyzed by DAOCS (a ring expansion) and DACS (hydroxylation)

Deacetoxycephalosporin C Synthase (DAOCS)

The first few steps in the biosynthesis of penicillins and cephalosporins are shared, and subsequently, the pathways branch with DAOCS catalyzing the first reaction in the cephalosporin biosynthetic pathway (Fig. 1).[7] This reaction causes the expansion of the five-membered thiazolidine ring of penicillin N into the six-membered dihydrothiazine ring of deactoxycephalosporin C (DAOC), with the concomitant oxidation of 2-oxoglutarate to succinate. In prokaryotic cephalosporin synthesizing organisms, there is a second 2-ODD, deacetylcephalosporin C synthase (DACS) that catalyzes the subsequent hydroxylation of DAOC to form deacetylcephalosporin (DAC).[20] In eukaryotic organisms, a single bifunctional enzyme (DAOCS/DACS) catalyzes both steps.[21] The amino acid sequences of DAOCS and DACS from the same species are 59% identical, and both show limited bifunctionality, with DAOCS having a low level of hydroxylase activity and DACS having a low level of expandase activity.[22]

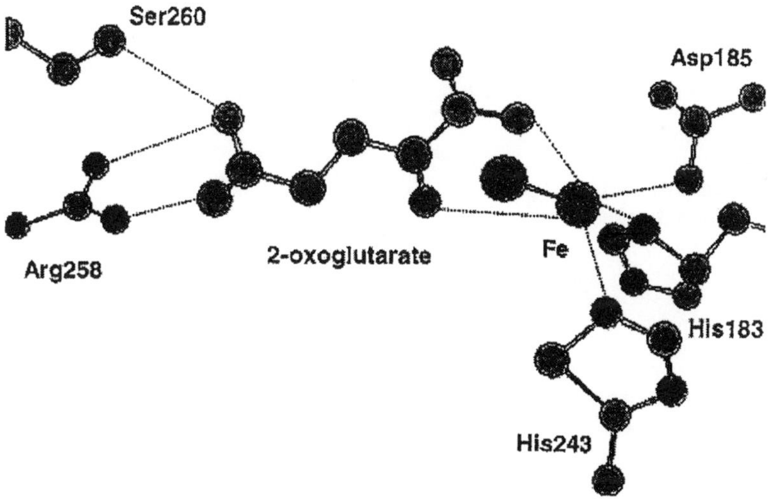

Figure 2. The active site of DAOCS, showing the iron and 2-oxoglutarate ligands.

The crystal structure and active site of DAOCS from *Streptomyces clavuligerus* have been studied in detail.[13, 23] The crystal structure of DAOCS is similar to that of IPNS, despite the primary amino acid sequence of the two enzymes being only 19% identical. The jelly-roll structural motif is maintained, and the ferrous iron is ligated by His183, Asp185, His243, and three water molecules in the DAOCS-Fe(II) complex (Fig. 2).[13] Valegård *et al.*, (1998)[13] noted that the only accessible site around the iron in the DAOCS-Fe(II)-2-oxoglutarate complex is located in a hydrophobic pocket lined by residues Ile192, Phe225, Val262, Phe264, and the methyl group of Thr190. They suggest that this pocket represents the binding site for dioxygen in DAOCS and other 2-ODDs. The crystal structures of the IPNS-Mn(II) and IPNS-Fe(II)-ACV complexes indicate that the C-terminus of the protein serves to protect the active site; Lloyd *et al.*, (1999) proposed that the C-terminus of DAOCS may stabilize the core of the resting enzyme and assist in the ordered binding of substrates to the active site.[23]

Mechanistic Hypotheses for 2-ODDs

Early studies on 2-ODDs indicated that there was sequential binding of co-substrates with 2-oxoglutarate binding first, followed by dioxygen and the primary substrate.[24] Crystallographic and spectroscopic data support this order of binding, and a mechanistic model for 2-ODDS that unifies all of the data has been proposed (Fig. 3).[13, 23]

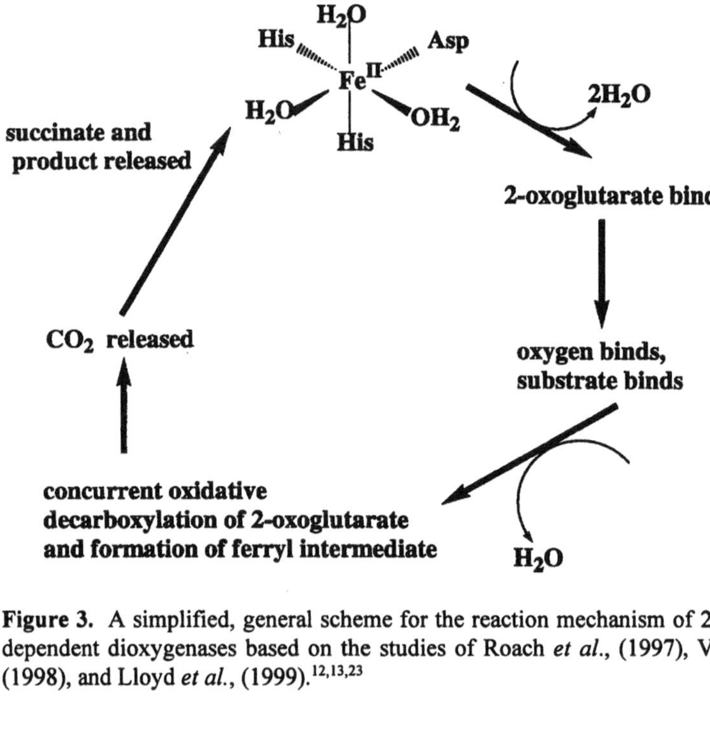

Figure 3. A simplified, general scheme for the reaction mechanism of 2-oxoglutarate-dependent dioxygenases based on the studies of Roach *et al.*, (1997), Valegård *et al.*, (1998), and Lloyd *et al.*, (1999).[12,13,23]

2-ODDS IN PLANTS

The study of 2-ODDs in plants has lagged behind those done on microorganisms and animals; this difference is largely due to the interest of the pharmaceutical and medical industries in the role of 2-ODDs in the synthesis of antibiotics and as markers or causative agents of human diseases. Plant molecular biology has enabled many of the 2-ODDs that were originally identified biochemically to be cloned and has also indicated that this family of enzymes has a far more extensive role in plant metabolism than was originally thought.

Flavonoid Biosynthesis

Four 2-ODDs have been identified as part of the anthocyanin biosynthetic pathway (Fig. 4); the flavone synthase type I (FSI), flavanone 3β-hydroxylase (F3H), flavonol synthase (FLS), and anthocyanidin synthase (ANS). Genes encoding three of these enzymes have been cloned and studied in detail. The 2-ODD with FSI activity has been found only in parsley suspension culture cells and in related species;[25] the majority of flavone synthases studied have been identified as P450 monooxygenases, (flavone synthases type II).[26-28] This suggests that flavone synthases have evolved independently twice.

F3H was originally identified as a 2-ODD in 1980;[29] the enzyme was subsequently partially purified, and the reaction studied in detail by Britsch and Grisebach (1986).[30] The F3H gene from *Petunia hybrida* has been expressed as a recombinant protein in *Escherichia coli* and purified to near-homogeneity.[31] This work identified three iron ligands of F3H by site-directed mutagenesis (His220, His278, and Asp222) and showed that these residues were those previously predicted as being iron ligands by sequence alignment to IPNS.[4] Mutagenesis of the conserved arginine residue (Arg288), proposed to be a ligand for 2-oxoglutarate, to a lysine or glutamine residue resulted in enzymes with extremely low specific activities and increased K_m values for 2-oxoglutarate, supporting the apparent role of this arginine residue in catalysis.[31]

FLS was identified as a 2-ODD in 1981,[32] a gene encoding this activity was isolated from *P. hybrida*;[33] the cDNA was expressed in yeast, and the recombinant protein catalyzed the conversion of dihydroquercetin to quercetin.[33] A detailed study of the biochemical characterization of a purified FLS has not yet been published.

The identification of ANS as a 2-ODD was not originally based on study of the biochemical characteristics of the enzyme. The gene presumed to encode the function of ANS was isolated by transposon-tagging of the gene encoding the

A2 locus of maize.[34] The *a2* mutant of maize was known to be blocked at a late stage of anthocyanin biosynthesis.[35] A mutation of a putative ANS gene was shown to be the cause of the acyanic floral phenotype of the *candica* mutant of *Antirrhinum majus,* and feeding of 2,3-*cis*-leucocyanidin to excised flowers produced no anthocyanin.[36] This work supported the suggestion that the *A2* and *Candica* gene products encoded an ANS. This assumption was further clarified by the isolation of a cDNA homologue of *A2* from *Perilla frutescens*.[37] The cDNA was expressed in *E. coli*, and the recombinant protein catalyzed the 2-oxoglutarate-dependent oxidation of leucoanthocyanidins to products presumed to be 2-flaven-3,4-diols that could be converted to anthocyanidins on acidification. There remain questions as to the identity of the products formed by the *Perilla* ANS and how the plant converts these products to anthocyanidins.

Alkaloid Biosynthesis

Two 2-ODDs have been identified as playing a role in alkaloid biosynthesis; hyoscyamine 6β-hydroxylase (H6H) in the synthesis of tropane alkaloids, and desacetoxyvindoline 4-hydroxylase (D4H) in the synthesis of indole alkaloids (Fig. 5).[38,39]

H6H catalyzes the hydroxylation of hyoscyamine to 6β-hydroxyhyoscyamine and the subsequent epoxidation of the hydroxy intermediate to scopolamine, an important anticholinergic agent.[38,40] The gene encoding H6H was originally isolated from *Hyoscyamus niger*,[41] and has also been cloned from *Atropa belladonna*.[42] The two genes encode proteins that are 91% identical at the amino acid level. *A. belladonna* contains predominantly hyoscyamine, with only low levels of scopolamine present in young plants.[43,44] *A. belladonna* has been used in transgenic experiments to demonstrate that the introduction and expression of the *H. niger* H6H gene in this species leads to elevated levels of the more valuable alkaloid, scopolamine.[45] Currently, a *Duboisia* hybrid is the most important commercial source of scopolamine; but recent results expressing the *H. niger* H6H gene in *Hyoscyamus muticus* have demonstrated that it is possible to get significant increases in scopolamine production in transgenic hairy root cultures, indicating that an *in vitro* source of scopolamine may become available.[46]

D4H is involved in the synthesis of vinblastine and vincristine, two cytotoxic dimeric indole alkaloids used to treat leukaemia and other cancers. The gene encoding D4H was cloned from *Catharanthus roseus*, the commercial source of vinblastine and vincristine.[47] Attempts to synthesize these alkaloids in cell cultures have failed because some of the necessary biosynthetic enzymes are not

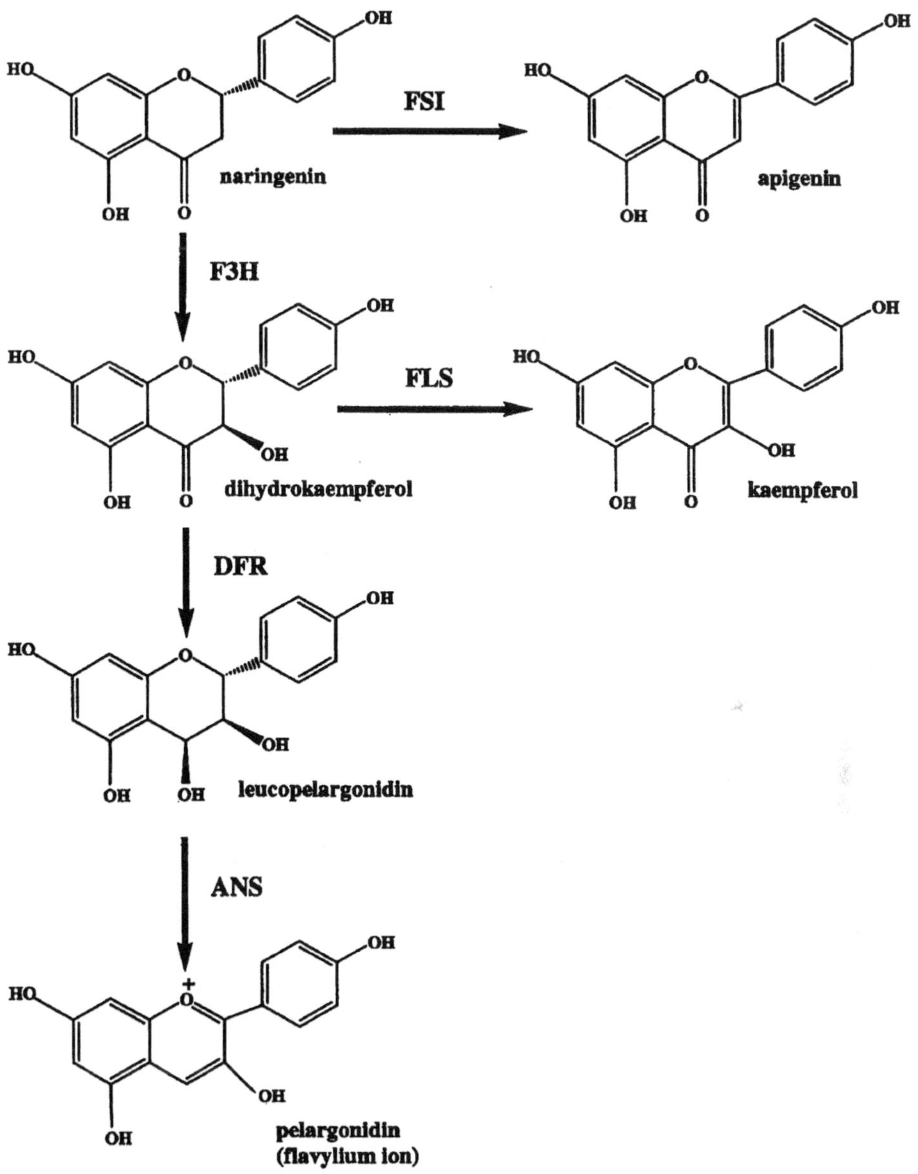

Figure 4. A generalized picture of the part of the flavonoid biosynthetic pathway involving 2-ODDs. The only reaction shown that is not catalyzed by a dioxygenase is that of the dihydroflavonol 4-reductase (DFR). FSI, FLS and ANS catalyze desaturations, while F3H catalyzes a hydroxylation.

expressed in these systems. D4H catalyzes the hydroxylation of desacetoxyvindoline to form deacetylvindoline, an important intermediate in the synthesis of vinblastine and vincristine.[39] Expression of recombinant D4H in *E. coli* has confirmed the identity of the gene,[47] and it is anticipated that manipulation of this and other genes involved in the biosynthesis of indole alkaloids in transgenic systems may aid in increasing the levels of the pharmaceutically important products.

Gibberellin Biosynthesis

Most of the genes encoding enzymes involved in the gibberellin biosynthetic pathway have been cloned over the last few years.[8] Gibberellin (GA) biosynthesis involves both P450-monooxygenases and 2-ODDs, with the monooxygenases catalyzing early hydroxylation reactions and 2-ODDs catalyzing later steps in the pathway (reviewed by Hedden and Kamiya, 1997).[8]

A 2-ODD has been identified as catalyzing the oxidation of the C-7 group of GAs from an aldehyde to a carboxylic acid in the endosperm of *Cucurbita maxima* and the embryos of *Hordeum vulgare* (Fig. 6).[48,49] Curiously, both species also have a P450-dependent monooxygenase capable of catalyzing the same reaction. In *C. maxima*, the two enzyme types have different pH optima and different substrate specificities, suggesting that they have different roles in the plant.[48] The gene encoding the dioxygenase form of the GA 7-oxidase of *C. maxima* has been cloned.[50] Recombinant GA 7-oxidase catalyzed oxidation of GA_{12}-aldehyde to GA_{12} but also produced a range of other products, one of which may have been 12α- or 12ß-hydroxy-GA_{12}, although the other products could not be identified.[50]

Monooxygenase and dioxygenase forms of both GA 13-hydroxylase and GA 12α-hydroxylase have been described (Fig. 6).[51,52] In the case of the 12α-hydroxylase, both types occur in *C. maxima* endosperm and embryos, and, as for the 7-oxidase, the two enzymes differ in terms of their substrate specificities and pH optima.[51,52] The only dioxygenase-type GA 13-hydroxylase isolated came from spinach leaves.[53]

The GA 20-oxidases catalyze a series of successive oxidations of carbon-20 from a methyl group to an alcohol, then an aldehyde, at which point the C atom is lost from the molecule to form the C_{19}-GA skeleton (Fig. 7). The enzyme can also convert the C-20 aldehyde to a carboxylic acid in some systems (reviewed in Hedden and Kamiya, 1997).8 The first cDNA encoding a GA 20-oxidase was cloned from *C. maxima*; a recombinant form of this enzyme was capable of

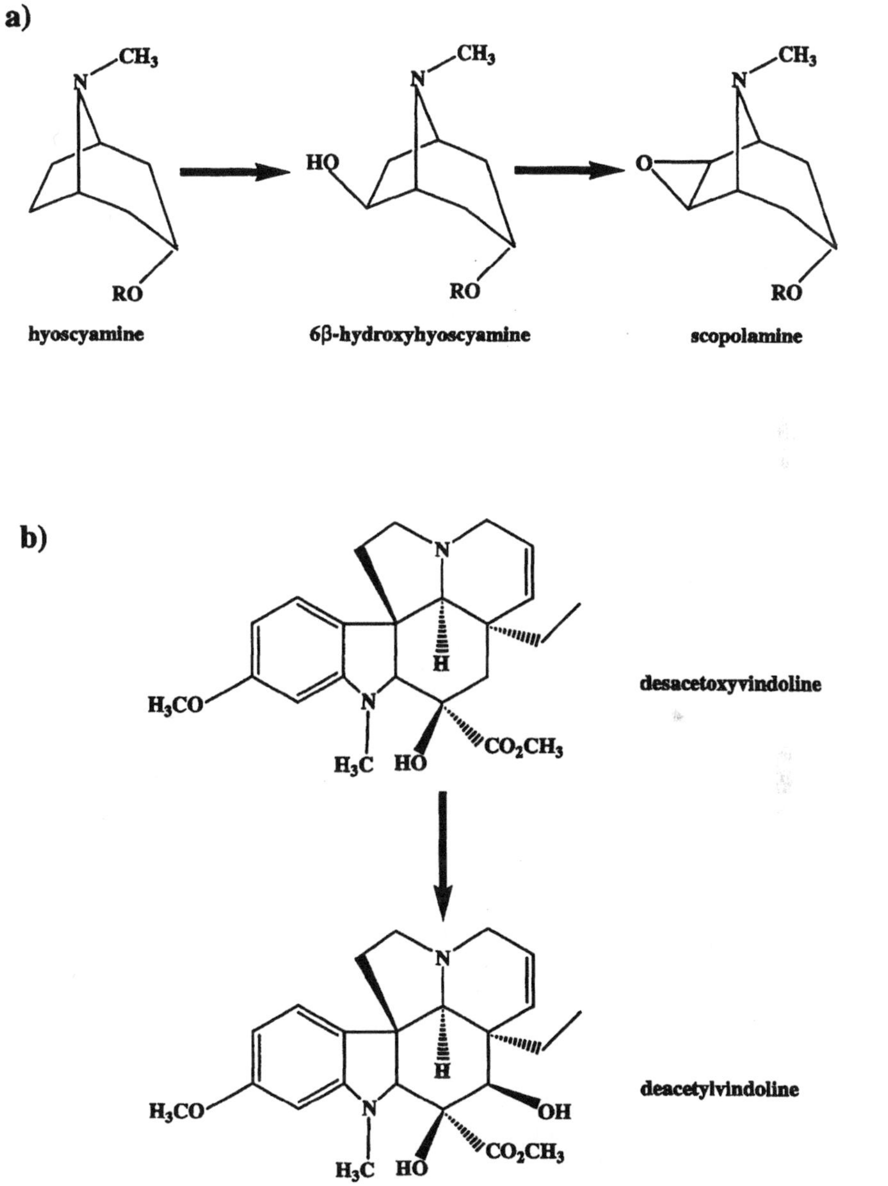

Figure 5. a) The transformations catalyzed by hyoscyamine 6β-hydroxylase (hydroxylation and epoxidation); b) D4H catalyzes the hydroxylation of desacetoxyvindoline.

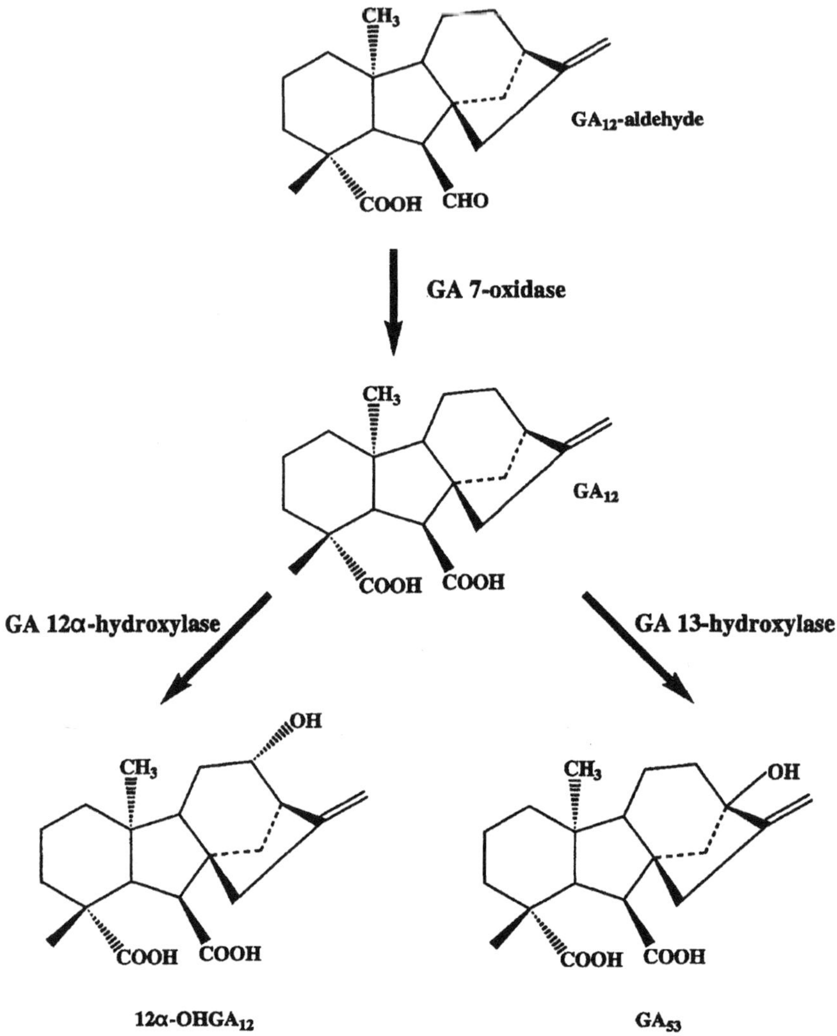

Figure 6. The oxidations catalyzed by GA 7-oxidase, and the GA 12α- and 13-hydroxylases. The pathway from GA_{12}-aldehyde is shown as an example.

catalyzing the successive oxidations of C-20 from the methyl group to the carboxylic acid or to the C_{19}-GA (at a low level) in an identical manner to the enzyme purified directly from the plant.[54]

GA 20-oxidases produce different ratios of the C_{19}-GAs and C_{20}-carboxylic acid GAs; the C_{19}-GAs are metabolized further by hydroxylation to produce the

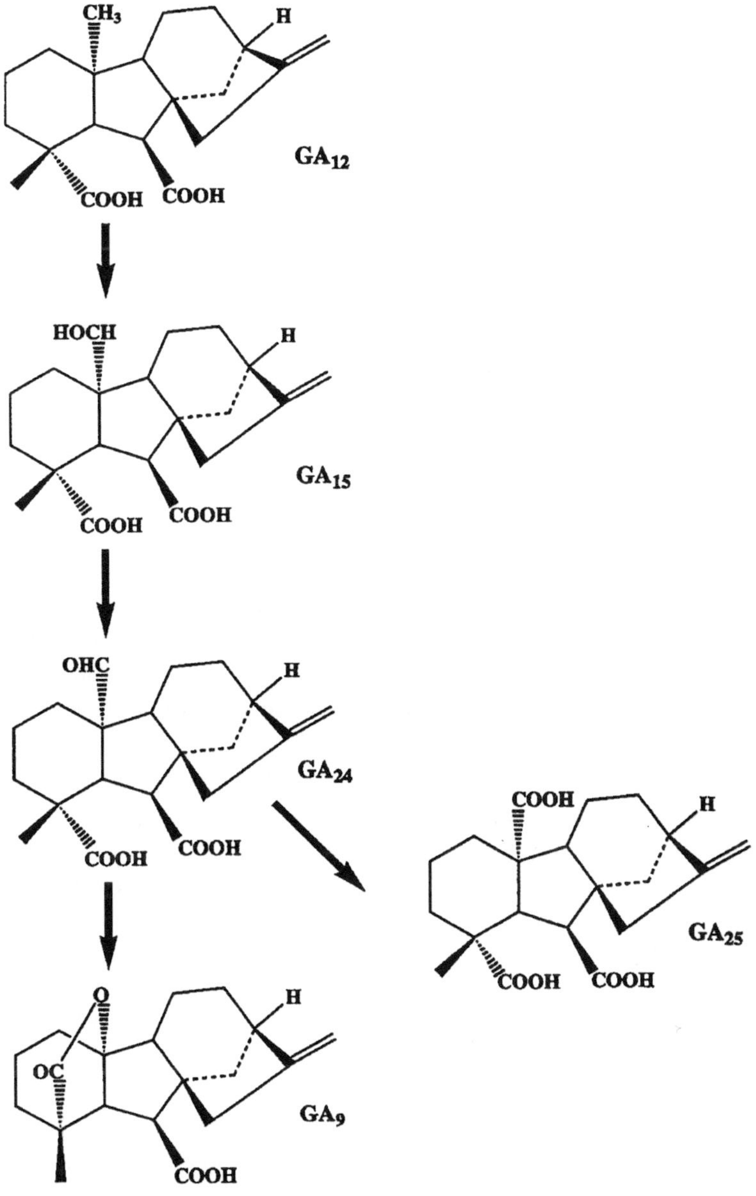

Figure 7. GA 20-oxidase catalyzes a sequence of reactions involving carbon-20. The function of the tricarboxylic acid GAs is unknown. The reaction cascade from GA_{12} is shown as an example.

GAs known to play a role in plant development, whereas the function of the C_{20}-carboxylic acid GAs is unknown. An attempt to define which parts of the enzyme may control the ratio of products generated was made by comparing the *C. maxima* enzyme, which synthesizes mainly tricarboxylic acid GAs, and the enzyme from the closely related species, *Marah macrocarpus*, which makes mainly C_{19}-GAs.[55] Recombinant chimeric proteins were produced, and the ratio of the different products made were analyzed for each construct. The results indicated that the carboxy-terminal region of the enzyme exerted the most influence on determining the ratio of the products formed.

The GA 3β-hydroxylase converts the C_{19}-GAs produced by the 20-oxidase to physiologically active GAs by the stereospecific addition of a hydroxyl group to the carbon-3 atom of the basic GA skeleton (Fig. 8).[56,57] The first gene encoding a GA 3 -hydroxylase to be cloned was isolated by T-DNA tagging and was shown to be the equivalent of the *GA4* locus in *A. thaliana*.[58] Two different GA 3β-hydroxylases from *A. thaliana* have been expressed as recombinant enzymes in *E. coli*.[59,60] Recombinant GA4 enzyme preferentially used non-13-hydroxylated GAs as substrates, and C_{19}-GAs were better substrates than C_{20}-GAs.[59] In addition, the recombinant enzyme also converted 2,3-didehydro-GAs to their corresponding epoxides, although these conversions were not very efficient. The homolog of the GA4 enzyme studied by Yamaguchi *et al.*, (1998)[60] was not characterized in such detail, but the results indicate that the two enzymes are broadly similar in their biochemical features.

A cDNA encoding a gibberellin 2β,3β-hydroxylase has been cloned from pumpkin endosperm.[61] This cDNA was expressed as a fusion protein in *E. coli* and was shown to be capable of catalyzing 3-hydroxylations of both C_{19}-GAs and C_{20}-GAs (with C_{20}-GAs being preferred as substrates) and 2-hydroxylations of GAs containing a carboxylic acid group at carbon atom-20. The recombinant enzyme had identical properties to partially purified GA 2β,3β-hydroxylase from *C. maxima* endosperm.[61]

Three cDNAs encoding GA 2β-hydroxylases (now called GA 2-oxidases) have been cloned from *A. thaliana*.[62] The three cDNAs were expressed as recombinant enzymes in *E. coli*; each of the three enzymes converted C_{19}-GAs to the corresponding 2 β-hydroxylated GAs as expected. A surprising result was the conversion of some 2 β-hydroxylated GAs (GA_4, GA_9, and GA_1) to their respective catabolites by further oxidation at carbon-atom-2 by two of the recombinant enzymes.[62] The 3 β-hydroxylated GAs are the physiologically active members of this group of plant growth regulators, while 2 β-hydroxylated GAs are physiologically inactive. Thus, the balance between the activities of the GA 3 β-hydroxylases, GA 2 β, 3 β-hydroxylases (where present), and GA 2-oxidases is crucial for the control of plant development.

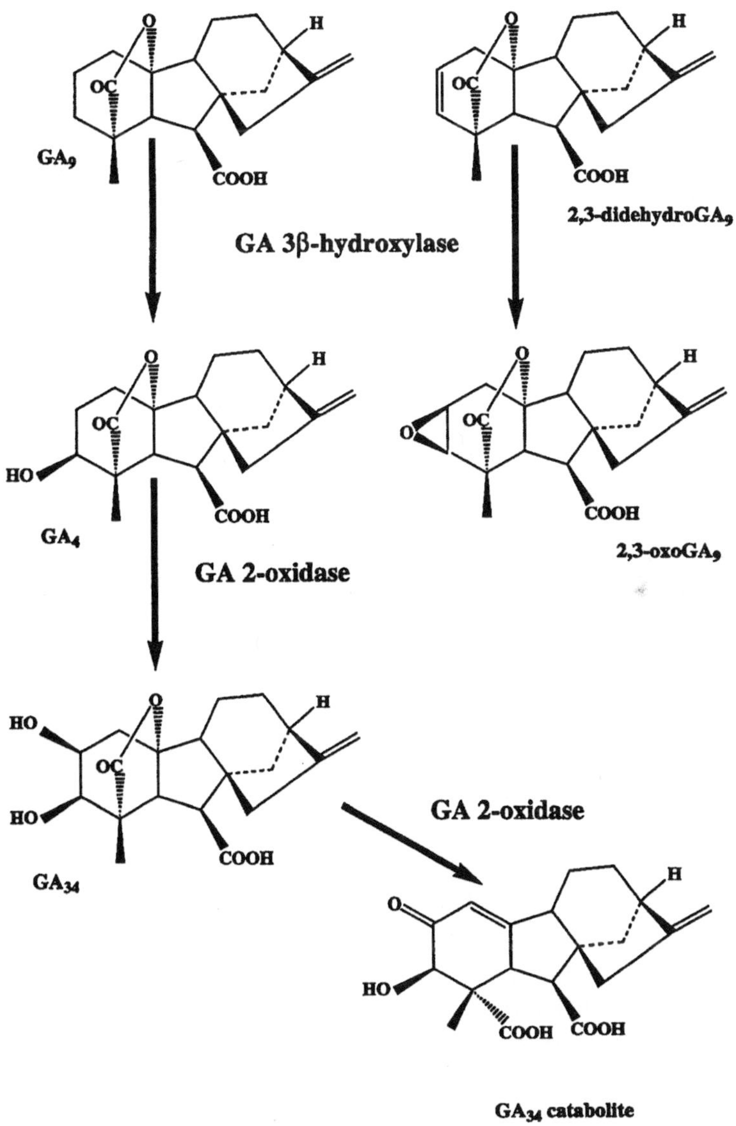

Figure 8. GA3β-hydroxylase can catalyze both an hydoxylation and the synthesis of 2,3-oxoGAs from 2,3-didehydroGAs. GA 2-oxidase can also catalyze two reactions, hydroxylation at carbon atom-2 and the synthesis of GA catabolites.

Ethylene Biosynthesis

The biosynthesis of the plant growth regulator ethylene has received a lot of attention, largely due to the simplicity of the pathway and the importance of ethylene in fruit ripening. The terminal step in ethylene biosynthesis is catalyzed by 1-aminocyclopropane carboxylic acid oxidase (ACCO), which converts 1-aminocyclopropane carboxylic acid (ACC) to ethylene, carbon dioxide, water, and hydrogen cyanide (Fig. 9).[63] ACCO is an unusual member of the 2-ODD family as it does not require 2-oxoglutarate as a co-substrate. This role is carried out by ascorbate.[64,65] ACCO is included within the 2-ODD superfamily due to strong sequence similarity to other 2-ODDs. ACCO is also unusual as it is the only 2-ODD that requires carbon dioxide as an activator.[64,66,67] The activation of ACCO by carbon dioxide was reviewed by Prescott and John (1996).[6]

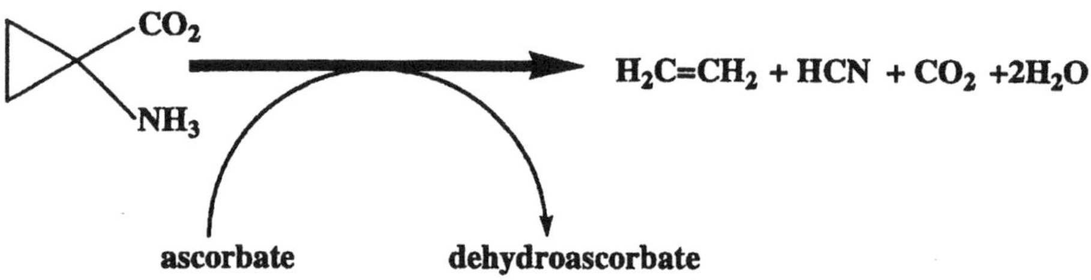

Figure 9. The terminal step in ethylene biosynthesis catalyzed by ACCO.

Many plant species contain multiple copies of genes encoding ACCO; several studies have shown that these genes are differentially expressed, reflecting the diverse roles of ethylene in plant growth and development.[68-71] Heterologous expression of 3 cDNAs encoding ACCOs from tomato in yeast showed that the 3 recombinant enzymes differed in their apparent K_m values for ACC, pI, and specific activities;[72] however, these studies were carried out on crude extracts, and, thus, the expression levels of the different recombinant ACCOs may have contributed to these differences.

An interesting aspect of ACCO concerns its location within plant tissues. ACCO activity has been localized to both the cytosol and the apoplastic space.[73-76] Although it is now certain that at least some ACCO is located on the external face of the plasmalemma in some tissues,[76] it is not known how the enzyme is transported to this location.

Several groups have investigated the role of conserved amino acid residues of ACCOs in catalysis.[77-79] Diethylpyrocarbonate (DEPC)-inactivation studies have indicated that two histidine residues form part of the active site of ACCO.[79,80] Mutagenesis studies show that His177 and His234 are likely to be iron ligands;[77-79] these residues also align with the known histidine iron ligands of IPNS and DAOCS. A third histidine (His211) is also proposed to have an important structural or catalytic role.[79] The third iron ligand identified by mutagenesis is Asp179, which corresponds to the ligands, Asp185 and Asp216 in DAOCS and IPNS, respectively.[77,78]

Purified ACCO from either plant sources or from a recombinant source has proved to be extremely labile, with a half-life under optimal conditions of as little as 8 minutes.[80] Several studies have investigated the cause of the labile nature of ACCO. Three different mechanisms of inactivation have been reported, non-oxidative inactivation and two different methods of oxidative inactivation.[81] A study of the inactivation of ACCO from pear fruit reported that catalase had an inhibitory effect on activity of ACCO *in vitro*.[82] These workers also studied the mechanism of ACCO inactivation on crude enzyme extracts and concluded that inactivation was the consequence of an enzyme, ACC, Fe^{2+}, and ascorbate complex.

A third study of ACCO inactivation that used a number of ACCO mutants suggested that purified apo-ACCO under normal reaction conditions can undergo an alteration in conformation resulting in an enzyme capable of binding Fe^{2+} that is only partially active.[78] Apo-ACCO may also bind Fe^{2+} incorrectly such that one of the iron ligands (His234) is not ligated. This conformer can undergo a change to give a fully active enzyme. Zhang *et al.*, (1997)[78] proposed that the reaction catalyzed by ACCO proceeds *via* a radical mechanism with ACC binding to the conserved RXS motif found in 2-ODDS. A second group has proposed a similar mechanism by investigating the ACCO reaction using an ACC analog; they also proposed that the reaction proceeds by a radical mechanism and provided evidence against an *N*-hydroxylation mechanism.[83]

Recently, an alternative mechanism has been proposed; electron paramagnetic resonance (EPR) spectroscopy of a recombinant ACCO containing Fe^{2+} in an anaerobic environment demonstrated that ascorbate excludes the oxygen analog, NO, from the active site.[84] This suggests that it is unlikely that ascorbate binding to the iron is the initial step in the dioxygen activation process as was proposed previously.[78,83] Rocklin *et al.*, (1999) showed that ACC and NO bind simultaneously to the iron at the active site of ACCO and have suggested a different catalytic mechanism (Fig. 10).[84]

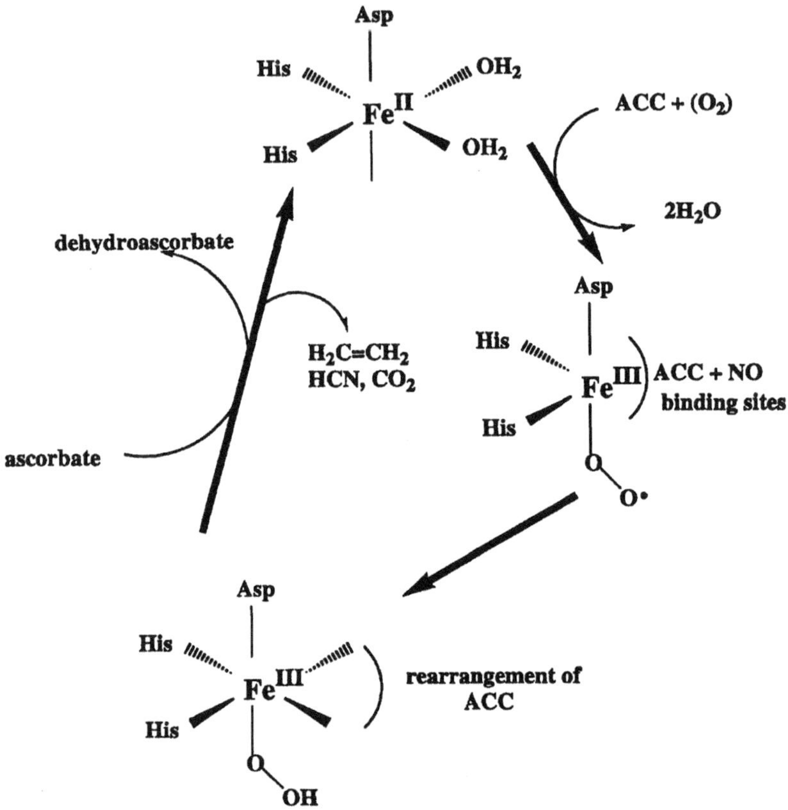

Figure 10. A simplified diagram of the reaction mechanism of ACCO proposed by Rocklin *et al.*, (1999).[84]

The ACCO of higher plants is related by both sequence, structure, and function to the ethylene-forming enzyme (EFE) from *Pseudomonas syringae*, a plant pathogen (reviewed by Fukuda *et al.*, 1993).[85] The enzyme requires 2-oxoglutarate, L-arginine, Fe^{2+}, and dioxygen to catalyze the reaction.[86] EFE catalyzes the synthesis of ethylene from 2-oxoglutarate and the oxidative decarboxylation of 2-oxoglutarate to succinate, with the concomitant formation of guanidine and L-δ-pyrroline-5-carboxylate from L-arginine.[87] A recent study investigated the roles of histidine residues in EFE by site-directed mutagenesis and found two histidines (His189 and His233) that appeared to be essential for catalytic activity.[88] This lends support to the idea that EFE has a similar structure to other 2-ODDs.

Hydroxylation of Proline Residues

Prolyl 4-hydroxylase (P4H) was first identified in mammals and catalyzes the introduction of a hydroxyl group into some of the prolyl residues of collagen.[89] The 4-hydroxyproline residues are essential for the correct folding of collagen polypeptide chains.[90] P4H was the first 2-ODD to be described, and thus, in the past, it has been considered to be the type member of the group. Mammalian P4H has been investigated in detail. For a recent review see Kivirikko and Pihlajaniemi (1998).[91] Mammalian P4H has an $\alpha_2\beta_2$ structure. The tetramer has no inter-chain disulphide bonds, but intra-chain bonds between α subunits appear to be essential.[91-93] However, the α subunits of P4H show limited sequence similarity to other 2-ODDs, with only a small number of critical residues in the active site being recognizable as conserved residues within the majority of members of the super-family.[3]

The α subunit contains the active site, and five residues have been identified by site-directed mutagenesis as having a critical role in catalysis. Two histidines and an aspartate residue have been identified as being the iron ligands, while a lysine residue and a third histidine residue are implicated in binding 2-oxoglutarate.[94,95] The β subunit of mammalian P4H is identical to protein disulfide isomerase (PDI).[96,97] The β subunit has PDI activity when present in the P4H tetramer but also exists and is active as a monomer.[98]

The role of the β subunit is partly to prevent the α subunits forming insoluble aggregates,[3,99-101] and also to anchor the enzyme in the lumen of the endoplasmic reticulum.[102] The disulfide isomerase activity of the β subunit is not required for P4H assembly or activity.[102] The PDI protein also has other roles in the cell, including being a subunit of the microsomal triglyceride transfer protein, acting as a chaperone, being the major cellular thyroid hormone-binding protein, and being a calcium binding protein, a dehydroascorbate reductase, and r-cognin (reviewed in Kivirikko and Pihlajaniemi, 1998).[91]

The α subunit of P4H in mammals is heterogeneous, occurring as α(I) and α(II) isoforms.[99,103,104] Genes encoding the α subunits of P4H have also been cloned from *Caenorhabditis elegans*, *Drosophila melanogaster*, and from *Paramecium bursaria Chlorella* virus-1 (PBCV-1). The amino acid sequence of the α subunit of the *Drosophila* P4H is 34 and 35% identical to the human α(I) and α(II) subunit types, and 31% identical to the *C. elegans* α subunit.[105] The amino acid sequence of the *C. elegans* α subunit is 45 and 42% identical to the human α(I) and α(II) subunit types.[100] The viral α subunit encodes a smaller polypeptide (242 residues as opposed to 500+ residues) with homology to the C-terminus of α subunits from other species.[106]

As yet, no genes encoding α subunits have been cloned from a higher plant source, although the enzyme has been studied in some detail (reviewed by Wojtaszek *et al.*, 1999).[107] The plant enzyme appears to be an $\alpha_2\beta_2$ tetramer and is associated with the endoplasmic reticulum and Golgi apparatus. Hydroxyproline residues are found in a variety of proteins in plants including extensins, arabinogalactans, proline-rich proteins, and hydroxyproline-rich glycoproteins (reviewed in Cassab, 1998).[108] It is likely that higher plant P4Hs also have broad substrate specificities, as there are multiple peptide "motifs" that are known to be hydroxylated in plant cell wall proteins. A putative β subunit has been cloned from alfalfa, but there is no evidence to suggest that this PDI forms part of a P4H tetramer.[109,110]

Other Plant 2-ODDs

A number of other genes encoding 2-ODDs have been cloned from a variety of species. As yet, the function of these genes remains to be determined, although it is likely that the *Ids2* and *Ids3* genes are involved in the synthesis of mugineic acid.[111, 112] Another 2-ODD of unknown function is *E8*; *E8* expression is regulated by ethylene during tomato fruit ripening, and the expression of this protein appears to regulate negatively the synthesis of ethylene in fruit.[113] The function of *E8* may be to influence elements of the signal transduction pathway, but it is most likely that this enzyme acts on a precursor of ethylene biosynthesis, such as *S*-adenosylmethionine, so that reduction of *E8* expression results in an increase in the availability of the precursor to the ethylene biosynthetic pathway.

2-ODDS IN ARABIDOPSIS THALIANA

The main thrust of this review is the evolution of 2-ODDs in plants. Thus, having reviewed known members of the 2-ODD super-family, the focus of the discussion now changes to the abundance of 2-ODDs and their functions in the model plant, *A. thaliana*. The sequencing of the *Arabidopsis* genome, although not yet completed, has revealed the abundance of different gene families in this species. It is currently estimated that there are approximately 313 P450 monooxygenases encoded in the *Arabidopsis* genome (source of information http://drnelson.utmem.edu/Arabfam.html).P450 monooxygenases catalyze similar reactions to 2-ODDs.[114] A survey of the sequencing output from the *Arabidopsis* genome, together with unpublished data, suggests that the number of 2-ODD genes found in *Arabidopsis* will be around 100 (A. Prescott, personal observation). At present, there are 64 complete gene sequences known; a further 2 gene sequences from the genome sequencing project contain frame shifts, suggesting that these may either be sequencing errors or genuine mutations in the ecotype Columbia. In addition to the complete genes, there are 4 pseudogenes that encode partial sequences of 2-ODDs and 20 genes for which there is only partial sequence information from ESTs (expressed sequence tags) or BAC (bacterial artificial chromosome) end sequences (A. Prescott, personal observation).

Members of the 2-ODD superfamily are not evenly distributed across the *Arabidopsis* genome. Chromosomes 2 and 3 contain only 11 genes encoding 2-ODDs in total, while each of the other three chromosomes have 18+ 2-ODD gene sequences. It should be noted that chromosomes 2 and 4 have been almost completely sequenced, while the other 3 chromosomes have been 45-70% sequenced at the time of writing. In a few instances, genes encoding a sub-family of 2-ODDs are clustered together as direct or inverted repeats, indicative of relatively recent gene duplications, but the majority of the members of most sub-families are not located in clusters (A. Prescott, personal observation). An example of the lack of clustering occurs within the *Arabidopsis* GA 20-oxidase gene family, where the 3 genes are found on chromosome IV (*c.* 80cM) and chromosome V (*c.*18cM and 104cM). In several of the cases studied to date, members of the same gene family also show low levels of sequence similarity despite catalyzing the same reaction. For example, the 3 GA 20-oxidases protein sequences are 63-74% identical, while the 3 GA 2-oxidases are 52-69% identical.

Phylogenetic studies of the 64 complete gene sequences from *Arabidopsis* indicate that the superfamily consists of a number of different subgroups (Fig. 11). Current evidence suggests that in some cases, 2-ODDs involved in a single biosynthetic pathway, such as those involved in flavonoid biosynthesis or the synthesis of gibberellins, are in related sub-groups, which indicates that these pathways have evolved partially through gene duplication and divergence of 2-ODDs. It should also be noted that *Arabidopsis* contains a number of sequences with homology to the C-terminus of the α subunit of P4H, a 2-ODD with little sequence similarity to other 2-ODDs, and, therefore, not included in the phylogenetic analysis or in the total of 2-ODDs, predicted to be found in this species.

A previous analysis of intron position in 2-ODD genomic sequences suggested that 2-ODDs contain up to 3 introns at 4 varying positions.[6] Current evidence shows that the *Arabidopsis* superfamily of 2-ODDs contains genes where the number of introns varies between 1 and 10 and that there are at least 8 different intron number/position variants (A. Prescott, personal observation). In general, members of the same sub-group have identical numbers of introns situated at the same place in each sub-group member; an exception to this occurs among the three genes encoding ACCOs in *Arabidopsis*, where 1 gene has 2 introns while the other two contain 3 introns (A. Prescott, personal observation).

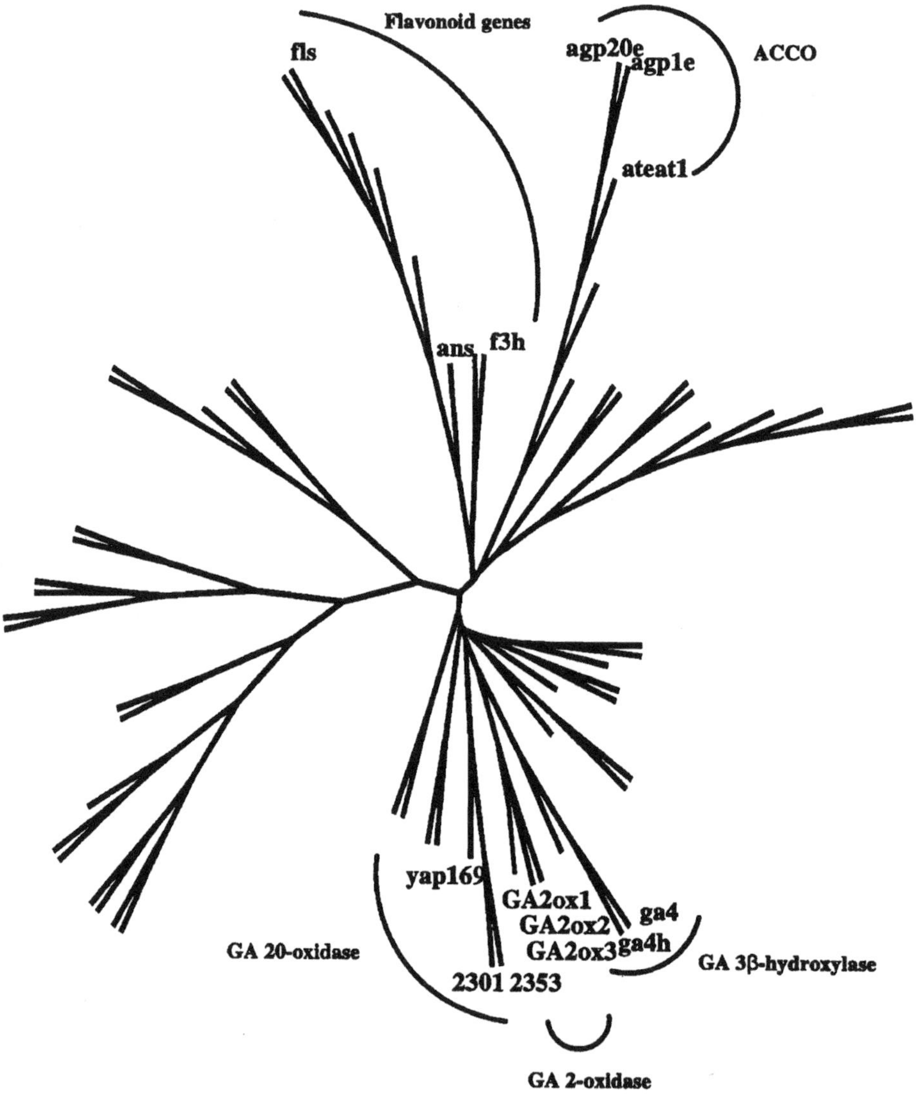

Figure 11. A phylogenetic tree of 2-ODD protein sequences found in the *Arabidopsis* genome to date. Each branch of the tree represents a single 2-ODD gene, the branches containing genes involved in GA, flavonoid, and ethylene biosynthesis are highlighted.

CONCLUSION

One of the questions that remains about the prevalence of 2-ODDs in *Arabidopsis* is what are the functions of these dioxygenases? Currently, 3 ACCOs, 8 genes involved in gibberellin biosynthesis, and 3 genes involved in flavonoid biosynthesis have been identified. In the case of the gibberellin 3β-hydroxylases, two genes have been functionally expressed in *E.coli* but phylogenetic analysis suggests that there is a third gene that is closely related to the 2 *bona fide* 3β-hydroxylases that is likely to encode either a third 3β-hydroxylase or another enzyme involved in GA synthesis/metabolism. Similarly, within the subgroup containing enzymes associated with flavonoid synthesis, a further 6 genes appear to be related to 1 or more of the 3 genes already known to have a role in flavonoid synthesis and are candidates for involvement in this biosynthetic pathway. However, even if these suppositions are largely correct, this still leaves approximately 80 genes/enzymes of completely unknown function. In addition, there are probably several hundred P450 monooxygenases capable of catalyzing similar types of reactions to 2-ODDs for which the functions are currently unknown.

A comparison of P450 monooxygenases and 2-ODDs reveals both similarities and differences. Most 2-ODDs are soluble enzymes found in the cytoplasm. In contrast, most P450 monooxygenases are membrane-bound and associated with microsomes and the endoplasmic reticulum. (For reviews of plant P450 monooxygenases, see Halkier, 1996 and Chapple, 1998;[114,115] for reviews on 2-ODDs, see Prescott, 1993 and Prescott and John, 1996).[1,6] Both types of enzymes require iron for activity, but 2-ODDs utilize Fe^{2+}, while the P450-dependent enzymes contain heme. Like 2-ODDs, monooxygenases catalyze a range of different reaction types, including hydroxylations and desaturations and are found in a large number of different biosynthetic pathways. In the case of 2-ODDs, the reaction mechanism is thought to involve an $Fe{=}O^{IV}$ species, while in P450-dependent enzymes $Fe{=}O^{V}$ appears to be the oxidizing species.[114,116]

Two intriguing questions are why organisms require both types of oxygenase enzyme and what are the advantages/disadvantages of using either an monooxygenase or a dioxygenase to catalyze a particular reaction? Where both types of enzyme occur in the same biosynthetic pathway, *e.g.* gibberellin or flavonoid synthesis, usually the monooxygenases precede the dioxygenases, indicating that dioxygenase-catalyzed reactions may involve more polar substrates. It is notable that some reactions may be catalyzed either by a 2-ODD or a P450-dependent monooxygenase, for example flavone synthase and the GA 7-, 12α-, and 13 hydroxylases.

The principal difference between P450-dependent enzymes and 2-ODDs lies in the structure of their active sites and the mechanistic implications of these structures. A number of mononuclear non-heme iron(II) enzyme families exist in addition to 2-ODDs, and a common feature in their active sites is a 2-His-1-carboxylate facial triad.[10] The triad provides the iron ligands, leaving 3 other coordination sites, orientated *cis* to each other, that can bind the substrate, molecular oxygen, and a co-factor. This allows an intramolecular mechanism of action of 2-ODDs. In contrast, P450-dependent monooxygenases have only one free coordination site on the heme that binds and activates molecular oxygen, while the substrate is bound at a separate site on the enzyme.[10] This indicates that 2-ODDs are likely to be more versatile than monooxygenases, and this is shown by the number of 2-ODDs that are capable of catalyzing multiple reactions in a pathway and can utilize a wide range of substrates *e.g.* GA 20-oxidases. Most 2-ODDs, when examined in detail, are capable of catalyzing several reactions, *e.g.* ACCOs are also known to be capable of utilizing some amino acids.[117] Some minor catalytic activities, as in the case of ACCO, may reflect the evolutionary origin of these enzymes.

Thus, there are some reactions which can be catalyzed equally well by either 2-ODDs or P450-dependent monooxygenases, while others that require *cis* coordination of substrate and molecular oxygen cannot be catalyzed by a heme enzyme that uses *trans* coordination chemistry.[118] This leads back to the title of this article. Are 2-ODDs inefficient enzymes or have they served a key role in the evolution of biochemical pathways? The flexibility of this class of enzyme in terms of the number of enzymes that can catalyze more than one reaction and/or utilize multiple substrates, including ones not found in their plant of origin, (A. Prescott, unpublished data) strongly suggests that 2-ODDs have aided the elaboration of various biosynthetic pathways, including those of flavonoid and gibberellin biosynthesis. It is also true that 2-ODDs can be termed as inefficient enzymes. Some such as ACCO, have short half-lives, and, due to the highly reactive oxidizing species at the active site, it is likely that catalytic inactivation (as shown for ACCOs) may occur in many members of the family. Most 2-ODDs are also capable of turning over 2-oxoglutarate in the absence of their true substrate, and it is generally thought that ascorbate acts as an alternative oxygen acceptor in these cases.[119]

The question still remains as to the functions of the 80 or so 2-ODDs in *Arabidopsis* for which there is currently no biochemical or genetic data. *A. thaliana* is not known for producing a wealth of unusual secondary metabolites; the only products that are found in this species and that are not widespread in higher plants are glucosinolates (restricted to the Brassicaceae) and the phytoalexin, camalexin.[120] This suggests that the majority of 2-ODDs found in this species will have homologs in most higher plants. There are possible roles for some of these enzymes as amino acid hydroxylases or in the later stages of brassinosteroid synthesis/metabolism, since in this pathway early hydroxylations are catalyzed by P450-dependent monooxygenases, and it is possible that later steps may be catalyzed by 2-ODDs, as happens in the GA biosynthetic pathway.

One of the challenges of the future is to see if plant 2-ODDs can be harnessed to produce new phyto-pharmaceuticals or plant growth regulators. It may even be possible, by mutagenesis, to create novel 2-ODDs to utilize the flexibility of these enzymes to synthesize entirely new products or to degrade metabolites/chemicals that currently pose problems in biological systems.

ACKNOWLEDGEMENTS

I thank Dr. Phil Mullineaux for critical reading of this manuscript and Dr. Rob Briddon for assistance with the phylogenetic analysis of 2-ODDs. I also acknowledge the financial assistance of the John Innes Centre, the Biotechnology and Biological Sciences Research Council and Zeneca Agrochemicals for funding research on 2-ODDs.

REFERENCES

1. PRESCOTT, A.G. 1993. A dilemma of dioxygenases (or where biochemistry and molecular biology fail to meet). J. Exp. Bot. 44: 849-861.
2. MARSH, E.N., DAH-YSYR CHANG, M., TOWNSEND, C.A. 1992. Two isozymes of clavaminate synthase central to clavulanic acid formation: cloning and sequencing both genes from *Streptomyces clavuligerus*. Biochemistry 31: 12648-12657.
3. MYLLYLÄ, R., GÜNZLER, V., KIVIRIKKO, K.I., KASKA, D.D. 1992. Modification of vertebrate and algal prolyl 4-hydroxylases and vertebrate lysyl hydroxylase by diethyl pyrocarbonate. Biochem. J. 286: 923-927.
4. ROACH, P.L., CLIFTON, I.J., FÜLÖP, V., HARLOS, K., BARTON, G.J., HAJDU, J., ANDERSSON, I., SCHOFIELD, C.J., BALDWIN, J.E. 1995. Crystal structure of isopenicillin *N* synthase is the first from a new structural family of enzymes. Nature 375: 700-704.
5. VERVERIDIS, P., JOHN, P. 1991. Complete recovery *in vitro* of ethylene-forming enzyme activity. Phytochemistry 30: 725-727.

6. PRESCOTT, A.G., JOHN, P. 1996. Dioxygenases: Molecular structure and role in plant metabolism. Annu. Rev. Plant Physiol. Plant Mol. Biol. 47: 245-271.
7. BALDWIN, J.E., ABRAHAM, E. 1988. The biosynthesis of penicillins and cephalosporins. Nat. Prod. Rep. 5: 129-145.
8. HEDDEN, P., KAMIYA, Y. 1997. Gibberellin biosynthesis: Enzymes, genes and their regulation. Annu. Rev. Plant Physiol. Plant Mol. Biol. 48: 431-460.
9. KOES, R.E., QUATTROCCHIO, F., MOL, J.N.M. 1994. The flavonoid biosynthetic pathway in plants: Function and evolution. Bioessays 16: 123-132.
10. HEGG, E.L., QUE, L., JR. 1997. The 2-His-1-carboxylate facial triad. Eur. J. Biochem. 250: 625-629.
11. FUKUMORI, F., HAUSINGER, R.P. 1993. *Alcaligenes eutrophus* JMP134 2,4-dichlorophenoxyacetate monooxygenase is an α-ketoglutarate-dependent dioxygenase. J. Bacteriol. 175: 2083-2086.
12. ROACH, P.L., CLIFTON, I.J., SHIBATA, N., HAJDU, J., SCHOFIELD, C.J., BALDWIN, J.E. 1997. Structure of isopenicillin N synthase complexed to substrate and the mechanism of penicillin formation. Nature 387: 827-830.
13. VALEGÅRD, K., TERWISSCHA VAN SCHELTINGA, A.C., LLOYD, M.D., HARA, T., RAMASWAMY, S., PERRAKIS, A., THOMPSON, A., LEE, H.-J., BALDWIN, J.E., SCHOFIELD, C.J., HAJDU, J., ANDERSSON, I. 1998. Structure of a cephalosporin synthase. Nature 394: 805-809.
14. BOROVOK, I., LANDMAN, O., KREISBERG-ZAKARIN, R., AHARONOWITZ, Y., COHEN, G. 1996. Ferrous active site of isopenicillin N synthase: Genetic and sequence analysis of the endogenous ligands. Biochemistry 35: 1981-1987.
15. TAN, S.H.D., SIM, T.S. 1996. Functional analysis of conserved histidine residues in *Cephalosporium acremonium* isopenicillin N synthase by site-directed mutagenesis. J. Biol. Chem. 271: 889-894.
16. LOKE, P., SIM, J., SIM, T.S. 1997. Functional analysis of a conserved aspartate D218 in *Cephalosporium acremonium* isopenicillin N synthase. FEMS Microbiol. Lett. 157: 137-140.
17. LANDMAN, O., BOROVOK, I., AHARONOWITZ, Y., COHEN, G. 1997. The glutamine ligand in the ferrous active site of isopenicillin N synthase of *Streptomyces jumonjinensis* is not essential for catalysis. FEBS Lett. 405: 172-174.
18. LOKE, P., SIM, T.S. 1998. Analysis of glutamines in catalysis in *Cephalosporium acremonium* isopenicillin N synthase by site-directed mutagenesis. Biochem. Biophys. Res. Commun. 252: 472-475.

19. SAMI, M., BROWN, T.J.N., ROACH, P.L., SCHOFIELD, C.J., BALDWIN, J.E. 1997. Glutamine-330 is not essential for activity in isopenicillin N synthase from *Aspergillus nidulans*. FEBS Lett. 405: 191-194.
20. TURNER, M.K., FARTHING, J.E., BREWER, S.J. 1978. The oxygenation of [3-*methyl*-^{3}H]desacetoxycephalosporin C [7 β-(5-D-aminoadipamido)-3-methylceph-3-em-4-carboxylic acid] to [3-*hydroxymethyl*-^{3}H]desacetylcephalosporin C by 2-oxoglutarate-linked dioxygenases from *Acremonium chrysogenum* and *Streptomyces clavuligerus*. Biochem. J. 173: 839-850.
21. SAMSON, S.M., DOTZLAF, J.E., SLISZ, M.L., BECKER, G.W., VAN FRANK, R.M., VEAL, L.E., YEH, W.-K., MILLER, J.R., QUEENER, S.W., INGOLIA, T.D. 1987. Cloning and expression of the fungal expandase/hydroxylase gene involved in cephalosporin biosynthesis. Bio/technology 5: 1207-1214.
22. BAKER, B.J., DOTZLAF, J.E., YEH, W.-K. 1991. Deacetoxycephalosporin C hydroxylase of *Streptomyces clavuligerus*. J. Biol. Chem. 266: 5087-5093.
23. LLOYD, M.D., LEE, H.-J., HARLOS, K., ZHANG, Z.-H., BALDWIN, J.E., SCHOFIELD, C.J., CHARNOCK, J.M., GARNER, C.D., HARA, T., TERWISSCHA VAN SCHELTINGA, A.C., VALEGÅRD, K., VIKLUND, J.A.C., HAJDU, J., ANDERSSON, I., DANIELSSON, Å., BHIKHABHAI, R. 1999. Studies on the active site of deactoxycephalosporin C synthase. J. Mol. Biol. 287: 943-960.
24. MYLLYLÄ, R., TUDERMAN, L., KIVIRIKKO, K.I. 1977. Mechanism of the prolyl hydroxylase reaction. 2. Kinetic analysis of the reaction sequence. Eur. J. Biochem. 80: 349-357.
25. BRITSCH, L. 1990. Purification and characterisation of flavone synthase I, a 2-oxoglutarate-dependent desaturase. Arch. Biochem. Biophys. 282: 152-160.
26. KOCHS, G., GRISEBACH, H. 1997. Induction and characterization of a NADPH-dependent flavone synthase from cell cultures of soybean. Z. Naturforsch. 42c: 343-348.
27. MARTENS, S., FORKMANN, G. 1998. Genetic control of flavone synthase II activity in flowers of *Gerbera* hybrids. Phytochemistry 49: 1953-1958.
28. STOTZ, G., FORKMANN, G. 1981. Oxidation of flavanones to flavones with flower extracts of *Antirrhinum majus* (snapdragon). Z. Naturforsch. 36c: 737-741.
29. FORKMANN, G. HELLER, W., GISEBACH, G. 1980. Anthocyanin biosynthesis in flowers of *Matthiola incana*: Flavanone 3-and flavonoid 3'-hydroxylases. Z. Naturforsch. 35c: 691-695.

30. BRITSCH, L., GRISEBACH, H. 1986. Purification and characterization of (2*S*)-flavanone 3-hydroxylase from *Petunia hybrida*. Eur. J. Biochem. 135: 569-577
31. LUKACIN, R., BRITSCH, L. 1997. Identification of strictly conserved hisidine and arginine residues as part of the active site in *Petunia hybrida* flavanone 3β-hydroxylase. Eur. J. Biochem. 249: 748-757.
32. BRITSCH, L., HELLER, W., GRISEBACH, H. 1981. Conversion of flavanone to flavone, dihydroflavonol and flavonol with an enzyme system from cell cultures of parsley *Petroselinum hortense*. Z. Naturforsch. 36c: 742-750.
33. HOLTON, T.A., BRUGLIERA, F., TANAKA, Y. 1993. Cloning and expression of flavonol synthase from *Petunia hybrida*. Plant J. 4: 1003-1010.
34. MENSSEN, A., HÖHMANN, S., MARTIN, W., SCHNABLE, P.S., PETERSON, P.A., SAEDLER, H., GIERL, A. 1990. The En/Spm transposable element of *Zea mays* contains splice sites at the termini generating a novel intron from a dSpm element in the *A2* gene. EMBO J. 9: 3051-3057.
35. REDDY, A.R., COE, E.H. 1962. Inter-tissue complementation: A simple technique for direct analysis of gene action sequence. Science 138: 149-150.
36. MARTIN, C., PRESCOTT, A., MACKAY, S., BARTLETT, J., VRIJLANDT, E. 1991. Control of anthocyanin biosynthesis in flowers of *Antirrhinum majus*. Plant J. 1: 37-49.
37. SAITO, K., KOBAYASHI, M., GONG, Z., TANAKA, Y., YAMAZAKI, M. 1999. Direct evidence for anthocyanidin synthase as a 2-oxoglutarate-dependent oxygenase: Molecular cloning and functional expression of cDNA from a red forma of *Perilla frutescens*. Plant J. 17: 181-189.
38. HASHIMOTO, T., YAMADA, Y. 1986. Hyoscyamine 6β-hydroxylase, a 2-oxoglutarate-dependent dioxygenase, in alkaloid-producing root cultures. Plant Physiol. 81: 619-625.
39. DE CAROLIS, E., CHAN, F., BALSEVICH, J., DE LUCA, V. 1990. Isolation and characterization of a 2-oxoglutarate dependent dioxygenase involved in the second-to-last step in vindoline biosynthesis. Plant Physiol. 94: 1323-1329.
40. HASHIMOTO, T., MATSUDA, J., YAMADA, Y. 1993. Two-step epoxidation of hyoscyamine to scopolamine is catalyzed by bifunctional hyoscyamine 6β-hydroxylase. FEBS Lett. 329: 35-39.
41. MATSUDA, J., OKABE, S., HASHIMOTO, T., YAMADA, Y. 1991. Molecular cloning of hyoscyamine 6β-hydroxylase, a 2-oxoglutarate-

dependent dioxygenase, from cultured roots of *Hyoscyamous niger*. J. Biol. Chem. 266: 9460-9464.

42. SUZUKI, K.-I., YUN, D.-J., CHEN, X.-Y., YAMADA, Y., HASHIMOTO, T. 1999. An *Atropa belladonna* hyoscyamine 6β-hydoxylase gene is differentially expressed in the root pericycle and anthers. Plant Mol. Biol. 40: 141-152.
43. HARBORNE, J.B., KHAN, M.B. 1993. Variations in the alkaloidal and phenolic profiles in the genus *Atropa* (Solanaceae). Bot. J. Linnean Soc. 111: 47-53.
44. HARTMANN, T., WITTE, L., OPRACH, F., TOPPEL, G. 1986. Reinvestigation of the alkaloid composition of *Atropa belladonna* plants, root cultures and cell suspension cultures. Planta Med. 52: 390-395.
45. YUN, D.-J., HASHIMOTO, T., YAMADA, Y. 1992. Metabolic engineering of medicinal plants: Transgenic *Atropa belladonna* with an improved alkaloid composition. Proc. Natl. Acad. Sci. USA 89: 11799-11803.
46. JOUHIKAINEN, K., LINDGREN, L., JOKELAINEN, T., HILTUNEN, R., TEERI, T.H., OKSMAN-CALDENTEY, K.-M. 1999. Enhancement of scopolamine production in *Hyoscyamus muticus* L. hairy root cultures by genetic engineering. Planta 208: 545-551.
47. VAZQUEZ-FLOTA, F., DE CAROLIS, E., ALARCO, A.-M., DE LUCA, V. 1997. Molecular cloning and characterization of desacetoxyvindoline 4-hydroxylase, a 2-oxoglutarate dependent-dioxygenase involved in the biosynthesis of vindoline in *Catharanthus roseus* (L.) G. Don. Plant Mol. Biol. 34: 935-948.
48. LANGE, T., SCHWEIMER, A., WARD, D.A., HEDDEN, P., GRAEBE, J.E. 1994. Separation and characterisation of three 2-oxoglutarate-dependent dioxygenases from *Cucurbita maxima* L. endosperm involved in gibberellin biosynthesis. Planta 195: 98-107.
49. GROSSELINDEMANN, E., LEWIS, M.J., HEDDEN, P., GRAEBE, J.E. 1992. Gibberellin biosynthesis from gibberellin A_{12}-aldehyde in a cell-free system from germinating barley (*Hordeum vulgare* L., cv. Himalaya) embryos. Planta 188: 252-257.
50. LANGE, T. 1997. Cloning gibberellin dioxygenase genes from pumpkin endosperm by heterologous expression of enzyme activities in *Escherichia coli*. Proc. Natl. Acad. Sci. USA 94: 6553-6558.
51. HEDDEN, P., GRAEBE, J.E., BEALE, M.H., GASKIN, P., MACMILLAN, J. 1984. The biosynthesis of 12α-hydroxylated gibberellins in a cell-free system from *Cucurbita maxima* endosperm. Phytochemistry 23: 569-574.

52. LANGE, T., HEDDEN, P., GRAEBE, J.E. 1993. Gibberellin biosynthesis in cell-free extracts from developing *Cucurbita maxima* embryos and the identification of new endogenous gibberellins. Planta 189: 350-358.
53. GILMOUR, S.J., ZEEVAART, J.A.D., SCHWENEN, L., GRAEBE, J.E. 1986. Gibberellin metabolism in cell-free extracts from spinach leaves in relation to photoperiod. Plant Physiol. 82: 190-195.
54. LANGE, T., HEDDEN, P., GRAEBE, J.E. 1994. Expression cloning of a gibberellin 20-oxidase, a multifunctional enzyme involved in gibberellin biosynthesis. Proc. Natl. Acad. Sci. USA 91: 8552-8556.
55. LANGE, T., KEGLER, C., HEDDEN, P., PHILLIPS, A.L., GRAEBE, J.E. 1997. Molecular characterisation of gibberellin 20-oxidases. Structure-function studies on recombinant enzymes and chimaeric proteins. Physiol. Plant. 100: 543-549.
56. KAMIYA,Y., KWAK, S.S. 1990. Partial characterisation of the 3β-hydroxylase from immature seeds of *Phaseolus vulgaris*. pp. 72-81 *in* Gibberellins (N. Takahashi, B.O. Phinney and J. MacMillan eds.), Springer Verlag, New York.
57. SMITH, V.A., GASKIN, P., MACMILLAN, J. 1990. Partial purification and characterisation of the gibberellin A_{20} 3β-hydroxylase from seeds of *Phaseolus vulgaris*. Plant Physiol. 94: 1390-1401.
58. CHIANG, H.-H., HWANG, I., GOODMAN, H.M. 1995. Isolation of the *Arabidopsis GA4* locus. Plant Cell 7: 195-201.
59. WILLIAMS, J., PHILLIPS, A.L., GASKIN, P., HEDDEN, P. 1998. Function and substrate specificity of the gibberellin 3β-hydroxylase encoded by the *Arabidopsis GA4* gene. Plant Physiol. 117: 559-563.
60. YAMAGUCHI, S., SMITH, M.W., BROWN, R.G.S., KAMIYA, Y., SUN, T. 1998. Phytochrome regulation and differential expression of gibberellin 3β-hydroxylase genes in germinating Arabidopsis seeds. Plant Cell 10: 2115-2126.
61. LANGE, T., ROBATZEK, S., FRISSE, A. 1997. Cloning and expression of a gibberellin 2β, 3β-hydroxylase cDNA from pumpkin endosperm. Plant Cell 9: 1459-1467.
62. THOMAS, S.G., PHILLIPS, A.L., HEDDEN, P. 1999. Molecular cloning and functional expression of gibberellin 2-oxidases, multi-functional enzymes involved in gibberellin deactivation. Proc. Natl. Acad. Sci. USA 96: 4698-4703.
63. YANG, S.F., HOFFMAN, N.E. 1984. Ethylene biosynthesis and its regulation in higher plants. Annu. Rev. Plant Physiol. 35: 155-189.

64. DONG, J.G., FERNÁNDEZ-MACULET, J.C., YANG, S.F. 1992. Purification and characterisation of 1-aminocyclopropane-1-carboxylate oxidase from apple fruit. Proc. Natl. Acad. Sci. USA 89: 9789-9793.
65. SMITH, J.J., VERVERIDIS, P., JOHN, P. 1992. Characterisation of the ethylene-forming enzyme partially purified from melon. Phytochemistry 31: 1485-1494.
66. PONELEIT, L.S., DILLEY, D.R. 1993. Carbon dioxide activation of 1-aminocyclopropane-1-carboxylate (ACC) oxidase in ethylene synthesis. Postharvest Biol. Technol. 3: 191-199.
67. SMITH, J.J., JOHN, P. 1993. Activation of 1-aminocyclopropane-1-carboxylate oxidase by bicarbonate/carbon dioxide. Phytochemistry 32: 1381-1386.
68. BARRY, C.S., BLUME, B., BOUZAYEN, M., COOPER, W., HAMILTON, A.J., GRIERSON, D. 1996. Differential expression of the 1-aminocyclopropane-1-carboxylate oxidase gene family of tomato. Plant J. 9: 525-535.
69. LASSERRE, E., BOUQUIN, T., HERNANDEZ, J.A., BULL, J., PECH, J.-C., BALAGUÉ, C. 1996. Structure and expression of three genes encoding ACC homologs from melon (*Cucumis melo* L.). Mol. Gen. Genet. 251: 81-90.
70. TANG, X., GOMES, A.M.T.R., BHATIA, A., WOODSON, W.R. 1994. Pistil-specific and ethylene-regulated expression of 1-aminocyclopropane-1-carboxylate oxidase genes in petunia flowers. Plant Cell 6: 1227-1239.
71. KIM, Y.S., CHOI, D., LEE, M.M., LEE, S.H., KIM, W.T. 1998. Biotic and abiotic stress-related expression of 1-aminocyclopropane-1-carboxylate oxidase gene family in *Nicotiana glutinosa* L. Plant Cell Physiol. 39: 565-573.
72. BIDONDE, S., FERRER, M.A., ZEGZOUTI, H., RAMASSAMY, S., LATCHÉ, A., PECH, J.-C., HAMILTON, A.J., GRIERSON, D., BOUZAYEN, M. 1998. Expression and characterization of three tomato 1-aminocyclopropane-1-carboxylate oxidase cDNAs in yeast. Eur. J. Biochem. 253: 20-26.
73. BOUZAYEN, M., LATCHÉ, A., PECH, J.C. 1990. Subcellular localization of the sites of conversion of 1-aminocyclopropane-1-carboxylic acid into ethylene in plant cells. Planta 180: 175-180.
74. ROMBALDI, C., LELIÈVRE, J.-M., LATCHÉ, A., PETITPREZ, M., BOUZAYEN, M., PECH, J.-C. 1994. Immunocytolocalisation of 1-aminocyclopropane-1-carboxylic acid oxidase in tomato and apple fruit. Planta 192: 453-460.

75. REINHARDT, D., KENDE, H., BOLLER, T. 1994. Subcellular localisation of 1-aminocyclopropane-1-carboxylate oxidase in tomato cells. Planta 195: 142-146.
76. RAMASSAMY, S., OLMOS, E., BOUZAYEN, M., PECH, J.-C., LATCHÉ, A. 1998. 1-aminocyclopropane-1-carboxylate oxidase of apple fruit is periplasmic. J. Exp. Bot. 49: 1909-1915.
77. LAY, V.J., PRESCOTT, A.G., THOMAS, P.G., JOHN, P. 1996. Heterologous expression and site-directed mutagenesis of the 1-aminocyclopropane-1-carboxylate oxidase from kiwi fruit. Eur. J. Biochem. 242: 228-234.
78. ZHANG, Z., BARLOW, J.N., BALDWIN, J.E., SCHOFIELD, C.J. 1997. Metal-catalyzed oxidation and mutagenesis studies on the iron (II) binding site of 1-aminocyclopropane-1-carboxylate oxidase. Biochemistry 36: 15999-16007.
79. TAYEH, M.A., HOWE, D.L., SALLEH, H.M., SHEFLYAN, G.Y., SON, J.-K., WOODARD, R.W. 1999. Kinetic and mutagenic evidence for the role of histidine residues in the *Lycopersicon esculentum* 1-aminocyclopropane-1-carboxylic acid oxidase. J. Protein Chem. 18: 55-68.
80. ZHANG, Z., SCHOFIELD, C.J., BALDWIN, J.E., THOMAS, P., JOHN, P. 1995. Expression, purification and characterisation of 1-aminocyclopropane-1-carboxylate oxidase from tomato in *Escherichia coli*. Biochem. J. 307: 77-85.
81. BARLOW, J.N., ZHANG, Z., JOHN, P., BALDWIN, J.E., SCHOFIELD, C.J. 1997. Inactivation of 1-aminocyclopropane-1-carboxylate oxidase involves oxidative modifications. Biochemistry 36: 3563-3569.
82. VIOQUE, B., CASTELLANO, J.M. 1998. *In vivo* and *in vitro* 1-aminocyclopropane-1-carboxylic acid oxidase activity in pear fruit: role of ascorbate and inactivation during catalysis. J. Agric. Food Chem. 46: 1706-1711.
83. PIRRUNG, M.C., CAO, J., CHEN, J.L. 1998. Ethylene biosynthesis: Processing of a substrate analog supports a radical mechanism for the ethylene-forming enzyme. Chem. Biol. 5: 49-57.
84. ROCKLIN, A.M., TIERNEY, D.L., KOFMAN, V., BRUNHUBER, N.M.W., HOFFMAN, B.M., CHRISTOFFERSEN, R.E., REICH, N.O., LIPSCOMB, J.D., QUE, L., JR. 1999. Role of the non-heme Fe(II) center in the biosynthesis of the plant hormone ethylene. Proc. Natl. Acad. Sci. USA 96: 7905-7909.
85. FUKUDA, H., OGAWA, T., TANASE, S. 1993. Ethylene production by micro-organisms. Adv. Microb. Physiol. 35: 275-306.

86. NAGAHAMA, K., OGAWA, T., FUJII, T., TAZAKI, M., TANASE, S., MORINO, Y., FUKUDA, H. 1991. Purification and properties of an ethylene-forming enzyme from *Pseudomonas syringae* pv. *phaseolicola* PK2. J. Gen. Microbiol. 137: 2281-2286.
87. FUKUDA, H., OGAWA, T., TAZAKI, M., NAGAHAMA, K., FUJII, T., TANASE, S., MORINO, Y. 1992. Two reactions are simultaneously catalyzed by a single enzyme: The arginine-dependent simultaneous formation of two products, ethylene and succinate, from 2-oxoglutarate by an enzyme from *Pseudomonas syringe*. Biochem. Biophys. Res. Commun. 188: 483-489.
88. NAGAHAMA, K., YOSHINO, K., MATSUOKA, M., TANASE, S., OGAWA, T., FUKUDA, H. 1998. Site-directed mutagenesis of histidine residues in the ethylene-forming enzyme from *Pseudomonas syringae*. J. Ferment. Bioeng. 85: 255-258.
89. HUTTON, J.J., TAPPEL, A.L., UDENFRIEND, S. 1967. Cofactor and substrate requirements of collagen prolyl hydroxylase. Arch. Biochem. Biophys. 118: 231-240.
90. VUORELA, A., MYLLYHARJU, J., NISSI, R., PIHLAJANIEMI, T., KIVIRIKKO, K.I. 1997. Assembly of human prolyl 4-hydroxylase and type III collagen in the yeast *Pichia pastoris*: Formation of a stable enzyme tetramer requires coexpression with collagen and assembly of a stable collagen requires coexpression with prolyl 4-hydroxylase. EMBO J. 16: 6702-6712.
91. KIVIRIKKO, K.I., PIHLAJANIEMI, T. 1998. Collagen hydroxylases and the protein disulphide isomerase subunit of prolyl 4-hydroxylases. Adv. Enzymol. Relat. Areas Mol. Biol. 72: 325-398.
92. KIVIRIKKO, K.I., MYLLYLÄ, R. 1980. The hydroxylation of prolyl and lysyl residues. pp. 53-104 *in* The Enzymology of Post-translational Modifications of Proteins (R.B. Freedman and H.C. Hawkins, eds.), Academic Press, London.
93. JOHN, D.C.A., BULLEID, N.J.1994. Prolyl 4-hydroxylase: Defective assembly of α-subunit mutants indicates that assembled α-subunits are intramolecularly disulphide bonded. Biochemistry 33: 14018-14025.
94. LAMBERG, A., PIHLAJANIEMI, T., KIVIRIKKO, K.I. 1995. Site-directed mutagenesis of the a subunit of human prolyl 4-hydroxylase. Identification of three histidine residues critical for catalytic activity. J. Biol. Chem. 270: 9926-9931.
95. MYLLYHARJU, J., KIVIRIKKO, K.I. 1997. Characterization of the iron- and 2-oxoglutarate-binding sites of human prolyl 4-hydroxylase. EMBO J. 16: 1173-1180.

96. PARKKONEN, T., KIVIRIKKO, K.I., PIHLAJANIEMI, T. 1988. Molecular cloning of a multifunctional chicken protein acting as the prolyl 4-hydroxylase β-subunit, protein disulphide-isomerase and a cellular thyroid-hormone-binding protein. Biochem. J. 256: 1005-1011.
97. PIHLAJANIEMI, T., HELAAKOSKI, T., TASANEN, K., MYLLYLÄ, R., HUHTALA, M.-J., KOIVU, J., KIVIRIKKO, K.I. 1987. Molecular cloning of the β-subunit of human prolyl 4-hydroxylase. This subunit and protein disulphide isomerase are products of the same gene. EMBO J. 6: 643-649.
98. KOIVU, J., MYLLYLÄ, R., HELAAKOSKI, T., PIHLAJANIEMI, T., TASANEN, K., KIVIRIKKO, K.I. 1987. A single polypeptide acts both as the β subunit of prolyl 4-hydroxylase and as a protein disulphide-isomerase. J. Biol. Chem. 262: 6447-6449.
99. HELAAKOSKI, T., ANNUNEN, P., VUORI, K., MACNEIL, I.A., PIHLAJANIEMI, T., KIVIRIKKO, K.I. 1995. Cloning, baculovirus expression, and characterization of a second mouse prolyl 4-hydroxylase α-subunit isoform: Formation of an $\alpha_2\beta_2$ tetramer with the protein disulphide isomerase/β subunit. Proc. Natl. Acad. Sci. USA 92: 4427-4431.
100. VEIJOLA, J., KOIVUNEN, P., ANNUNEN, P., PIHLAJANIEMI, T., KIVIRIKKO, K.I. 1994. Cloning, baculovirus expression, and characterization of the α subunit of prolyl 4-hydroxylase from the nematode *Caenorhabditis elegans*. J. Biol. Chem. 269: 26746-26753.
101. VUORI, K., PIHLAJANIEMI, T., MARTTILA, M., KIVIRIKKO, K.I. 1992. Characterisation of the human prolyl 4-hydroxylase tetramer and its multi-functional protein disulfide-isomerase subunit synthesized in a baculovirus expression system. Proc. Natl. Acad. Sci. USA 89: 7467-7470.
102. VUORI, K., PIHLAJANIEMI, T., MYLLYLÄ, R., KIVIRIKKO, K.I. 1992. Site-directed mutagenesis of human protein disulfide isomerase: Effect on the assembly, activity and endoplasmic reticulum retention of human prolyl 4-hydroxylase in *Spodoptera frugiperda* cells. EMBO J. 11: 4213-4217.
103. HELAAKOSKI, T., VUORI, K., MYLLYLÄ, R., KIVIRIKKO, K.I., PIHLAJANIEMI, T. 1989. Molecular cloning of the α-subunit of human prolyl 4-hydroxylase: The complete cDNA-derived amino acid sequence and evidence for alternative splicing of RNA transcripts. Proc. Natl. Acad. Sci. USA 86: 4392-4396.
104. ANNUNEN, P., HELAAKOSKI, T., MYLLYHARJU, J., VEIJOLA, J., PIHLAJANIEMI, T., KIVIRIKKO, K.I. 1997. Cloning of the human prolyl 4-hydroxylase α-subunit isoform α(II) and characterization of the type II enzyme tetramer. The α(I) and α(II) subunits do not form a mixed $\alpha(I)\alpha(II)\beta_2$ tetramer. J. Biol. Chem. 272: 17342-17348.

105. ANNUNEN, P., KOIVUNEN, P., KIVIRIKKO, K.I. 1999. Cloning of the α subunit of prolyl 4-hydroxylase from *Drosophila* and expression and characterization of the corresponding enzyme tetramer with some unique properties. J. Biol. Chem. 274: 6790-6796.
106. ERIKSSON, M., MYLLYHARJU, J., TU, H., HELLMAN, M., KIVIRIKKO, K.I. 1999. Evidence for 4-hydroxyproline in viral proteins. Characterization of a viral prolyl 4-hydroxylase and its peptide substrates. J. Biol. Chem. 274: 22131-22134.
107. WOJTASZEK, P., SMITH, C.G., BOLWELL, G.P. 1999. Ultrastructural localisation and further biochemical characterisation of prolyl 4-hydroxylase from *Phaseolus vulgaris*: comparative analysis. Int. J. Biochem. Cell Biol. 31: 463-477.
108. CASSAB, G.I. 1998. Plant cell wall proteins. Annu. Rev. Plant Physiol. Plant Mol. Biol. 49: 281-309.
109. SHORROSH, B.S., DIXON, R.A. 1991. Molecular cloning of a putative plant endomembrane protein resembling vertebrate protein disulphide-isomerase and a phosphatidylinositol-specific phospholipase C. Proc. Natl. Acad. Sci. USA 88: 10941-10945.
110. SHORROSH, B.S., SUBRAMANIAM, J., SCHUBERT, R.A., DIXON, R.A. 1993. Expression and localisation of plant protein disulfide isomerase. Plant Physiol. 103: 719-726.
111. NAKANISHI, H., OKUMURA, N., UMEHARA, Y., NISHIZAWA, N.-K., CHINO, M., MORI, S. 1993. Expression of a gene specific for iron deficiency (*Ids3*) in roots of *Hordeum vulgare*. Plant Cell Physiol. 34: 401-410.
112. OKUMURA, N., NISHIZAWA, N.-K., UMEHARA, Y., OHATA, T., NAKANISHI, H., YAMAGUCHI, T., CHINO, M., MORI, S. 1994. A dioxygenase gene (*Ids2*) expressed under iron deficiency conditions in the roots of *Hordeum vulgare*. Plant Mol. Biol. 25: 705-719.
113. KNEISSL, M., DEIKMAN, J. 1996. The tomato *E8* gene influences ethylene biosynthesis in fruit but not in flowers. Plant Physiol. 112: 537-547.
114. HALKIER, B.A. 1996. Catalytic reactivities and structure/function relationships of cytochrome P450 enzymes. Phytochemistry 43: 1-21.
115. CHAPPLE, C. 1998. Molecular-genetic analysis of plant cytochrome P450-dependent monooxygenases. Annu. Rev. Plant Physiol. Plant Mol. Biol. 49: 311-343.
116. QUE, L., JR., HO, R.Y.N. 1996. Dioxygen activation by enzymes with mononuclear non-heme iron active sites. Chem. Rev. 96: 2607-2624.

117.GIBSON, E.J., ZHANG, Z., BALDWIN, J.E., SCHOFIELD, C.J. 1998. Substrate analogs and inhibition of ACC oxidase: conversion of D-valine to iso-butanal. Phytochemistry 48: 619-624.

118.SCHOFIELD, C.J., BALDWIN, J.E., BYFORD, M.F., CLIFTON, I., HADJU, J., HENSGENS, C., ROACH, P. 1997. Proteins of the penicillin biosynthesis pathway. Curr. Opin. Struct. Biol. 7: 857-864.

119.MYLLYLÄ, R., MAJAMAA, K., GÜNZLER, V., HANAUSKE-ABEL, H.M., KIVIRIKKO, K.I. 1984. Ascorbate is consumed stoichiometrically in the uncoupled reactions catalysed by prolyl 4-hydroxylase and lysyl hydroxylase. J. Biol. Chem. 259: 5403-5405.

120.CHAPPLE, C.C.S., SHIRLEY, B.W., ZOOK, M., HAMMERSCHMIDT, R. SOMERVILLE, S.C. 1994. Secondary metabolism in *Arabidopsis*. pp. 989-1030 *in Arabidopsis* (E.M. Meyerowitz and C.R. Somerville, eds.), Cold Spring Harbor Laboratory Press, New York.

Chapter Nine

EVOLUTION OF ACYLTRANSFERASE GENES: Origin and Diversification of the BAHD Superfamily of Acyltransferases Involved in Secondary Metabolism

Benoit St-Pierre[1,2] and Vincenzo De Luca[1,3]

Institut de Recherche en Biologie Végétale
Université de Montréal[1]
Canada

Laboratoire de Physiologie Végétale
Université de Tours
France[2]

Novartis Agribusiness Biotechnology Research, Inc.
3054 Cornwallis Road
Research Triangle Park, North Carolina, 27709-2257[3]

INTRODUCTION

Enzymatic acylation is a widespread substitution reaction that is notably involved in the diversification of natural substances produced by plant secondary metabolism (Fig. 1). A large number of small molecules, covering most classes of secondary metabolites, are subject to either *O*-, *N*-, *C*-, or *S*- acyl group transfer.

Evolution of Metabolic Pathways – J.T. Romeo et al. (Editors)

Acylation may also trigger non-enzyme-catalyzed reactions in natural product biosynthesis. For instance, acetylation of the isoquinoline alkaloid salutaridinol by a stereospecific acetyltransferase is followed by spontaneous allylic elimination of acetate and non-enzymatic oxide bridge closure to yield thebaine (Fig. 2). In addition to small molecules, acylation of polysaccharides and proteins yields macromolecules with various regulatory properties.

Acyltransferases (EC 2.3.1.x) catalyze transfer of acyl groups from a donor molecule to the hydroxyl, amino, or thiol group of an acceptor molecule to yield an acyl ester derivative. Different acyltransferase classes have been identified that use 1-*O*-acylglucosides, acyl coenzyme A, acylated-acyl carrier protein, or quinic acid ester (*e.g.* chlorogenic acid) as high energy activated acyl- donors. Acyltransferases that utilize 1-*O*-acylglucoside as an acyl- donor in plants, including galloyltransferase (1-*O*-galloyl-β-D-glucose:1,2,3,6-tetra-*O*-galloylglucose 4-*O*-galloyltransferase),[1] glucose fatty acyltransferase (1-*O*-fatty acyl-β-D-glucose:acylglucose/ glucose fatty acyltransferases),[2] and indole-3-acetyl- β-1-D-glucoside:*myo*-inositol indoleacetyltransferase,[3] 1-*O*-(sinapoyl)- β-D-glucoside:1-(sinapoyl)- β-D-glucose 2-*O*-sinapoyl-transferase,[4] as well as various others,[5] have been described.[2] Acylated acyl carrier protein may act as an acyl- donor for fatty acid and phospholipid biosynthesis. Chlorogenic acid may act as a caffeic acid donor in the formation of isochlorogenic acid (3,5-bis-*O*-caffeoylquinic acid)[6] and of caffeoylglucaric acid.[7] Finally, coenzyme A esters act as acyl donors in the most abundant group of acyltransferases that catalyze diverse reactions involved in primary and secondary metabolism.

CoA-dependent acyltransfer reactions are involved in the biosynthesis of acetyl-CoA, cysteine, arginine, phospholipids, fatty acids, and polyamines, among others. In animals, CoA-dependent *O*- and *N*-acylation plays an important role in the biosynthesis of acetylcholine neurotransmitters,[8] of cholesterol esters,[9] in the detoxification/bioactivation of arylamine carcinogens,[10,11] and in the regulation of gene transcription through histone acetylation.[12] In prokaryotes, *N*- and *O*-acetyltransferases are major determinants for the detoxification of antibiotics such as chloramphenicol, aminoglycosides, puromycin, streptothricine, and phosphinotricin.[13,14] In plants, CoA thioesters play important roles in *O*- and *N*-acylation of quinolizidine-,[15] benzylisoquinoline-,[16] monoterpenoid indole-,[17] and tropane alkaloids,[18] as well as in the formation of dianthramides,[19] hydroxycinnamic acid esters of flavonoids and anthocyanins,[20] terpenes,[21,22] and aromatic esters.[23]

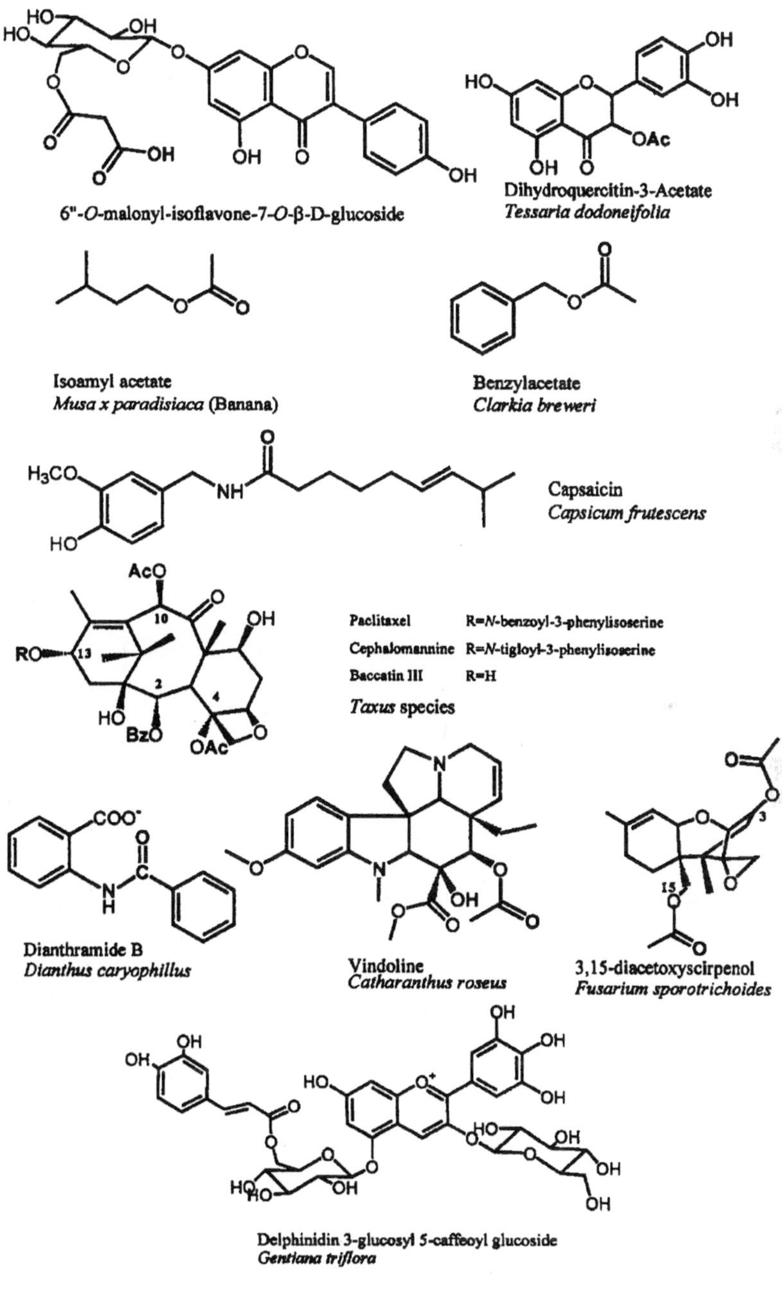

Figure 1. Some mono- and polyacylated plant secondary products.

The CoA-dependent acyltransferases are also much diversified with respect to their molecular mass, oligomeric structure, and amino acid sequences. Several distinct classes of enzymes have been characterized among prokaryotes and animals, and many do not show any sequence conservation with other acyltransferases described above. Futhermore, even within one class of enzyme, *e.g.* the chloramphenicol acetyltransferases, two families of evolutionarily unrelated genes have been characterized.[14] In addition, crystallographic studies have indicated a diversity of folds in CoA binding proteins, and no conserved sequence motifs for loops involved in CoA binding.[24,25] This review describes a recently characterized acyltransferase superfamily from higher plants and fungi. Evolutionary aspects of this gene superfamily will be discussed and special emphasis will be given to the acyltransferases of *Arabidopsis thaliana* whose genome is being completely sequenced.

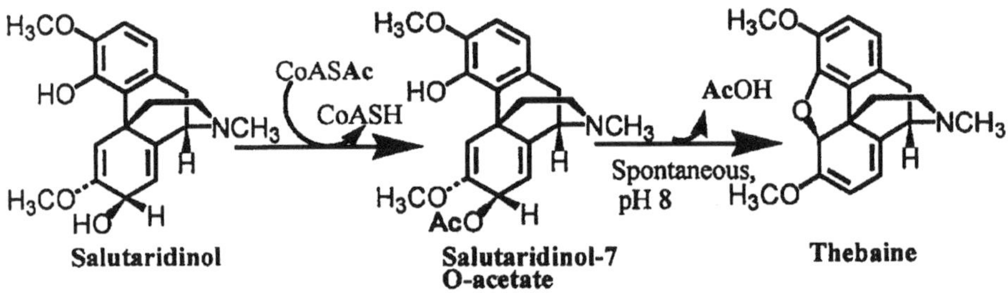

Figure 2. Salutaridinol 7-*O*-acetyltransferase catalyzed reaction and spontaneous allylic elimination of acctat yielding thebaine.

BIOCHEMISTRY OF COA ESTER-DEPENDENT ACYLTRANSFERASES

Plant CoA-dependent acyltransferases show diverse structural properties, most notably in their molecular mass and oligomeric composition. Their subunit sizes range from 15 kD for the trimeric quinate *O*-hydroxycinnamoyltransferase,[26] 25 kD for the dimeric hydroxycinnamoyl-CoA:tyramine hydroxycinnamoyltranferase,[27,28] 34 kD for the tetrameric serine acetylransferase,[29] 38 kD for the monomeric 1-aminocyclopropane-1-carboxylate (ACC) *N*-malonyltransferase,[30,31] up to 71 kD for the monomeric 10-hydroxytaxane *O*-acetyltransferase.[32] However, monomeric CoA- dependent acyltranferases with Mrs close to 50 kD have been the most frequently identified class in plants (Table 1).

Table 1. Molecular Size of Denatured and Native Enzymes for Selected CoA Ester Dependent Acyltransferases

Source	Pathway	Acyl group transfered	Subunits Mr (kD)	Native Mr (kD)
Catharanthus roseus	Indole Alkaloid	Acetyl-[33]	**50**[a]	45
Papaver	Isoquinoline alkaloid	Acetyl-[16]	**50**	50
Clarkia breweri	Benzyl Acetate	Acetyl-[23]	**58 (48)**[b]	N.D.
Taxus canadensis	Taxanes	Acetyl-[22] (5-*O*-acylation)	**N.D.**	50
Taxus chinensis	Taxanes	Acetyl-[32] (10-*O*-acylation)	**71**	72
Mentha piperita	Monoterpene (menthol acetate)	Acetyl-[21]	**N.D.**	37
Musa sapientum	Isoamylacetate	Acetyl-[34]	**N.D.**	40
Lupinus termis	Quinolizidine alkaloids	Tigloyl-[15]	**50**	40
Datura stramonium	Tropane alkaloid	Tigloyl-[18]	**65**	65
Dianthus caryophyllus	Dianthramides	Benzoyl/hydro xycinnamoyl-[19]	**53**	44
Nicotiana tabacum	Suberin (ω-hydroxyfatty acid esters)	Hydroxy-cinnamoyl-[35]	**52**	55
Equisetum arvense	Caffeic acid esters (tartrate, shikimate)	Hydroxycinna moyl-[36]	**N.D.**	45 50, 52
Gentiana triflora *Perilla frutescens*	Anthocyanins	Hydroxycinna moyl-37, 38	**52**[38] **50**[37]	49[38] 49[37]
L. esculentum *Vigna radiata*	Ethylene (ACC *N*-conjugate)	Malonyl-[30, 31, 39]	**38**[30] **40**[39]	38,[30] 40,[31] 54[39]
Arachis hypogea	Anthranilic acid *N*-conjugate	Malonyl-[40]	**51**	50
Arachis hypogea	? (2-methoxyethanol *O*-malonylation)	Malonyl-[40]	**54**	55
Petroselinum hortense	Flavonoids	Malonyl-[41]	**50**	50

[a] *C. roseus* acetyltransferase was originaly purified as a heterodimer of 33 and 21 kD subunits,[42] however immunoblot analysis showed that the enzyme exists as single 50 kD subunit *in vivo* that appears to be sensitive to proteolysis.[33] *Dianthus* benzoyl/hydroxycinnamoyltransferase was also sensitive to partial proteolysis during the course of purification.[19]

[b] Although the predicted molecular mass of BEAT was 48.2 kD, the purified protein migrated as a 58 kD band on SDS-PAGE. The discrepancy may be caused, at least in part, by an unusual concentration of charged residues on the protein.[23]

These enzymes catalyze acetyl-, malonyl-, tigloyl-, benzoyl-, and hydroxycinnamoyl transfer reactions, which are mostly associated with secondary metabolic pathways. In addition to a monomeric structure and a similar molecular mass, several other common features of this acyltransferase family can be described. Diethylpyrocarbonate has been used to inactivate deacetylvindoline 4-*O*-acetyltransferase (DAT), taxane 5-*O*-acetyltransferase, and other acyltransferases,[22,33,43] which suggests that histidine residues are important for catalytic activity.

The reactivation or enhancement of enzyme activity with sulfhydryl protecting reagents (DTT or DTE) was described,[15,17,21,30,44] and inactivation by cysteine specific reagents,[15,21,37,38,43] suggested that reduced sulfhydryl groups are also required for activity. Two monomeric acyltransferases also appear to be subject to partial proteolysis,[19,33] and this could be a common feature of this class of enzyme. However, more biochemical analyses and sequence information are required to clarify the relationship between these acyltransferases (Table 1), with particular respect to the smaller (38 kD[40]) and larger (71 kD[32]) proteins.

THE BAHD SUPERFAMILY OF ACYLTRANSFERASES

During the past few years, structural genes for several monomeric acyltransferases have been isolated from a number of plant (Table 2) and fungal (Table 3) species. They encode proteins with known biochemical functions whose predicted molecular masses of 45 to 61 kD are consistent with the Mr of monomeric acyltransferases described in Table 1.

Genes encoding three of these acetyltransferases and two hydroxycinnamoyltransferases are all associated with secondary metabolism. A gene from the Madagascar periwinkle encodes DAT, which is involved in the terminal step in vindoline biosynthesis.[33] A second gene encoding a homologous protein (78% similarity) appears to encode minovincinine 19-*O*-acetyltransferase, which has a different substrate specificity (unpublished data) from DAT.[17,45] A more distant acetyltransferase gene (35% similarity) from *Clarkia breweri* encodes a benzylalcohol *O*-acetyltransferase involved in the formation of benzylacetate, which is a major floral scent component in this species.[23]

A gene from carnation encodes benzoyl/hydroxycinnamoyl CoA:anthranilate *N*-benzoyl/hydroxycinnamoyltransferase (HCBT) involved in the formation of dianthramide phytoalexins, which are characteristic of this species.[19] This enzyme, like most aromatic acyltransferases of known function, shows a broad substrate specificity for the acyl donor. Hydroxycinnamoyl

CoA:anthocyanin *O*-hydroxycinnamoyltransferase genes that catalyze *O*-acylation of the 3-glucosyl or 5-glucosyl residue of anthocyanidin 3,5-diglucosides[20,46] have been isolated from five different plant species. These genes all share low but significant sequence similarity (25-26 %) to DAT type acetyltransferases (Table 2). Multiple protein sequence alignment identified several conserved regions,[33] suggesting that there is an evolutionary relationship within this group of enzymes. Hence, this class of proteins constitutes a superfamily of acyltransferases for which we coined the name BAHD, based on the names of the first genes (<u>B</u>EAT <u>A</u>HCT <u>H</u>CBT1 <u>D</u>AT) isolated from plant species.

Table 2. Plant Acyltransferase Genes of the BAHD Superfamily

Gene	Reaction Catalyzed	Source	Predicted Size	Sequence Similarity (Identity)
DAT[33]	Deacetylvindoline 4-*O*-acetyltransferase	*Catharanthus roseus*	439 aa (50 kD)	100 %
MAT*	Minovincinine 19-*O*-acetyltransferase	*Catharanthus roseus*	443 aa (50 kD)	82% (75 %)
BEAT[23]	Benzylalcohol *O*-acetyltransferase	*Clarkia breweri*	433 aa (48 kD)	35% (23 %)
HCBT[19]	Anthranilate *N*-hydroxycinnamoyl/ benzoyltransferase	*Dianthus caryophyllus*	442-446 aa (50 kD)	26 % (18%)
AHCT[20, 46]	Anthocyanin *O*-hydroxy-cinnamoyltransferase	*Petunia hybrida, Senecio cruentus, Gentiana triflora, Perilla ocimoides, Lavandula angustifolia*	405-480 aa (45-54 kD)	24-27 % (14-18%)

* Unpublished data.

Apart from higher plant species, only eukaryotic fungal acyltransferase genes with significant sequence similarity have been identified (Table 3). Trichothecene 3-*O*-acetyltransferase, which is involved in the biosynthesis of sesquiterpenoid trichothecene mycotoxins in phytopathogenic *Fusarium* species, has significant similarity to DAT.[47] Decalonectrin 15-*O*-acetyltransferase,[48] which is also involved in trichothecene biosynthesis, and brewer's yeast alcohol *O*-acetyltransferase[49] have low sequence similarity to DAT (16-17%), but they have a conserved sequence motif for the proposed active site of BAHD acyltransferases (see next section). The predicted size of fungal acyltransferase proteins (50-61 kD) falls within the range of sizes for plant monomeric acyltransferases.

In addition to acyltransferases with known biochemical functions, a large number of other unknown sequences can be retrieved from sequence databases by similarity searches. The remaining cDNAs and predicted gene sequences with significant similarity to the BAHD acyltransferases are all from plant species, except for a putative acyltransferase gene from yeast.[33]

Table 3. Fungal Acyltransferase Genes Related to the BAHD Superfamily

Gene	Reaction Catalyzed	Source	Predicted Size	Sequence Similarity (Identity) (DAT=100%)
Tri 101[47]	Trichothecene 3-*O*-acetyltransferase	*Fusarium graminearum*	451 aa (50 kD)	20 % (11%)
Tri 3[48]	Trichothecene (decalonectrin) 15-*O*-acetyltransferase	*Fusarium sporotrichoides*	513 aa (57 kD)	16 % (8%)
ATF1[49]	Alcohol *O*-acetyltransferase (Isoamyl alcohol)	*Saccharomyces cerevisae*	525 aa (61 kD)	17 % (9%)

In *Arabidopsis thaliana*, 42 putative BAHD acyltransferases have so far been identified by a sequence similarity search of the nucleotide databases (our unpublished results). Only one of these, *Cer2*, has a known function. *Cer2* and its maize homolog, known as *Glossy2*, are epicuticular wax mutants that appear to be defective in the two-carbon elongation of C28 and C30 fatty acids, respectively.[50-52] No homologies of these mutant proteins to those with known functions, such as condensing enzymes involved in the elongation of very long chain fatty acids (FAE1), have previously been identified.[53] The homology of wax mutant genes to BAHD acyltransferases suggests that these genes might indeed catalyze acyl CoA-dependent fatty acid elongations. The elongation of long chain fatty acids up to C22 normally requires four reactions: decarboxylative condensation of malonyl CoA with an acyl CoA to form a β-ketoacyl derivative, reduction, dehydration, and a further reduction.[54,55] Although it is not clear how such an acyltransferase could participate in elongation of epicuticular wax fatty acids, these results could help to elucidate the poorly understood biochemistry involved in the process.

The homology of DAT to both TOM36 and CMPMEL2, two genes associated with ripening in tomato and muskmelon fruits,[56,57] respectively, might also give insight into the biochemistry of ripening. Although no known biochemical function has been assigned to TOM36, its antisense expression in transgenic tomato fruits is claimed to increase the content of solids and reducing sugars (US Patent 5,569,829).

CONSERVED BAHD DOMAINS AMONG ACYL-TRANSFERASES

Multiple protein sequence alignments of all plant acyltransferases revealed several regions of high similarity.[33] These include three highly conserved amino acids (Fig. 3) in a **His**-XXX-[Ile, Leu, Val, Met]-**Asp**-[**Gly**, Ala] motif (HXUXDG), with perfectly conserved histidine and aspartate residues, as well as a conserved glycine residue (60 %) among the plant and fungal BAHD acyltransferases. Another region of high similarity within the plant/fungal acyltransferases consists of a 7-amino acid motif near the C-terminal [**Asp** or Glu]-**Phe**-**Gly**-[**Trp** or Phe, Tyr, Leu, Met]-**Gly**-[**Lys** or Arg]-[**Pro** or Ala, Gly, Ser] (DFGWGKP motif), which is highly conserved in plant acyltransferases, except for Cer2 (Fig. 3) and Glossy2. Within the acyltranferases of fungal origin, a similar motif with a 2-aa insertion (DFGFGLGKP) was also found in the trichotecene 3-*O*-acetyltransferase (Fig. 3) but not in trichothecene (decalonectrin) 15-*O*-acetyltransferase or in alcohol acetyltransferase.[48, 49]

A comparison of conserved sequences reveals interesting similarities to some CoA ester-dependent acyltransferase families (Table 4), but not to others (Table 5). The HXUXDG motif is similar to conserved sequence motifs in several families of acyltransferase, which includes the variations observed in those involved in phospholipid and glycerolipid biosynthesis (Table 4).

The position of this motif within the carboxy terminus of the chloramphenicol acetyltransferase (CAT)[58] and the dihydrolipoamide acetyltransferase (DHLAT) families is divergent.[59] For the choline acetyltransferase (ChAT) and the carnitine *O*-acyltransferases (COAT) families,[60, 61] this motif appears in the middle of the protein sequence, and the size of the polypeptide chain C-terminal to this motif is equivalent to that found in the BAHD acyltransferases (Fig. 3).[33]

Eubacteria and filamentous fungi possess another interesting class of enzymes that also contain an analogous conserved motif (H**H**UU**X**D**G**).[62] These enzymes are involved in the production of antibiotics or siderophore peptide molecules. The non-ribosomal peptide synthases may have several copies of this motif in peptide elongation and epimerization domains. Interestingly, the position of the HHUUXDG and the size of the elongation/epimerization domain (450-550 aa) appear to be similar to the BAHD acyltransferases (Table 4). By analogy to CAT, the second histidine in this motif is thought to catalyze acyl transfer of the activated amino acid between successive 4'-phosphopantetheinyl serine-bound cofactors along with the multiple domains of these multienzymatic proteins.[62,65] These residues may also participate in peptide epimerization reactions of amino

Table 4. Characteristics of Selected CoA Ester-dependent Acyltransferase Protein Families and 4'-Phosphopantetheinyl Cofactor-dependent Acyltransfer Domain of Peptide Synthetases with a Histidine Residue at the Active Site

Enzyme family	Conserved sequences	Size (aa)	Oligomeric state
Dihydrolipoamide Acyltransferases:	**DHRUUDG**	404-637	trimer
-Dihydrolipoamide Acetyltransferase -Dihydrolipoamide Succinyltransferase - Dihydrolipoamide Transacylase			
Chloramphenicol Acetyltransferases	H**H**[A,S]VC	~210	trimer
-type I, II and III	**DG**		
ChoAT / COT / CPT Acyltransferases	E**H**SXX**DG**	620-670	monomer/
- Choline *O*-Acetyltransferase - Carnitine *O*-Acetyltransferase - Carnitine *O*-Octanoyltransferase - Carnitine *O*-Palmitoyltransferases			oligomer ?
Glycerolipid Acyltranferase	N**H**XSXU**D**	G3PAT 396-827	monomer/
- *sn*-Glycerol-3-phosphate Acyltransferase - Lysophosphatidic acid Acyltransferase - 2-Acyl-*sn*-glycerophosphoethanolamine Acyltransferase		LPAAT 245-308	oligomer ?
Non-ribosomal Peptide Synthethases	H**H**UUX**DG**	450-550	multi-domain
- Elongation domain [62] - Epimerization domain			enzyme

Information sources: Swiss-Prot, Prosite, Brenda and ProDom databases. U stands for alkyl residues Leu, Ile, Val, Met

acids bound to the 4'-phosphopantetheinyl arm. It is interesting that the reaction mechanism involved in non-ribosomal peptide biosynthesis may be similar to that of CoA-dependent acylation, as both involve the use of a 4'-phosphopantetheinyl group. Low, but statistically significant, sequence similarities in the first 180 aa of DAT to the N-terminus of the elongation domain of many non-ribosomal peptide synthases were detected by an iterative search (PSI-BLAST)[66] of the database that used position-specific score matrices generated from sequence alignements (results not shown). In contrast, only a short region of 40 aa centered on the putative active site of DAT showed significant sequence similarity to CAT. This suggests that BAHD acyltransferase shows evolutionary relationships to the elongation domain of the non-peptide synthases and to CAT.

X-ray diffraction studies have shown that the HXXXDG motif resides in the active sites of CAT, DHLAT, and dihydrolipoamide succinyl-transferase (DHLST).[67-69] In addition, mutagenic analysis showed that the conserved histidine residue was essential for the catalytic activity of CAT,[70] DHLAT,[71]

Table 5. Selected CoA Ester-dependent Acyltransferase Protein Families not Related to the Superfamily of BAHD Acyltransferase

Enzyme family	Conserved sequences	size	Oligomeric state
Hexapeptide-repeat containing-transferases	repeats of [LIV]-G-x(4)	200-220 aa	monomer
- Serine *O*-Acetyltransferase - Galactoside Acetyltransferase - UDP-*N*-acetylglucosamine Acyltransferase - Nodulation protein L (nod L) - Chloramphenicol Acetyltransferase type IV - Streptogramin A Acetyltransferase			
Spermidine/Spermine N-Acetyltransferases	RKLGMGS		
- Tyramine N-hydroxycinnamoyl-Transferase (Potato)[63]		248 aa	dimer
Polyketide ß-ketoacyl synthases	[QHG]-x-G-C-[FYNA]-[GA]-G-[GA]	390 aa	dimer
- Chalcone Synthase - Stilbene Synthase -Acridone Synthase			
Homoserine *O*-Acetyltransferases family		400-520	monomer[64]
- Deacetylcephalosporin C *O*-Acetyltransferase			

Information sources: Swiss-Prot, Prosite, Pfam and ProDom databases.

dihydrolipoamide transacylase,[72] choline acetyltransferase,[61] carnitine palmitoyltransferase II,[60] and for glycerolipid acyltransferase (where the histidine residue is found in a slightly different region NHXSXUD).[73] It is interesting to note that, in spite of the low amino acid sequence similarity among the plant/fungal acyltransferases and CAT, DHLAT, and ChAT/COAT, the region containing the active site histidine is conserved. Furthermore, chemical modification analysis with the histidine-specific reagent diethylpyrocarbonate suggested that a catalytically important histidine residue is present at the active site of DAT.[33] The conservation of the HXUXDG domain histidine residue and the inhibitor analysis both strongly suggest that this conserved histidine is part of the active site of this class of enzymes.[33]

The CAT reaction mechanism, shown in Fig. 4 is a paradigm for acyltransferases with a conserved histidine active site domain (HXXXDG).[58, 60, 61,68,72,74] The imidazole Nε2 of the active site histidine (His-195) abstracts a proton from the hydroxyl group of chloramphenicol. The nucleophilic oxygen, strengthened by the proton abstraction, attacks the carbonyl carbon of acetyl CoA and forms a putative tetrahedral intermediate. The energy of this intermediate is lowered by an active site serine (Ser-148, not shown), which provides a hydrogen bond to the charged transition state. The acyl transfer reaction is then completed

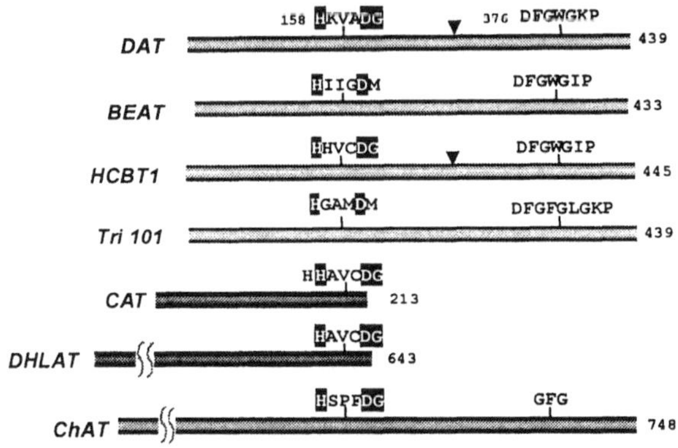

Figure 3. Structures and sequence of conserved regios within plant and fungal BAHD acyltransferases, CAT, DHLAT and ChAT

following breakdown of this tetrahedral intermediate.

In addition to the HXXXDG motif, there may be additional functional residues conserved between CAT and BAHD acyltransferase. For instance, a highly conserved threonine/serine residue, characteristic of BAHD acyltransferases,[33] corresponding to Ser-143 of DAT, Ser-148 of CAT, Ser-558 of DHLAT, and Thr-323 of DHLST, might play a role in stabilization of the transition state.[67-69] In addition, secondary structure prediction for DAT showed similarity to the well conserved secondary structures found in CAT, DHLAT, and DHLST.[67-69] For instance, the β-strand and α-helix flanking the HXXXDG residues in CAT, DHLAT, and DHLST were also predicted near DAT putative active site residues, and correspond to a highly conserved region in BAHD acyltransferases (result not shown). These conserved secondary structures and amino acid residues further argue for an evolutionary relationship within this large group of acyltransferases and against the possiblity that this family evolved separately through convergent evolution.

However, one major characteristic of the CAT/DHLAT proteins apparently has not been conserved in the BAHD acyltransferases. CAT, DHLAT, and DHLST adopt a homotrimeric structure whose interface forms a long channel that provides the binding pocket for substrates (Fig. 4).[67-69] The active site channel is

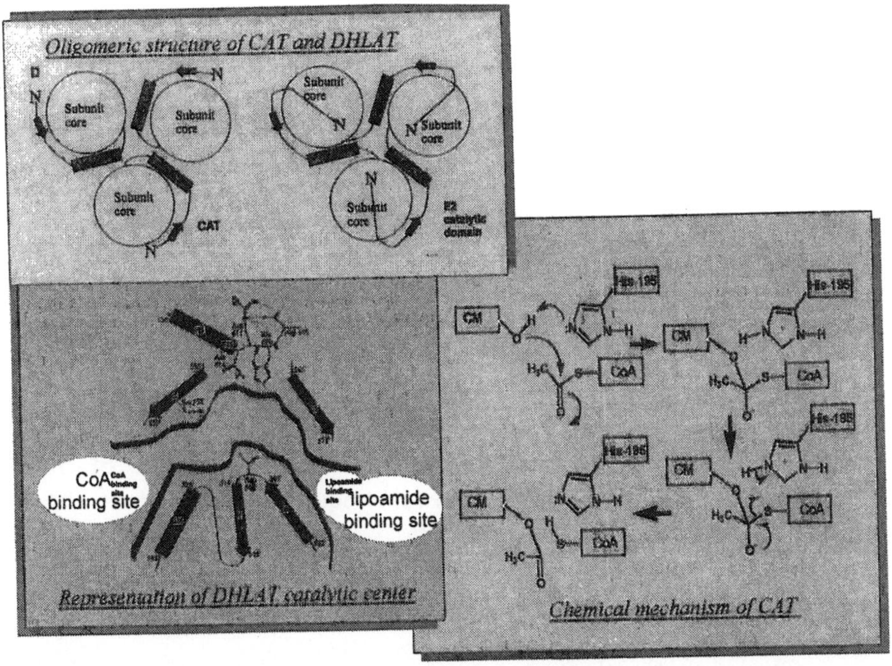

Figure 4. Oligomeric structure, catalytic domain representation of DHLAT and mechanism of the CAT catalyzed acyltransfer reaction. (Left panels are reproduced with permission from reference #68. Right panel is reproduced with permission from reference #74).

located at the interface of two 3-fold related subunits such that each trimer has three active-site channels. In view of the conservation of the active site motif and of some of the secondary structure elements between DAT, CAT, and DHLAT, a homotrimeric structure would have been expected for DAT. However, the size of DAT (45 kD), deduced from its sequence and as determined by gel filtration chromatography, does not agree with this hypothesis. This is also true for other members of the family (Table 1).

Although the monomeric composition of BAHD acyltransferases differs from the oligomeric composition of CAT/DHLAT enzymes, it corresponds to that of ChAT/COAT. The accomodation of HXXXDG active site residues within a functional monomer may have required changes in protein structure. Alternatively, BAHD acyltransferase may have a composite origin where the N-terminus corresponding to the active site domain would have been fused to a C-terminal domain of unknown origin. Interestingly, a DFGWGKP motif is highly conserved in the carboxy terminus of BAHD acyltransferases and may correspond to a region (PVVXXG[FY]G) that is highly conserved in the C-terminus of carnitine and choline acyltransferases60 (Fig. 3). These C-terminal motifs also are similar to motif A (UXXUOUXPXUQXXGUO$XUU, where U represents a bulky hydrophobic residue; O, a small residue; and $, an S, T, R, or K residue) that is conserved in a variety of N-acetyltransferases.[75,76] It has been suggested that these conserved regions bind the nucleotide moiety of coenzyme A, based on their sequence similarity to the adenine binding loop (RVVPGYG) of CoA dependent citrate synthase and to mononucleotide triphosphate binding sites (GXXGXGK).[60,75] Mutations in these region were shown to affect the activity of protein *N*-acetyltranferase (MAK3),[77,78] and rat carnitine palmitoyltransferase II (CPTII).[60] However, mutagenic analysis of the CPT II conserved region suggested it did not bind CoA, but may more likely interact with the acyl aceptor carnitine.[60] The similar C-terminal position and conserved DFGWGKP amino acid sequence also suggests this region may serve a critical function in the activity of the BAHD acyltransferases. Definite conclusions about the origins of functional acyltransferase monomers will, however, require knowledge of the tertiary structure of a member of this superfamily.

DIVERSIFICATION OF THE BAHD SUPERFAMILY- IN *ARABIDOPSIS*

The international effort to sequence the complete genome of *Arabidopsis thaliana* has created a remarkable opportunity to study the evolution and molecular mechanisms leading to the diversification of this superfamily in a higher plant species. Although the *Arabidopsis* genome project remains to be completed, over 60% of it has already been sequenced (June, 12 1999). Within this dataset, 43 genes, which show 21-38% similarity (12-26% identity) to DAT, have so far been identified by sequence similarity progams (Blast 2.0).[66] If the frequency of BAHD-like genes is constant in the genome, the total number should reach close to 65. They all contain conserved HXUXDG and DFGWGKP regions, although one sequence that diverged at the potential active site histidine residue may be a pseudogene. The BAHD-type acyltransferases, thus, represent an important family of genes that may be involved in the diversification of secondary metabolism in higher plants.

The amino acid sequence identities within *Arabidopsis* acyltransferases vary from 10 to 86% and average 19%, suggesting significant diversification of this family. In addition, there would be 25 distinct families, if a 40% sequence identity were used to define different family members. This compares with the 40 cytochrome P-450 families and the 211 full length sequences, which are currently recognized in *A. thaliana* (http://drnelson.utmem.edu/Arabfam.html).

Phylogenetic Analysis

Following alignment and exclusion of alignment-ambiguous sites, the sequences (*ca.* 280 amino acids) of bacterial elongation domain-like sequences (2), BAHD acyltransferases from *A. thaliana* (43), as well as those from other higher plant (14) and fungal (6) species were submitted to phylogenetic analysis by using a neighbor-joining method (Fig. 5). This analysis placed, with high statistical confidence (88.3% bootstrap value), the plant AT sequences within a monophyletic group derived from non-plant genes. Plant BADH ATs may, therefore, have diverged from a common ancestral gene, through gene duplication and mutation to yield the various functional enzyme groups that have been characterized in plants. Cladistic analysis identified 6 monophyletic groups among the plant sequences.

Group A contains 11 sequences including *Gentiana triflora* Anthocyanin 5-hydroxycinnamoyltransferase (A5-HCT), *Perilla ocimoides* Anthocyanin 3-hydroxycinnamoyltransferase, and seven *Arabidopsis* sequences that display 25 to 29 % amino acid sequence identity to A5-HCT. Clade B regroups 8 sequences from *Arabidopsis* and a *Petunia hybrida* AHCT-like sequence, which has not been biochemically characterized.[46] Clade C contains 8 sequences from *Arabidopsis* associated with *C. roseus* DAT,[33] MAT,[45] and *Clarkia breweri* BEAT.[23] The level of sequence identity between DAT and clade C *Arabidopsis* sequences (24 to 27%) is approximately equivalent to that occurring between DAT and BEAT (25%), suggesting that they may all catalyze acetylations. Three *Arabidopsis* sequences of clade F display 17 to 19% identity to clade C DAT, but they may not be sisters to clade C since a low bootstrap value was assigned to the C-F branch. Clade D regroups 15 *Arabidopsis* acyltransferases together with *Dianthus caryophyllus* anthranilate *N*-hydroxycinnamoyl/benzoyltransferase (HCBT1),[79] ripening-associated CMPMEL2[56] and TOM36,[57] *N. tabacum* hypersensitive-response HSR201,[80] and anther-specific ASP2547. The sequence identity between *A. thaliana* sequences and HCBT1 may be as low as 18%, indicating a high level of sequence diversity within this clade. However, *A. thaliana* clone AB017061 has 44% sequence identity to HCBT1, suggesting it may encode a reaction similar to anthranilate *N*-hydroxycinnamoyl/ benzoyltransferase. Clade E includes *A. thaliana* Cer2 gene and its *Z. mays* homolog Glossy2 (28% identity).[50-52] This phylogenetic analysis may help in the biochemical identification of *A. thaliana* acyltransferases. For example, further experiments will determine whether the members of clades A and C are hydroxycinnamoyltransferases and acetyltransferases, respectively.

The phylogenetic analysis also resolved the fungal trichothecene acetyltransferases (Tri3 and Tri101) in two different clades. Sister to the plant clade is a group containing four sequences, including *Fusarium* trichothecene 3-*O*-acetyltransferase *Tri101*s. Paraphyletic to the plant and Tri101 clade are regrouped *Fusarium* trichothecene 15-*O*-acetyltransferase Tri3 and *S. cerevisae* alcohol acetyltransferase ATF1,[48,49] but with low statistical confidence. The sequence identity (6%) between Tri3 and Tri101 is low, whereas DAT aligns more closely to Tri101 (11% identity) than to Tri3 (7% identity), both with respect to the overall sequence relatedness and to the conservation of the C-terminal motif (Fig. 3). A horizontal gene transfer (from unknown species) has been proposed for the acquisition of *Tri101* by trichothecene producing organisms, since this gene is not on the biosynthetic gene cluster where *Tri3* is found.[81] It has been speculated that *Tri101* further protects the fungus against its own toxin. Thus, the BADH family of plant acyltransferases is organized into at least six evolutionarily conserved groups. Plant genes are separated into six clades, and each clade but one contains sub-clades composed of genes from distantly related species. The two available monocot sequences (*Z. mays Glossy2* and ASP2547) fall into different clades together with dicots. These observations suggest that each group originated from lineages that split off before species divergence, and perhaps after fungal-plant divergence. The division into six distinct groups of plant acyltransferases, therefore, may predate the monocot/dicot divergence estimated to have occured between 135 and 300 million years ago.

Figure 5. Phylogenetic tree for BAHD acyltransferase amino acid sequences and mycobacterial elongation domain-like sequences inferred from the neighbor-joining distance analysis. Bootstrap values (1000 replicates) indicate statistical validity of the branches. Chromosomal localization of the *Arabidopsis* sequences on the RI genetic map and structure of the respective genes are also shown. Sequences are identified with the taxon name (or AT for *A. thaliana*) and the gene names in Tables 2 and 3 or the DNA accession number followed, when required, by order of appearance of the gene in the Genbank accession. Chromosomal localization of genes on the *Arabidopsis* RI genetic map was obtained from the *Arabidopsis* database (http://genome-www.stanford.edu/*Arabidopsis*/). Predicted exons and introns are defined by open boxed and thick lines, respectively.

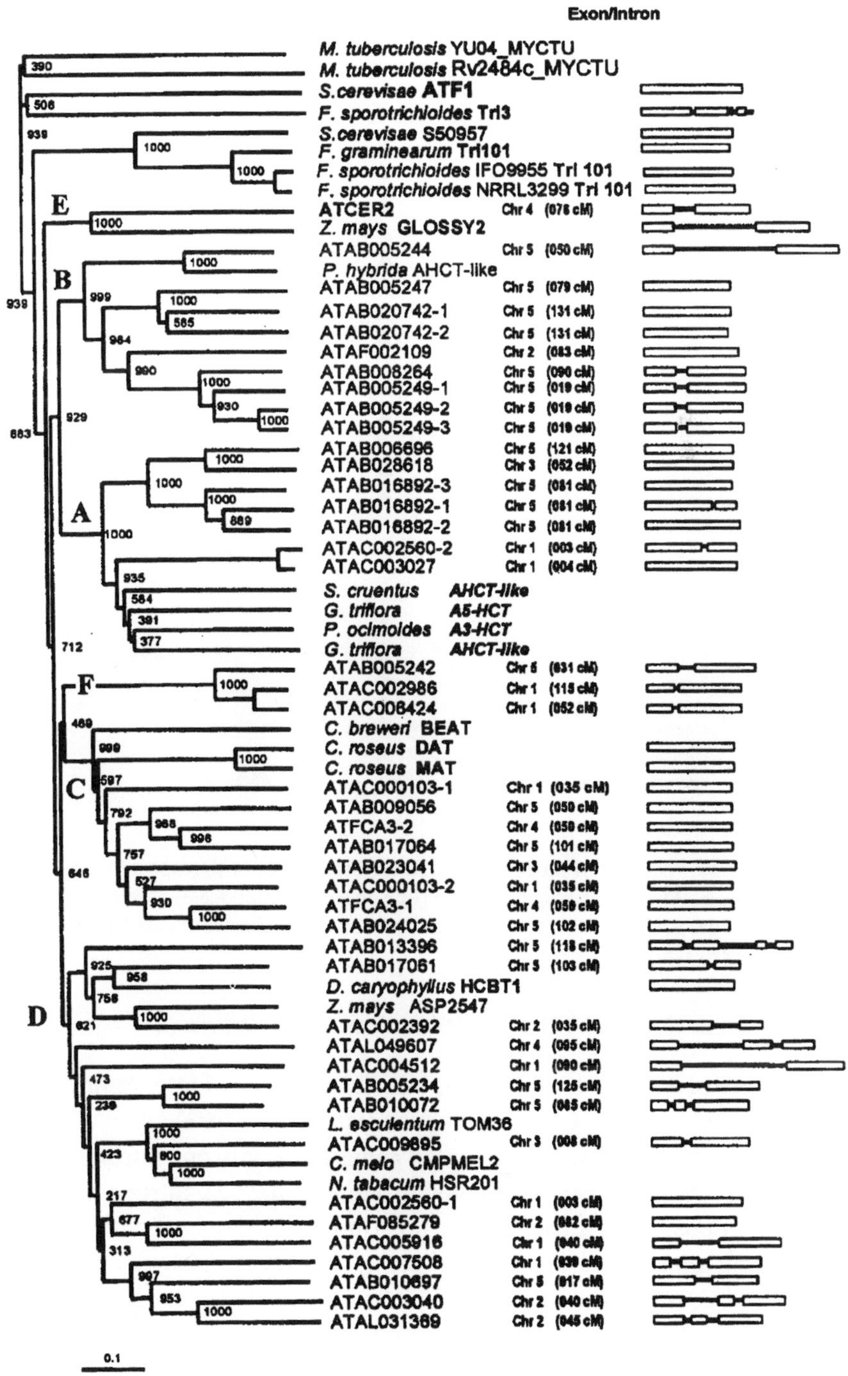
Exon/Intron
M. tuberculosis YU04_MYCTU
M. tuberculosis Rv2484c_MYCTU
S.cerevisae ATF1
F. sporotrichioides Tri3
S.cerevisae S50957
F. graminearum Tri101
F. sporotrichioides IFO9955 Tri 101
F. sporotrichioides NRRL3299 Tri 101
E
ATCER2 Chr 4 (078 cM)
Z. mays GLOSSY2
ATAB005244 Chr 5 (050 cM)
B
P. hybrida AHCT-like
ATAB005247 Chr 5 (079 cM)
ATAB020742-1 Chr 5 (131 cM)
ATAB020742-2 Chr 5 (131 cM)
ATAF002109 Chr 2 (083 cM)
ATAB008264 Chr 5 (090 cM)
ATAB005249-1 Chr 5 (019 cM)
ATAB005249-2 Chr 5 (019 cM)
ATAB005249-3 Chr 5 (019 cM)
ATAB006696 Chr 5 (121 cM)
ATAB028618 Chr 3 (052 cM)
ATAB016892-3 Chr 5 (081 cM)
A
ATAB016892-1 Chr 5 (081 cM)
ATAB016892-2 Chr 5 (081 cM)
ATAC002560-2 Chr 1 (003 cM)
ATAC003027 Chr 1 (004 cM)
S. cruentus AHCT-like
G. triflora A5-HCT
P. ocimoides A3-HCT
G. triflora AHCT-like
ATAB005242 Chr 5 (031 cM)
F
ATAC002986 Chr 1 (115 cM)
ATAC006424 Chr 1 (052 cM)
C. breweri BEAT
C. roseus DAT
C. roseus MAT
C
ATAC000103-1 Chr 1 (035 cM)
ATAB009056 Chr 5 (050 cM)
ATFCA3-2 Chr 4 (058 cM)
ATAB017064 Chr 5 (101 cM)
ATAB023041 Chr 3 (044 cM)
ATAC000103-2 Chr 1 (035 cM)
ATFCA3-1 Chr 4 (058 cM)
ATAB024025 Chr 5 (102 cM)
ATAB013396 Chr 5 (118 cM)
ATAB017061 Chr 5 (103 cM)
D. caryophyllus HCBT1
D
Z. mays ASP2547
ATAC002392 Chr 2 (035 cM)
ATAL049607 Chr 4 (095 cM)
ATAC004512 Chr 1 (090 cM)
ATAB005234 Chr 5 (125 cM)
ATAB010072 Chr 5 (085 cM)
L. esculentum TOM36
ATAC009895 Chr 3 (008 cM)
C. melo CMPMEL2
N. tabacum HSR201
ATAC002560-1 Chr 1 (003 cM)
ATAF085279 Chr 2 (082 cM)
ATAC005916 Chr 1 (040 cM)
ATAC007508 Chr 1 (039 cM)
ATAB010697 Chr 5 (017 cM)
ATAC003040 Chr 2 (040 cM)
ATAL031369 Chr 2 (045 cM)
0.1

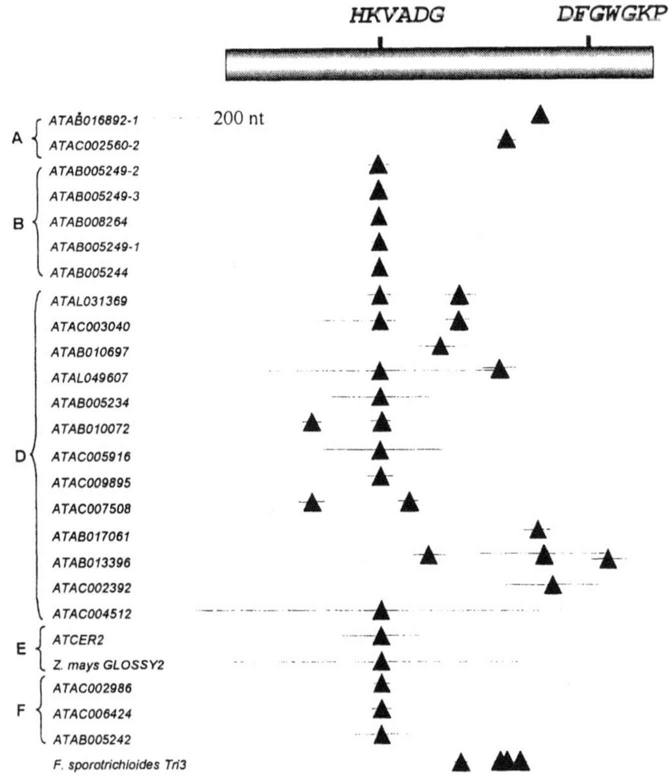

Figure 6. Positions of introns in BAHD acyltranferases with respect to DAT amino acid sequence. Position (triangle) and size (line on a nucleotide scale) of introns found in plant and fungal BAHD genes (accession numbers or gene names found on left) are shown with respect to DAT protein.

Molecular Genetics of BAHD Acyltransferase Evolution

Gene structure prediction analysis suggests the presence of intervening sequences within 24 of the 43 BAHD acyltransferase-like genes (56%) currently identified in *Arabidopsis* (Fig. 5). Most interrupted genes have only one intron, fewer genes (6) have more. With respect to aligned protein sequences, introns may be found at 12 different positions, however, one position was conserved within 75% of the intron-bearing genes (Fig. 6). The intron "Q" occurs in the sequence encoding a highly conserved glutamine residue, which is found 17-aa towards the fewer genes (6) have more. With respect to aligned protein sequences, introns may be found at 12 different positions, however, one position was conserved within 75% of the intron-bearing genes (Fig. 6). The intron "Q" occurs in the sequence encoding a highly conserved glutamine residue, which is found 17-aa towards the N-terminus from the active site motif of DAT.[33] Despite the sequence diversity within this superfamily, conservation of an intron position within members of several plant clades (B, D, E, and F) is in agreement with the hypothesis that all plant sequences are derived from a common ancestor (Fig. 6). The phylogenetic groupings obtained also seem to correlate with their gene structures. This is illustrated by the genes of clade C that contain no introns, those of clades B, E, and F that have a single "Q" intron, those of clade A where only C-terminal introns are observed, and those of clade D that contain single or multiple intron structures. Interestingly, the divergent trichothecene acetyltransferase *Tri3* from fungal species is unique, with four introns that occur at non-conserved positions.

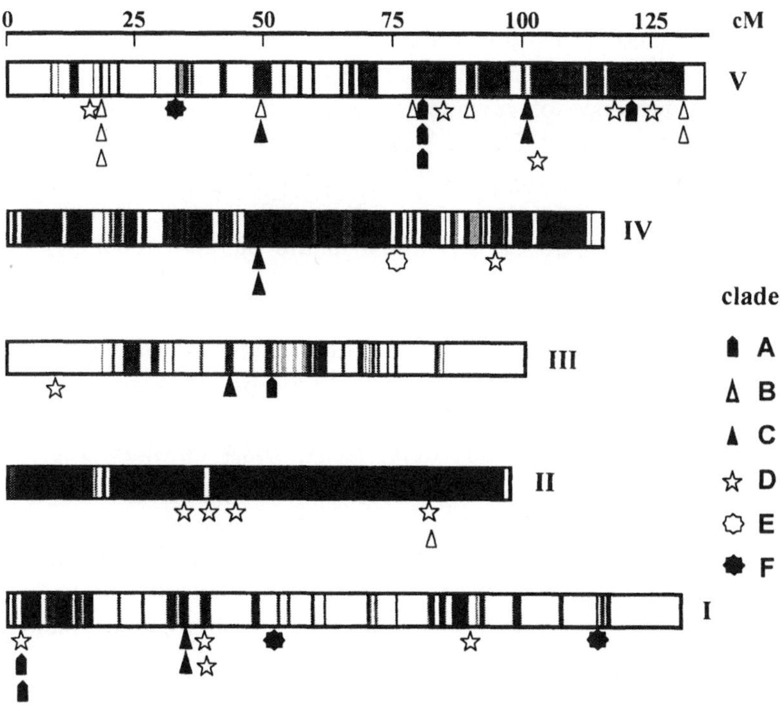

Figure 7. Chromosomal localization of Acyltransferase genes in *A. Thaliana* The association of BAHD acyltransferase genes to each clade is shown on the right.

The clustering of acyltransferase genes on *Arabidopsis* chromosomes reveals the role played by sequence duplication in diversification of this superfamily. The majority of acyltransferase sequences (54%) are localized in ten loci on chromosome V of *Arabidopsis*, which includes two or three gene clusters occurring within 1 cM (Fig. 7). Furthermore, most of the clusters (80%) include arrays of phylogenetically related genes that are also associated with their sister sequences (14/23) or close relatives (4/23) on the sequence tree (Fig. 6). An ancient event involving duplication of a two-gene cluster was apparently followed by a chromosomal translocation event that led to the shuffling of the sister sequences of ATFCA3-1 and ATFCA3-2 from chromosome 4 to chromosome 5, or vice versa (Fig. 5 and 7). Close inspection of genomic sequences in the vicinity of these clusters reveals that the presumed duplication event also involved other unrelated genes, which was most likely followed by a significant reorganization of the genes within these loci (Fig. 8). In addition to the duplication of ATFCA3-1/ATAB024025 and ATFCA3-2/ATAB017064 ancestors, close members of the CYP705A family of cytochrome P450 and a triterpene synthase gene also appear to have been included in the duplication/translocation event (Fig. 8). Following translocation, the number and position of genes apparently may have been further modified. The presence of reverse transcriptase sequences in the genomic fragments covered by AB009056 and AC000103 accession, where additional clade C sequences are found, may suggest a transposition-based mechanism responsible for these chromosomal translocations.

A striking coincidence was noted in the putative functions of genes associated with these loci. Several secondary metabolic pathway genes that catalyze distinct reactions (acyltransferase, monooxygenase, terpene cyclase) are clustered within a single locus, and some observations suggest that they may be involved in the same biosynthetic pathway. One or two genes for triterpene synthases and two BAHD acyltransferases from clade C are embedded within a large cluster of cytochrome P450 702A, 705A, 708A families (Fig. 8). It is interesting that the CYP702A and CYP708A families are phylogeneticaly close to hydroxylases, which use di- and triterpene substrates (http://drnelson.utmem.edu/CytochromeP450.html). Brassinolide biosynthesis CYP90A1 (steroid 23-hydroxylase) and CYP90B1 (steroid 22α-hydroxylase) genes,[82,83] together with gibberellin biosynthesis CYP88A1 (dwarf3) and CYP85 genes,[84,85] are phylogeneticaly close to the CYP702A and CYP708A families.

Close inspection of BAHD acyltransferase regions revealed the presence of several other secondary metabolism biosynthetic genes. At 50 cM on chromosome 5, the clade C AB009056 gene is linked with a 5-epiaristolochene synthase-like sequence (Fig. 9), which is similar to the linkage of its sister clade

C AB017064 gene (Fig. 5) at 101 cM on Chr5 (Fig. 8) with another sesquiterpene cyclase-like sequence. Furthermore, clade F AB005242 and AC002986 genes, which may be related to clade C (Fig. 5), are also linked to the terpenoid metabolic genes, isopentenyl pyrophosphate isomerase on Chr 5 (Fig. 9) and a cluster of three triterpene cyclase-like genes on Chr 1 (not shown), respectively. The clade D AB013396 gene, which is a phylogenetically close relative of the anthranilate *N*-acyltransferase HCBT, is clustered with anthranilate synthase β-like sequence on

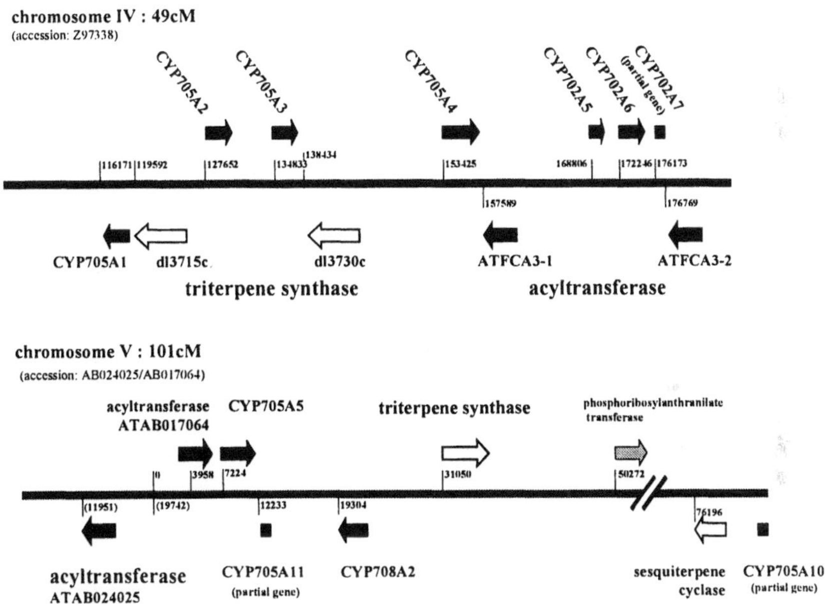

Figure 8. Physical map of the loci for phylogeneticaly related acyltransferases on chromosome IV and V of *A. thaliana*.
Physical map and predicted genes are shown for a section of chromosome IV at 49cM and a section of chromosome V at 101 cM. The phylogenetic relationships between acyltransferases are shown as solid arrows, and dotted solid arrows, with the closest sequences bearing the same pattern. Hatched arrows identify cytochrome P450 genes and include their family number. Incomplete pseudogenes are also shown (hatched boxes). Predicted genes for tritepene synthases and sesquiterpene cyclase (open arrows) are also indicated. The gene coordinate indicated refers to those of the genbank accession Z97338 for chromosome IV, and of the two contiguous genomic sequences in accession AB024025 (coordinate in bracket) and in accession AB017064 for chromosome V.

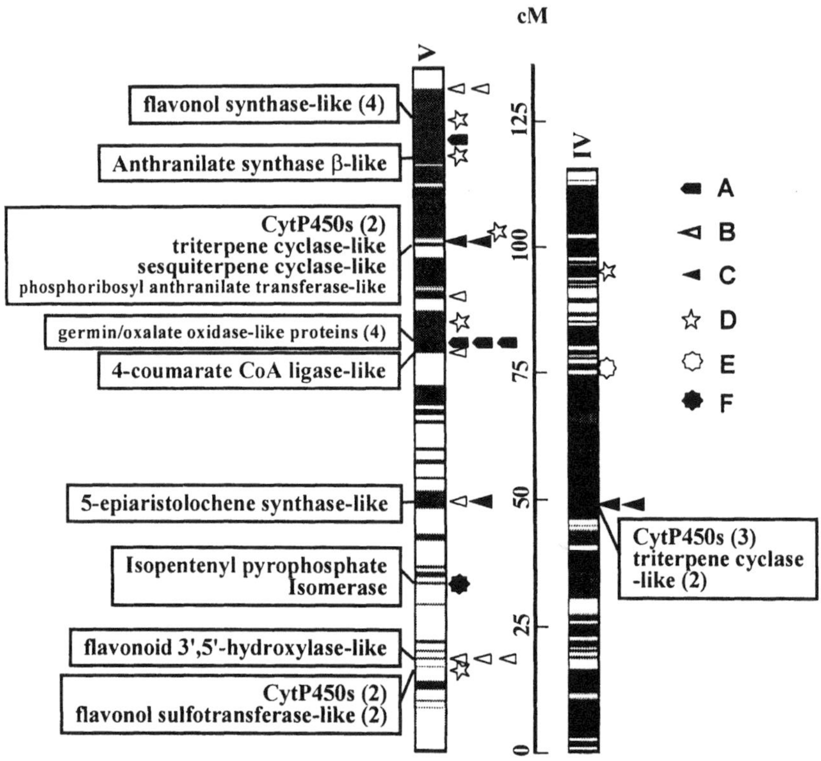

Figure 9. Genes closely linked (less than 100 Kb) with BAHD acyltranferases on *Arabidopsis* chromosomes 4 and 5.

Arabidopsis Chromosome V (Fig. 9). The clade D AB010697 and AB005234 genes may be linked to flavonoid biosynthesis through their association to two flavonol sulfotransferase-like genes and two cytochrome P450s (CYP712A2 and CYP93D1) at 17 cM, and to four flavonol synthase-like genes at 125 cM of Chr 5, respectively (Fig. 9). The associated cytochrome P450s (CYP712A2 and CYP93D1) may also be involved in flavonoid biosynthesis, as indicated by the function of other genes of the CYP93 family. CYP93A1 is a flavanone 2-hydroxylase of licorice, CYP93C1v2 catalyzes aryl migration of isoflavonoid biosynthesis, in addition to CYP712, which is a member of a sister family to CYP93 genes (http://drnelson.utmem.edu/BiblioD.html). The three acyltransferase sequences of the clade B AB005249 gene are linked to CYP75B1, which is a flavonoid 3',5'-hydroxylase (CYP75A)-like gene. This clustering and the presence of a *P. hybrida* anthocyanin HCT-like sequence (Fig. 5) may indicate that clade B sequences are, like clade A, associated with the anthocyanin pathway.

This survey is certainly far from complete since the functions of most *Arabidopsis* predicted acyltransferase-coding genes are not known. Nevertheless, the chromosomal linkage of several distinct clases of genes to secondary metabolism raises questions about the origin of these gene clusters and the possibility that they may be functionally organized.

Clustering of genes that participate in a common metabolic pathway, a general feature of prokaryotic gene organization, was observed in filamentous fungi for many dispensable metabolic pathways (rare nutrient utilization and natural product biosynthesis).[86] This is notably the case in *Fusarium sporotrichoides* where nine trichothecene biosynthetic genes are clustered, including the acyltranferase gene *Tri3*.[48,86] Our observations suggest that this unusual functional organization of genes in some eukaryotes may still have been partially conserved in higher plant genomes.

CONCLUSION

The present review suggests that plants contain a superfamily of BAHD acyltransferases, which may be related to the ancient CAT family of genes. The presence of a HXXXDG motif, which outlines the active sites of CAT, DHLAT, DHLST, the elongation domain of non-peptide synthases, and the BAHD superfamily, suggests that a common reaction mechanism has been used during evolution to promote the formation of a diversity of natural products. In contrast to previous assumptions, the association of BAHD genes with various natural product pathways in the *Arabidopsis* genome suggests that secondary metabolism may be functionally organized, rather than randomly distributed throughout the genome. The association of acyltransferases with particular pathways will be useful, together with phytochemical investigations, functional genomics, proteomics, and metabolic profiling to elucidate the biochemistry of this versatile superfamily of genes.

ACKNOWLEDGEMENTS

This work was supported by the Natural Sciences and Engineering Research Council of Canada and Le Fonds pour la Formation de Chercheurs et l'Aide à la Recherche.

REFERENCES

1. NIEMETZ, R., GROSS, G. G., 1998. Gallotannin biosynthesis - purification of β-glucogallin - 1,2,3,4,6-pentagalloyl-β-D-glucose galloyltransferase from Sumac leaves. Phytochemistry 49: 327-332.
2. LI, A.X., EANNETTA, N., CHANGAS, G.S., STEFFENS, J., 1999. Glucose polyester biosynthesis. Purification and characterization of a glucose acyltransferase. Plant Physiol. 121: 453-60.
3. KESY, J. M., BANDURSKI, R. S., 1990. Partial purification and characterization of indol-3-ylacetylglucose: *myo*-inositol indol-3-ylacetyltransferase (indoleacetic acid-inositol synthase). Plant Physiol. 94: 1598-1604.
4. DAHLBENDER, B., STRACK, D., 1986. Purification and properties of 1-(hydroxycinnamoyl)-glucose: 1-(hydroxycinnamoyl)-glucose hydroxycinnamoyl-transferase from radish seedlings. Phytochemistry 25: 1043-1046.
5. SCHÜTTE, H. R., 1992. II. Secondary plant substances. Special topics in phenylpropanoid metabolism. Prog. Bot. 53: 78-98.
6. VILLEGAS, R. J. A., SHIMOKAWA, T., OKUYAMA, H., KOJIMA, M., 1987. Purification and characterization of chlorogenic acid: chlorogenate caffeoyl transferase in sweet potato roots. Phytochemistry 26: 1577-1581.
7. STRACK, D., GROSS, W., 1990. Properties and activity changes of chlorogenic acid:glucaric acid caffeoyltransferase from tomato (*Lycopersicon esculentum*). Plant Physiol. 92: 41-47.
8. WU, D., HERSH, L. B., 1994. Choline acetyltransferase: celebrating its fiftieth year. J. Neurochem. 62: 1653-1663.
9. STURLEY, S. L., 1997. Molecular aspects of intracellular sterol esterification: The acyl coenzyme A: cholesterol acyltransferase reaction. Curr. Opin. Lipidol. 8: 167-173.
10. HEIN, D. W., RUSTAN, T. D., DOLL, M. A., BUCHER, K. D., FERGUSON, R. J., FENG, Y., FURMAN, E. J., GRAY, K., 1992. Acetyltransferases and susceptibility to chemicals. Toxicol. Lett. 64-65: 123-130.
11. HENGSTLER, J. G., ARAND, M., HERRERO, M. E., OESCH, F., 1998. Polymorphisms of *N*-acetyltransferases, glutathione *S*-transferases, microsomal epoxide hydrolase and sulfotransferases: influence on cancer susceptibility. Rec. Res. Cancer Res 154: 47-85.
12. KUO, M. H., ALLIS, C. D., 1998. Roles of histone acetyltransferases and deacetylases in gene regulation. Bioessays 20: 615-626.

13. DAVIES, J., 1994. Inactivation of antibiotics and the dissemination of resistance genes. Science 264: 375-382.
14. MURRAY, I. A., SHAW, W. V., 1997. *O*-Acetyltransferases for chloramphenicol and other natural products. Antimicrob. Agents Chemother. 41: 1-6.
15. SUZUKI, H., MURAKOSHI, I., SAITO, K., 1994. A novel *O*-tigloyltransferase for alkaloid biosynthesis in plants. Purification, characterization, and distribution in *Lupinus* plants. J. Biol. Chem. 269: 15853-15860.
16. LENZ, R., ZENK, M. H., 1995. Acetyl coenzyme A:salutaridinol-7-*O*-acetyltransferase from *Papaver somniferum* plant cell cultures. The enzyme catalyzing the formation of thebaine in morphine biosynthesis. J. Biol. Chem. 270: 31091-31096.
17. DE LUCA, V., BALSEVICH, J., KURZ, W. G. W., 1985. Acetyl coenzyme A: deacetylvindoline *O*-acetyltransferase, a novel enzyme from *Catharanthus*. Plant Physiol. 121: 417-428.
18. RABOT, S., PEERLESS, A. C. J., ROBINS, R. J., 1995. Tigloyl-Coa: pseudotropine acyl transferase - an enzyme of tropane alkaloid biosynthesis. Phytochemistry 39: 315-322.
19. YANG, Q., REINHARD, K., SCHILTZ, E., MATERN, U., 1997. Characterization and heterologous expression of hydroxycinnamoyl/benzoyl-CoA:anthranilate N-hydroxycinnamoyl/benzoyltransferase from elicited cell cultures of carnation, *Dianthus caryophyllus* L. Plant Mol. Biol. 35: 777-789.
20. FUJIWARA, H., TANAKA, Y., YONEKURA-SAKAKIBARA, K., FUKUCHI-MIZUTANI, M., NAKAO, M., FUKUI, Y., YAMAGUCHI, M., ASHIKARI, T., KUSUMI, T., 1998. cDNA cloning, gene expression and subcellular localization of anthocyanin 5-aromatic acyltransferase from *Gentiana triflora*. Plant J. 16: 421-431.
21. CROTEAU, R., HOOPER, C. L., 1978. Metabolism of monoterpenes. Acetylation of (-)-menthol by a soluble enzyme preparation from peppermint (*Mentha piperita*) leaves. Plant Physiol. 61: 737-742.
22. WALKER, K., KETCHUM, R. E. B., HEZARI, M., GATFIELD, D., GOLENIOWSKI, M., BARTHOL, A., CROTEAU, R., 1999. Partial purification and characterization of acetyl coenzyme A: taxa-4(20),11(12)-dien-5 alpha-ol *O*-acetyl transferase that catalyzes the first acylation step of taxol biosynthesis. Arch. Biochem. Biophys. 364: 273-279.
23. DUDAREVA, N., DAURIA, J. C., NAM, K. H., RAGUSO, R. A., PICHERSKY, E., 1998. Acetyl-CoA-benzylalcohol acetyltransferase - an enzyme involved in floral scent production in *Clarkia breweri*. Plant J. 14:

297-304.
24. MODIS, Y., WIERENGA, R., 1998. Two crystal structures of *N*-acetyltransferases reveal a new fold for coA-dependent enzymes. Structure 6: 1345-1350.
25. ENGEL, C., WIERENGA, R., 1996. The diverse world of coenzyme A binding proteins. Curr. Opin. Struct. Biol. 6: 790-797.
26. RHODES, M. J. C., WOOLTORTON, L. S. C., LOURENÇO, E. J., 1979. Purification and properties of hydroxycinnamoyl CoA quinate hydroxycinnamoyl transferase from potatoes. Phytochemistry 18: 1125-1129.
27. HOHLFELD, H., SCHEEL, D., STRACK, D., 1996. Purification of hydroxycinnamoyltransferase-CoA-tyramine hydroxycinnamoyltransferase from cell-suspension cultures of *Solanum tuberosum* L cv Datura. Planta 199: 166-168.
28. NEGREL, J., JAVELLE, F., 1997. Purification, characterization and partial amino acid sequencing of hydroxycinnamoyl-Coa-tyramine *N*-(hydroxycinnamoyl)transferase from tobacco cell-suspension cultures. Eur. J. Biochem. 247: 1127-1135.
29. DROUX, M., RUFFET, M. L., DOUCE, R., JOB, D., 1998. Interactions between serine acetyltransferase and *O*-acetylserine (thiol) lyase in higher plants--structural and kinetic properties of the free and bound enzymes. Eur. J. Biochem. 255: 235-245.
30. MARTIN, M. N., SAFTNER, R. A., 1995. Purification and characterization of 1-aminocyclopropane-1-carboxylic acid *N*-malonyltranferase from tomato fruit. Plant Physiol. 108: 1241-1249.
31. CHICK, W. S. H., LEUNG, P. C., 1997. Immunopurification and characterization of a 40-kD 1-aminocyclopropane-1-carboxylic acid *N*-malonyltransferase from Mung bean seedling hypocotyls. Plant Physiol. 113: 119-126.
32. MENHARD, B., ZENK, M. H., 1999. Purification and characterization of acetyl coenzyme A: 10- hydroxytaxane *O*-acetyltransferase from cell suspension cultures of *Taxus chinensis*. Phytochemistry 50: 763-774.
33. ST-PIERRE, B., LAFLAMME, P., ALARCO, A. M., DE LUCA, V., 1998. The terminal *O*-acetyltransferase involved in vindoline biosynthesis defines a new class of proteins responsible for coenzyme A-dependent acyl transfer. Plant J. 14: 703-713.
34. HARADA, M., UEDA, Y., IWATA, T., 1985. Purification and some properties of alcohol acetyltransferase from banana fruit. Plant Cell Physiol. 26: 1067-1074.

35. LOTFY, S., JAVELLE, F., NEGREL, J., 1996. Purification and characterization of hydroxycinnamoyl-Coenzyme A: ω-hydroxypalmitic acid *O*-hydroxycinnamoyltransferase from tobacco (*Nicotiana tabacum* L.) cell-suspension cultures. Planta 199: 475-480.
36. HOHLFELD, M., VEIT, M., STRACK, D., 1996. Hydroxycinnamoyltransferases involved in the accumulation of caffeic acid esters in gametophytes and sporophytes of *Equisetum arvense*. Plant Physiol. 111: 1153-1159.
37. FUJIWARA, H., TANAKA, Y., FUKUI, Y., ASHIKARI, T., YAMAGUCHI, M., KUSUMI, T., 1998. Purification and characterization of anthocyanin 3-aromatic acyltransferase from *Perilla frutescens*. Plant Sci. 137: 87-94.
38. FUJIWARA, H., TANAKA, Y., FUKUI, Y., NAKAO, M., ASHIKARI, T., KUSUMI, T., 1997. Anthocyanin 5-aromatic acyltransferase from *Gentiana triflora* - Purification, characterization and Its role in anthocyanin biosynthesis. Eur. J. Biochem. 249: 45-51.
39. GUO, L., ARTECA, R., PHILIPS, A., Y, L., 1992. Purification and characterization of 1-aminocyclopropane-1-carboxylate *N*-malonyltranferase from etiolated mung bean hypocotyls. Plant Physiol. 100: 2041-2045.
40. MATERN, U., FESER, C., HELLER, W., 1984. N-malonyltransferases from peanut. Arch. Biochem. Biophys. 235: 218-227.
41. MATERN, U., FESER, C., HAMMER, D., 1983. Further characterization and regulation of malonyl-coenzyme A flavonoid glucoside malonyltransferases from parsley cell suspension cultures (*Petroselinum hortense*). Arch. Biochem. Biophys. 226: 206-217.
42. POWER, R., KURZ, W. G. W., DE LUCA, V., 1990. Purification and characterization of acetylcoenzyme A: deacetylvindoline 4-*O*-acetyltransferase from *Catharanthus roseus*. Arch. Biochem. Biophys. 279: 370-376.
43. LOTFY, S., 1995. Inactivation and kinetic characterization of hydroxycinnamoyl-Coa - hydroaromatic acid *O*-hydroxycinnamoyltransferases from *Cichorium endivia* and *Phoenix dactylifera*. Plant Physiol. Biochem. 33: 423-431.
44. WINK, M., HARTMANN, T., 1982. Enzymatic synthesis of quinolizidine alkaloid esters: a tigloyl-CoA: 13-hydroxylupanine *O*-tigloltransferase from *Lupinus albus* L. Planta 156: 560-565.
45. FAHN, W. GRUNDLACH, H. DEUR-NEUMANN, B. STOCKIGT, J., (1985) Late enzymes in vindoline biosynthesis. Acetyl-CoA: 17-*O*-acetyltransferase. Plant Cell Rep. 4: 333-336.
46. ASHIKARI, T., TANAKA, Y., FUJIWARA, H., NAKAO, M., FUKUI, Y., SAKAKIBARA, K., MIZUTANI, M., KUSUMI, T. (1997). Genes coding for

proteins having acyl transfer activity European Patent Office. Suntory Limited, Japan, Japan.
47. KIMURA, M., KANEKO, I., KOMIYAMA, M., TAKATSUKI, A., KOSHINO, H., YONEYAMA, K., YAMAGUCHI, I., 1998. Trichothecene 3-*O*-acetyltransferase protects both the producing organism and transformed yeast from related mycotoxins - cloning and characterization of Tri101. J. Biol. Chem. 273: 1654-1661.
48. MCCORMICK, S. P., HOHN, T. M., DESJARDINS, A. E., 1996. Isolation and characterization of Tri3, a gene encoding 15-*O*-acetyltransferase from *Fusarium sporotrichioides*. Appl. Environ. Microbiol. 62: 353-359.
49. FUJII, T., NAGASAWA, N., IWAMATSU, A., BOGAKI, T., TAMAI, W., HAMACHI, M., 1994. Molecular cloning, sequence analysis, and expression of the yeast alcohol acetyltransferase gene. Appl. Environ. Microbiol 60: 2786-2792.
50. NEGRUK, V., YANG, P., SUBRAMANIAN, M., MCNEVIN, J. P., LEMIEUX, B., 1996. Molecular cloning and characterization of the CER2 gene of *Arabidopsis thaliana*. Plant J. 9: 137-145.
51. TACKE, E., KORFHAGE, C., MICHEL, D., MADDALONI, M., MOTTO, M., LANZINI, S., SALAMINI, F., DORING, H. P., 1995. Transposon tagging of the maize *Glossy2* locus with the transposable element En/Spm. Plant J. 8: 907-917.
52. XIA, Y., NIKOLAU, B. J., SCHNABLE, P. S., 1996. Cloning and characterization of *CER2*, an *Arabidopsis* gene that affects cuticular wax accumulation. Plant Cell 8: 1291-1304.
53. JAMES, D. W. J., LIM, E., KELLER, J., PLOOY, I., RALSTON, E., DOONER, H. K., 1995. Directed tagging of the *Arabidopsis FATTY ACID ELONGATION1 (FAE1)* gene with the maize transposon activator. Plant Cell 7: 309-319.
54. FEHLING, E., MUKHERJEE, K. D., 1991. Acyl-CoA elongase from a higher plant (*Lunaria annua*): Metabolic intermediates of very-long-chain acyl-CoA products and substrate specificity. Biochim. Biophys. Acta 1082: 239-246.
55. FEHLING, E., LESSIRE, R., CASSAGNE, C., MUKHERJEE, K. D., 1992. Solubilization and partial purification of constituents of acyl-CoA elongase from *Lunaria annua*. Biochim. Biophys. Acta 1126: 88-94.
56. AGGELIS, A., JOHN, I., KARVOUNI, Z., GRIERSON, D., 1997. Characterization of two cDNA clones for mRNAs expressed during ripening of melon (*Cucumis melo* L.) fruits. Plant Mol. Biol. 33: 313-322.
57. SLATER, A., MAUNDERS, M. J., EDWARDS, K., SCHUCH, W., GRIERSON, D., 1985. Isolation and characterization of cDNA clones for

tomato polygalacturonase and other ripening-related proteins. Plant Mol. Biol. 5: 137-147.
58. SHAW, W. V., LESLIE, A. G. W., 1991. Chloramphenicol acetyltransferase. Annu. Rev. Biophys. Chem. 20: 363-386.
59. RUSSELL, G. C., GUEST, J. R., 1991. Sequence similarities within the family of dihydrolipoamide acyltransferases and discovery of a previously unidentified fungal enzyme. Biochim. Biophys. Acta 1076: 225-232.
60. BROWN, N. F., ANDERSON, R. C., CAPLAN, S. L., FOSTER, D. W., MCGARRY, J. D., 1994. Catalytically important domains of rat carnitine palmitoyltransferase II as determined by site-directed mutagenesis and chemical modification. Evidence for a critical histidine residue. J. Biol. Chem. 269: 19157-19162.
61. CARBINI, L. A., HERSH, L. B., 1993. Functional analysis of conserved histidines in choline acetyltransferase by site-directed mutagenesis. J. Neurochem. 61: 247-253.
62. DE CRÉCY-LAGARD, V., MARLIÈRE, P., SAURIN, W., 1995. Multienzymatic non ribosomal peptide biosynthesis: identification of the functional domains catalysing peptide elongation and epimerisation. C. R. Acad. Sci. III 318: 927-936.
63. SCHMIDT, A., GRIMM, R., SCHMIDT, J., SCHEEL, D., STRACK, D., ROSAHL, S., 1999. Cloning and expression of a potato cDNA encoding hydroxycinnamoyl-CoA:tyramine N-(hydroxycinnamoyl)transferase. J. Biol. Chem. 274: 4273-80.
64. VELASCO, J., GUTTIERREZ, S., CAMPOY, S., MARTIN, J. F., 1999. Molecular characterization of the *Acremonium chrysogenum cefG*gene product: the native deacetylcephalosporin C acetyltransferase is not processed into subunits. Biochem. J. 337: 379-385.
65. STACHELHAUS, T., MOOTZ, H. D., BERGENDAHL, V., MARAHIEL, M. A., 1998. Peptide bond formation in nonribosomal peptide biosynthesis. Catalytic role of the condensation domain. J. Biol. Chem. 273: 22773-22781.
66. ALTSCHUL, S. F., MADDEN, T. L., SCHAFFER, A. A., ZHANG, J., ZHANG, Z., MILLER, W., LIPMAN, D. J., 1997. Gapped BLAST and PSI-BLAST: a new generation of protein database search programs. Nucl. Acids Res. 25: 3389-3402.
67. KNAPP, J. E., MITCHELL, D. T., YAZDI, M. A., ERNST, S. R., REED, L. J., HACKERT, M. L., 1998. Crystal structure of the truncated cubic core component of the *Escherichia coli* 2-oxoglutarate dehydrogenase multienzyme complex. J. Mol. Biol. 280: 655-668.

68. MATTEVI, A., OBMOLOVA, G., SCHULZE, E., HALK, K. H., WESTPHAL, A. H., DE KOK, A., HOL, W. G. J., 1992. Atomic structure of the cubic core of the pyruvate dehydrogenase multienzyme complex. Science 255: 1544-1550.
69. LESLIE, A. G., 1990. Refined crystal structure of type III chloramphenicol acetyltransferase at 1.75 Å resolution. J. Mol. Biol. 213: 167-186.
70. LEWENDON, A., MURRAY, I. A., SHAW, W. V., GIBBS, M. R., LESLIE, A. G., 1994. Replacement of catalytic histidine-195 of chloramphenicol acetyltransferase: evidence for a general base role for glutamate. Biochemistry 33: 1944-1950.
71. RUSSELL, G. C., GUEST, J. R., 1990. Overexpression of restructured pyruvate dehydrogenase complexes and site-directed mutagenesis of a potential active-site histidine residue. Biochem. J. 269: 443-450.
72. MENG, M. H., CHUANG, D. T., 1994. Site-directed mutagenesis and functional analysis of the active-site residues of the E2 component of bovine branched-chain alpha-keto acid dehydrogenase complex. Biochemistry 33: 12879-12885.
73. HEATH, R. J., ROCK, C. O., 1998. A conserved histidine is essential for glycerolipid acyltransferase catalysis. J. Bacteriol. 180: 1425-1430.
74. SHAW, W. V., 1992. Chemical anatomy of antibiotic resistance: chloramphenicol acetyltransferase. Sci. Progr. 76: 565-580.
75. COON, S. L., ROSEBOOM, P. H., BALER, R., WELLER, J. L., NAMBOODIRI, M. A., KOONIN, E. V., KLEIN, D. C., 1995. Pineal serotonin *N*-acetyltransferase: expression cloning and molecular analysis. Science 270: 1681-1683.
76. TERCERO, J. C., RILES, L. E., WICKNER, R. B., 1992. Localized mutagenesis and evidence for post-transcriptional regulation of MAK3. A putative N-acetyltransferase required for double-stranded RNA virus propagation in *Saccharomyces cerevisiae*. J. Biol. Chem. 267: 20270-20276.
77. TERCERO, J. C., RILES, L. E., WICKNER, R. B., 1992. Localized mutagenesis and evidence for post-transcriptional regulation of MAK3. A putative N-acetyltransferase required for double-stranded RNA virus propagation in Saccharomyces cerevisiae. J. Biol. Chem. 267: 20270-6.
78. TERCERO, J. C., DINMAN, J. D., WICKNER, R. B., 1993. Yeast MAK3 N-acetyltransferase recognizes the N-terminal four amino acids of the major coat protein (gag) of the L-A double-stranded RNA virus. J. Bacteriol. 175: 3192-4.
79. YANG, Q., GRIMMIG, B., MATERN, U., 1998. Anthranilate *N*-hydroxycinnamoyl/benzoyltransferase gene from carnation: rapid elicitation of transcription and promoter analysis. Plant Mol. Biol. 38: 1201-1214.

80. CZERNIC, P., HUANG, H. C., MARCO, Y., 1996. Characterization of hsr201 and hsr515, two tobacco genes preferentially expressed during the hypersensitive reaction provoked by phytopathogenic bacteria. Plant Mol. Biol. 31: 255-265.
81. KIMURA, M., MATSUMOTO, G., SHINGU, Y., YONEYAMA, K., YAMAGUCHI, I., 1998. The mystery of the trichothecene 3-*O*-acetyltransferase gene - Analysis of the region around Tri101 and characterization of its homologue from *Fusarium sporotrichioides*. FEBS Lett. 435: 163-168.
82. SZEKERES, M., NEMETH, K., KONCZ-KALMAN, Z., MATHUR, J., KAUSCHMANN, A., ALTMANN, T., REDEI, G. P., NAGY, F., SCHELL, J., KONCZ, C., 1996. Brassinosteroids rescue the deficiency of CYP90, a cytochrome P450, controlling cell elongation and de-etiolation in *Arabidopsis*. Cell 85: 171-182.
83. CHOE, S., DILKES, B. P., FUJIOKA, S., TAKATSUTO, S., SAKURAI, A., FELDMANN, K. A., 1998. The DWF4 gene of Arabidopsis encodes a cytochrome P450 that mediates multiple 22alpha-hydroxylation steps in brassinosteroid biosynthesis. Plant Cell 10: 231-243.
84. WINKLER, R. G., HELENTJARIS, T., 1995. The maize Dwarf3 gene encodes a cytochrome P450-mediated early step in Gibberellin biosynthesis. Plant Cell 7: 1307-1317.
85. BISHOP, G. J., HARRISON, K., JONES, J. D., 1996. The tomato Dwarf gene isolated by heterologous transposon tagging encodes the first member of a new cytochrome P450 family. Plant Cell 8: 959-969.
86. KELLER, N. P., HOHN, T. M., 1997. Metabolic pathway gene clusters in filamentous fungi. Fungal Genet. Biol. 21: 17-29.

Chapter Ten

GLYCOSYLTRANSFERASES INVOLVED IN PLANT SECONDARY METABOLISM

Thomas Vogt

Leibniz-Institut für Pflanzenbiochemie
Weinberg 3
D-06120 Halle/Saale, Germany

INTRODUCTION

Glycosylation is considered as one of the final modifications late in the biosynthesis of many compounds of plant secondary metabolism. Glycosides of all major classes of secondary compounds, *i.e.* phenylpropanoids, terpenoids, alkaloids, glucosinolates, cyanogenic glycosides, and steroids have been identified for many plants. The large number of naturally occurring glycosides[1] does not correlate with a detailed knowledge of the corresponding enzymes leading to their formation, the glycosyltransferases. In addition - with a few exceptions - research is restricted nearly exclusively to glycosyltransferases involved in flavonoid metabolism.[2] Except for their well-documented role in detoxification of xenobiotics[3,4] and their ability to increase the

Evolution of Metabolic Pathways – J.T. Romeo et al. (Editors)

hydrophilicity of hydrophobic or amphiphilic aglycones, the function of the variety of glycosyltransferases in plant secondary metabolism remains substantially uncharacterized.

The vast majority of plant glycosyltransferases are soluble enzymes, therefore, isolation and purification should be straightforward. However, often present in low amounts, purification of individual glycosyltransferases to apparent homogeneity is a challenging task. This, however, is essential to assign defined substrate specificities to an individual enzyme.

In recent years, recombinant DNA technology has entered the arena of research on enzymes involved in plant secondary metabolism, and both gene cloning and heterologous gene expression have proven to be powerful tools for purification and characterization of various recombinant plant glucosyltransferases. In addition, computational analysis based on amino acid and nucleotide sequence comparisons[5,6] has identified new candidates from rapidly growing databases. This increases the number of members of the glucosyltransferase family, but fails to promote the functional identification of individual genes and the corresponding proteins.

This review gives an overview of our current knowledge of plant glycosyltransferase proteins and genes. The first part includes an up to date summary of those enzymes involved in plant secondary metabolism. The second part on biochemical and molecular data is strengthened by a section on the physiological significance of plant glucosyltransferases. Based on amino acid sequence similarity data, substrate specificities, and phylogenetic aspects, a classification of plant secondary product glucosyltransferases will be proposed.

GLYCOSYLTRANSFERASES INVOLVED IN FLAVONOID BIOSYNTHESIS

The nucleotide diphosphoglycoside:glycosyltransferases of plant secondary metabolism represent a superfamily of enzymes that catalyze the addition of a sugar residue by use of energy rich nucleotide sugars to mostly small hydrophobic molecules (Fig. 1). Most plant glycosyltransferases use UDP-glucose as the sugar donor to catalyze *O*-glycoside formation and are nearly exclusively regarded as soluble enzymes, whereas the corresponding animal UDP-glucuronosyltransferases are usually membrane bound.[7] C-glycoside formation is also known, but the corresponding glucosyltransferases have only been described once,[8] and our current knowledge of these enzymes is poor.

Glycosyltransferases involved in 3-*O*-glycosylation of flavonols and anthocyanidins are those most frequently reported in the literature.[1,9] Recent reports on anthocyanidin 3-*O*-glycosyltransferases include cell suspension cultures of carrot (*Daucus carota*)[10] and flower buds of *Senecio x hybridus*.[11] Two enzymes from carrot that catalyze the subsequent formation of cyanidin-3-*O*-galactoside and cyanidin 3-*O*-xylosylgalactoside have been purified to apparent homogeneity, whereas in the case of *S. x hybridus,* the state of purification for the cyanidin 3-*O*-glucosyltransferase was not indicated. Furthermore, a 3-*O*-glucosyltransferase specific for anthocyanidins also has been purified from grapes of *Vitis vinifera,*[12] and the presence of multiple flavonol- and anthocyanidin glycosylating enzymes in several grape varieties recently has been postulated.[13] Several enzyme activities for flavonol and anthocyanidin 3-*O*-glycosyltranferases also have been described for apple (*Pyrus malus*) peel.[14,15] No attempts, however, have been made to purify the respective enzymes from grapevine and apple. Two tissue-specific flavonol glycosyltranferases have been described from pollen of *Petunia hybrida.*[16] The first of those, a flavonol 3-*O*-galactosyltransferase, is unusual in terms of apparently being membrane bound, whereas the second is soluble and specifically transfers glucose to the 2'-position of galactose. Seedlings of mung bean (*Vigna mungo)* contain distinct flavonol 3-*O*-glucosyl- and 3-*O*-galactosyltransferases,[17] whereas for cassava (*Manihot esculenta),* only the flavonol specific glucosyltransferases[18] have been described. In alfalfa (*Medicago sativa*), flavonoid and isoflavonoid glucosyltranferases have been reported with distinct specificities either towards flavonols or the isoflavonoids, medicarpin and coumestrol, respectively.[19]

Glycosylation of the A- or B-ring of flavonoids has been demonstrated only for a few plant species. Flavanone 7-*O*-glucosyltransferase activity in several *Citrus* species was described.[20-22] A corresponding flavanone 7-*O*-glucoside rhamnosyltransferase was subsequently purified from *Citrus maxima.*[23] In this case, both glycosyltransferases are involved in the synthesis of the bitter tasting flavanone glycoside naringin. A partially purified glucosyltransferase from ripening strawberry (*Fragaria ananassa*) displayed considerable 7-*O*-glucosylating activity with several flavonoids as substrates.[24] A flavonol 7-*O*-glucosyl-transferase was also detected in flowers of *Chrysanthemum segetum,*[25] with quercetin and 8-hydroxyquercetin (gossypetin) as the preferred glucose acceptors. Since, in these cases, the enzymes were not purified to homogeneity, it is difficult to decide if strawberry or *C. segetum* contain one unspecific 7-*O*-glucosyltransferase or a set of specific flavonoid glycosyltransferases. In *Euonymus europaeus,* flavonol 7-*O*-glucosyltranferase activity was determined in addition to flavonol 3-*O*-glucosyl- and xylosyltransferases.[26] Although the enzymes were not purified, the authors were able to separate individual activities by column chromatography. A regiospecific flavone 7-*O*-glucosyltransferase was purified from *Silene latifolia.*[27] In petals of *Dahlia variabilis,* Stich *et al.* reported for the first time a glucosyltransferase with specificity to the 4 -hydroyl group of a 6 -deoxychalcone, isoliquiritigenin.[28] This glucosylation is biosynthetically equivalent to the A-ring glucosylation of heterocyclic flavonoids. The first isoflavone 4 -O-glucosyltransferase has been described from leaves of *Prunus x yedoensis,*[29] whereas a corresponding 7-*O*-glucosyltransferase was already known from chick pea, *Cicer arietinum.*[30]

Flavonol ring B-specific glucosyltransferases have been purified and characterized only from onion (*Allium cepa*) bulbs[31,32] or from a semi-aquatic weed, *Chrysosplenium americanum* (Saxifragaceae).[33] Both the onion and *C. americanum* glucosyltransferases are highly substrate and regiospecific. The enzyme from *A. cepa* glucosylates the 4 -OH group of flavonols, whereas two *C. americanum* glucosyltransferases transfer glucose either to the 2 - or the 5 -hydroxyl group of partially methylated flavonols. In *A. cepa*, only quercetin, but not the corresponding 3-*O*-glucoside, was further glucosylated at the 4 -hydroxyl group.[32] In the case of *C. americanum*, purification of both enzymes to homogeneity revealed their regiospecificity for one hydoxyl group only, stressing again the importance of accurate purification before assigning any detailed substrate specificities to one enzyme. Soluble glucuronosyltransferase activities have only been reported so far for primary leaves of rye (*Secale cereale*), where they are involved in the modification of the flavone C-glucoside isovitexin.[34]

GLYCOSYLTRANSFERASES INVOLVED IN OTHER BIOSYNTHETIC PATHWAYS OF PLANT SECONDARY METABOLISM

Pigmentation of most families of the Caryophyllales is not due to anthocyanins, but to betalains, the yellow betaxanthins and the red to violet betacyanins.[35] Two soluble regiospecific glucosyltransferases were purified and identified from *Dorotheanthus bellidiformis* (Aizoaceae) cell suspension cultures.[36-38] The enzymes were shown to effectively glucosylate flavonols, flavones, and anthocyanidins, in addition to the natural substrate betanidin. Betanidin 5-*O*-glucosyltransferase showed specificity towards 4'- and, to a lesser extent, towards the 7-hydroxyl groups with a strong preference for ortho 3', 4'-dihydroxylated flavonols (Fig. 2), whereas the corresponding betanidin 6-*O*-glucosyltransferase glucosylated only the 3-hydroxyl group of either cyanidin or quercetin with high affinity.

Glucosyltransferases with hydroxycinnamic acids and other simple phenolics as their metabolites include: soluble enzymes involved in lignin biosynthesis in spruce,[39] sinapine formation in *Brassica napus*,[40] 4-hydroxy benzoic acid glucosylation in *Lithospermum erythrorhizon*,[41] scopolin biosynthesis in Solanaceae and cassava,[18,42] and glucosylation of salicylic acid in oat (*Avena sativa*) roots.[43] It should be noted that in the case of sinapic acid carboxylglucosyltransferase, involved in sinapine biosynthesis, an energy rich 1-*O*-acylglucoside is formed, with a high group-transfer potential to further biosynthetic reactions.[44] These kinds of activated glucose esters are also involved in the biosynthesis of gallotannins and other simple phenolic compounds.[45] Similar reactions also have been described for primary metabolism, where glucosyltransferases catalyze the activation of fatty acids such as 1-*O*-acylglucoses.[46] Formation of energy rich glycoside bonds are also known for flavonoids, where a reversible transfer of glucose to the 3-

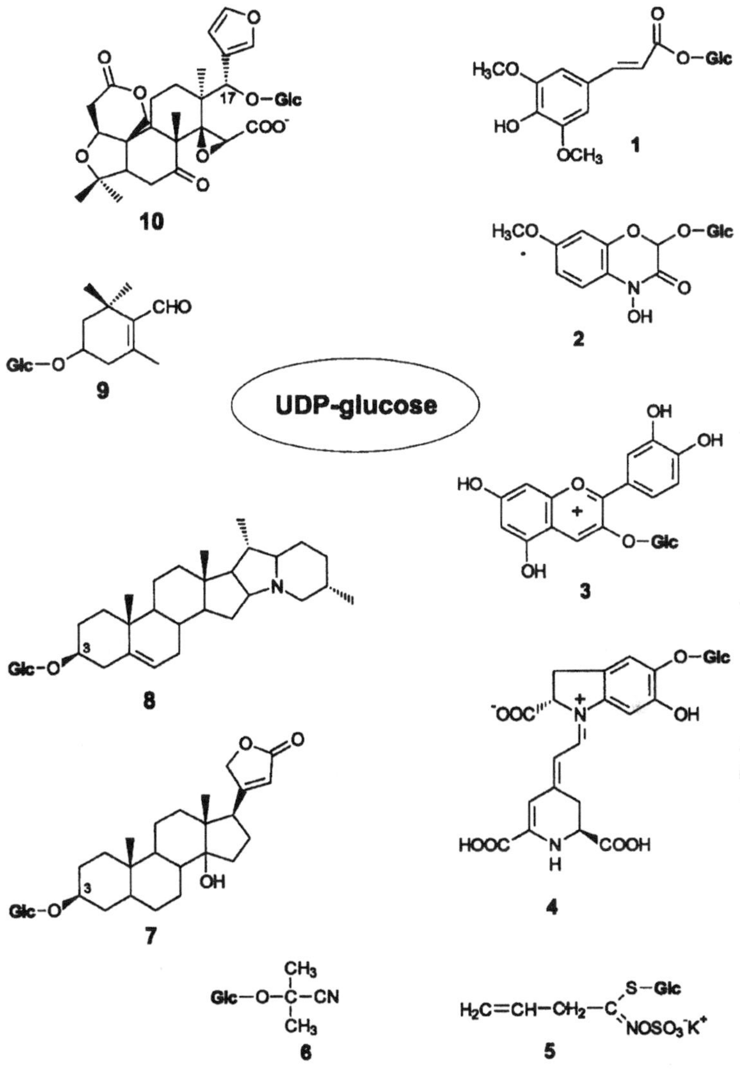

Figure 1. Natural products of plant UDP-glucose glucosyltransferases using UDP-glucose as donor. 1. hydroxycinnamic acids, *e.g.* sinapic acid; 2. hydroxamic acid glucosides, *e.g.* DIMBOA-2-*O*-glucoside, 3. anthocyanins, *e.g.* cyanidin 3-*O*-glucoside; 4. betacyanins, *e.g.* betanidin-5-*O*-glucoside; 5. glucosinolates, e.g.sinigrin; 6. cyanogenic glucosides, *e.g.* linamarin; 7. cardiac glucosides, *e.g.* digitoxigenin-glucoside; 8. steroidal glycoalkaloids, *e.g.* solanidine glucoside; 9. glycosydically bound volatiles, *e.g.* picrocrocin; 10. triterpenoid glucosides, *e.g.* limonoate-17-*O*-glucoside.

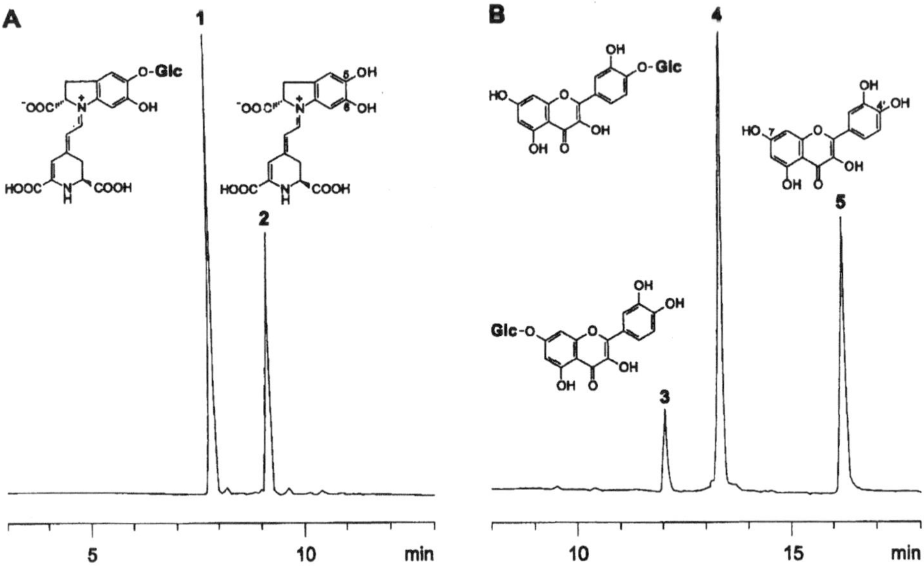

Figure 2. HPLC profiles of enzyme reaction mixtures in the presence of recombinant 5-GT, UDP-glucose and betanidin. (A. detection at 540 nm) or quercetin (B. detection at 360 nm). Peak identification: 1. betanin; 2. betanidin, 3. quercetin-7-*O*-glucoside; 4. quercetin 4 -*O*-glucoside; 5. quercetin. (Reproduced from Vogt *et al.*(1999); Plant J. 19, 509-521, ßios Scientific Publishers).

hydroxyl group of flavonols has been observed.[47]

Substrate specificity of glucosyltransferases may also include stereospecificity. Coniferyl alcohol glucosyltransferase of *Fagus grandiflora* (american beech) is specific for *Z*-coniferyl alcohol and yields *Z*-coniferin in monolignol formation; no *E*-coniferin is formed.[48] Recently a stereospecific 4-coumaric acid glucosyltransferase from *Sphagnum fallax* was highly purified[49] and shown to exhibit a pronounced specificity to catalyze the transfer of glucose to give 4 -*Z*-*O*-ß-D-glucosyl-*p*-coumaric acid.

Soluble glucosyltransferases involved in the glucosylation of *a*-hydroxynitriles forming cyanogenic glycosides have been described for cassava[18] and *Sorghum bicolor*.[50] The penultimate step in glucosinolate biosynthesis in rape-seed (*Brassica napus*) and in *Arabidopsis thaliana* is catalyzed by a soluble UDPG:thiohydroximate glucosyltransferase.[51,52] A homogenous preparation of this enzyme was obtained by immunopurification from discolored *Brassica oleracea* leaves.[53]

Several terpenoid glucosyltransferases have been reported to catalyze the transfer of sugars from UDP-glucose or UDP-glucuronic acid to terpenoid aglycones in plants. UDP-glucose:limonoid glucosyltransferase, purified from the albedo of orange,[54] is of special interest. It reduces the formation of bitter triterpenoid limonoid aglycones in *Citrus* varieties. A microsomal triterpenoid glucuronosyltranferase has been detected in suspension cultures of licorice, *Glycyrrhiza glabra*.[55] Glycyrrhizin is a triterpenoid saponin in the roots of licorice, and is used as a natural sweetener. Other natural sweeteners include the glycosylated steviol derived diterpenoid glycosides, stevioside and rebaudioside A, which are glucosylated by two soluble enzymes from *Stevia rebaudiana*.[56]

Steroidal glycoalkaloid biosynthesis requires the involvement of UDP-glucose solanidine glucosyltransferase, present only in trace amounts in potato (*Solanum tuberosum*).[57] This enzyme glycosylates the related glycoalkaloids solasodine, tomatidine, and solanidine with decreasing affinity[58] (see also Figure 5). Related enzymes also have been partially purified from the eggplant, *Solanum melongena*,[59] where a stepwise glycosylation was postulated for higher glycosylated solanidine derivatives.[60] These cytosolic enzymes can be distinguished from the membrane-bound UDP-glucose:sterol ß-D-glucosyltransferase.[61,62] The enzyme, which has highest affinity for cholesterol and ß-sitosterol, was purified from oat seedlings. It has the unique property of being active after extraction with diethylether.[63] Additional soluble glucosyltransferases have been reported from oat, and are involved in the glucosylation of the 3-ß hydroxylated derivatives of pregnane and androstane.[64]

Leaves of *Digitalis* species contain a complex mixture of cardiac glycosides, another important group of steroid derived natural compounds. Glycosylation of cardiac glycosides has been studied in shoot cultures of foxglove (*Digitalis lanata*),[65] and at least three different glucosyltransferases have been identified. Virtually no information is available on the subsequent transfer of the rare 6-deoxy or methylated sugars, such as ß-D-digitoxose and D-fucose, to the cardiac aglycones, cardenolides, or bufadienolides.

Cyclic hydroxamic acids from several Poaceae, probably involved in the defense response against a variety of bacteria, fungi, and insects, are glycosylated by soluble glucosyltransferases. These exhibit a profound substrate specificity towards 2,4-dihydroxy-1, 4-benzoxazin-3-one (DIBOA) and its 7-methoxy analogue DIMBOA, but not against the 2-hydroxy-1,4-benzoxazin-3-one.[66]

In addition to their natural substrates, plant glycosyltransferases have the capacity to glycosylate herbicides and other xenobiotics.[67] Two constitutive glucosyltransferases, one soluble and one membrane associated enzyme, have been characterized from seedlings of soybean (*Glycine max*).[68] Both were able to glucosylate the herbicide bentazone. However, the affinity for bentazone was much lower than that for the natural substrates, kaempferol and 4-hydroxyphenylpyruvate, respectively. In addition, cell cultures of soybean have been shown to contain other, probably unspecific glucosyltransferases involved in detoxification of a variety of xenobiotics, including further bentazone derivatives and the insecticide DDT (1,1,1-trichloro-2, 2-(bis-4-chlorophenyl)-ethane).[3,69]

The existence of this "green liver" concept apparently is an evolutionary old phenomenon and has been documented for lower plants, such as algae and mosses.[70]

Apparently, some glucosyltransferases are able to effectively glycosylate different substrates, if they share some structural similarities. This is consistent with the observation that most glycosyltransferases, like *O*-methyltransferases or cytochrome P450-dependent mono-oxygenases, are position specific rather than substrate specific.[71,72] In contrast, the key enzymes of plant secondary metabolism are highly substrate specific. The superfamily of polyketide synthases, for example, start with malonyl-CoA and either *p*-coumaric acid or N-methyl anthranilate, and form chalcones, stilbenes, or acridones, respectively.[73] In the sense of ecology and evolution, the broad substrate specificity of the late, modifying enzymes might be advantageous to the plant. The resulting structural diversity of plant secondary products ensures a high degree of plasticity towards rapidly changing environmental conditions,[74] without the need to change or to shut down a fundamental pathway essential for vital growth.

DETECTION AND PROPERTIES

Although the individual substrates are heterogenous, the overall biochemical properties of the glycosyltransferases are similar and consistent among the subclasses described. In most cases, the enzymes are monomers sharing a molecular mass between 50 and 60 kDa, with a neutral to weakly basic pH optimum (pH 7.0 - 8.5). Affinity for the sugar acceptor is high, with a K_m-value mostly in the low micromolar range, whereas the affinity for the sugar donor is lower, with a K_m-value in the high micromolar to low millimolar range.[10-32] There is no dependence on divalent cations, although millimolar concentrations of Ca^{2+} and Mg^{2+} stimulate the activity of some partially purified glycosyltransferases.[24,25,34] The requirement for reducing agents is not consistent and may depend on the reactivity of individual SH-groups in the native protein conformation. Many glycosyltransferases are irreversibly inactivated by the SH-blockers iodacetamide or *p*-chloro-mercuribenzoesulfonic acid.[25,26,38] Micromolar concentrations of Zn^{2+} and Cu^{2+} also completely inactivate the majority of glycosyltransferases.[31,38,40,52] In alfalfa, this might have physiological significance since alfalfa plants treated with Cu^{2+} to elicit the phytoalexin response were unable to accumulate flavonoid conjugates, and an increase in the corresponding aglycones was observed that stimulates *Rhizobium* growth.[19] Sensitivity towards diethylpyrocarbonate[38] may point towards the involvement of one or more histidine residues in the catalytic mechanism.[75] Several histidine residues are conserved in the amino acid sequences of all glycosyltransferases (see for example Figure 4). Sensitivity towards 4, 4'-diisothiocyanostilbenedisulfonic acid at low micromolar concentrations may indicate involvement of lysine in the catalytic site of a betanidin 5-*O*-glucosyltransferase.[38] However, these results should always be interpreted with caution since not all inhibitors tested are absolutely specific for individual amino acids,[75] and a 50 % inhibition with millimolar concentrations of any inhibitor might indicate protein denaturation rather than interference with the catalytic mechanism of the enzymes.

PURIFICATION STRATEGIES

Major advances in resolution and performance of liquid column chromatography- techniques have revolutionized the purification of low abundance enzymes. Conventional procedures used to purify glycosyltransferases involve ammonium sulfate precipitation, ion (usually anion) exchange, and hydroxyapatite- or gel filtration chromatography. These techniques mostly lead to only partially purified enzymes.[12,17] Hydrophobic interaction chromatography, on phenyl-based matrices, has been applied successfully in the purification of several glycosyltransferases.[10,27] Under high salt conditions, the enzymes bind to the hydrophobic matrices and are usually eluted under highly stringent low millimolar salt concentrations. Without further desalting, the eluted enzymes often can be directly applied to further chromatographic steps. This is especially important if a multi-step purification procedure is performed, since instability and loss of activity is often associated with low protein concentration and dilution of the glucosyltransferases, although after concentration, activity can often be restored (Vogt, unpublished).

More sophisticated chromatographic techniques use affinity purification. Chromatography on UDP-glucuronic acid agarose has been successfully applied in the purification of flavanone 7-*O*-glucosyltransferase,[21] 4 -OH benzoate glucosyltransferase,[41] and limonoid glucosyltransferase.[54] Unfortunately, not all glycosyltransferases exhibit a high affinity for this matrix. One effective alternative is dye-ligand affinity chromatography. This technique has been successfully used to purify a variety of nucleotide binding proteins, including coenzyme A-dependent enzymes or nucleotide dependent oxidoreductases.[76] Likewise, it is also effective in the purification of UDP-sugar dependent glycosyltransferases, since the dye-ligand is considered to interact with the nucleotide binding fold of the UDP-sugar.[77] Among hundreds of dyes available, the task of selecting a suitable immobilized one remains empirical, involving the testing of a number of dyes to bind the protein in question. Fortunately, elution from the ligand, rather than binding, is a problem, offering the unique possibility to highly enrich glycosyltransferases from most plant protein extracts.[10,33,39] Such a purification step may often lead to a 10-40 fold enrichment.

Unspecific elution of glycosyltransferases with high salt can be further improved by affinity elution with UDP alone or with UDP-glucose.[23,38] For example, nearly homogeneous betanidin 6-*O*-glucosyl-transferase, present in minute amounts in cell suspension cultures of *D. bellidiformis*, was purified from a crude plant extract in two chromatographic steps and simultaneously separated from the more abundant 5-*O*-glucosyltransferases (Fig. 3).[38] Combination of both affinity techniques is the method of choice to differentiate between individual enzymes in plant tissues, and to purify even low abundant glycosyltransferases in amounts sufficient for protein sequence analysis or antibody generation.

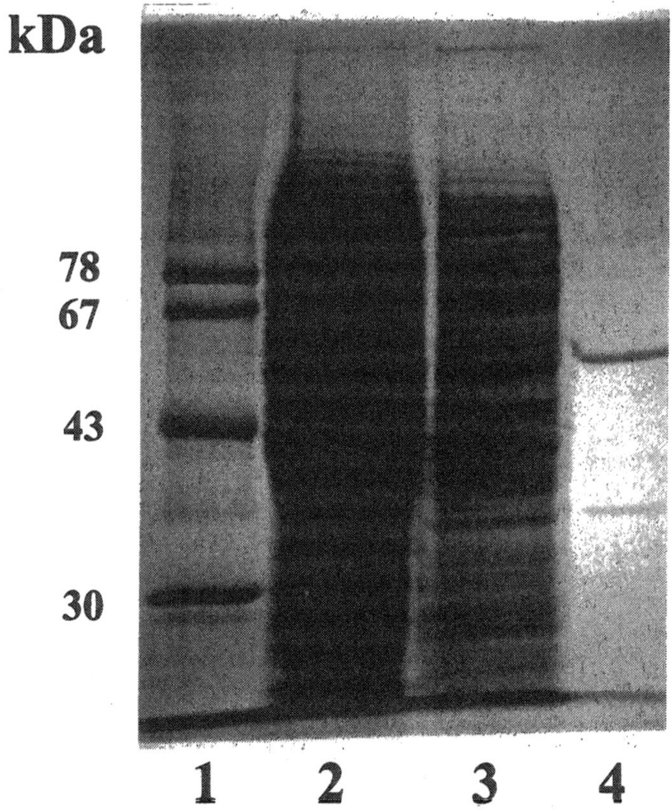

Figure 3. SDS-PAGE illustrating the two step purification of betanidin 6-*O*-glucosyltransferase from cell suspension cultures of *D. bellidiformis*. Lane 1. molecular weight marker; lane 2. crude protein extract; 3. after anion exchange chromatography; 4. after dye ligand chromatography on *Reactive Yellow 3 Agarose* (Sigma, Deisenhofen, Germany). Reproduced from Vogt et al.(1997); Planta 203: 349-361, Springer Verlag).

MOLECULAR CLONING AND HETEROLOGOUS EXPRESSION

Several approaches were taken to clone glucosyltransferases of plant secondary metabolism. The first glucosyltransferase gene, corresponding to the *bronze 1* (*bz1*) locus of maize (*Zea mays*), was cloned by transposon tagging.[78] The corresponding cDNA of *bz1* was used as a template to characterize GT-alleles in maize[79] and in *Antirrhinum majus*.[80] The *A. majus* gene was cloned by differential screening of cDNA from wild-type and mutant flowers. The resulting cDNA clones were used during the following years as probes to screen cDNA libraries constructed from a variety of other plants. Sources include grapes of *Vitis vinifera*[81] and leaves of *Perilla frutescens*.[82] By homology to the *A. majus* sequence, several glycosyltransferase cDNAs were isolated from a cassava cDNA library.[83] After combining their own data with an existing database of protein sites and patterns,[84] and performing a detailed sequence analysis, the authors proposed a "plant secondary product glucosyltransferase consensus sequence" ("PSPG-box"), which is considered to contain the UDP-sugar binding motif (Fig. 4). The box is marked by a highly conserved "W X P Q" amino acid sequence at the left (the N-terminal) site and an absolutely conserved glutamine at the right the (C-terminal) site. In the center of this

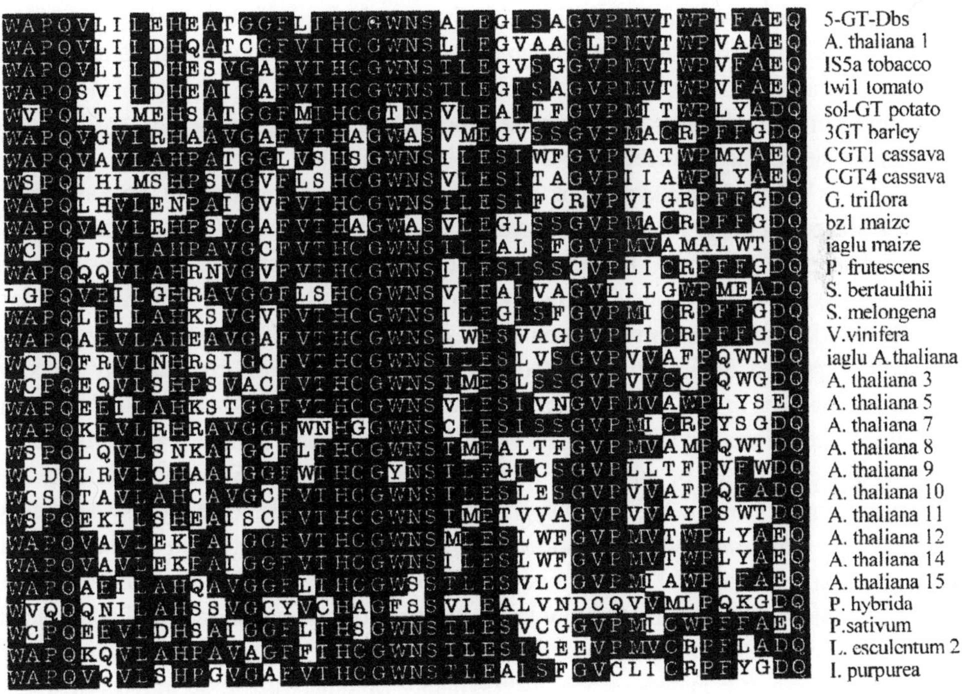

Figure 4. "PSPG"-box of thirty plant glycosyltransferase amino acid sequences from the GENBANK- and EMBL-databases. (For accession numbers in GENBANK see Figure 7).

"PSPG" box, the amino acid sequence "H C G W N S" is present in 95 % of all glucosyltransferases of plant secondary metabolism. Perfect to design primers of low degeneracy, this conserved area is the major reason why many attempts to purify new glycosyltransferases by PCR and sequence similarity cloning are successful.

In *Petunia hybrida*, certain flower phenotypes can be correlated with the presence or absence of transcripts for structural genes of the flavonoid pathway. By a differential screening approach of several *P. hybrida* varieties, the *Rt* locus was verified as coding for an anthocyanidin 3-*O*-glucoside rhamnosyltransferase.[85,86] Cloning and sense/antisense expression of the RT-gene in *P. hybrida* in both cases led to a slight change in flower color from pink to purple to variegated purple/pink transgenic flowers. Although not investigated in detail, these color variations may correlate with the accumulation or repression of the corresponding anthocyanidin glycosides. Heterologous expression of the *A. majus* 3-*O*-glucosyltransferase in the cut flower lisianthus (*Eustoma grandiflorum*) resulted in a transgenic line with accumulation of new flavonol and anthocyanin 3-*O*-glucosides, in addition to the 3-*O*-galactosides present in untransformed control plants.[87,88] Flower color, however, was unaffected in the transgenic plants, even though the transgene caused an excessive alteration in pigment composition.[88] These examples again indicate the possibility that chromogenic pigment pattern can be changed in plants by transformation with glucosyltransferase genes, but also show that the glycosylation pattern cannot directly be correlated with a change in flower color.

With the availability of high efficiency expression cloning vectors, substrate specificities of many recombinant plant enzymes are comparable to data obtained with the original plant proteins. However, only a few glycosyltransferase proteins have been expressed in heterologous systems and checked carefully for their substrate specificities. Kinetic data from the purified recombinant enzymes from *V. vinifera* or *Gentiana triflora*, both expressed in *E. coli*, indicated that both *bz1* homologs preferentially glucosylate anthocyanidins *in vivo*, although flavonols were also accepted as substrates.[89,90] A putative glucosyltransferase cDNA, obtained by differential display from *P. frutescens* green and red leaves,[82] could be identified as a specific anthocyanin 5-*O*-glucosyltransferase by expression in yeast.[91] An elegant approach to identify the solanidine glycosylating enzyme from *S. tuberosum* was also undertaken by performing expression studies in yeast.[58] Selection of individual clones was based on the ability of a recombinant (alkaloid glucosylating) yeast strain to grow in the presence of micromolar alkaloid concentrations, which were lethal to the glucosyltransferase (-) strain (Fig. 5). This example clearly demonstrates that increasing the hydrophilicity of otherwise toxic aglycones might be extremely beneficial for plant and animal cells. The classical way from protein to gene (reverse genetics approach), involving identification and purification of the enzyme, obtaining protein sequence data to design degenerate oligonucleotides, and cloning and expression of the corresponding gene by screening of a cDNA library with a PCR-

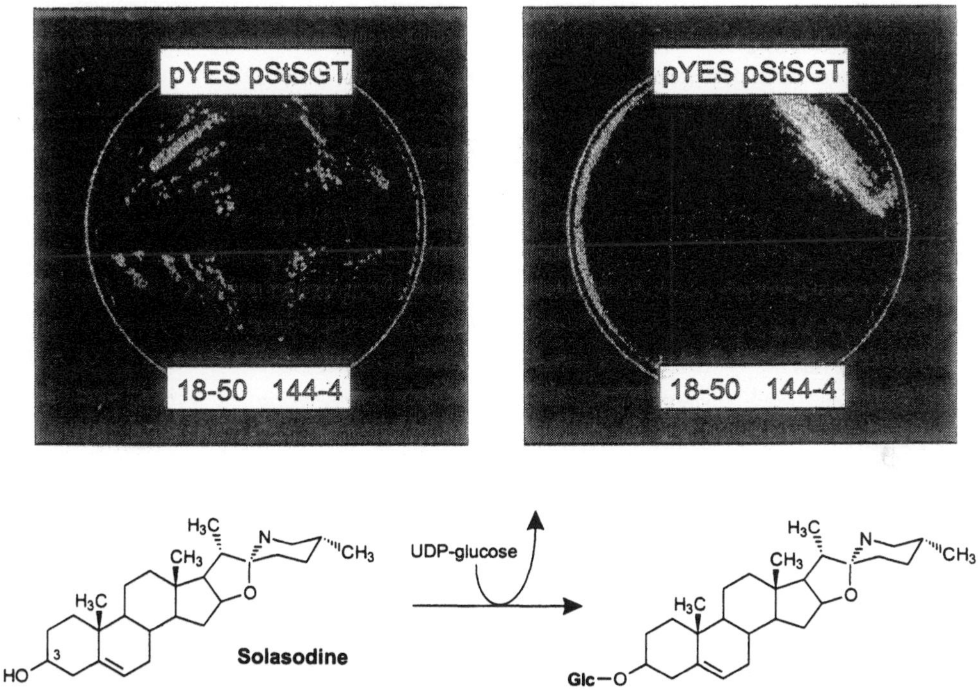

Figure 5. Growth of yeast transformants in the absence and presence of 12.5 μM solasodine. *S. cerivisiae* strain KT1115 was transformed with the empty vector (pYES2) and three potato cDNA containing plasmids (SGT, 18-50, 144-4). Growth on the aglycone was only observed with the SGT-containing vector. (reproduced with kind permission from Moehs et al. (1997); Plant J. 11: 227-236, ßios Scientific Publishers).

generated cDNA fragment has been successfully achieved in the case of betanidin 5-*O*-glucosyltranferase from cell cultures of *D. bellidiformis*.[92]

Most recently the UDP-glucose:p-hydroxymandelonitrile-*O*-glucosyltransferase that catalyses the last step in the biosynthesis of the cyanogenic glucoside dhurrin in *S. bicolor* was cloned and expressed in *E. coli*.[93] Other cloned, non flavonoid biosynthetic glucosyltransferase genes include the *Iaglu* gene involved in the conjugation of the phytohormone indole 3-acetic acid. The corresponding cDNA from maize was cloned by screening of an expression library with a polyclonal antibody.[94] The plasma membrane-associated sterol glucosyltransferase was obtained from a cDNA library of oat by screening with a PCR product, amplified with degenerate primers deduced from amino acid sequences. The recombinant protein was expressed in *E. coli* and displayed substrate specificities identical to the plant enzyme.[95]

The cloned and heterologously expressed plant glycosyltransferases are summarized in Table 1.

Table 1. Recombinant and Heterologously Expressed Plant Glycosyltransferases

Source	Preferred Substrates	Expressed in	References
A. majus	flavonols/ anthocyanidins	lisianthus	87, 88
G. triflora	anthocyanidins	*E. coli*	90
V. vinifera	anthocyanidins	*E. coli*	89
P. frutescens	anthocyanins	*S. cerevisiae*	91
D. bellidiformis	betanidin/flavonols	*E. coli*	92
S. tuberosum	steroidal glycoalkaloids	*S. cerevisiae*	58
A. sativa	sterols	*E. coli*	95
S. bicolor	cyanogenic glycosides	*E. coli*	93

TISSUE SPECIFICITY AND CELLULAR LOCALIZATION

Enzymatic and molecular data indicate that individual flavonoid glycosyltransferases are expressed in a tissue specific manner. Expression of a specific anthocyanidin glucosyltransferase together with other structural genes is associated with anthocyanin accumulation and coordinately controlled in light and dark in *V. vinifera* and *P. frutescens.*[81,82] Anthocyanin formation in *V. vinifera* is also strongly correlated with the red color of grape varieties. Since all other genes of flavonoid biosynthesis are present in white and red grape varieties, one specific anthocyanidin 3-*O*-glucosyltransferase might determine the red color of grapes.[13] Expression of the *P. hybrida Rt*-gene is strictly tissue-specific, detectable only in young petals, and to a lower extent in styles, and is controlled by the regulatory *An1* and *An2* genes.[85] Accumulation of rare flavonol glucosylgalactosides is associated with the presence of two glycosyltransferases, only active in the gametophyte, in the pollen grains of *P. hybrida*, and not detectable in the sporophyte.[16] Tissue specificity of an enzyme usually is correlated with product formation in the same tissue. In contrast, localization of flavone glucuronides and the corresponding glucuronosyltransferases in different tissues of rye seedlings has been postulated.[96]

Intracellular localization of glycosyltranferases has not been investigated in detail. Glycosyltransferases are considered to be soluble proteins, with the exception of a membrane bound galactosyltransferase from *P. hybrida* pollen, the plasma membrane-bound sterol glucosyltransferases, and a 4-hydroxy phenylpyruvate glucosyltransferase.[16,62,68] However, based on cell fractionation studies, the whole flavonoid metabolic pathway, including the modifying enzymes also has been proposed to be associated with cellular membranes.[97] Immunogold labelling suggested association of flavonol B-ring specific glycosyltransferases with the endoplasmic reticulum.[33] In a detailed study on rye mesophyll, however, a cytosolic localization was suggested for initial glucuronosyltranferases, followed by a final glucuronidation step in the vacuoles.[96] This requires transport of flavonoid conjugates across the tonoplast, which involves glutathione S-transferase activity in some cases.[98] However, in rye, uptake of the endogenous flavone C-glucoside, isovitexin, is not dependent on conjugation with glutathione, and can also be distinguished from a strictly energy dependent transport of xenobiotics.[99]

INDUCIBLE GLUCOSYLTRANSFERASES

Enhanced expression of glucosyltransferase genes coordinates with other flavonoid structural genes and has been described in detail for *P. frutescens* and *V. vinifera* in response to increased light intensity.[81,82] Induction of glycosyltransferase transcripts in grape seedlings upon exposure to light is already observed after six hours.[81] Even faster responses after exposure to various stressors point to a more diverse role of other putative glucosyltransferases than formerly anticipated. From differential display and sequence similarity data, two putative glycosyltransferase sequences, *IS5a* and *IS10a*, were identified and cloned from tobacco (*N. tabacum*) cell suspension cultures.[100] The corresponding *IEGT* (immediate early-induced glucosyltransferase) gene was rapidly and transiently induced by salicylic acid and tobacco mosaic virus (TMV) infection (Fig. 6). This induction occurs within minutes, much faster than the induction of other pathogen-related proteins, and can be stimulated by cycloheximide, a potent inhibitor of protein biosynthesis. Preliminary data suggest flavonols as substrates of both glycosyltransferases.[101] A similar scenario was described for tomato in which a rapid induction of the putative glycosyltransferase transcript *twi1* was observed after wounding and elicitor treatment.[102] The authors proposed involvement of the corresponding gene in the oxidative burst,[103] however, no identification of the corresponding substrates was performed. The similarity of the response, as compared to tobacco, as well as the high level of amino acid sequences identity of the deduced twi1 protein and the IEGT protein, may suggest identical, or at least similar, substrates in both cases. Two *IS5a* and *Is10a* homologues from the tobacco variety *Samsun* could effectively glucosylate the coumarins, esculetin and scopoletin among several other phenylpropanoids and phenolic acids, whereas salicylic acid itself was a poor substrate.[104] Therefore, it is unlikely that the phytohormone is the actual substrate of these induced glucosyltransferases and the reported cause of increased levels of the glucosides. A corresponding salicylic acid glucosyltransferase that is induced in tobacco leaves after infection with TMV may correspond to yet another unidentified sequence.[105]

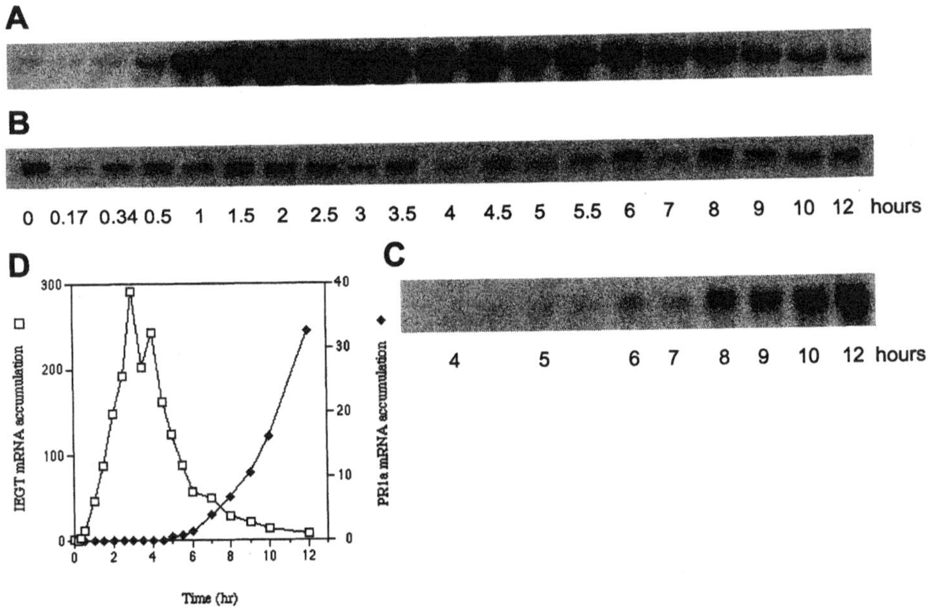

Figure 6. Rapid and transient accumulation of a putative glucosyltransferase IEGT after treatment of BY-2 tobacco cell suspension cultures with 0.2 mM salicylic acid. Autoradiogram of 20 μg of RNA blot probed with IEGT (A), pathogen related PR-1 protein (B) and ß-ATPase (C). D. quantification of IEGT and PR-1 by absolute average phosphorimager units after normalization with the ß-ATPase signal. (reproduced with kind permission from Horvath and Chua (1996); Plant Mol. Biol. 31: 1061-1072; Kluver Academic Press).

The above described transcripts are not induced by jasmonic acid (JA) and, therefore, are clearly different from a recently published glycosyltransferase sequence, *JIGT* (jasmonate induced glucosyltransferase) obtained also from tobacco cell cultures by differential cDNA screening.[106] This putative glycosyltransferase can be induced by nanomolar concentrations of methyl jasmonate or coronatine, a phytotoxin from *Pseudomonas syringae*, which has similar effects on plant tissues as JA-derivatives. No expression studies of *JIGT* were performed and its natural substrate(s) await elucidation.

SEQUENCE ANALYSIS AND PHYLOGENETIC TREES

Plant glycosyltransferase genes encode a superfamily of proteins, comparable to that of the *O*-methyltransferase superfamily that also act late in the modification of secondary metabolites.[107] Sequence alignments of thirty, mostly full length glucosyltransferase sequences involved in flavonoid biosynthesis show that the corresponding proteins share a uniform length of 400 to 489 amino acids. Several putative *A. thaliana* glycosyltransferase genes, identified in the *Arabidopsis* genome, were included in the sequence alignment. Except for the *A. thaliana iaglu* gene,[108] all other putative *A. thaliana* glycosyltransferase genes have not been functionally assigned. In addition to sequences from *A. thaliana*, the corresponding dendrogram (Fig. 7) is dominated by sequence data from the Solanaceae, with only a few sequences from other plant families.

Although several subclusters can be detected, the dendrogram cannot be directly correlated to phylogenetic distances of the individual plant families. However, if we assume substrate specificity as one principle of cluster formation, several arrangements are plausible. One cluster apparently represents the anthocyanidin glucosyltransferase. It consists of already expressed genes and the corresponding recombinant proteins from *G. triflora*, *V. vinifera*, and *P. frutescens*. [13,90,91] We postulate that the enzymes of *S. melongena*, (Toguri, unpublished) and morning glory (*Ipomea purpurea*) (Tiffin *et al.*, unpublished) represent anthocyanidin- or anthocyanin glucosyltransferases and not flavonol glycosyltransferases. Both monocot sequences, from barley and maize, are placed adjacent to each other in one phylogenetic branch, very close to the anthocyanin cluster. In this case, both sequences are considered putative flavonol 3-*O*- glucosyltransferases. Structural similarity of the glucose acceptors, together with a regiospecificity for the 3-hydroxyl group, may reflect the placement close to the glycosyltransferases involved in anthocyanin formation.

Unresolved clusters of other glucosyltransferases, in most cases from *A. thaliana*, reflect our fragmentary knowledge of most cloned glycosyltransferase genes. Characterized by significant sequence identities to the *iaglu* gene from maize, sequence No. 8 from *A. thaliana* (Fig. 7) might identify the corresponding gene from this plant, rather than the previously suggested sequence.[108] Other genes from *A. thaliana* are placed randomly and may represent individual subclusters. The substrates and the functions of all these sequences are unknown. This includes the cassava sequences CGT1 and CGT4, respectively.

Another cluster comprises glucosyltranferases characterized from three unrelated plant families, Aizoaceae, Brassicaceae, and Solanaceae. However, the similarity of inducible putative glycosyltransferases from Solanaceae *IS5a*[100] and *twi1*[102] with the betanidin 5-*O*-glucosyltransferase of *D. bellidiformis* and one *A. thaliana* gene of unknown function, is of particular interest. A sequence alignment was performed and illustrates the high degree of sequence identities at the amino acid level in this cluster (Fig. 8). This is difficult to explain since classical[109] as well as molecular systematics[110] show quite distant correlations of the Aizoaceae, the Solanaceae, and the Brassicaceae. A similar degree of sequence identity within one cluster is also observed among the better characterized anthocyanin glucosyltransferase group (data not shown).

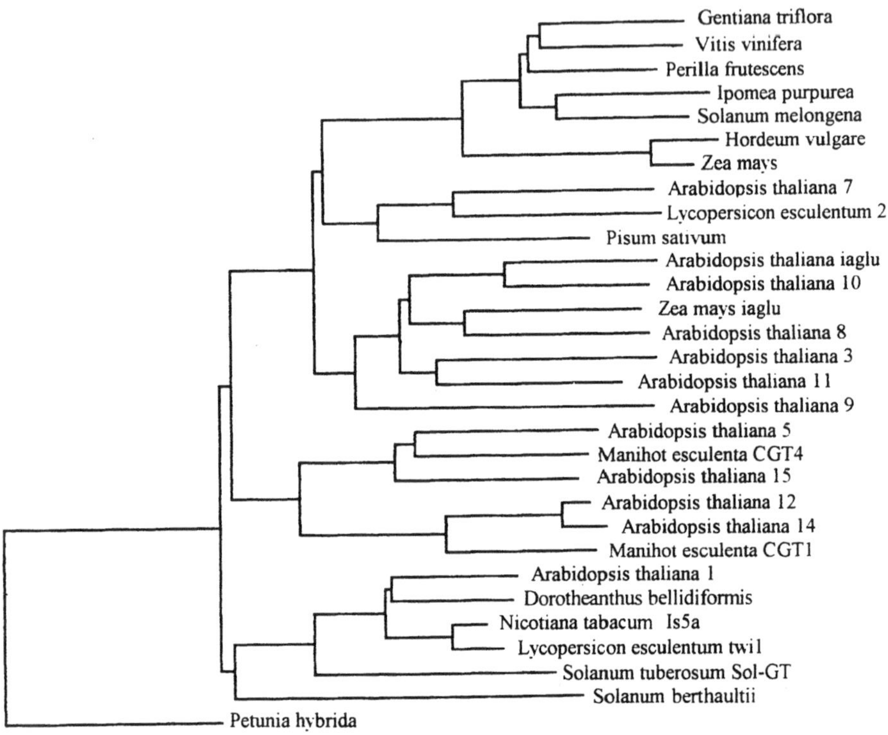

Figure 7. Phylogenetic dendrogram of thirty glycosyltransferases of plant secondary metabolism. The dendrogram was created using the "clustal sequence alignment" program of the *Lasergene* software package (DNASTAR, Madison, U.S.A.). Accession numbers of glucosyltransferases in GENBANK: *Gentiana triflora (D85186), Vitis vinifera (AF000371), Perilla frutescens (AB002818), Ipomea purpurea (AF028237), Solanum melongena (X77369), Hordeum vulgare (X15694), Zea mays (X13501), Arabidopsis thaliana 7 (AC002505), Lycopersicon esculentum (S39507), Pisum sativum (AF034743), Arabidopsis thaliana iaglu (U81293), Arabidopsis thaliana 10 (Z97335), Zea mays iaglu (L34847), Arabidopsis thaliana 8 (AC002333), Arabidopsis thaliana 3 (Z97339), Arabidopsis thaliana 11 (AC002391), Arabidopsis thaliana 9 (AC004165), Arabidopsis thaliana 5 (Z99708), Manihot esculenta CGT4 (X77462), Arabidopsis thaliana 15 (AF077407), Arabidopsis thaliana 12 (Z97338), Arabidopsis thaliana 14 (Z97338), Manihot esculenta CGT1 (X77459), Arabidopsis thaliana 1 (AL021961), Dorotheanthus bellidiformis 5-GT (Y18871), Nicotiana tabacum Is5a (U32644), Lycopersicon esculentum twi1 (X85138), Solanum tuberosum Sol-GT (U82367), Solanum bertaulthii (AF006081), Petunia hybrida (Z25802).* (Reproduced from Vogt et al. (1999); Plant J. 19: 509-519; ßios Scientific Publishers).

This may suggest that in both cases substrate specificity is a consequence of sequence identity, with the betanidin glucosylating 5-*O*-glycosyltransferase exhibiting identical or at least similar substrate specificities to the salicylic acid inducible tobacco genes *IS5a* and *Is10a*.[100] In fact, recent data suggest that these overexpressed glycosyltransferases from the Solanaceae most efficiently glucosylate flavonols, especially quercetin,[101] and quercetin is the second best substrate for the betanidin glucosylating enzyme. Whether the regiospecificity of the enzyme plays a role in this sequence alignment cannot be answered yet, since no flavonoid A- or B-ring specific glucosyltransferase has been cloned. However, this cluster is clearly separate from the flavonol 3-*O*-glucosyltransferases of maize and barley, which may be evidence for different regiospecificity and phylogenetic origin, respectivley.

If these assumptions are valid, one might suggest flavonols as possible substrates for the unknown *A. thaliana 1* glycosyltransferase sequence. Unfortunately, in this most thoroughly investigated higher plant, neither the glucosyltransferases nor the flavonoid glycosides are known in detail.[111] The solanidine glucosyltransferase involved in the biosynthesis of potato steroidal glycoalkaloids[57] also showed considerable homology to this subfamily (Fig. 8), which may be based on some structural similarities of the flavonoid and the steroid molecules.

Significant identities of the betanidin glucosylating 5-*O*-glucosyltransferase to flavonoid glycosylating proteins may also shed light on the phylogenetic origin of betalain biosynthesis compared to the flavonoid pathway. As stated earlier, betalain pigments are characteristic of the plant order Caryophyllales, whose systematic classification in the plant kingdom is still in question.[112] Nothing is known about the evolution of the corresponding pigments and the mutual exclusion of betalains and anthocyanins in the plant kingdom.[113] The existence of glucosyltransferases, which share substrates from both betalain and flavonoid biosynthesis, and which show significant identities to enzymes glycosylating flavonoids or certain hydroxycoumarins,[101,104] points towards a phylogenetically late origin of the betalain biosynthetic pathway. A second enzyme, the betanidin-6-*O*-glucosyltransferase, also present in *D. bellidiformis*, showed, except for betanidin, its highest activity towards the 3-hydroxyl group of cyanidin.[38] Preliminary sequence data from this enzyme show only little sequence similarity to the corresponding 5-*O*-glucosyltransferase, however, significant identities to some unknown glycosyltransferases from cassava, suggesting a polyphyletic origin of the betanidin glucosylating enzymes (Vogt, unpublished).

The only rhamnosyltransferase, cloned from *P. hybrida*,[85,86] appears as an outgroup separated from the rest of the glycosyltransferases of the Solanaceae. This is due to a rather unique "PSPG-box", designed to bind UDP-rhamnose instead of UDP-glucose. It will be interesting to compare this sequence with that of the corresponding 3-*O*-glucosyltranferase, the initial enzyme in flavonol and anthocyanidin glucosylation in *P.*

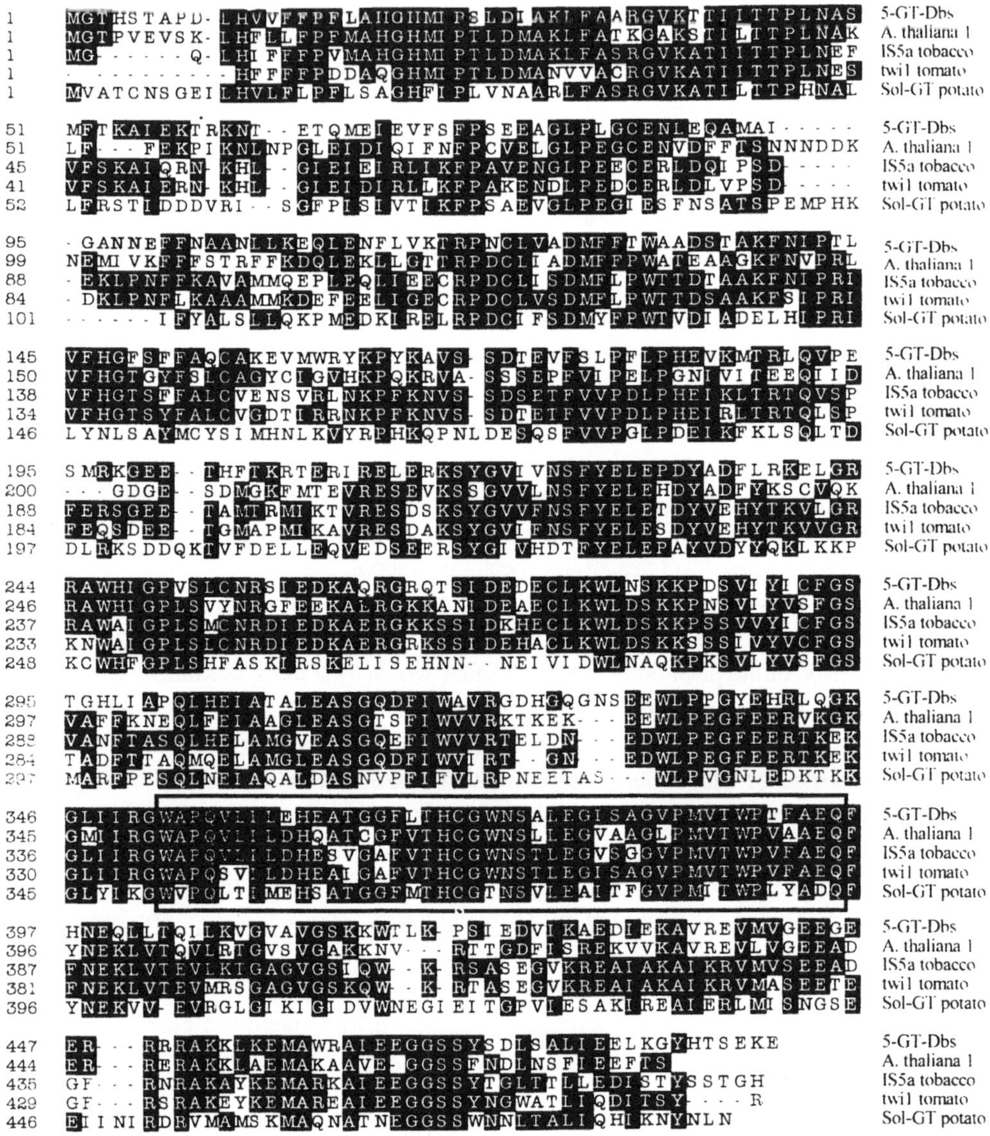

Figure 8. Sequence identities of betanidin glucosyltransferase (5-GT-Dbs) to inducible glucosyltransferases of *N. tabacum and L. esculentum* (Is5a; twi 1), solanidine glucosyltransferase from *S. tuberosum* (Sol-GT potato) and an unknown sequence from *A. thaliana* (A. thaliana 1). The "PSPG-box" is marked (Reproduced from Vogt et al. (1999); Plant J. 19: 509-519; ßios Scientific Publishers).

hybrida. Both substrates apparently are accepted by this enzyme with a preference for flavonols,[114] but the corresponding gene has not been cloned yet. Once the gaps in the dendrogram are filled with data from other plant families, it should help to assign substrate- and even regiospecificity to those cloned but functionally uncharacterized glucosyltransferases.

Highly conserved amino acids of all glycosyltransferases may point to important structural or functional domains. A significant portion of the protein sequences displayed in the multiple alignment start with a hydrophobic *N*-terminus.[92] This may suggest that they are part of a larger family of the "type II" class of eukaryotic membrane-bound glycosyltransferases that are defined by a hydrophobic *N*-terminal membrane anchor.[115] *In vivo,* these enzymes may be loosely associated with membrane structures, consistent with the data obtained by immunoaffinity labelling.[33] Two histidine residues are highly conserved in this part of all glycosyltransferase genes analyzed.[92] In contrast, amino acids conserved in some, but not in other proteins, deserve special attention in order to assign the binding domains of the different sugar acceptors. Based on unique sequence identities of the hydrophobic domain of the potato solanidine glucosyltransferase with steroid-specific animal UDP-glucuronosyltransferase, a steroidal alkaloid binding domain for amino acids 109-143 of the respective plant protein was suggested.[57] Hydrophilicity plots[116] from a variety of glycosyltransferases also show one conserved hydrophobic area from amino acid 120 to 140, consistent with the proposed putative role in substrate binding (Vogt, unpublished).

SUMMARY

The glycosyltranferases of plant secondary metabolism constitute a superfamily of modifying enzymes involved in late steps of the biosynthesis of natural products. Purification and characterization revealed several conserved properties of all glycosyltransferases. These include similar molecular weights and a tendency to bind to hydrophobic and dye ligand matrices, both successfully used in the purification of individual enzymes. Heterologous cloning and expression studies of various glycosyltransferase genes unequivocally defined high substrate specificities, even regiospecificities towards the sugar acceptor. This is reflected in a dendrogram deduced from amino acid sequence data, which can be interpreted by substrate similarities, rather than taxonomic classification of the corresponding plant source. While purification of glycosyltransferases from *A. thaliana* is a challenging task due to the limited amount of plant tissue, functional analysis may start at the nucleotide sequence level by sequence similarity search and heterologous expression of the deduced proteins. These functional studies will merit further investigations, since the complexity of potential glucosyltransferase targets including xenobiotics, plant and insect hormones, and putative drugs has only been partly realized.[117] With these goals in mind, a refined classification of the glucosyltranferase superfamily in plants, similar to the analysis presented for glucuronosyltransferases of animal tissues,[7] might be within reach over the next few years, and probably will enable the design of new enzymes with defined substrate specificities and new properties.

ACKNOWLEDGEMENTS

I thank Dieter Strack for his advice and helpful discussions during preparation of this work. I also thank Willibald Schliemann and Martina Rauscher for critical reading of the manuscript. I am grateful to Diana Horvath and Bill Belknap for sharing some information prior to publication and kind permission to use Figures 5 and 6 of this manuscript, respectively. The Deutsche Forschungsgemeinschaft is acknowledged for financial support.

REFERENCES

1. IKAN, R. 1998. Naturally Occurring Glycosides. John Wiley & Sons, Chichester, 444p.
2. HELLER, W., FORKMANN, G. 1994. Biosynthesis of flavonoids. pp. 499-535 *in* The Flavonoids, Advances in Research Since 1986. (Harborne, J.B., ed.), Chapman & Hall, London.
3. WETZEL, A., SANDERMANN JR., H. 1994. Plant biochemistry of xenobiotics: Isolation and characterization of a soybean *O*-glucosyltransferase of DTT metabolism. Arch. Biochem. Biophys. 314: 323-328.
4. COLEMANN, J.O.D., BLACKE-KALFF, M.M.A., DAVIES, T.G.E. 1997. Detoxification of xenobiotics by plants: Chemical modification and vacuolar compartmentation. Trends. Pl. Sci. 2: 144-151.
5. ALTSCHUL, S.F., GISH, W., MILLER, W., MYERS, E.W., LIPMAN, D.J. 1990. Basic local alignment search tool. J. Mol. Biol. 215: 403-10.
6. GISH, W., STATES, D.J. 1993. Identification of protein coding regions by database similarity search. Nat. Genet. 3: 266-72.
7. MACKENZIE, P.I., OWENS, I.S., BURCHELL, B., BOCK, K.W., BAIROCH, A., BELANGER, A., FOURNEL-GIGLEUX, S.,GREEN, M., HUM, D.W., IVANAGI, T., LANCET, D., LOUISOT, P., MAGDALOU, J., CHOWDHURY, J.R., RITTER, J.K., SCHLACHTER, H., TEPHLY, T.R., TIPTON, K.F., NEBERT, D.W. 1997. The UDP glycosyltransferase gene superfamily: Recommended nomenclature update based on evolutionary divergence. Pharmacogenetics 7: 255-269.

8. KERSCHER, F., FRANZ, G. 1988. Isolation and some properties of an UDP-glucose: 2-hydroxflavanone-6(or 8)-C-glucosyltransferase from *Fagopyrum esculentum*. J. Plant. Physiol. 132: 110-115.
9. HÖSEL, W. 1981. Glycosylation and glycosidases. pp.725-753 *in* The Biochemistry of Plants Vol. 7, (Stumpf, P.K., Conn, E.E., eds.), Academic Press, New York.
10. ROSE, A., GLÄßGEN, W.E., HOPP. W., SEITZ, H.U. 1996. Purification and characterization of glycosyltransferases involved in anthocyanin biosynthesis in cell-suspension cultures of *Daucus carota*. Planta 198: 397-403.
11. OGATA, J., TERAMOTO, S., YOSHITAMA, K. 1998. Isolation and characterization of UDP-glucose:cyanidin 3-*O*-glucosyltransferase from the flower buds of *Senecio x hybridus*. J. Plant Res. 111: 213-216.
12. DO, C.B., CORMIER, F., NICOLAS, Y. 1995. Isolation and characterization of a UDP-glucose:cyanidin 3-*O*-glucosyltransferase from grape cell suspension cultures (*Vitis vinifera* L.). Plant Sci. 112: 43-51.
13. FORD, C.M., HØJ, P.B. 1998. Multiple glucosyltransferase activities in the grapevine *Vitis vinifera* L. Austr. J. Grape and Wine Res. 4: 48-58.
14. LISTER, C.E., LANCASTER, J.E., WALKER, J.R.L. 1996. Developmental changes in enzymes of flavonoid biosynthesis in the skins of red and green apple cultivars. J. Sci. Food and Agriculture 71: 313-320.
15. LISTER, C.E., LANCASTER, J.E., SUTTON, K.H., WALKER, J.R.L. 1997. Aglycone and glycoside specificity of apple skin flavonoid glycosyltransferase. J. Sci. Food and Agriculture 75: 378-382.
16. VOGT, T., TAYLOR, L.P. 1995. Flavonol 3-*O*-glycosyltransferases associated with *Petunia* pollen produce gametophyte specific flavonol diglycosides. Plant Physiol. 108: 903-911.
17. ISHIKURA N., MATO, M. 1993. Partial purification and some properties of flavonol 3-*O*-glucosyltransferases from seedlings of *Vigna mungo*, with special reference to the formation of kaempferol 3-*O*-galactoside and 3-*O*-glucoside. Plant Cell Physiol. 34: 329-335.
18. MEDERACKE, H., SELMAR, D., BIEHL, B. 1995. Glucosyltransferases of the cyanogenic plant cassava (*Manihot esculenta*, Crantz). Angew. Bot. 69: 119-124.
19. PARRY, A.D., EDWARDS, R. 1994. Characterization of *O*-glucosyltransferases with activities toward phenolic substrates in alfalfa. Phytochemistry 37: 655-661.
20. LEWINSOHN, E., BRITSCH, L., MAZUR, Y., GRESSEL, J. 1989. Flavanone glycoside biosynthesis in *Citrus*. Chalcone synthase, UDP-glucose:flavanone-7-*O*-glucosyltransferase and rhamnosyltransferase activities in cell free extracts. Plant Physiol. 91: 1323-1328.
21. MCINTOSH, C.A., LATCHINIAN, L., MANSELL, R.L. 1990. Flavanone-specific 7-*O*-glucosyltransferase activity in *Citrus paradisii* seedlings: purification and

characterization. Arch. Biochem. Biophys 282: 50-57.

22. BERHOW, M.A., SMOLENSKY, D. 1995. Developmental and substrate specificity of hersperetin-7-*O*-glucosyltransferase activity in *Citrus limon* tissues using high-performance liquid chromatographic analysis. Plant Sci. 112: 139-147.
23. BAR-PELED, M., LEWINSOHN, E., FLUHR, R., GRESSEL, J. 1991. UDP-rhamnose:flavanone-7-*O*-glucoside 2"-*O*-rhamnosyltransferase. Purification and characterization of an enzyme catalyzing the production of bitter compounds in *Citrus*. J. Biol. Chem. 31: 20953-20959.
24. CHENG, G.W., MALENICK, D.A., BREEN, P.J. 1994. UDP-glucose:flavonoid-*O*-glu-cosyltransferases from strawberry fruit. Phytochemistry 35: 1435-1439.
25. STICH, K., HALBWIRTH, H., WURST, F., FORKMANN, G. 1997. UDP-glucose: flavonol 7-*O*-glucosyltransferase activity in flower extracts of *Chrysanthemum segetum*. Z. Naturforsch. 52c: 153-158.
26. ISHIKURA, N., YANG, Z-Q. 1994. Multiple forms of flavonol *O*-glucosyltransferases in young leaves of *Euonymus alatus f. ciliato-centatus*. Phytochemistry 36: 1139-1145.
27. VELLEKOOP, P., LUGONES, L., VAN BREDERODE, J. 1993. Purification of an UDP-glucose:flavone 7-*O*-glucosyltransferase, from *Silene latifolia* using a specific interaction between the enzyme and phenyl-sepharose. FEBS 330: 36-40.
28. STICH, K., HALBWIRTH, H. WURST, F., FORKMANN, G. 1994. Formation of a 6 -deoxychalcone 4 -glucosides by enzyme extracts from petals of *Dahlia variabilis*. Z. Naturforsch. 49c: 737-741.
29. ISHIKURA, N., YAMAMOTO, K. 1990. Detection and characterization of UDP-glucose: flavonoid *O*-glucosyltranferases in the leaves of *Prunus x yedoensis Matsum*. Plant Cell Physiol. 31: 1109-1115.
30. KÖSTER, J., BARZ, W. 1981. UDP-glucose:isoflavone 7-*O*-glucosyltransferase from roots of chick pea (*Cicer arietinum L.*) Arch. Biochem. Biophys. 212: 98-104.
31. LATCHINIAN-SADEK, L., IBRAHIM, R. 1991a. Partial purification and characterization of a ring B-*O*-glucosyltransferase from onion bulbs. Phytochemistry 30: 1767-1771.
32. TSUSHIDA, T., SUZUKI, M. 1996. Content of flavonol glucosides and some properties of metabolizing the glucosides in onion. Nippon Shokuhin Kagaku Kogaku Kaishi 43: 642-649.
33. LATCHINIAN-SADEK, L., IBRAHIM, R. 1991b. Flavonol ring B-specific *O*-glucosyltransferases: Purification, production of monoclonal antibodies and immunolocalization. Arch. Biochem. Biophys. 289: 230-236.
34. SCHULZ, M., WEISSENBÖCK, G. 1988. Three specific UDP-glucuronate:flavone-glucuronosyltransferases from primary leaves of *Secale cereale*. Phytochemistry 27: 1261-1267.

35. MABRY, T.J. 1980. Betalains pp. 513-533 *in* Encyclopedia of Plant Physiology Vol. 8, (Bell, E.A., Charlwood, B.V., eds.) Springer, Berlin-Heidelberg-New York.
36. HEUER, S., STRACK, D. 1992. Synthesis of betanin from betanidin by a protein preparation from cell cultures of *Dorotheanthus bellidiformis* (Burm.f.) N.E.Br. Planta 186: 626-628.
37. HEUER, S., VOGT, T., BÖHM, W., STRACK, D. 1996. Partial purification and characterization of UDP-glucose:betanidin-5-*O*- and 6-*O*-glucosyltransferases from cell cultures of *Dorotheanthus bellidiformis* (Burm.f.) N.E.Br. Planta 199: 244-250.
38. VOGT, T., ZIMMERMANN, E., GRIMM, R., MEYER, M., STRACK, D. 1997. Are the characteristics of betanidin glucosyltransferases from cell suspension cultures of *Dorotheanthus bellidiformis* indicative for their phylogenetic relationship with flavonoid glucosyltransferases? Planta 203: 349-361.
39. SCHMID, G., GRISEBACH, H. 1982. Enzymic synthesis of lignin precursors. Purification and properties of UDPglucose:coniferyl alcohol glucosyltransferases from cambial sap of spruce (*Picea abies* L.) Eur. J. Biochem. 123: 363-370.
40. WANG, S., ELLIS, B. 1998. Enzymology of UDP-glucose:sinapic acid glucosyltransferase from *Brassica napus*. Phytochemistry 49: 307-318.
41. LI, S.M., WANG, Z.X., HEIDE, L. 1997. Purification of UDP-glucose: 4-hydroxy-benzoate glucosyltransferases from cell cultures of *Lithospermum erythrorhizon*. Phytochemistry 46: 27-32.
42. BETRY, P., FLINIAUX, M.A., MACKOVA, M., GILLET, F., MACEK, T., JACQUIN-DUBREUIL, A. 1995. Scopoletin glucosyltransferase activity from *Duboisia myoporoids*; improvement of cultivation conditions and crude extract preparation procedure. J. Plant Physiol. 146: 503-507.
43. YALPANI, N., SCHULZ, M., DAVIS, M.P., BALKE, N.E. 1992. Partial purification and properties of an inducible uridine 5 -diphospho-glucose:salicylic acid glucosyltransferase from oat roots. Plant Physiol. 100: 457-463.
44. MOCK, H.-P., STRACK, D. 1993. Energetics of the Uridine5 -diphosphoglucose:hy-droxycinnamic acid acyl-glucosyltransferase reaction. Phytochemistry 32: 575-579.
45. GROSS, G.G. 1992. Enzymatic synthesis of gallotannins and related compounds pp. 297-324 *in* Phenolic Metabolism in Plants. (Stafford, H. and Ibrahim, R.K., eds.) Plenum Press, New York
46. KUAI, J.-P., GHANGAS, G.S., STEFFENS, J.C. 1997. Regulation of triacylglucose fatty acid composition[1]. Uridine diphosphate glucose:fatty acide glucosyltransferases with overlapping chain-length specificity. Plant. Physiol. 115: 1581-1587.
47. SUTTER, A., GRISEBACH, H. 1975. Free reversibility of the UDP-glucose.flavonol 3-*O*-glucosyl-transferase Arch. Biochem. Biophys. 167: 444-447.

48. YAMAMOTO, E., INCIONG, M.E.J., DAVIN, L.B., LEWIS, N.G. 1990. Formation of *cis*-coniferin in cell-free extracts of Fagus grandiflora Ehrh bark. Plant Physiol. 94, 209-213.
49. RASMUSSEN, S., RUDOLPH, H. 1997. Isolation, purification and characterization of UDP-glucose:*cis-p*-coumaric acid-ß-D-glucosyltransferase from *Sphagnum fallax*. Phytochemistry 46: 449-453.
50. KAHN, R.A., BAK, S., SVENDSON, I., HALKIER, B.A., MØLLER, B.L. 1997. Isolation and reconstitution of cytochrom P450ox and in vitro reconstitution of the entire biosynthetic pathway of the cyanogenic glucoside dhurrin from *Sorghum*. Plant Physiol. 115: 1661-1670.
51. REED, D.W., DAVIN, L., JAIN, J.C., DELUCA, V., NELSON, L., UNDERHILL, E.W. 1993. Purification and characterization of UDP-glucose:thiohydroximate glucosyltransferase from *Brassica napus L.* seedlings. Arch. Biochem. Biophys. 305: 526-532.
52. GUO, L., POULTON, J. 1994. Partial purification and characterization of *Arabidopsis thaliana* UDPG:thiohydroximate glucosyltransferase. Phytochemistry 36: 1133-1138.
53. GROOTWASSINK, J.W.D., REED, D.W., LOLENOWSKY, A.D. 1994. Immunoprification and immunocharacterization of the glucosinolate biosynthesis enzyme thiohydrximate:S-glucosyltranferase. Plant Physiol. 105: 425-433.
54. HASEGAWA, S., SUHAYDA, G.S., HSU, W.J., ROBERTSON, G.H. 1997. Purification of limonoid glucosyltransferase from navel orange albedo tissues. Phytochemistry 46: 33-37.
55. HAYASHI, H., NISHIYAMA, Y., TOMIZAWA, N., HIROAKA, N., IKESHIRO, Y. 1996. UDP-glucuronic acid:triterpene glucuronsosyltransferase activity in cultured licorice cells. Phytochemistry 42: 665-666.
56. SHIBATA, H., SAWA, Y., TKA-AKI, O., SONOKE, S., KIM, K.K., YOSHIOKA, M. 1995. Steviol and steviol-glycoside:glucosyltransferase activities in *Stevia rebaudiana Bertoni* - Purification and partial characterization. Arch. Biochem. Biophys. 321: 390-396.
57. STAPLETON, A., ALLEN, P.V., TAO, H.P., BELKNAP, W.R., FRIEDMAN, M. 1992. Partial amino acid sequence of potato solanidine UDP-glucose glucosyltransferase purified by anion-exchange chromatography and size exclusion media. Protein Expr. Purif. 3: 85-92.
58. MOEHS, C.P., ALLEN, P.V., FRIEDMAN, M., BELKNAP, W.R. 1997. Cloning and expression of solanidine UDP glucose glucosyltransferase from potato. Plant J. 11: 227-236.
59. PACZKOWSKI, C., KALINOWSKA, M., WOJCIECHOWSKI, Z.A. 1997. UDP-glucose: solasodine glucosyltranserase from eggplant (*Solanum melongena L.*) leaves: partial purification and characterization. Acta Biochem. Pol. 44: 43-53.

60. PACZKOWSKI, C., WOJCIECHOWSKI, A. 1994. Glucosylation and galactosylation of diosgenin and solasidine by soluble glycosyltransferases from *Solanum melongena*. Phytochemistry 35: 1429-1434.
61. ZIMOWSKI, J. (1992) Specificity and some other properties of cytosolic and membraneous UDPGlc:3-*beta*-hydroxysteroid glucosyltransferases from *Solanum tuberosum* leaves. Phytochemistry 31: 2977-2981.
62. ULLMANN, P., URY, A., RIMMELE, D., BENISTE, P., BOUVIER-NAVE, P. 1993. UDP-glucose:sterol ß-D-glucosyltransferase, a plasma membrane-bound enzyme of plants: Enzymatic properties and lipid dependence. Biochimie 75: 713-723.
63. WARNECKE, D.C., HEINZ, E. 1994. Purification of a membrane-bound UDP-glucose:sterol-ß-D-glucosyltransferase based on its solubility in diethylether. Plant Physiol. 105: 1067-1073.
64. KALINOWSKA, M. 1994. Glucosylation of C_{19} and C_{21}-hydroxysteroids by soluble and membrane-bound glucosyltransferase from oat (*Avena sativa*) seedlings. Phytochemistry 36: 617-622.
65. THEURER, C., TREUMANN, H.J., FAUST, T., MAY, U., KREIS, W. 1994. Glycosylation in cardenolide biosynthesis. Plant Cell, Tissue Organ Cult. 38: 327-335.
66. LEIGHTON, V., NIEMEYER, H.M., JONSSON, L.M.V. 1994. Substrate specificities of a glucosyltransferase and an N-hydroxlase involved in the biosynthesis of cyclic hydroxamic acids in Gramineae. Phytochemistry 36: 887-892.
67. KREUZ, K., TOMMASINI, R., MARTINOIA, E. 1996. Old enzymes for a new job: herbicide detoxification in plants. Plant Physiol. 111: 349-353.
68. LEAH, J.M., WORRALL, T.L., COBB, A.H. 1992. Isolation and characterization of two glucosyltransferases from *Glycine max* associated with bentazone metabolism. Pestic. Sci. 34: 81-87.
69. WETZEL, A., SANDERMANN, JR., H. 1994. Plant biochemsitry of xenobiotics and characterization of a soybean *O*-glucosyltransferase of DDT metabolism. Arch. Biochem. Biophys. 314: 323-328.
70. PFLUGMACHER, S., SANDERMANN, JR., H. 1998. Taxonomic distribution of plant glucosyltransferases acting on xenobiotics. Phytochemistry 49: 507-511.
71. IBRAHIM, R.K., DE LUCA, V., KHOURI, H., LATCHINIAN, L., BRISSON, L., CHAREST, P.M. 1997. Enzymology and compartmentation of polymethylated flavonol glucosides in *Chrysosplenium americanum*. Phytochemistry 26: 1237-1245.
72. CHAPPLE, C.C.S. 1998. Molecular-genetic analysis of plant cytochrome P450-dependent monooxygenases pp. 311-343 *in* Annual Review of Plant Physioloy and Plant Molecular Biology Vol. 49, (Jones, L.J., Somerville, C.R., Walbot, V., eds.) Annual Reviews, Palo Alto, California, U.S.A.

73. SCHROEDER, J. 1997. A family of plant specific polyketide synthases: Facts and predictions. Trends Plant Sci.: 2, 373-378.
74. HARTMANN, T. 1996. Diversity and variability of plant secondary metabolism: A mechanistic view. Entomol. Exp. Applicata 80: 177-188.
75. BATTAGLIA E., PRITCHARD M., OUZZINE, M., FOURNEL, S., RADOMINSKA, A., SIEST, G., MAGDALOU, J. 1994. Chemical modification of human UDP-glucuronosyl transferase UGT1*6 by diethyl pyrocarbonate: Possible involvement of a histidine residue in the catalytic process. Arch. Biochem. Biophys. 309: 266-272.
76. LOWE, C.R., PEARSON, J.C. 1984. Affinity chromatography on immobilised dyes. Methods in Enzymology 104: 97-113.
77. HATZFELD, W.D., MOESINGER, E. 1995. Active-site-directed inhibition of sucrose-phophate synthase by cibacron blue F3G-A and 1-deoxynojirimycin. Planta 197: 296-300.
78. FEDOROFF, N.V., FURTEK, D., NELSON, O.E. 1984. Cloning of the bronze locus in maize by a simple and generalizable procedure using the transposable controlling element Activator (Ac). Proc. Natl. Acad. Sci. U.S.A. 81: 3825-3829.
79. FURTEK, D., SCHIEFELBEIN, J.W., JOHNSTON, F., NELSON, O.E. 1988. Sequence comparisons of three wild-type Bronze-1 alleles from *Zea mays*. Plant Mol. Biol. 11: 473-481.
80. MARTIN, C., PRESCOTT, A., MACKAY, S., BARTLETT, J., VRIJLANDT, E. 1991. The control of anthocyanin biosynthesis in flowers of *Anthirrhinum majus*. Plant J. 1: 37-49.
81. SPARVOLI, F., MARTIN, C., SCIENZA., A., GAVAZZI, G., TONELLI, C. 1994. Cloning and molecular analysis of structural genes involved in flavonoid and stilbene biosynthesis in grape (*Vitis vinifera L.*) Plant Mol. Biol. 24: 743-755.
82. GONG, Z., YAMAZAKI, M., SUGIYAMA, M., TANAKA, Y., SAITO, K., 1997. Clon-ing and molecular analysis of the structural genes involved in anthocyanin biosynthesis and expressed in a forma-specific manner in *Perilla frutescens*. Plant Mol. Biol. 35: 915-927.
83. HUGHES, J., HUGHES, M.A. 1994. Multiple secondary plant product UDP-glucose glucosyltransferase genes expressed in cassava. DNA Sequence. 5: 41-49.
84. BAIROCH, A. 1991. PROSITE: A dictionary of sites and patterns in proteins. Nucl. Acid. Res. 19: 2241-2245.
85. BRUGLIERA, F., HOLTON, T.A., STEVENSON, T.W., FARCY, E., LU, C.Y., COR-NISH, E.C. 1994. Isolation and characterization of a cDNA clone corresponding to the *Rt* locus of *Petunia hybrida*. Plant J. 5: 81-92.
86. KROON, J., SOUER, E., DE GRAAF, A., XUE, Y., MOL, J., KOES, R. 1994. Cloning and structural analysis of the anthocyanin pigmentation locus Rt of *Petunia*

hybrida: Characterization of insertion sequences of two mutant alleles. Plant J. 5: 69-80.

87. MARKHAM, K.R. 1996. Novel anthocyanins produced in petals of genetically transformed lisianthus. Phytochemistry 42: 1035-1038.
88. SCHWINN, K.L., DAVIES, K.M., DEROLES, S.C., MARKHAM, K.R., MILLER, R.M., BRADLEY, J.M., MANSON, D.G., GIVEN, N.K. 1997. Expression of an *Anthirrhinum majus* UDP-glucose:flavonoid-3-*O*-glucosyltransferase transgene alters flavonoid glycosylation and acylation in lisianthus (*Eustoma grandiflorum* Grise.). Plant Science 125: 53-61.
89. FORD, C.M., BOSS, P.K., HØJ, P.B. 1998. Cloning and characterization of *Vitis vinifera* UDP-glucose: flavonoid 3-*O*-glucosyltransferase, a homologue of the enzyme encoded by the maize Bronze-1 locus that may primarily serve to glycosylate anthocyanidins in vivo. J. Biol. Chem. 273: 9224-9233.
90. TANAKA, T., YONEKURA, K., FUKUCHI-MIZUTANI., M., I., FUKUI, Y., FUJI-WARA, H., ASHIKARI, T., KUSUMI, T., 1996. Molecular and biochemical characterization of three anthocyanin synthetic enzymes from *Gentiana triflora*. Plant Cell Physiol. 37: 711-716.
91. YAMAZAKI, M., ZHIZHONG, G., FUKUCHI-MIZUTANI, M., FUKUI, Y., TANAKA, Y., KUSUMI, T. AND SAITO, K. 1999. Molecular cloning and biochemical characterization of a novel anthocyanin 5-*O*-glucosyltransferase by mRNA differential display for plant forms regarding anthocyanin. J. Biol. Chem. 274: 7405-7411.
92. VOGT, T., GRIMM, R., STRACK, D. 1999. Cloning and expression of a cDNA encoding betanidin 5-*O*-glucosyltransferase, a betanidin- and flavonoid-specific enzyme with high homology to inducible glucosyltransferases from the Solanaceae. Plant J. 19, 509-519.
93. JONES, P.R., MØLLER, B.L., HØJ, P.L. 1999. The UDP-glucose:*p*-hydroxymandelo-nitrile-*o*-glucosyltransferase which catalyzes the last step in synthesis of the cyanogenic glucoside dhurrin in *Sorghum bicolor*: isolation, cloning, heterologous expression and substrate specificity. J. Biol. Chem.. In press.
94. SZERSON, J.B., SZCZYGLOWSKI, K., BANDURSKI, R.S. 1994. *iaglu*, a gene from *Zea mays* involved in conjugation of growth hormone indole-3-acetic-acid. Science 265: 1699-1701.
95. WARNECKE, D.C., BALTRUSCH, M., BUCK, F., WOLTER, F.P., HEINZ, E. 1997. UDP-glucose:sterol glucosyltransferase: Cloning and functional expression in *Escherichia coli*. Plant Mol. Biol. 35: 597-603.
96. ANHALT, S., WEISSENBÖCK, G. 1992. Subcellular localization of luteolin glucuronides and related enzymes in rye mesophyll. Planta 187: 83-88.

97. HRAZDINA, G., WAGNER, G.J. 1985. Metabolic pathways as enzyme complexes: Evidence for the synthesis of phenylpropanoids and flavonoids on membrane-associated enzyme complexes. Arch. Biochem. Biophys. 237: 88-100.
98. MARSS, K.A., ALFENITO, M.R., LLOYD, A.M., WALBOT, V. 1995. Gluthathione S-transferase involved in vacuolar transfer encoded by the maize gene *bronze-2*. Nature 375: 397-400.
99. KLEIN, M., WEISSENBÖCK, G., DUFAUD, A., GAILLARD, C., KREUZ, K., MAR-TINOIA, E. 1996. Different energization mechanisms drive the vacuolar uptake of a flavonoid glucoside and a herbicide glucoside. J. Biol. Chem. 47: 29666-29671.
100. HORVATH, D., CHUA, N.-H. 1996. Identification of an immediate-early salicylic acid-inducible tobacco gene and characterization of induction by other compounds. Plant Mol. Biol. 31: 1061-1072.
101. HORVATH, D., HUANG, D.J., CHUA, N.-H. 1998. Four classes of salicylate-induced tobacco genes. Mol. Plant-Microbe Interact. 11: 895-905.
102. TRUESDALE, M.R., DOHERTY, H.M., LOAKE, G.J., MCPHERSON, M.J., ROBERTS, M.R., BOWLES, D.J. 1996. Molecular cloning of a novel wound inducible gene from tomato: *Twi1*. Plant Physiol. 112: 446.
103. O DONNELL, P.J., TRUESDALE, M.R., CALVERT, C.M., DORANS, A., ROBERTS, M.R., BOWLES, D.J. 1998. A novel gene that rapidly responds to wound- and pathogen-related signals. The Plant J. 14: 137-142.
104. FRASSINET-TACHET L., BALTZ, R., CHONG, J., KAUFFMAN, S., FRITIG, B., SAINDRE-NAN, P. 1998. Two tobacco genes induced by infection, elicitor and salicylic acid encode glucosyltransferases acting on phenylpropanoids and benzoic acid derivatives, including salicylic acid. FEBS Letters 437: 319-323.
105. ENYEDI, A., RASKIN, I. 1993. Induction of UDP-glucose:salicylic acid glucosyltransferase activity in tobacco mosaic virus-inoculated tobacco (*Nicotiana tabacum*) leaves. Plant Physiol. 101: 1375-1380.
106. IMANISHI, S., HASHIZUME, K., KOJIMA, H., ICHIHARA, A., NAKAMURA, K. 1998. An mRNA of tobacco cell, which is rapidly inducible by menthyl jasmonate in the presence of cycloheximide, codes for a putative glycosyltransferase. Plant Cell Physiol. 39: 202-211.
107. IBRAHIM, R.K. 1997. Plant *O*-methyltransferase signatures. Trends. Pl. Sci. 2: 249-250.
108. GRAHAM, R., THORNBURG, R. 1997. DNA sequence of UDP glucose:indole-3-acetate ß-D-glucosyltransferase from *Arabidopsis thaliana*. Accession No.(81293) (PGR 97-044). Plant Physiol. 113: 1004.
109. CRONQUIST, A. 1981. An Integrated System of Classification of Flowering Plants. Columbia University Press, New York.

110. SOLTIS, D.E., SOLTIS, P.S., NICKRENT, D.L., JOHNSON, L.A., HAHN, W.J., HOOT, S.B., SWEERE, J.A., KUZOFF, R.K., KRON, K.A., CHASE, M.W., SWENSON, S.M., ZIMMER, E.A., SHU-MIAW, C., GILLESPIE, L., KRESS, W.J., SYTSMA, K.J. 1997. Angiosperm phylogeny inferred from 18S ribosomal DNA sequences. Ann. Mo. Bot. Gard. 84: 1-49.
111. GRAHAM, T.L. 1998. Flavonoid and flavonol glycoside metabolism in *Arabidopsis*. Plant Physiol. Biochem. 36: 135-144.
112. CLEMENT, J.S., MABRY, T.J. 1996. Pigment evolution in the Caryophyllales: A systematic overview. Bot. Acta 109: 360-367.
113. STAFFORD, H. 1994. Anthocyanins and betalains: Evolution of two mutually exclusive pathways. Plant Science 101: 91-98.
114. JONSSON, L.M.V., AARSMAN, M.E.G., BASTIAANEET, J., DONKER-KOOPMAN, W.E., GERATS, A.G.M., SCHRAM, A.W. 1984. Common identity of UDP-glucose:anthocyanidin 3-*O*-glucosyltransferase and UDP-glucose:flavonol 3-*O*-glucosyltransferase in flowers of *Petunia hybrida*. Z. Naturforsch. 39c: 559-567.
115. NARIMATSU, H. 1994. Recent progress in molecular cloning of glucosyltransferase genes from eucaryotes. Microbiol. Immunol. 38: 489-504.
116. DYTE, J., DOOLITTLE, F.R. 1982. A simple method for displaying the hydropathic character of proteins. J. Mol. Biol. 157: 105-132.
117. O'REILLY, D.R., MILLER, L.K. 1989. Abaculovirus blocks insect molting by producing ecdysteroid UDP-glucosyltransferase. Science 243: 1110-1112.

Chapter Eleven

THE METHYLTRANSFERASE GENE SUPERFAMILY: A TREE WITH MULTIPLE BRANCHES

Ragai K. Ibrahim and Ingrid Muzac

Plant Biochemistry Laboratory
Department of Biology
Concordia University
Montréal, Québec, Canada H3G 1M8

INTRODUCTION

S-Adenosyl-L-methionine (AdoMet)-dependent methyltransferases (MTs) [E.C. 2.1.1-] catalyze the transfer of the methyl group of AdoMet to an acceptor molecule, with the concomitant formation of the corresponding methylated

Evolution of Metabolic Pathways – J.T. Romeo et al. (Editors)

derivative and *S*-adenosyl-L-homocysteine (AdoHcy) as products. MTs are essential enzymes that play a variety of roles in the maintenance of biological activities of all organisms. In general, the enzymatic methylation of natural plant products inactivates the reactivity of their hydroxyl and/or carboxyl groups and, thus, alters their solubility and intracellular compartmentation. In plants, MTs are involved in the biosynthesis of a wide array of natural products related to the plant's survival. These include phenylpropanoids, lignins, flavonoids, alkaloids, and phytoalexins, to mention a few. In animals, on the other hand, MTs play an important role in the regulation of biological activity of both cholinergic and adrenergic neurotransmitters, as well as in the biosynthesis of several cellular components, such as phosphatidyl choline, various quinones, biological amines, and steroids. MTs are also involved in the biosynthesis of a variety of polyketide-derived aflatoxins and antibiotics that are produced by fungi and bacteria, respectively. Furthermore, the methylation of carbohydrates is well documented in bacteria by the ubiquitous accumulation of a number of methylated sugars, including 2-methylfucose as well as 2,4- and 3,4-dimethylrhamnose, all of which contribute to the stability of the mycoside fraction of several bacterial species. The natural occurrence of 4-methylglucuronic acid and 4-methylgalacturonic acid in the hemicellulose fraction of plant cell walls plays an important role in their plasticity, as well as in fruit ripening.

The fact that biological molecules are considered documents of evolutionary history,[1] has enticed biochemists and molecular biologists to search for the genes encoding various metabolic pathways. As a result, the basis for phylogenetic inference has now shifted from the level of cellular and organismal properties to that of molecular characteristics, *i.e.* molecular phylogeny. However, the phylogenetic relationships of MTs should also be discussed in conjunction with their biochemical properties, as to how and when their enzyme activities are likely to have evolved. This chapter reviews the most common AdoMet-dependent MTs that have been cloned in various organisms. It also describes the structural-functional relationships of their genes, and attempts to trace the evolution of this gene superfamily. The most interesting prospects will be the characterization of other genes encoding novel methyl transfer reactions, not only to increase the pool of genes for manipulation of plant secondary metabolism, but also to contribute to a better understanding of the phylogenetic relationships of these genes.

METHYLTRANSFERASES

The Methyl Donor

AdoMet is one of the most versatile compounds in the cell's biochemical inventory. It is formed from L-methionine and ATP by AdoMet synthase that is encoded by a highly conserved gene in plants. Little wonder that, except for ATP, no other molecule has been applied to more biological reactions than AdoMet. The fact that it participates in a myriad of transmethylations including those of nucleotides, proteins, lipids, and carbohydrates, as well as a variety of natural products, gives it a universal distribution in all organisms. Its role in regulatory functions and its specificity of assignments suggest that AdoMet appeared relatively early in biochemical evolution. In addition, the simplicity of the nonreactive, hydrophobic methyl group of AdoMet seems incongruous with the concomitant modification to the biochemical properties of the methylated metabolites and, hence, to the physiology of the organism. However, it is interesting to note that, whereas AdoMet is the methyl donor for the majority of enzymatic methylations, it receives that methyl group from N^5-methyl tetrahydrofolate by a cobalamine-independent methionine synthase.[2]

Conserved Domains

Although the X-ray structure of a plant MT is not yet available, the Mg^{2+} and AdoMet-binding domains and the active site have been identified from the recently resolved crystal structure of the rat liver catechol *O*-methyltransferase[3] (COMT, Fig. 1). These consist of a [xLExGxGxG] motif for AdoMet binding, and a [KGTVL] motif for Mg^{2+} binding. The fact that these domains are similar to those of the DNA-MT,[4] suggests that all AdoMet-dependent, functionally analogous enzymes have a common structure and have evolved from a common ancestor in spite of their different substrate specificities.

Whereas the consensus motif I of COMT (Fig. 1) seems to be common to almost all published MT sequences, albeit with some modification, attempts to designate additional AdoMet-binding conserved regions that are shared among all plant MTs have been controversial, either because of the limited number of plant MT sequences available at the time,[5] or that the proposed motifs are semi-conserved and contain a considerable number of mismatches.[6] In fact, three (B, C, and I, Fig. 1) of the four main motifs proposed by Joshi and Chiang[6] contain 10 mismatches in the 37 designated amino acid residues. Furthermore, the three (J, K, and L, Fig. 1) additional regions proposed by these authors are the same motifs (regions II, III, and V, Fig. 1) previously described as a signature for most, if not all, plant OMTs.[7] The latter motifs were deduced from the amino acid sequence alignment of some 30 plant OMTs with specificity for a variety of methyl acceptor molecules.[8] The consensus motifs[7] (Fig. 1) are mostly located within the third part of the OMT sequence distal to the carboxyl terminal. These motifs, which are rich in glycine (ca. 28%), are characterized by several positions where hydrophobic residues are bracketing 2 or 3 glycine residues as in regions I, II, and V, or the motif is entirely hydrophobic as in regions II and IV (Fig. 1). However, different protein sequences of the MT gene superfamily may harbor one or more of these regions, whereas others may contain the whole complement. Most often, the degree of sequence homology among different MTs is related to the degree of similarity/identity in the consensus motifs.

Vidgren Et al.[3]	XLEXGXGXG ***I***		KGTVL	
Kagan & Clarke[5]	VLDIGGGTG	PQFDAIFC	LLRPGGRLLI	
Joshi & Chiang[6]	LVDVGGGXG ***A***	VPXXDAXXMKWI ***B***	ALPXXGKVLIXXEXILP ***C***	LDRXLRLL ***I***
		IKGINFDLPHVI ***J***	PGVEHVGGDMF ***K***	GGKERTXXEFLA ***L***
Ibrahim[7]	LVDVGGGXG ***I***	GINFDLPHV ***II***	EHVGGDMF ***III***	GGKERT ***V***
			NGKVI ***IV***	

Figure 1. Conserved regions in methyltransferases including the AdoMet-binding motifs. X, any amino acid residue. Conserved regions are designated by bold letters or bold Roman numbers in italics.

STRUCTURAL-FUNCTIONAL FEATURES OF METHYLTRANSFERASES

MTs can be conveniently divided into four distinct families according to the nature of the target atom in the methyl acceptor molecule, and thus designated as O, C, N, and SMTs. Only those MTs whose genes have been cloned and characterized (Table 1) will be considered in this brief survey. Most MTs fall into two categories: one with a M_r of 38 to 43 kDa range that does not require Mg^{2+} for catalytic activity, and represents the majority of plant MTs; and the other, with a

Table 1. Representative members of the methyltransferase gene superfamily, their methyl acceptor molecules, and their functional significance.

Methyltransferase/ Methyl acceptor molecule	Source	Functional significance
I. *O*-Methyltransferases		
Phenylpropanoids		
Caffeic/5-hydroxyferulic	Several spp.	Lignification
Caffeoyl/5-hydroxyferuloyl CoA esters	Several spp.	Lignification
(Iso)eugenol	*Clarkia breweri*	Flower scent volatiles
Bergaptol*	*Petroselinum crispum*	Parsely phytoalexiin
Flavonoids		
Pinosylvin*	*Pinus sylvestris*	Stilbene phytoalexin
Isoliquiritigenin	*Medico sativa*	Nod-gene factor
Apigenin	*Hordeum vulgare*	Barley phytoalexin
Quercetin	*Arabidopsis thaliana*	Secondary metabolism
Trimethylquercetin	*Chrysosplenium americanum*	Me-flavonol synthesis
Daidzein	*M. sativa*	Alfalfa phytoalexin
6a-OH-maackiain	*Pisum sativa*	Pea phytoalexin
Alkaloids		
(Acetyl)serotonin	Human, bovine	Melatonin synthesis
Scoulerine	*Coptis japonicum*	Berberine biosynthesis
Norcoclaurine	*Thalictrum tuberosum*	Berberine biosynthesis
Polyketides		
Sterigmatocystin	*Aspergillus spp.*	Aflatoxin biosynthesis
Carminomycin, demethyl-puromycin, tetracenomycin	*Streptomyces spp.*	Antibiotic biosynthesis

Table 1. (cont.)

Carboxyl methylations		
Peptide-L-isoaspartyl	*A. thaliana*	Damaged protein repair
Benzoic acid*	*Antirrhinum majus*	Flower scent production
Salicylic acid	*C. breweri*	Flower scent production
II. *C*-Methyltransferases		
Cycloartenol	Several species	Steroid metabolism
γ-Tocopherol	*A. thaliana*	α-Tocopherol synthesis
Ubiquinone	*Saccharomyces cerevisiae*	Quinone biosynthesis
Uroporphyrinogen III	*A. thaliana*	Siroheme biosynthesis
III. *N*-Methyltransferases		
Putrescine	*Nicotiana tabacum*	Nicotine biosynthesis
RUBISCO-LSU	*N. tabacum, P. sativa*	Post-transl. Modification
Phosphatidylethanolamine	*S. cerevisiae*	Choline biosynthesis
IV. *S*-Methyltransferases		
Betaine/homocysteine	*A. thaliana*	Methionine biosynthesis
Thiopurine	Mouse	Antilukemic, immunosuppr.
Thioether	Mouse	Detoxification of xenobiotics
V. Halide methyltransferase		
Sodium chloride	*Batis maritima*	Halophyte adaptation

*Unpublished sequence

M_r of 23 to 27 kDa that requires Mg^{2+} for activity. Both types of enzymes can be found in plants, microorganisms, and animals.

O-Methyltransferases (OMTs)

These constitute the most important family of MTs with several genes cloned in various organisms (Table 1). Their genes encode the methylation of the hydroxyl groups of a wide range of catecholic compounds, including phenylpropanoids, flavonoids, alkaloids, and polyketides, but with a significant degree of substrate- and position specificities. In addition, some OMTs are known to catalyze the carboxyl methylation of some acceptor molecules. In all the structures cited in this article (**I** to **XXVII**), the methyl groups shown in bold represent those introduced by the corresponding methyltransferase.

Hydroxycinnamic acid OMT. The methylation of hydroxycinnamic acids (HCAs, **Ia**) is an important enzymatic reaction in the biosynthesis of lignin in all vascular plants. Several HCA-OMT cDNAs have been isolated and characterized from different sources. They encode both caffeic acid and 5-hydroxyferulic acid OMTs, with a methylation ratio of 2:1 to 3:1 depending on the source of the enzyme, to give rise to ferulic and caffeic acids, the precursors of the monolignols, coniferyl and sinapyl alcohols, respectively. It is believed that the methylation of both acids is catalyzed by one 'bifunctional/ bispecific' OMT, rather than two distinct forms of the enzyme. Furthermore, it is presumed that the marked differences observed in the ratios of methylation of both substrates by the recombinant proteins may be due to the bulky methoxyl group on 5-hydroxyferulic acid, whose methylation may not be favored to the same extent as that of caffeic acid. Such a property favors the concept of two distinct enzymes rather than one bifunctional protein. Recently, however, two HCA-OMT isoforms I and II were purified, for the first time, from cabbage leaves to near homogeneity.[9] They exhibited different product ratios (sinapic to ferulic: 1.7-1.8 for isoform I and 2.6-2.8 for isoform II) that were consistent during enzyme purification, thus indicating distinct substrate preferences.[9] Considering the fact that cabbage accumulates sinapoyl esters as the predominant phenylpropanoid compounds, it is reasonable to assume that the differential activity of both isoforms is indicative of two distinct enzymes, each of which is specific for one substrate, and can accept the other substrate with a lower affinity. The cloning and characterization of both isoforms and their expression in a heterologous system may help clarify the structure-function relationship of the presumed 'bispecific' HCA OMT. A parallel situation in *Petunia hybrida*[10] has demonstrated that the flavonoid 3′ and 3′/5′ hydroxylases are two distinct enzymes that are encoded by two genes assigned to the loci *Hf1* and *Hf2*.

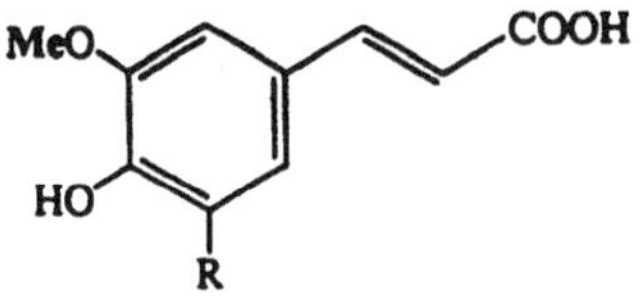

Ia

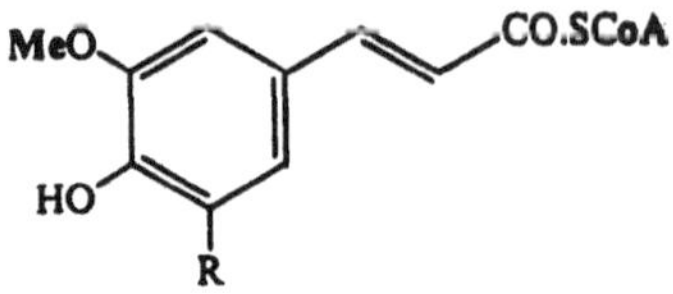

Ib

The first cloning of the HCA-OMT was reported from aspen,[11] which is considered the type-member of the group. This was followed by those of alfalfa,[12] poplar,[13] and several other species,[14-17] including gymnosperms.[18,19] These OMTs have similar primary structures (ca 365 aa, 40 kDa) and do not require Mg^{2+} for catalytic activity. Whereas many of these cDNA clones were identified on the basis of sequence similarities with other known lignin OMTs, most of the recombinant OMT preparations tested for enzyme activity exhibited a higher preference for 5-hydroxyferulic acid over caffeic acid, with a relative ratio of 3:1. These OMT sequences share an overall 80-95% similarity, and display almost all of the 5 conserved motifs, except for the TMV-induced tobacco OMT[20] (75%), the gymnosperm OMT[18] (60%), as well as a root-specific OMT involved in suberin biosynthesis,[21] and a jasmonate-induced[22] and a pathogen-induced[23] OMT from barley leaves, all of which share an overall 50% similarity with the type-member of the group.

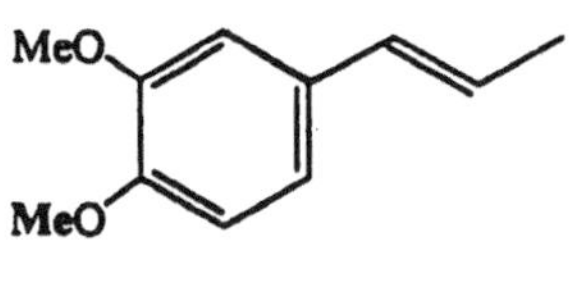

IIa

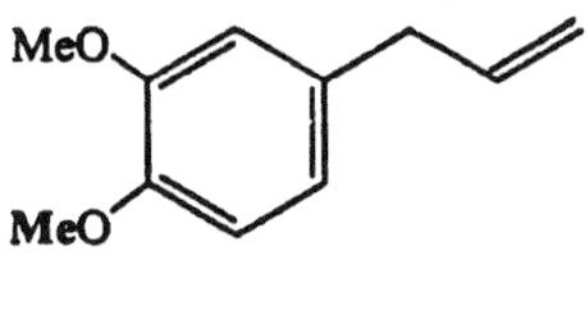

IIb

In contrast with the lignin cDNA clones that encode *meta*-specific methylations, that of *Clarkia breweri*[24] (368 aa, 40 kDa) methylates the phenylpropenes, eugenol (**IIa**) and isoeugenol (**IIb**) at the *para*-position to their 4-methyl derivatives, all of which are components of the floral scent. This OMT shares only a 65% amino acid sequence similarity with those involved in lignin methylation, possibly because of its regiospecificity for position 4, or due to the absence of a carboxyl group in these acceptor molecules, although it displays all of the five consensus motifs.

Hydroxycinnamoyl CoA OMTs. Another group of enzymes, the hydroxycinnamoyl CoA (HCCoA) OMTs, catalyzes the methylation of caffeoyl CoA and 5-hydroxyferuloyl CoA (**Ib**), but not of the free acids, to their feruloyl- and sinapoyl CoA esters, albeit with different ratios. The common occurrence of this methylation reaction seems to define an alternative pathway to that which utilizes the free acids for lignin biosynthesis. A number of HCCoA cDNA clones have been isolated from parsley,[25] *Zinnia elegans*,[26] and grape[27] cell cultures among several other species. The first reported cDNA clone[25] (241 aa, 27.1 kDa), the type-member of this group, was isolated from parsley (*Petroselinum crispum*) cells that were elicited with a fungal cell wall preparation. This clone encodes an OMT that catalyzes the specific methylation of only caffeoyl CoA ester, and exhibits no significant sequence homology to that of aspen OMT[11] (the only known sequence), nor does it harbor any of the conserved motifs common to other OMTs. It was proposed that the parsley caffeoyl CoA OMT plays a role in cell wall strengthening as part of the plant's response to elicitation or microbial attack.

On the other hand, the grape (*Vitis vinifera*) OMT cDNA[27] (242 aa, 27.2 kDa) encodes the methylation of both caffeoyl CoA and 5-hydroxyferuloyl CoA esters in a relative ratio of 3:1, the reverse ratio for those OMTs which utilize the free acids. Based on these activity ratios, it was proposed that the parsley OMT may be involved in the biosynthesis of guaiacyl residues, whereas that of grape is involved in the formation of both the guaiacyl and syringyl residues of lignin. The grape OMT shares 85-94% sequence identity with other members of the HCCoA OMT group, but it exhibits no significant homology to those which utilize the free acids.[27]

Speculation on the dominant role of HCCoA OMT in lignification was based on studies carried out with the *Zinnia* cDNA clone[26] (245 aa, 27.6 kDa). Tissue-print hybridization showed that expression of this gene is associated with lignification during xylogenesis as an alternative pathway in lignin biosynthesis. This OMT shares (% similarity/identity) 93/85 with parsley, 70/50 with members of this OMT group, and 45/21 with the rat liver COMT at the N-terminal residues 13-126, and 48/21 with the human COMT at the N-terminal residues 90-176, including the 9 amino acid residues (region I) involved in AdoMet binding. It appears, therefore, that the structural-functional properties of these three OMTs[25-27] define the general characteristics of the HCCoA-OMT group.

More recently, an OMT cDNA clone (381 aa, 42 kDa) isolated from the differentiating xylem of loblolly pine (*Pinus taeda*)[19] was reported to encode the methylation of both caffeic and 5-hydroxyferulic acids as well as their CoA esters to a similar extent, although the native xylem extracts exhibited higher caffeoyl CoA (1.7-fold) and 5-hydroxyferuloyl CoA (1.9-fold) activities than with the respective free acids. However, the fact that these data were obtained with crude tissue extracts, the insufficient evidence for the identity of the putative CoA products, and the low titration values for immunoprecipitation of the enzyme activity, all raise serious questions as to the "versatility" of this "novel" OMT. Whereas most characterized HCCoA OMTs[25-27] share 85-95% sequence identity with one another and exhibit no significant homology to those utilizing the free acids, that of loblolly pine shares only 15% identity with the former group and 60% similarity and 40% identity with the latter group.[19] In addition, this OMT habors all of the five conserved motifs characteristic of HCA OMTs that are usually absent in the enzymes[25-27] that utilize the CoA esters. These properties suggest that the versatile loblolly pine OMT may, in fact, be another member of the HCA-OMT group of enzymes which, surprisingly as a gymnosperm OMT, utilizes 5-hydroxyferulic acid and its CoA ester as substrates. However, until other gymnosperm OMTs are cloned and their gene products properly characterized, the versatility of this enzyme remains in question.

Flavonoid OMTs. A distinct group of OMT cDNAs encode the methylation of a variety of flavonoid compounds. These include one clone for a stilbene (unpublished), two for chalcones,[28,29] and one each for a flavone,[30] an isoflavone,[31] and a pterocarpan,[32] as well as three flavonol[33-35] cDNAs.

Stilbenes accumulate in several plants including gymnosperms where they are believed to function as phytoalexins.[36] They may be considered analogs of chalcones, since both groups of compounds are biosynthetically derived from the same precursors via their structurally related enzymes,[37] stilbene synthase and chalcone synthase. A cDNA clone, recently isolated from ozone-induced Scots pine needles (H. Chiron, pers. comm.), encodes a 364 aa, 40 kDa protein that catalyzes the methylation of pinosylvin (**III**), although it also accepts catechol, HCAs, and quercetin to a higher extent than pinosylvin. The stilbene OMT shares 30-50% sequence identity with most HCA OMTs, 40% identity with flavonol OMTs, and a 90% identity with the consensus motifs.

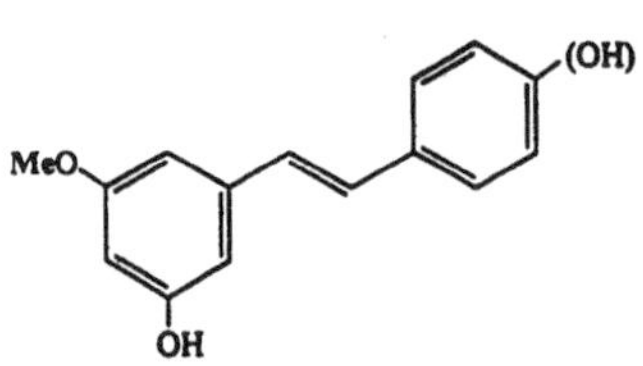

III

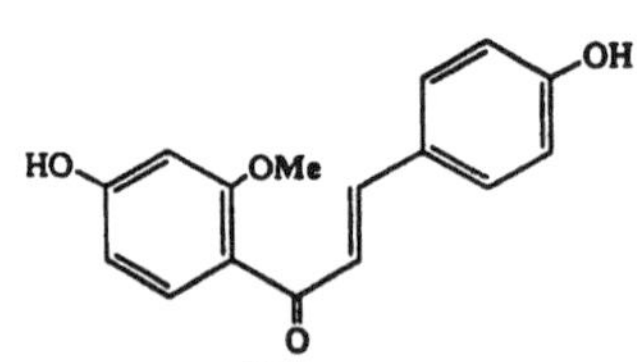

IV

The inducible chalcone cDNAs of alfalfa[28] and licorice[29,38] (375 aa, 41.2 kDa) encode OMTs for the 2′-methylation of the 5-deoxy chalcones isoliquiritigenin (**IV**) and licodione, respectively. Both enzymes exhibit 83% sequence similarity with one another, 50% with the *Chrysosplenium americanum* trimethylflavonol 3′/5′-OMT,[33] and 60% with the alfalfa HCA- OMT, as well as 80-85% identity with the consensus motifs.

The previously reported HCA-OMT cDNA clone expressed in barley in response to pathogen infection[23] has recently been characterized as a flavone 7-OMT,[30] with the highest affinity for apigenin (**V**) as the methyl acceptor molecule. This OMT (390 aa, 42.3 kDa) shares the highest (41%) overall sequence identity with the maize root-specific OMT involved in cell-wall suberin synthesis.[21] It occupies a distinct branch in a phylogenetic tree among the other HCA OMTs, since it exhibits only a 61% identity with the conserved motifs.

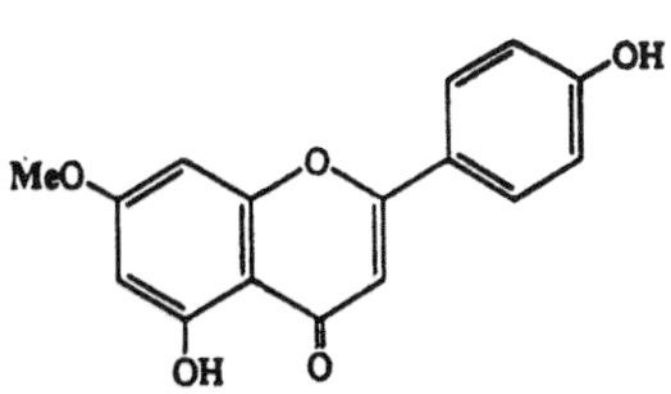

V

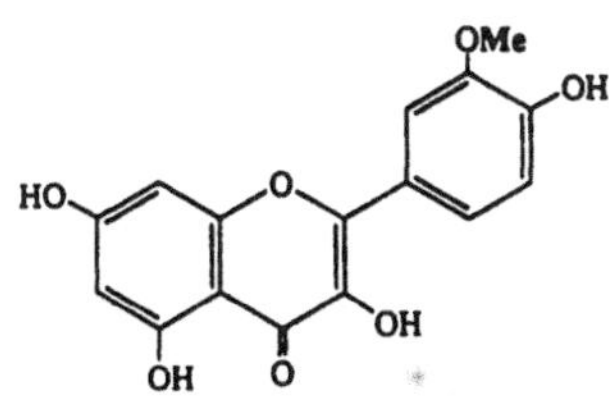

VI

Three, constitutively-expressed flavonol (**VI**) cDNA clones[33-35] share a high (ca 85%) amino acid sequence similarity and 95-100% identity with the consensus motifs, although their gene products exhibit different substrate preferences. One of these, isolated from *C. americanum*[33,39] (345 aa, 37.6 kDa), encodes the specific 3′/5′-methylation of 3,7,4′-trimethylquercetin and accepts neither the mono- or dimethyl derivatives nor the parent aglycone as substrates. The second cDNA clone (363 aa, 40.5 kDa), recently isolated from *Arabidopsis thaliana* by the yeast two-hybrid system and characterized as a HCA OMT based on high sequence similarity,[40] has now been identified as a quercetin 3′-OMT that does not accept HCAs as substrates.[34] In contrast, the third cDNA clone[35] (343 aa, 37.9 kDa) isolated from *C. americanum* encodes the 3′-methylation of luteolin and quercetin, as well as the 3/5-methylation of caffeic and 5-hydroxyferulic acids, respectively. However, the substrate preference of this OMT is not unexpected, since its sequence differs from that of HCA OMTs by only 32 invariant amino acid residues that are located in the regions proposed for substrate specificity (aa 20-120) and substrate binding (aa 120-200).[24,41] The variations in amino acid substitutions in both regions may well be responsible for the broad specificity of this OMT, as compared with the strict specificity of the former enzymes[33,34] for their substrates. In fact, the structural similarity of ring B and its contiguous 3-C side chain in flavonoids to that of phenylpropanoids may favor catalysis of both types of substrates, albeit to different degrees.

The stress-inducible isoflavone OMT cDNA clone[31] encodes a 352 aa, 39.6 kDa protein that mediates the 7-*O*-methylation of daidzein (**VII**, a 5-deoxyisoflavone), although it also accepts 6,7,4′-trihydroxyisoflavone (136% that of daidzein) and 6a-hydroxymaakiain (49%). This OMT shares the highest (50%) sequence identity with the pterocarpan 3-OMT of pea,[32] 40% with the maize suberin OMT,[21] 37% with the barley flavone 7-OMT,[29] and 21% with the *C. americanum* flavonol 3′/5′-OMT,[33] although it exhibits a fairly high (66%) identity with the consensus motifs.[7]

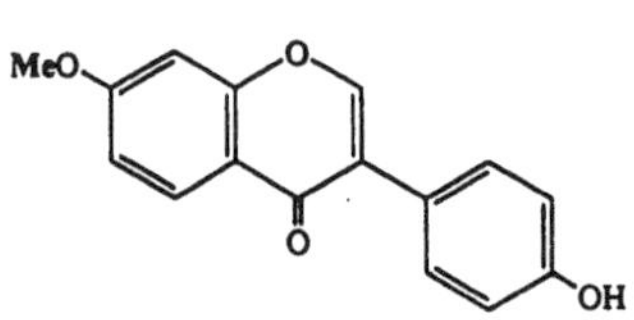

VII

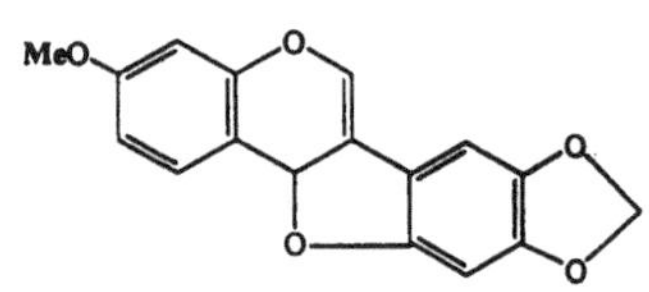

VIII

The pterocarpan cDNA[32] encodes a 360 aa, 40.4 kDa protein for the 3-*O*-methylation of 6a-hydroxymaackiain (**VIII**), the terminal step in the biosynthesis of the pea phytoalexin pisatin. This fungal-induced OMT shares a higher (50%) sequence identity with that of alfalfa than with the maize (40.6%), barley (39.1%), or other HCA (32%) OMTs, although it displays a significantly high (70%) identity with the AdoMet conserved motifs.[7]

Amino acid sequence alignment and visual comparison of the nine available flavonoid OMTs[28-35] indicate a significant degree of similarity (data not shown). In fact, these OMT sequences, which vary in length between 345 and 390 amino acids, share among themselves 39 identical residues, as well as 21 others with one mismatch and 19 with two mismatches. Otherwise, the variations among their sequences are mostly located in the regions proposed for substrate specificity and substrate binding, and seem to determine their specificity for the different flavonoid acceptor molecules. However, in spite of the differences in the structure of these flavonoid OMTs, the alignment of their conserved regions I-V[7] indicates a high degree of identity, especially among the flavonol-, stilbene- and chalcone OMTs (Fig. 2).

	I	II	III	IV	V	Identity with consensus* (%)
Chryso-Flv3´/5´	LVDVGGGTG	INFDLPHVI	EHVGGDMFVSVPKG	GKVIL	GGKERTEKE	100
Chryso-Flv3´/HCA	LVDVGGGTG	INFDLPHVI	EHVGGDMFVSVPKG	GKVIV	GGKERTEKE	100
Arabid-Flv3´	LVDVGGGIG	INFDLPHVI	EHVGGDMFVSVPKG	GKVIL	GGKERTEKE	95
Pinus-Stilbene	LVDVGGGVG	INFDMPHVV	KHVGGDMFDSVPSG	GKVIV	GGKERTEQE	90
Alfalfa-Chalc	LVDVGGGSG	INFDLPGVI	EHVGGDMFASVPQG	GKVII	GGRERTEQY	85
Glycyrrhiza-Chalc	LVDVGGGNG	INFDLPQVI	ELVGGDMFASVPQG	GKVIV	GGRERTQKQ	80
Pisum-Pterocarpan	LVDVAGGTG	TVFDQPQVV	NFVGGDMFKSVPSA	GKVII	LGKERTKKE	70
Alfalfa-Isoflav	IVDVGGGTG	IVFDRPQVV	TYVGGDMFTSIPNA	GKVII	NGKERNEEE	66
Hordeum-Flavone	LVDVAGGNG	SVLDLPQVI	EFVAGDMMEFVPPA	GKVII	NGKVREEQN	61
Consensus	LVDVGGGXG	INFDLPHVI	EHVGGDMFXSVPXG	GKVII	GGKERTEKE	
		Q V	A	L/V	Q	

Figure 2. Alignment of the conserved regions in flavonoid OMT sequences. Invariant amino acid residues are highlighted. X, represents any amino acid. *Taking into consideration the variant substitutions.

Alkaloid OMTs. The cDNA clones known for alkaloid methylation include several hydroxyindole OMTs from various organisms and two plant benzylisoquinoline alkaloid OMTs. The bovine cDNA[42] encodes a 350 aa, 77 kDa dimeric OMT for the methylation of serotonin (5-hydroxytryptamine, **IX**), as well as its *N*-acetyl derivative, to melatonin. This reaction is localized in the pineal gland of most animals, including birds and fish, and plays an important role in biological rhythms. Although plants accumulate a number of *O*-methylated arylamines, none of the genes encoding the enzymes catalyzing these methylations has yet been cloned. A prospective candidate would be *Chenopodium rubrum*, an obligate short-day plant, which was recently reported to contain melatonin, the agent responsible for regulating the photoperiodic behaviour of flowering in this plant.[43]

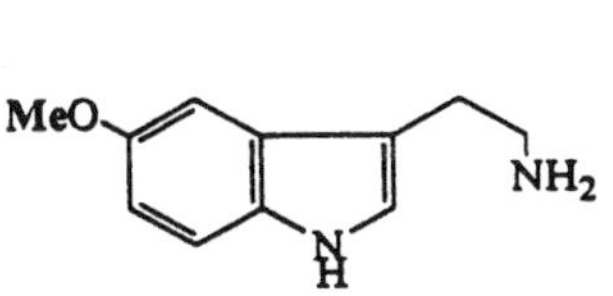

IX

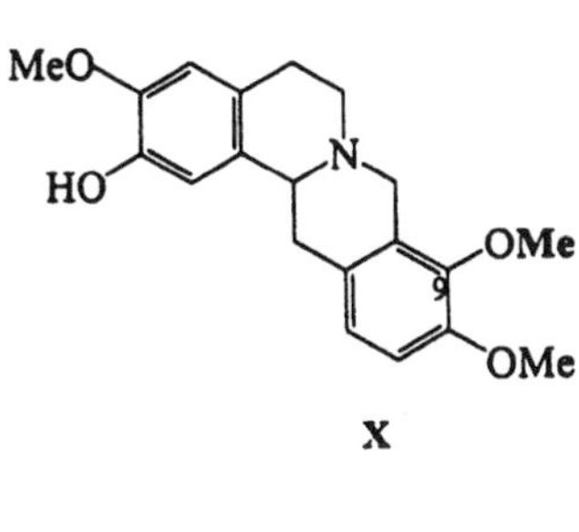

X

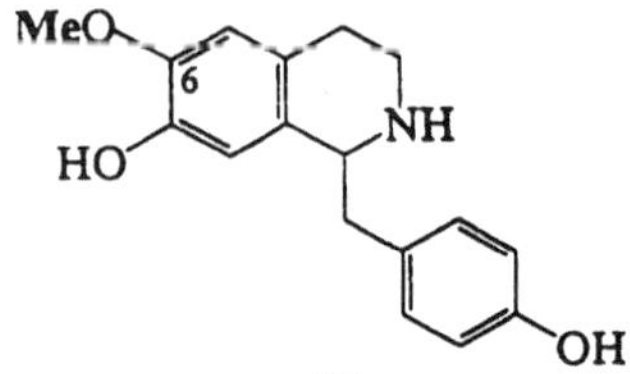

XI

The two benzylisoquinoline alkaloid OMT cDNA clones were isolated from *Coptis japonica*[44] and *Thalictrum tuberosum*[45] cell suspension cultures. The former clone[44] (381 aa, 38.4 kDa), which was isolated by a reverse-genetics technique, encodes the 9-*O*-methylation of scoulerine (**X**) to tetrahydrocolumbamine. The low M_r of this enzyme was attributed to a premature protein that is further processed post-translationally. This OMT shares a 55.6% similarity and 46.4% identity with the aspen HCA-OMT, as well as 80% identity with the consensus motifs.

The *Thalictrum* cDNA clone[45] (365 aa, 74 kDa dimer) was isolated as a member of a small gene family from a methyl jasmonate-elicited cell suspension culture, by using primer sequences based on conserved amino acid residues in lignin OMTs.[13] This cDNA clone encodes the 6-*O*-methylation of norcoclaurine to coclaurine (**XI**). Both coclaurine and tetrahydrocolumbamine are intermediates in the biosynthesis of the antimicrobial alkaloid, berberine.[46] The norcoclaurine OMT exhibits a broad substrate specificity, as it also methylates both caffeic acid and catechol to a higher extent than norcoclaurine, and was described as a 'multifunctional' OMT involved in secondary metabolite synthesis.[45] It is not surprising, therefore, that this OMT shares a significantly high (73.6%) identity with that of aspen HCA-OMT,[11] and 95% identity with the conserved motifs. However, it is interesting to note that whereas the scoulerine and norcoclaurine OMTs catalyze the methylation of two 'paralogous' hydroxyl groups on the benzylisoquinoline ring system, they only share 53.5% sequence similarity (44.8% identity). In fact, their sequences differ by 123 invariant amino acid residues, eight of which are located within the highly conserved regions of 46 amino acids.[44,45] The other invariant amino acids are mostly situated in the regions proposed for substrate specificity and substrate binding. Therefore, the cDNA cloning of other alkaloid OMTs (*e.g.* those of papaverine or reticuline) should allow for a better understanding of the structure-function relationship among the benzylisoquinoline alkaloid OMTs.

Carboxyl OMTs. Several enzymes catalyze the carboxyl methylation of a variety of substrates, including polypeptides, phenolic acids, and polyketides. α-

Carboxyl methylation of D-aspartyl/L-isoaspartyl residues of damaged proteins has been reported in plants, humans, and bacteria, and is considered an important enzymatic reaction in limiting protein deterioration. An *Arabidopsis thaliana* cDNA clone[47] (230 aa, 25.7 kDA) encodes a seed-specific, cytosolic protein that responds to abscisic acid treatment, but not to salt or drought stresses. This OMT shares a higher sequence similarity with the hydroxyindole-OMT and the RNA/DNA-MTs than with any known plant OMT.

Carboxyl methylation of naturally occurring aliphatic and aromatic acids is a common enzymatic reaction in plant secondary metabolism. The carboxymethylated products, such as those of acetic, propionic, butyric, benzoic, salicylic, and cinnamic acids, are volatile compounds that contribute to flower scent and fruit flavor. On the other hand, that of jasmonic acid acts as a signal molecule in a variety of plant interactions. Two cDNA clones encoding the carboxyl methylations of benzoic[48] and salicylic[49] acids have recently been isolated and characterized. They are involved in floral scent production in *Antirrhinum* and *Clarkia* species, respectively. Both cDNA clones may represent a new class of plant OMTs, since they exhibit no significant sequence similarity or conserved motif identity with any of the other plant OMTs. However, the salicylic acid carboxyl OMT[49] shares 20-40% identity with a few *A. thaliana* genes of unknown function, and whose sequences have recently been determined in the large-scale sequencing projects.

Myo-Inositol OMT. An unusual OMT cDNA clone[50], isolated from the facultative halophyte *Mesembryanthemum crystallinum*, encodes a 365 aa, 40.3 kDa protein that catalyzes the methylation of *myo*-inositol to ononitol, an intermediate in pinitol biosynthesis that plays an important role in osmotic stress tolerance. This OMT shares 50% overall sequence similarity with both the bovine pineal gland hydroxyindole[42]- and aspen lignin[11]-OMTs, as well as 70% identity with the conserved motifs. Its low sequence homology to other plant OMTs may be attributed to the alcoholic nature of the acceptor-molecule hydroxyl groups, as compared with the phenoxide ion of other catecholic substrates.

Polyketide OMTs. Polyketides represent an important family of metabolites produced mostly by bacteria (as antibiotics), fungi (as aflatoxins), and, to a lesser extent, by plants (as flavoring and pigment agents). In bacteria and some fungi, the genes encoding the biosynthesis of these compounds are usually clustered and can be cloned in DNA fragments and expressed as multifunctional polypeptides. Three cDNAs from *Streptomyces spp.* encode OMTs for the methylation of antibiotics, and one from *Aspergillus parasiticus* involved in aflatoxin biosynthesis represent the structural characteristics of this group of OMTs.

Except for *S. glaucescens* cDNA,[51] which encodes a 270-amino acid, 30-kDa protein, both the *S. alboniger*[52] and *S. peucetius*[53] gene products consist of 367 aa (40.3 kDa) and 356 aa (41 kDa), respectively; the latter being a homo-tetramer as the native protein. These bacterial OMTs catalyze the C-9 carboxyl methylation of the anthracycline-type antibiotic tetracenomycin tcmE to tcmA2 (**XII**), and of *O*-demethylpuromycin to puromycin,[52] as well as the 4-*O*-methylation of carminomycin to daunorubicin;[53] the latter is a widely used agent for chemotherapy. These proteins do not exhibit any significant homology to other plant OMTs, but they share the highest (25-35%) sequence identity with one another, as well as with the bovine hydroxyindole OMT and a bacterial hydroxyneurosporine OMT involved in carotenoid biosynthesis. These homologies can be traced to the 60% to 70% sequence identity they share with one or more of the conserved regions I, II, and V.[7]

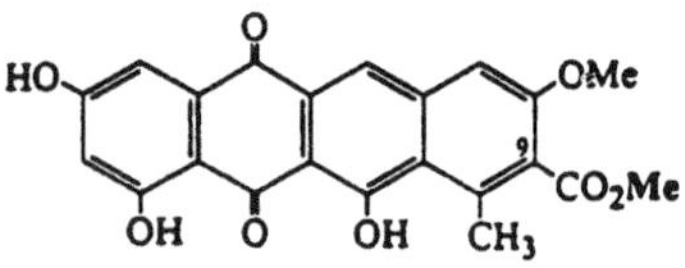

XII

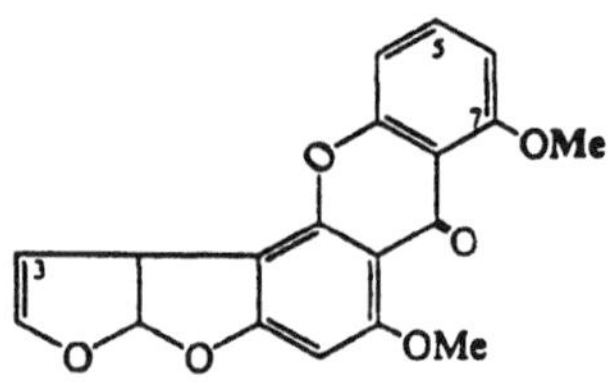

XIII

Aflatoxins (AF) are produced by the filamentous fungi *A. parasiticus* and *A. flavus* when they infect corn, cotton, peanuts and nut trees. These compounds are both toxic and carcinogenic to animals and humans. A cDNA clone encoding an OMT for the 7-*O*-methylation of sterigmatocystin (**XIII**) and its 2,3-dihydro derivative was isolated from *A. parasiticus*.[54] This 377 aa, 42 kDa protein (a homo-tetramer) shares 97% sequence identity with that of *A. flavus*,[55] and exhibits 66% identity with the three conserved regions I, II, and V.

C-Methyltransferases (CMTs)

These constitute a small family of heterogeneous MTs that catalyze the methylation of structurally unrelated molecules, such as steroids, ubiquinone/menaquinone, tocopherol, and uroporphyrinogen III. These enzyme proteins are mostly localized in the endoplasmic reticulum membranes of plant cells, and do not share significant sequence similarities with one another.

Sterol CMT. Three cDNA clones have been characterized from *Arabidopsis*[56] (361 aa, 41.2 kDa), soybean[57] (367 aa, 41.5 kDa), and *Zea mays*[58] (345 aa, 40 kDa). They encode a C-24 OMT for the conversion of cycloartenol (**XIV**) to phytosterols (*e.g.* stigmasterol), which function as structural components of cell membranes, and are required for cell proliferation and cell growth. They also serve as precursors of the recently discovered plant growth regulators, brassinosteroids. Whereas the *Arabidopsis*[56] and soybean[57] sterol CMT sequences are highly divergent, with only 46% similarity, that of corn[58] shares 66% similarity and 46% identity with the yeast sterol CMT,[59] and 75% and 37% identity with those of soybean and *Arabidopsis*, respectively. Hydropathy profiles of the four sequences indicate the presence of a 25-amino acid leader peptide in the *Arabidopsis*, but not in the corn, soybean, or yeast sterol CMTs.[56] Except for the soybean protein that displays the conserved motifs of regions I and V, those of *Arabidopsis* and corn exhibit only region I motif. In fact, alignment of the four sterol CMT sequences reveals two additional, highly conserved motifs; one consists of 11 amino acids starting at Phe[65] of the corn sequence (FYEYGWGESFHFA) and the other, of 8 amino acids beginning at Glu[179] (EATCHAPD); both motifs are proposed as 'finger-prints' for sterol CMTs.[58]

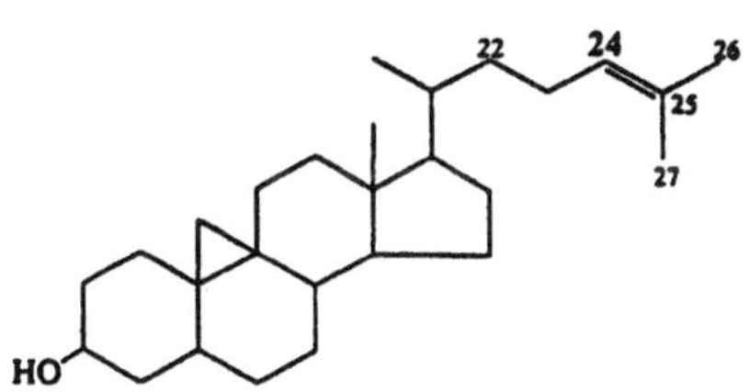

XIV

Tocopherol CMT. Tocopherols, collectively known as vitamin E, are classified into α-, β-, γ- and δ-isomers, according to the number and position of methyl substituents on the phenolic ring moiety. α-Tocopherol is an important vitamin to human health as an antioxidant, as a decreased-risk factor of cardiovascular diseases and of some cancers, as well as in improving immune function.[60] A cDNA encoding γ-tocopherol CMT (348 aa, 33 kDa) was isolated from *A. thaliana*,[61] by using a heterologous probe prepared from the photosynthetic bacterium *Synechocystis* PCC6803 by a genomics-based approach. This gene product catalyzes the C-5 methylation of γ- to α-tocopherol (**XV**), and results in a 9-fold increase in the seed-oil α-tocopherol content of transgenic *Arabidopsis* over the wild-type controls.[61] This cDNA shares a high sequence similarity with the sterol CMTs, including two of the conserved regions common to plant MTs, as well as a 47-amino acid transit peptide, consistent with the site of tocopherol synthesis.

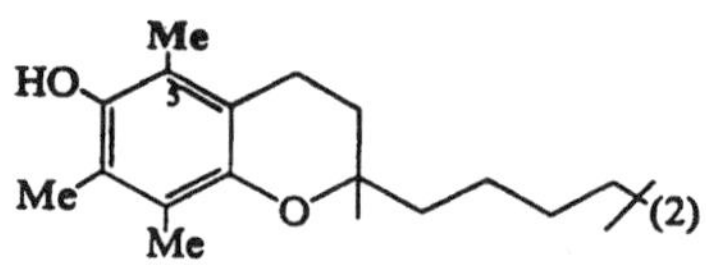

XV

Ubiquinone CMT. Ubiquinone (coenzyme Q) is a lipophilic metabolite that functions in the electron transport chain in the plasma membranes of prokaryotes, and the inner mitochondrial membranes of eukaryotes, apart from its roles as an antioxidant and in the regeneration of tocopherols. A nuclear clone was isolated from a yeast genomic library, based on its ability to restore growth of a representative mutant. Its gene product[62] (307 aa, 34.7 kDa) catalyzes the C-5 methylation of 2-methoxy-6-polyprenyl-1,4-benzoquinone to CoQ5 (**XVI**). This CMT shares 44% sequence identity over 262 amino acids with the *E. coli* UbiE analog, and exhibits two partially modified AdoMet-binding motifs.

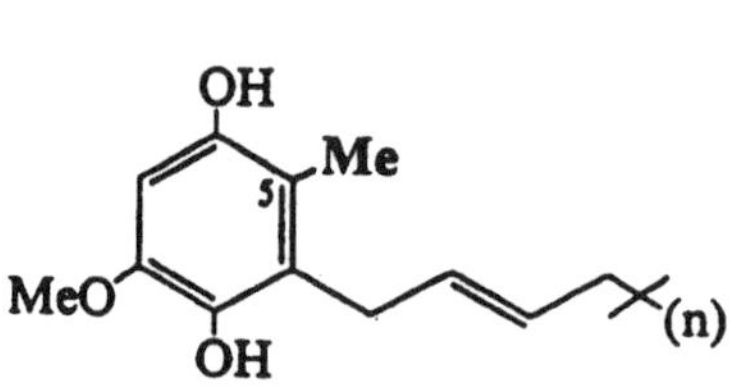

XVI

Uroporphyrinogen III CMT. Siroheme acts as the prosthetic group for both nitrite reductase and sulfite reductase. Its biosynthetic pathway involves a *C*-methylation of the tetrapyrrole uroporphyrinogen (UP)III, that is not common to either the heme or chlorophyll pathways. A cDNA cloned from *A. thaliana*,[63] that functionally complements an *E. coli cysG* mutant, encodes a 369 amino acid, 39.9 kDa protein that contains an *N*-terminal transit peptide for targeting the mitochondria or chloroplasts. This CMT shares significant homology with the yeast and bacterial analogs, and exhibits two highly conserved domains common to UPIII-CMTs that are located between amino acids 120-134 and 195-228.

N-Methyltransferases (NMTs)

In spite of the common occurrence in plants of *N*-methylated amines and *N*-methylated alkaloids, only a few NMT cDNAs have been isolated and characterized. These include three cDNAs involved in the methylation of putrescine,[64] of the lysyl residues of ribulose 1,5-bisphosphate carboxylase/oxygenase (RUBISCO) large subunit,[65] and of phosphatidylethanolamine in yeast.[66]

Putrescine NMT. The putrescine (**XVII**) NMT cDNA[64] (375 aa, 41 kDa), which was isolated from tobacco cell culture by substractive hybridization, catalyzes the first committed step in nicotine biosynthesis, and its down regulation increases the levels of polyamines at the expense of nicotine. This protein exhibits a much higher sequence similarity to the mammalian spermidine synthase than to the *E. coli* analog, and has no significant homology to plant MT sequences, except for region I of the AdoMet binding motif.

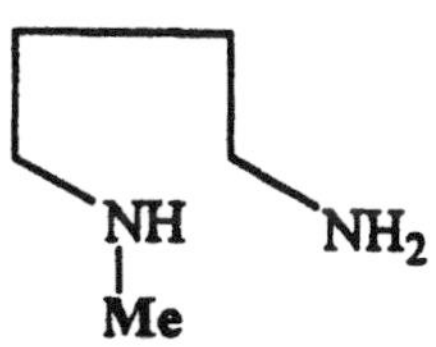

XVII

Protein-lysine NMT. Protein (lysine) NMT, also known as protein methylase III, represents a group of enzymes that catalyzes the methylation of ε-amino groups of lysine as a post-translation reaction, resulting in the formation of trimethyllysyl residues. This reaction is common to many, but not all, plant and animal species with no apparent specific function, although it appears to be species-related. The RUBISCO-NMT cDNA of tobacco[65] (491 aa, 56 kDa) exhibits a 75.3% overall sequence similarity to that of pea, although neither of these proteins displays any of the conserved motifs common to AdoMet-dependent MTs.[7]

Phospholipid NMT. Phosphatidyl choline (PC, phosphatidyltrimethylethanolamine, **XVIII**) is an important component of the cell membrane structure and function. Its biosynthesis is catalyzed by the stepwise *N*-methylation of phosphatidylethanolamine (PE) to its mono- and dimethyl derivatives as intermediates; a pathway that has been extensively studied in mammalian tissues, *Neurospora crassa*, and *S. cerevisiae*, among others. This pathway has been the subject of much dispute as to how many NMTs are involved in the conversion of PE to PC, and what are their substrate specificities. Two PE cDNA clones were isolated by genetic complementation using yeast mutants.[66] One clone (869 aa, 101.2 kDa), designated as PE-MT, encodes the methylation of PE to PMME, the monomethyl derivative; the other, as phospholipid (PL)-MT, encodes a protein (206 aa, 23.2 kDa) that catalyzes the synthesis of PC from PE, although both the monomethyl and dimethyl intermediates (PMME and PDME) are better substrates than PE. The homology between the PE-MT and PL-MT sequences suggests that gene duplication occurred during evolution of PL-MT, and that the two enzymes may have been derived from a common ancester.[66] Two partially conserved (50% similarity) motifs of regions I and V can be discerned in the PE-MT sequence, but none in that of PL-MT.

S-Methyltransferases (SMTs)

Besides the N^5-methyltetrahydrofolate-dependent methionine synthase (homocysteine SMT) known in plants,[2] three cDNAs that are involved in the methylation of S-containing compounds have been isolated from mammalian tissues; none of which exhibits any significant similarity to plant MTs. A betaine-homocysteine cDNA[67] encodes a bifunctional protein (406 aa, 45 kDa) that converts betaine to dimethylglycine, and homocysteine to methionine, and exhibits some homology to the bacterial methionine synthase.

A murine thiopurine-MT cDNA[68] encodes a 240 aa, 27.5 kDa protein that mediates the methylation of aromatic and heterocyclic compounds. Thiopurines and related mercapto compounds are important antileukemic and immunosuppressive agents.[68] This SMT, which has a high sequence similarity to its human analog, exhibits none of the conserved motifs common to AdoMet-dependent enzymes.

Another related enzyme, the murine thioether-SMT[69] (264 aa, 29 kDa), methylates sulfur as well as selenium and tellurium atoms of aliphatic thiols (*e.g.* dimethyl sulfide) to their trimethylsulfonium derivatives and, thus, plays an important role in the detoxification and elimination of xenobiotics. This SMT, which displays three of the five conserved motifs,[7] shares a high sequence identity with two mammalian cDNAs encoding the *N*-methylation of nicotinamide and phenylethylamine, but shows low homology with the thiopurine-MT isolated from the same organism.

Very recently, a cDNA clone (338 aa, 36.7 kDa) encoding the S-methylmethionine (SMM)-dependent methylation of selenocysteine has been isolated from *Astragalus bisulcatus* (Fabaceae), a Se-tolerant species which accumulates organoselenium compounds.[70] Expression of the selenocysteine MT in *E. coli* and functional analysis of the recombinant protein indicate that it increases Se tolerance and strongly reduces its unspecific incorporation into proteins, provided that SMM is present in the medium. However, none of these SMTs shares any significant sequence similarity to any of the known plant MTs.

Halide Methyltransferase

A putative halide/bisulfide MT was recently purified to near homogeneity from *Brassica oleracea* and partially sequenced.[71] However, based on the low rates of methyl chloride synthesis and the high K_m value for Cl^-, it was postulated that this enzyme functions in sulfur, rather than halide, metabolism. Nevertheless, the cloning of such a protein and the determination of its substrate specificity would improve our understanding of the biosynthesis of S-containing compounds in plants, especially both aliphatic and aromatic glucosinolates.

The production of methyl halides, found in the upper atmosphere, by several organisms has an important implication on the integrity of the stratospheric ozone layer. A cDNA (230 aa, 22.5 kDa), encoding a chloride MT, has recently been characterized from the halophyte *Batis maritima* (Bataceae).[72] The physiological role of this enzyme is believed to dispose of excess chloride ions by their elimination as a volatile methyl chloride. This cDNA exhibits no significant sequence similarity to other known MTs, except for the presence of three partially conserved regions that are common to the AdoMet-dependent enzymes.[7]

EVOLUTIONARY AND PHYLOGENETIC RELATIONSHIPS OF METHYLTRANSFERASES

In spite of the progress made in gene cloning, our state of knowledge of the phylogenetic relationships among MTs involved in secondary metabolite synthesis remains fragmentary. This is mainly due to the scarcity of cloning of their genes, except for those involved in lignin biosynthesis that have been cloned in several species. In contrast with the type II DNA-MTs, where 11 genes encoding the methylation of adenine (type IIA) and 12 genes for cytosine (type IIC) were already known by 1989,[4] it was only in 1991 that the first plant genes were reported to encode enzymes for the methylation of lignin precursors.[11,12,25] To date, there are more than 100 known AdoMet-dependent MT cDNAs in the databank encoding O-, C-, N-, and S-methyl transfers to a variety of metabolites in plants (70), fungi (9), bacteria (12), several animal species (11), and human (9). Although these MTs are functionally similar enzymes they are structurally different proteins, albeit for members of the OMT family that share a high amino acid sequence similarity/identity, as well as a high degree of conservation in the five motifs common to AdoMet-dependent MTs.[7]

The fact that the C-, N-, and SMT sequences cannot be confidently aligned, either with one another or with any of the OMT sequences, clearly indicates that they are probably no more related to one another than either is to the OMT family, except for one or two poorly conserved motifs of regions I, II, or V. Therefore, a close phylogenetic relationship among these three MT families is highly unlikely, and they may have evolved from unrelated ancestral genes. According to one hypothesis on enzyme evolution,[73] these MTs may have arisen by a selective, piece-by-piece assembly of pre-existing polypeptide subunits, where multiple amino acid substitutions had led to the acquisition of new enzyme functions (Fig. 3). These piece-meal assemblies, derived from functional units, must have retained some conserved regions, at least the AdoMet-binding motif (region I, Fig. 1) for their catalytic activity.

Recent views on the phylogeny of land plants recognize that lignins may have first emerged with the appearance of pteridophytes.[74,75] This biopolymer, which is found in all vascular plants, performs the dual function as a mechanical support of the vascular elements and in the transport of water and solutes. It can, therefore, be assumed that vascular plant lignin OMTs were among the earliest to evolve, with the HCCoA OMT that is likely responsible for the synthesis of the guaiacyl-type lignin as the ancestral gene.[26] Through duplication and evolutionary divergence, this gene may have given rise to the HCA OMT that is involved in

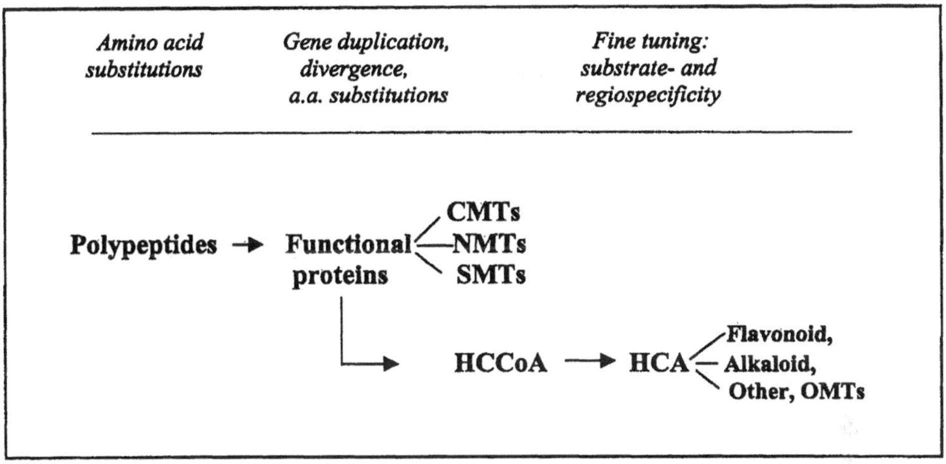

Figure 3. Proposed origin of methyltransferases.

the synthesis of both the guaiacyl- and syringyl-type lignins, as well as other genes encoding the methylation of various secondary metabolites, especially flavonoids that occur in some bryophytes, ferns, and gymnosperms. These OMTs must have acquired both the preference and regiospecificity for their methyl acceptors through a process of fine-tuning arrived at by several, selectively important amino acid substitutions (Fig. 3).

The predictions of this interesting hypothesis can be tested by using a phylogenetic analysis of the more ubiquitous OMTs. The HCCoA OMT should appear more ancestrally on the phylogenetic tree if it were the precursor of the HCA OMT. In addition, the flavonoid OMTs should be more recent innovations, and would be expected to cluster together to the exclusion of HCCoA OMT. The availability of several HCA-, HCCoA-, and flavonoid-OMT cDNAs enabled the alignment of their sequences, with the aim of detecting the phylogenetic relationships among these genes. Analysis of a selected number of these sequences, by using the PILEUP[76] of the GCG program and PROTPARS of the PHYLIP[77] program, resulted in a phylogram (Fig. 4) with three distinct groups I, II, and III that include a number of smaller clusters, which are supported by high bootstrap frequencies, except for the TMV-induced tobacco OMT.[20] These groups reflect the main differences in their OMT structure, molecular size, requirement for Mg^{2+}, and their substrate specificities.

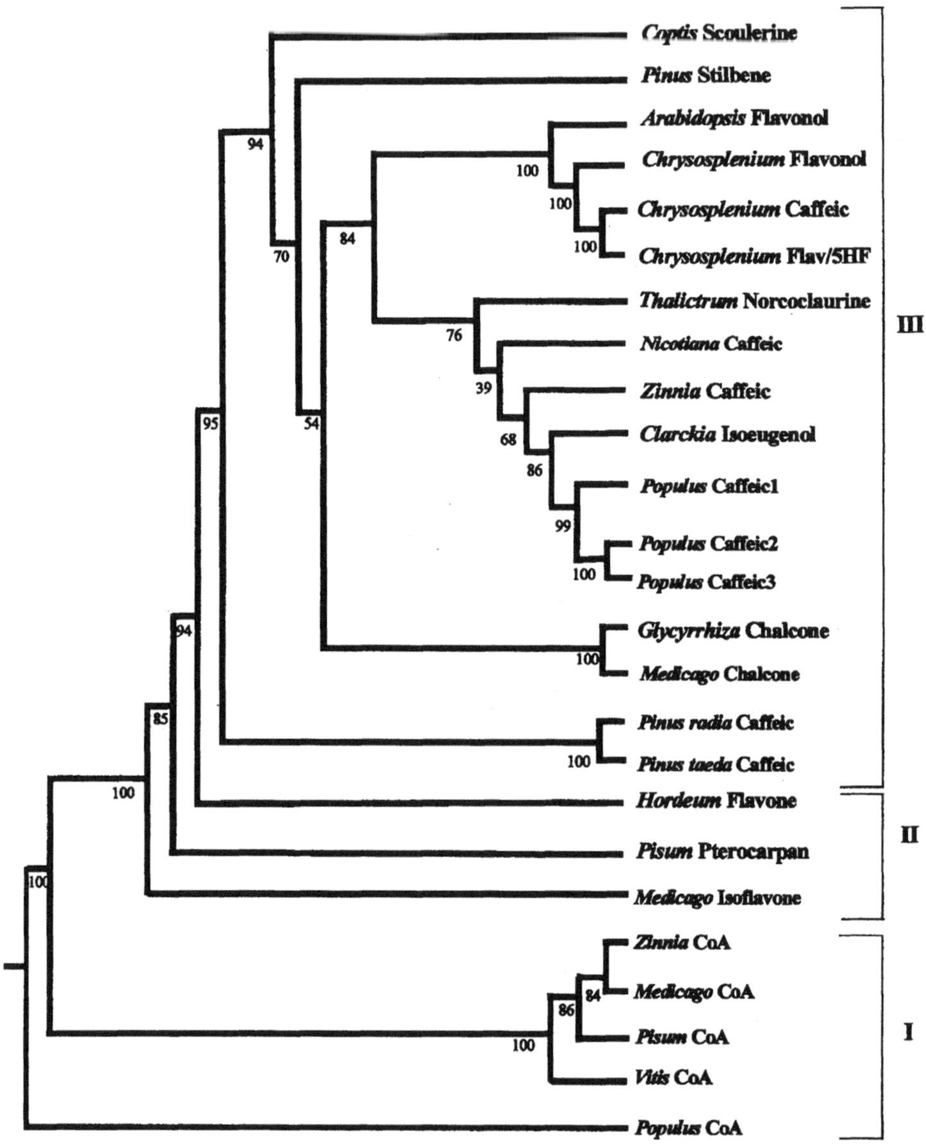

Figure 4. Phylogram of some OMT sequences representing the different methyl acceptor molecules, constructed using PILEUP[76] of the GCG program and PROTPARS of the PHYLIP[77] program. Bootstrap values are shown at the nodes.

One group (Group I, Fig. 4) that represents the HCCoA OMTs is phylogenetically the most distant from the two other groups, with the poplar OMT gene situated as an outgroup for the entire unrooted tree (Fig. 4). Evidently the HCCoA OMT group, which exhibits a high degree of sequence divergence, does not satisfy the requirements of the OMT gene family. On the other hand, both the HCA- and flavonol OMTs (Group III) are more closely related to one another, and appear in two distinct monophyletic clusters. However, the stilbene OMT and the other flavonoid (flavone, pterocarpan, and isoflavone) OMTs (in Group III) are situated peripherally to, and distantly from, these two clusters although they share with them a high level of conservation (Fig. 2). In addition, the *Thalictrum* (norcoclaurine) OMT seems to be more closely related to the HCA OMTs than to its paralog, *Coptis* (scoulerine) OMT, both being separated from one another by a significant distance (Fig. 4). Their location on the phylogram reflects a high divergence between the two genes, with a difference of 123 amino acid residues, and sharing only 40% sequence identity with one another (see Alkaloids OMT section). This corroborates with the preference of the norcoclaurine OMT for a variety of dihydroxyphenolic substrates rather than for benzylisoquinolines. Other clusters exhibit significant taxonomic relationships in addition to significant sequence similarities among their members, as evidenced by the association of *Chrysosplenium*, poplar, and gymnosperm-HCA OMTs and the legume-chalcone OMTs. However, it is interesting to note that whereas the flavone, pterocarpan, and isoflavone OMTs (Group II Fig. 4) appear to be phylogenetically related, their sequences are significantly distant from those in the flavonol cluster, in spite of the high level of conservation of their motifs (Fig. 2) and the structural similarity of their flavonoid acceptor molecules.

Similar results were obtained from a phylogram of the 9 available flavonoid OMT sequences (Fig. 5) that shows one distinct monophyletic cluster for the flavonol and stilbene OMTs, another that includes the isoflavone, pterocarpan and flavone OMTs, and a minor cluster for the legume-chalcone OMTs. Again, the two latter clusters are separated from one another by a significant distance from the flavonol-stilbene cluster. The neighbour-joining phylogram (not shown) also resulted in the same order of branching, indicating that the OMTs included in each of these clusters may have evolved as distinct groups. However, it will not be possible to draw any more inference to these preliminary findings unless more sequences become available for each of these flavonoid OMT groups. Nevertheless, the information generated from these analyses clearly indicates that plant OMTs have evolved from a common ancestral gene by duplication and divergence, with a remarkable level of conservation of the consensus motifs.

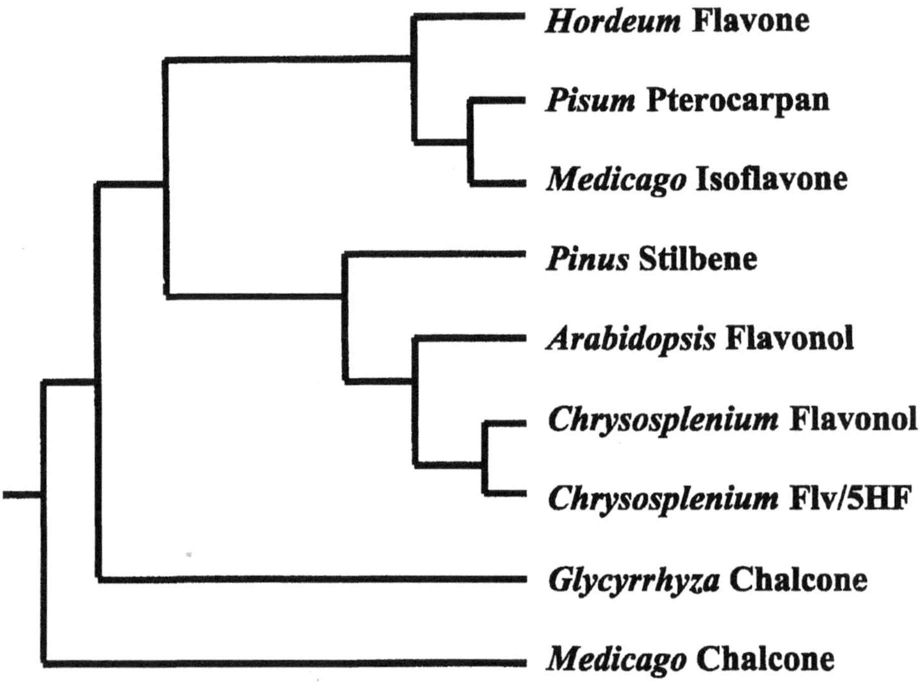

Figure 5. Phylogram of the 9 flavonoid OMT sequences, constructed as described in Figure 4.

CONCLUSIONS AND PERSPECTIVES

The brief inventory of the genes encoding the most common methyl transfer reactions indicates the ubiquity and universality of these essential enzymes among all organisms. The fact that the genes encoding C, N and S-MTs exhibit divergent sequences, both among themselves as well as among members of the same family, suggests that they have evolved from unrelated ancestral genes. In contrast, the more structurally related members of the OMT family appear to have evolved from an ancestral gene, with the HCCoA OMT as a progenitor of the different branches involved in the methylation of other metabolites, such as HCAs, flavonoids, and alkaloids. However, more genes need to be cloned and characterized, both at the molecular and biochemical level, before drawing any conclusions on the evolutionary and phylogenetic relationships of these enzymes. In addition, the characterization of the domains responsible for the catalytic and regiospecificity of these proteins should provide the tools for manipulation of the pathways involving the methylation of plant secondary metabolites and, hence, the biotechnological production of organisms with desirable traits.

In fact, plants represent a 'treasure-chest' of a rich variety of secondary metabolites of different biosynthetic origins, many of which harbor methylations that await the cloning and characterization of. the genes encoding their methylations. The availability of novel methylation genes will contribute to a better understanding of their phylogenetic relationship, as well as increase the pool of genes that can be used for the manipulation of the methylation pattern of secondary metabolites using transgenic technology.

O-Methylation of flavonoids reduces the reactivity of the phenolic hydroxyl groups and increases the antimicrobial activity of these compounds.[8] It also constitutes an important reaction in the biosynthesis of some flavonoid-based phytoalexins. To date, there are only a few genes that are known to encode single methylation steps of various flavonoid structures. The penta- to hepta-*O*-methylated flavones of *Chrysoplenium americanum* (**XIX**) (Saxifragaceae) and of some *Citrus spp.* represent a valuable source of flavonoid OMT genes that await cloning. Sequence comparisons of the primary structures of these genes will allow us to designate the regions, if not the specific amino acid residues, that are responsible for the regio- and position specificities of these enzymes. 3′/5′-*O*-Methylation of anthocyanins (**XX**) is responsible for the wide array of crimson to purple flower colors in peonies, petunias, and malvas. Apart from acquiring additional anthocyanin-specific OMTs for sequence comparison, the introduction of these genes into other species should broaden the color spectrum of their flowers, and enhance both their esthetic and commercial values.

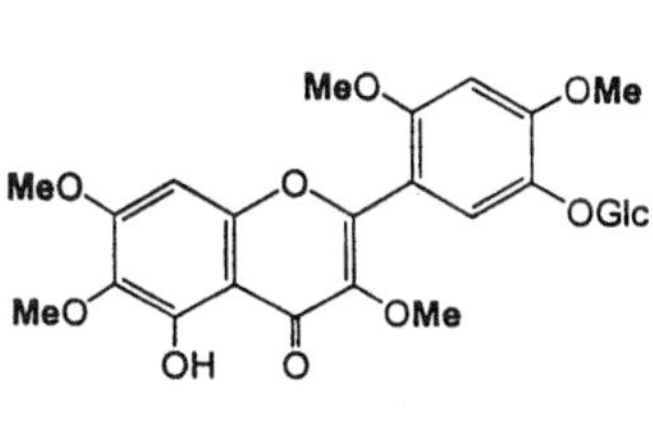

XIX

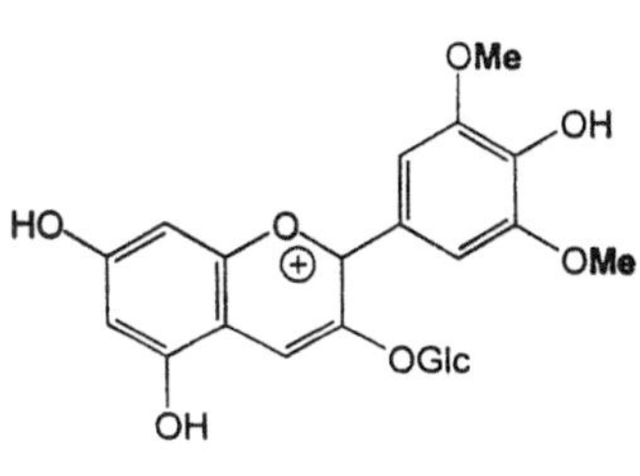

XX

Methylated polyketides and mevalonate-derived compounds are of common occurrence in plants, fungi, and bacteria. One exceptional coumarin (**XXI**), produced by *Aspergillus variecolor*, is derived from acetate, including its

5 methyl group; whereas the methyl groups at positions 4 and 7 are introduced after coumarin ring formation. The fact that this fungal coumarin as well as the isocoumarin 6-methoxymellein, the phytoalexin of carrot, are biogenetically different from the shikimate-derived furanocoumarin bergapten (whose OMT has been cloned, but not published), suggests that the amino acid sequence comparisons of the enzymes catalyzing their methylations would be of significant phylogenetic interest.

Another interesting acetate-derived plant anthraquinone is questin, which has CMe and OMe groups at positions 2 and 5, respectively. Fungal metabolism of this compound results in oxidative ring fission to its carboxy derivative, sulochrin (**XXII**), which is further carboxyl methylated. This metabolite represents a challenging combination of genes encoding C-, O-, and carboxyl methylations that are both structurally and phylogenetically interesting.

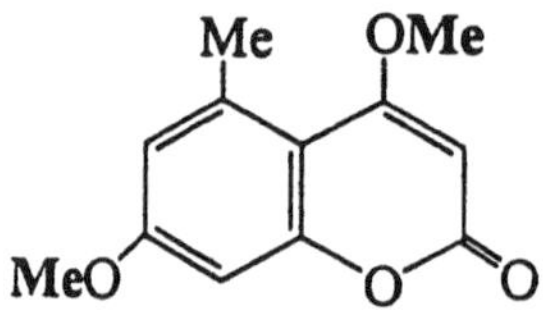

XXI

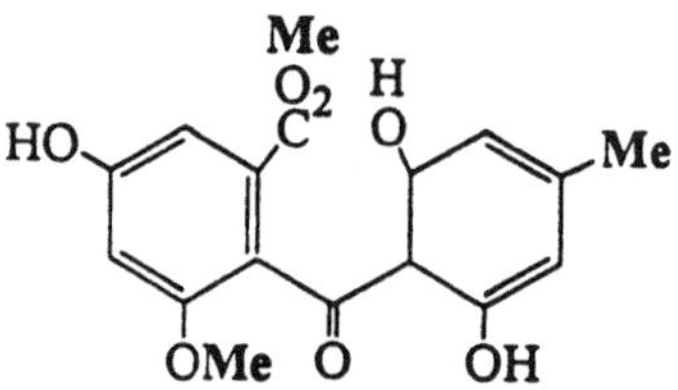

XXII

Except for the C-24 methylation of steroids, other naturally occurring isoprenoids are rarely methylated after their assembly. Yet, the sesquiterpenoid juvenile hormone of the giant silk worm (**XXIII**) is characterized by two AdoMet-dependent *C*-methylations and a carboxyl methylation that are encoded by genes of insect origin. In addition, other novel C-, N-, and O-methyl-transfer reactions are of common occurrence in ricinine (**XXIV**), in the toxic principle of castor bean oil, as well as in several phenylethylamines, such as mescaline (**XXV**) and pellotine (**XXVI**), that have unique 3,4-di and 3,4,5-tri-*O*-methylation patterns. The structural and phylogenetic analysis of the genes encoding these methylations can be rewarding from both the scientific and applied view points.

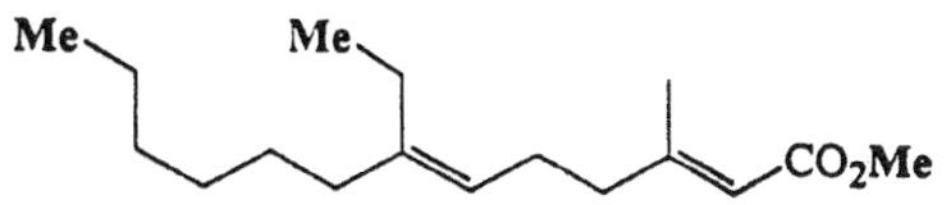

XXIII

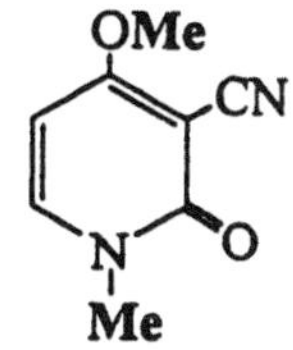

XXIV

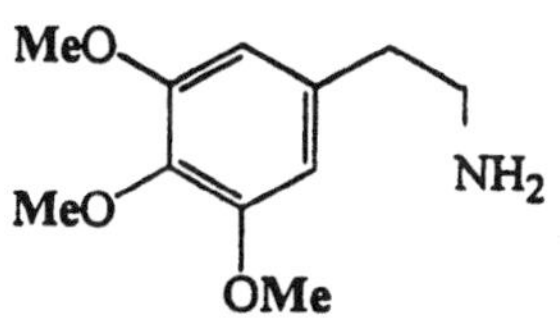

XXV

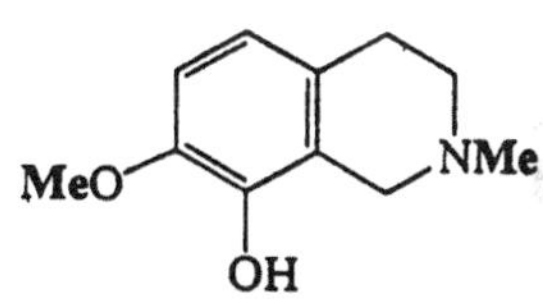

XXVI

Finally, the monoterpenoid indole alkaloid, vindoline (**XXVII**), which combines *in vivo* with catharanthine to form the antileukemic bis-alkaloids, vinblastine and vincristine, represents an interesting example of O-, N-, and carboxyl methyl transfer reactions. In addition to the recently cloned desacetylvindoline acetyltransferase,[77] the cloning of the genes encoding the latter methylations may allow the production of this valuable alkaloid by using semi-synthetic methods in combination with recombinant DNA technology.

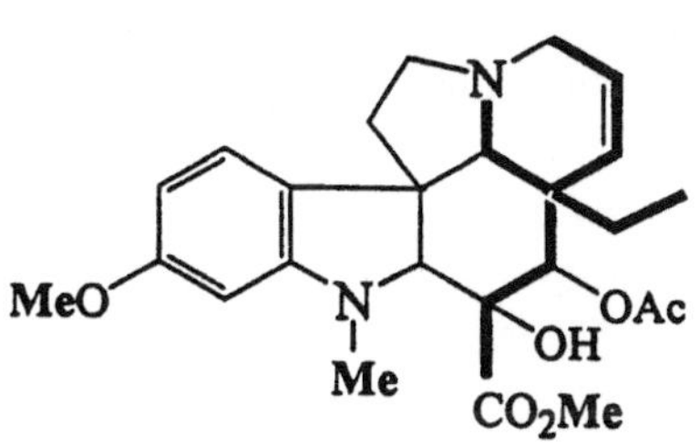

XXVII

ACKNOWLEDGEMENTS

We wish to thank Anne Bruneau, Université de Montréal and Ehab Abou-Heif, Duke University for their advise on the phylogenetic trees, and Patrick J. Gulick for helpful discussions. The research work reported from our laboratory is being supported by operating and equipment grants from the Natural Sciences and Engineering Research Council (NSERC) of Canada.

REFERENCES

1. ZUCKERKANDL, E., PAULING, L. 1965. Molecules as documents of evolutionary history. J. Theor. Biol. 8: 357-366.
2. RAVANEL, S., GAKIÈRE, B., JOB, D., DOUCE, R. 1998. The specific features of methionine biosynthesis and metabolism in plants. Proc. Natl. Acad. Sci. USA 95: 7805-7812.
3. VIGDREN, J., SVENSSON, L.A., LILJAS, A. 1994. Crystal structure of catechol *O*-methyltransferase. Nature 368: 354-357.
4. SCHLUCKEBIER, G., O'GARA, M., SAENGER, W., CHENG, X. 1995. Universal catalytic domain structure of AdoMet-dependent methyltransferases. J. Mol. Biol. 247: 16-20.
5. KAGAN, R.M., CLARKE, S. 1994. Widespread occurrence of three sequence motifs in diverse *S*-adenosyl-L-methionine-dependent methyltransferases suggests a common structure for these enzymes. Arch. Biochem. Biophys. 310: 417-427.
6. JOSHI, C.P., CHIANG, V.L. 1998. Conserved sequence motifs in plant *S*-adenosyl-L-methionine-dependent methyltransferases. Plant Mol. Biol. 37: 663-674.
7. IBRAHIM, R.K. 1997. Plant *O*-methyltransferase signatures. Trends Plant Sci. 2: 249-250.
8. IBRAHIM, R.K., BRUNEAU, A., BANTIGNIES, B. 1998. Plant *O*-methyltransferases: molecular analysis, common signature and classification. Plant Mol. Biol. 36: 1-10.
9. DE CAROLIS, E., IBRAHIM, R.K. 1989. Purification and kinetics of phenylpropanoid *O*-methyltransferase activities from *Brassica oleracea*. Biochem. Cell Biol. 67: 763-769.
10. HOLTON, T.A., CORNISH, E.C. 1995. Genetics and biochemistry of anthocyanin biosynthesis. Plant Cell 7: 1071-1083.
11. BUGOS, R.C., CHIANG, V.L., CAMPBELL, W.H. 1991. cDNA cloning, sequence analysis and seasonal expression of lignin bispecific caffeic acid/5-hydroxyferulic acid *O*-methyltransferase of aspen. Plant Mol. Biol. 17: 1203-1215.
12. GOWRI, G., BUGOS, R.C., CAMPBELL, W.H., DIXON, R.A. 1991. Stress responses in alfalfa (*Medicago sativa* L.). X. Molecular cloning and expression of *S*-adenosyl-L-methionine: caffeic acid 3-*O*-methyltransferase, a key enzyme

of lignin biosynthesis. Plant Physiol. 97: 7-14.
13. DUMAS, B., VAN DOORSSELAERE, J., GIELEN, J., LEGRAND, M., FRITIG, B., VAN MONAGU, M., INZÉ, D. 1992. Nucleotide sequence of a cDNA encoding *O*-methyltransferase from poplar. Plant Physiol. 98: 796-797.
14. COLLAZO, P., MONTOLIU, L., PUIGDOMÈNECH, P., RIGAU, J. 1992. Structure and expression of lignin *O*-methyltransferase gene from *Zea mays* L. Plant Mol. Biol. 20: 857-867.
15. MCINTYRE, C.L., RAE, A.L., CURTIS, M.D., MANNERS, J.M. 1995. Sequence and expression of a caffeic acid *O*-methyltransferase cDNA homolog in the tropical forage legume *Stylosanthus humilis*. Aust. J. Plant Physiol. 22: 474-478.
16. GARCIA-MAS, J., MESSEGUER, R., ARUS, P., PUIGDOMENECH, P. 1995. The caffeic acid *O*-methyltransferase from *Prunus amygdalus*. Plant Physiol. 108: 1341, PGR 95-006.
17. HAYAKAWA, T., NATO, K., KAWAI, S., KATAYAMA, Y., MOROHOSHI N. 1996. Molecular cloning and tissue-specific expression of two genes that encode caffeic acid *O*-methyltransferase from *Populus kitakamiensis*. Plant Sci. 113: 157-165.
18. MOYLE, R., WAGNER, A., WALTER, C. 1999. Nucleotide sequence of a caffeic acid *O*-methyltransferase gene from *Pinus radiata*. Plant Physiol. 119: 1147 (PGR 99-026).
19. LI, L., POPKO, J.L., ZHANG, X.H., OSAKABE, K., TSAI, C.J., JOSHI, C.P., CHIANG, V.L. 1997. A novel multifunctional *O*-methyltransferase implicated in a dual pathway associated with lignin biosynthesis. Proc. Natl. Acad. Sci. USA 94: 5461-5466.
20. PELLIGRINI, L., GEOFFROY, P., FRITIG, B., LEGRAND, M. 1993. Molecular cloning and expression of a new class of *o*-diphenol *O*-methyltransferases induced in tobacco leaves by infection or elicitor treatment. Plant Physiol. 103: 509-517.
21. HELD, B.M., WAND, H., JOHN, I., WURTELE, E.S., COLBERT, J.T. 1993. An m-RNA putatively encoding for an *O*-methyltransferase preferentially in maize roots and is located predominantly in the region of the endodermis. Plant Physiol. 102: 1001-1008.
22. LEE, J.E., VOGT, T., HAUSE, B., LÖBLER, M. 1997. Methyl jasmonate induces an *O*-methyltransferase in barley. Plant Cell Physiol. 38: 851-862.
23. GREGERSEN, P.L., CHRISTENSSEN, A.B., SOMMER-KNUDSEN, J., COLLINGE D.B. 1994. A putative *O*-methyltransferase from barley is induced by fungal pathogens and UV light. Plant Mol. Biol. 26: 1797-1806.

24. WANG, J., PICHERSKY, E. 1998. Characterization of *S*-adenosyl-L-methionine: (iso)eugenol *O*-methyltransferase involved in floral scent production in *Clarkia breweri*. Arch. Biochem. Biophys. 349: 153-160.
25. SCHMITT, D., PAKUSCH, A.-E., MATERN, U. 1991. Molecular cloning, induction and taxonomic distribution of caffeoyl CoA 3-*O*-methyltransferase, an enzyme involved in disease resistance. J. Biol. Chem. 266: 17416-17423.
26. YE, Z.-H., KNEUSEL, R.E., MATERN, U., VARNER, J.E. 1994. An alternative methylation pathway in lignin biosynthesis in *Zinnia*. Plant Cell 6: 1427-1439.
27. BUSAM, G., JUNGHANNS, K.T., KNEUSEL, R.E., KASSEMEYER, H.-H., MATERN, U. 1997. Characterization and expression of caffeoyl CoA 3-*O*-methyltransferase proposed for the induced resistance response of *Vitis vinifera* L. Plant Physiol. 115: 1039-1048.
28. MAXWELL, C.A., HARRISON, M.J., DIXON, R.A. 1993. Molecular characterization and expression of alfalfa isoliquiritigenin 2′-*O*-methyltransferase, an enzyme specifically involved in the biosynthesis of an inducer of *Rhizobium meliloti* nodulation genes. Plant J. 4: 971-981.
29. HAGA, M., AKASHI, T., AOKI, T., AYABE, S. 1997. A cDNA for *S*-adenosyl-*L*-methionine:isoliquiritigenin/licodione 2′-*O*-methyltransferase from cultured licorice (*Glycyrrhiza echinata* L.) cells. Plant Physiol. 113: 663.
30. CHRISTENSSEN, A.B., GREGERSEN, P.L., OLSEN, C.E., COLLINGE, D.B. 1998. Flavonoid 7-*O*-methyltransferase is expressed in barley leaves in response to pathogen attack. Plant Mol. Biol. 36: 219-227.
31. HE, X.-E., REDDY, J.T., DIXON, R.A. 1998. Stress responses in alfalfa (*Medicago sativa* L.). XXII. cDNA cloning and characterization of an elicitor-inducible isoflavone 7- *O*-methyltransferase. Plant Mol. Biol. 36: 43-54.
32. WU, Q., PREISIG, C.L., VANETTEN, H.D. 1997. Isolation of the cDNAs encoding (+)6a-hydroxymaackiain 3-*O*-methyltransferase, the terminal step for the biosynthesis of the phytoalexin pisatin in *Pisim sativum* L. Plant Mol. Biol. 35: 551-560.
33. GAUTHIER, A., GULICK, P.J., IBRAHIM, R.K. 1996. cDNA cloning and characterization of a 3′/5′-*O*-methyltransferase for partially methylated flavonols from *Chrysosplenium americanum*. Plant Mol. Biol. 32: 1163-1169.
34. MUZAC, I., NOCE, A., ZHANG, H., IBRAHIM, R.K. 1999. Purification and characterization of a novel flavonol 3′-*O*-methyltransferase cDNA clone from *Arabidopsis thaliana*. PSNA Conf. Proc., Poster Abstract No. 29, p. 46.
35. GAUTHIER, A., GULICK, P.J., IBRAHIM, R.K. 1998. Characterization of two cDNA clones which encode *O*-methyltransferases for the methylation of both flavonoid and phenylpropanoid compounds. Arch. Biochem. Biophys.

351: 243-249.
36. GORHAM, J. 1995. The Biochemistry of Stilbenoids. Chapman & Hall, London.
37. SCHRÖDER, G., BROWN, J.W.S., SCHRÖDER, J. 1988. Molecular analysis of resveratrol synthase: cDNA clones and relationship with chalcone synthase. Eur. J. Biochem. 172: 161-169.
38. ISHIMURA, M., FURUNO, T., TAKAHASHI, T., DIXON, R.A., AYABE, S. 1997. Enzymatic *O*-methylation of isoliquiritigenin and licodione in alfalfa and licorice cultures. Phytochemistry 44: 991-995.
39. SÉGUIN, J., MUZAC, I., IBRAHIM, R.K. 1998. Purification and immunological characterization of a recombinant trimethylflavonol 3′-*O*-methyltransferase. Phytochemistry 49: 319-325.
40. ZHANG, H. WANG, J., GOODMAN, H.M. 1997. An *Arabidopsis* gene encoding a putative 14-3-3-interacting protein, caffeic acid/5-hydroxyferulic acid *O*-methyltransferase. Biochim. Biophys. Acta 1353: 199-202.
41. MENG, H., CAMPBELL, W.H. 1998. Biochemistry and molecular biology of lignin-specific *O*-methyltransferases from woody species. pp. 55-64. *in* Lignin and Lignan Biosynthesis (N.G. Lewis and S. Sarkanen, eds.). ACS Symposium Series No. 697, American Chemical Society, Washington D.C.
42. DONOHUE, S.J., ROSEBOOM, P.H., KLEIN, D.C. 1992. Bovine hydroxyindole *O*-methyltransferase. J. Biol. Chem. 267: 5184-5185.
43. KOLAR, J., MACHACKOVA, I., EDER, J., PRINSEN, E., VAN DONGEN, W., VAN ONCKELEN, H., ILLNEROVA, H. 1997. Melatonin: occurrence and daily rhythm in *Chenopodium rubrum*. Phytochemistry 44: 1407-1413.
44. TAKESHITA, N., FUJIWARA, H., MIMURA, H., FITCHEN, J., YAMADA, Y., SATO, F. 1995. Molecular cloning and characterization of *S*-adenosyl-L-methionine:scoulerine 9- *O*-methyltransferase from cultured cells of *Coptis japonica*. Plant Cell Physiol. 46:29-36.
45. FRICK, S., KUTCHAN, T.M. 1999. Molecular cloning and functional expression of *O*-methyltransferase common to isoquinoline alkaloid and phenylpropanoid biosynthesis. Plant J. 17: 329-339.
46. KUTCHAN, T.M. 1998. Molecular genetics of plant alkaloid biosynthesis. pp. 257-316. *In* The Alkaloids, (G. Cordell, ed.), Vol. 50, Academic Press, San Diego, CA,
47. MUDGETT, M., CLARKE, S. 1996. A distinctly regulated protein repair L-isoaspartyl methyltransferase from *Arabidopsis thaliana*. Plant Mol. Biol. 30: 723-737.
48. BUSHUE, L., MANN, C., GORENSTEIN, N, DUDAREVA, N. 1999. New class of plant methyltransferases:benzoic acid carboxyl methyltransferase from

snapdragon flowers. PSNA Conf. Proc., Oral Abstract No. 17, p. 32.
49. ROSS, J.R., NAM, K.H., D'AURIA, J.C., PICHERSKY, E. 1999. *S*-Adenosyl-L-methionine: salicylic acid carboxyl methyltransferase, an enzyme involved in floral scent production and plant defense, represents a new class of plant methyltransferases. Arch. Biochem. Biophys. 367: 9-16.
50. VERNON, D., BOHNERT, H.J. 1992. A novel methyltransferase induced by osmotic stress in the facultative halophyte *Mesembryanthemum crystallinum*. EMBO J. 11: 2077-2085.
51. DECKER, H., MOTAMEDI, H., HUTCHINSON, C.R. 1993. Nucleotide sequence and heterologous expression of *tcmG* and *tcmP* biosynthetic genes for tetracenomycin C in *Strepromyces glaucescens*. J. Bacteriol. 175: 3876-3886.
52. LACALLE, R.A., RUIZ, D., JIMÉNEZ, A. 1991. Molecular analysis of *dmpM* gene encoding *O*-demethylpuromycin *O*-methyltransferase from *Streptomyces alboniger*. Gene 109: 55-61.
53. MADDURI, K., TORTI, F., COLOMBO, A.L., HUTCHINSON, C.R. 1993. Cloning and sequencing a gene encoding carminomycin 4- *O*-methyltransferase from *Streptomyces peucetius* and its expression in *Escherichia coli*. J. Bacteriol. 175: 3900-3904.
54. YU, J., CARY, J.W., BHATNAGAR, D., CLEVELAND, T.E., KELLER, N.P., CHU, F.S. 1993. Cloning and characterization of a cDNA from *Aspergillus parasiticus* encoding an *O*-methyltransferase involved in aflatoxin biosynthesis. Appl. Environ. Microbiol. 59: 3564-3571.
55. YU, J., CHANG, P.-K., PAYNE, G.A., CARY, G.W., BHATNAGAR, D., CLEVELAND, T.E. 1995. Comparison of the *omtA* genes encoding *O*-methyltransferases involved in aflatoxin biosynthesis from *Aspergillus parasiticus* and *A. flavus*. Gene 163: 121-125.
56. HUSSELSTEIN, T., GACHOTTE, D., DESPEREZ, T., BARD, M., BENVENISTE, P. 1996. Transformation of *Saccharomyces cerevisiae* with a cDNA encoding a sterol *C*-methyltransferase from *Arabidopsis thaliana* results in the synthesis of 24-ethyl sterols. FEBS Lett. 381: 87-92.
57. SHI, J., GONZALES, R.A., BATTACHARYYA, M.K. 1996. Identification and characterization of an *S*-adenosyl-*L*-methionine:@^{24}-sterol *C*-methyltransferase cDNA from soybean. J. Biol. Chem. 271: 9384-9389.
58. GREBENOK, R.J., GALBRAITH, D.W., DELLAPENNA, D. 1997. Characterization of *Zea mays* endosperm C-24 sterol methyltransferase: one of two types of sterol methyltransferase in higher plants. Plant Mol. Biol. 34: 891-896.

59. GARBER, R.F., COPPLE, D.M., KENNEDY, B.K., VIDAL, M., BARD, M. 1989. The yeast gene *ERG6* is required for normal membrane function, but is not essential for biosynthesis of the cell-cycle-sparking sterol. Mol. Cell Biol. 9: 3447-3456.
60. TABER, M.G., SIES, H. 1996. Vitamin E in humans: demand and delivery. Annu. Rev. Nutr. 16:321-346.
61. SHINTANI, D., DELLAPENNA, D. 1998. Elevating the vitamin E content of plants through metabolic engineering. Science 282: 2098-2100.
62. BARKOVICH, R.J., SHTANKO, A., SHEPHERD, J.A., LEE, P.T., MYLES, D.C., TZAGOLOFF, A., CLARKE, C.F. 1997. Characterization of the *COQ5* gene from *Saccharomyces cerevisiae*: evidence for a *C*-methyltransferase in ubiquinone biosynthesis. J. Biol. Chem. 272: 9182-9188.
63. LEUSTEK, T., SMITH, M., MURILLO, M., SINGH, D.P., SMITH, A.G., WOODCOCK, S.C., AWAN, S.J., WARREN, M.J. 1997. Siroheme biosynthesis in higher plants: Analysis of an *S*-adenosyl-L-methionine-dependent uroporphyrinogen III methyltransferase from *Arabidopsis thaliana*. J. Biol. Chem. 272: 2744-2752.
64. HIBI, N., HIGASHIGUCHI, S., HASHIMOTO, T., YAMADA, Y. 1994. Gene expression in tobacco low-nicotine mutants. Plant Cell 6: 723-735.
65. YING, Z., JANNEY, N., HOUTZ, R.L. 1996. Organization and characterization of ribulose 1,5-bisphosphate caboxylase/oxygenase large subunit $^{\epsilon}$N-methyltransferase gene in tobacco. Plant Mol. Biol. 32: 663-671.
66. KODAKI, T., YAMASHITA, S. 1987. Yeast phosphatidyl ethanolamine methylation pathway: cloning and characterization of two distinct methyltransferase genes. J. Biol. Chem. 262: 15428-15435.
67. GARROW, T.A. 1996. Purification, kinetic properties, and cDNA cloning of mammalian betaine-homocyteine methyltransferase. J. Biol. Chem. 271: 22831-22838.
68. FESSING, M.Y., BELKOV, V.M., KRYNETSKI, E.Y., EVANS, W.E. 1998. Molecular cloning and functional characterization of the cDNA encoding murine thiopurine *S*-methyltransferase. FEBS Lett. 424: 143-145.
69. WARNER, D.R., MOZIER, N.M., PEARSON, J.D., HOFFMAN, J.L. 1995. Cloning and base sequence analysis of a cDNA encoding mouse lung thioether *S*-methyltransferase. Biochim. Biophys. Acta 1246: 160-166.
70. NEUHIER, B., THANBICHLER, M., LOTTSPIECH, F., BÖCK, A. 1999. A family of *S*-methylmethionine-dependent thiol/selenol methyltransferases. J. Biol. Chem. 274: 5407-5414.
71. ATTIEH, J.M., HANSON, A.D., SAINI, H.S. 1995. Purification and characterization of a novel methyltransferase responsible for biosynthesis of

halomethanes methanethiol in *Brassica oleracea*. J. Biol. Chem. 270: 9250-9257.

72. NI, X., HAGER, L.P. 1998. cDNA cloning of *Batis maritima* methyl chloride transferase and purification of the enzyme. Proc. Natl. Acad. Sci. USA 95: 12866-12871.
73. CROW, J.F., DOVE, W.F. 1989. Anecdotal, historical and critical commentaries on genetics. Genetics 122: 1-6.
74. QIU, Y.-L., PALMER, J.D. 1999. Phylogeny of early land plants: insights from genes and genomes. Trends Plant Sci. 4: 26-30.
75. LEWIS, N.G., DAVIN, L.B., SARKANEN, S. 1999. The nature and function of lignins, pp. 617-745, *in* Comprehensive Natural Products Chemistry (M. Pinto, series ed.). Vol. 3, Elsevier, Amsterdam.
76. PILEUP. 1994. Program Manual for the Wisconsin Package, version 8. Genetics Computer Group, Madison, WI.
77. FELSENSTEIN, J. 1993. PHYLIP (Phylogeny Inference Package) version 3.5c. Distrubuted by the author, Depatment of Genetics, University of Washington, Seattle, WA.
78. ST-PIERRE, B., LAFLAMME, P., ALARCO, A.-M., DE LUCA, V. 1998. The terminal *O*-acetyltransferase involved in vindoline biosynthesis defines a new class of proteins responsible for coenzyme A-dependent acyl transfer. Plant J. 14: 703-714.

Chapter Twelve

BRASSINOSTEROIDS - STRUCTURES, ANALYSIS AND SYNTHESIS

Jürgen Schmidt, Barbara Spengler, Brunhilde Voigt and Günter Adam

Institute of Plant Biochemistry
Weinberg 3, D-06120
Halle/S., Germany

INTRODUCTION

Brassinosteroids (BR's) represent highly active plant-growth-promoting substances with structures similar to animal steroidal hormones. The fascinating history of BR's started in 1979 when Grove *et al.* isolated from pollen of rape (*Brassica napus*) brassinolide (Fig. 1,(**1**), a plant growth regulator with high a biological activity in the bean second internode bioassay. Its structure was elucidated as (22*R*,23*R*,24*S*)-2α,3α,22,23-tetrahydroxy-24-methyl-β-homo-6a-oxa-5α-cholestan-6-one by using spectroscopic methods (EI-MS, FAB-MS, NMR) and X-ray analysis.[1]

The high biological activity of brassinolide (**1**) at very low concentrations

Evolution of Metabolic Pathways – J.T. Romeo et al. (Editors)

induced intense research activities in many laboratories, especially in Japan. Since the discovery of brassinolide, the natural occurrence of more than 40 members of this group has been detected. Today, brassinosteroids can be regarded as a class of plant hormones with ubiquitous occurrence in the plant kingdom. A series of reviews has been published covering the synthesis, biochemistry, and biological mode of action of BR's as well as their practical application in agriculture.[2-11] Recent molecular biological studies have demonstrated their essential role for normal plant growth and development.[7,12,13]

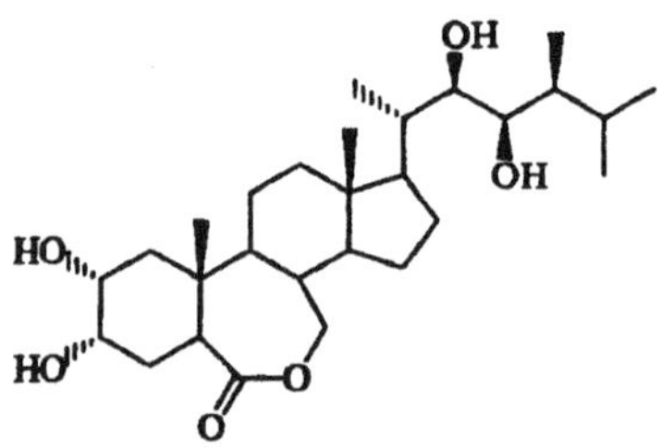

Brassinolide (1)

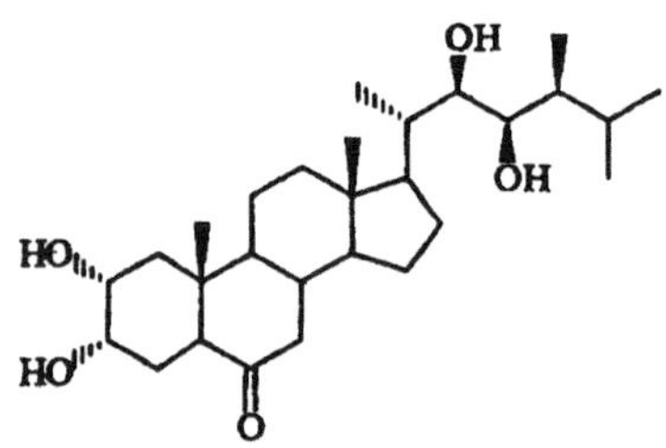

Castasterone (2)

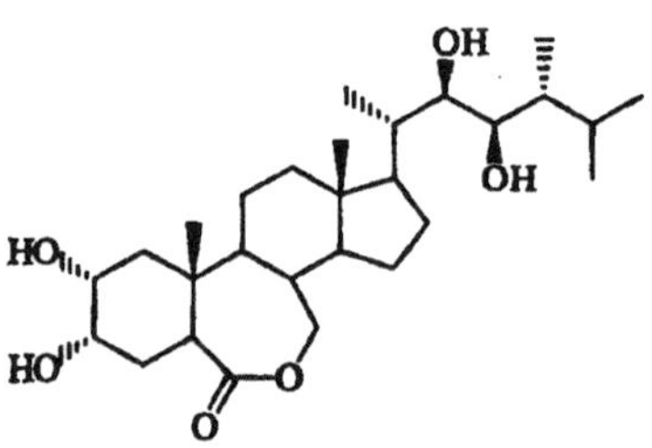

24-Epibrassinolide (3)

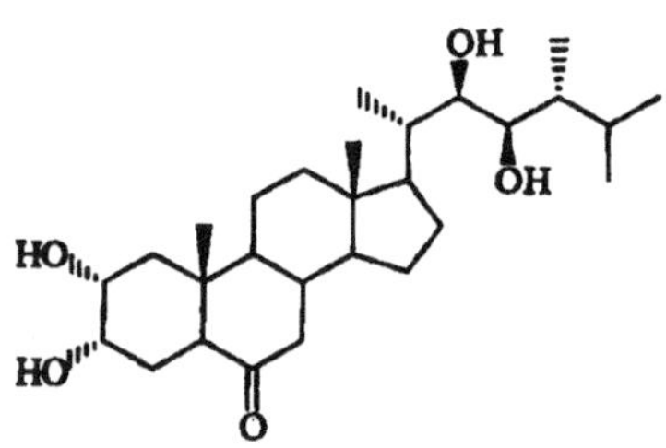

24-Epicastasterone (4)

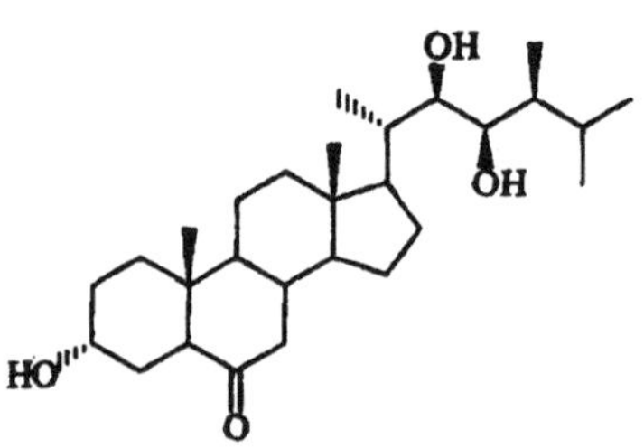

Typhasterol (5)

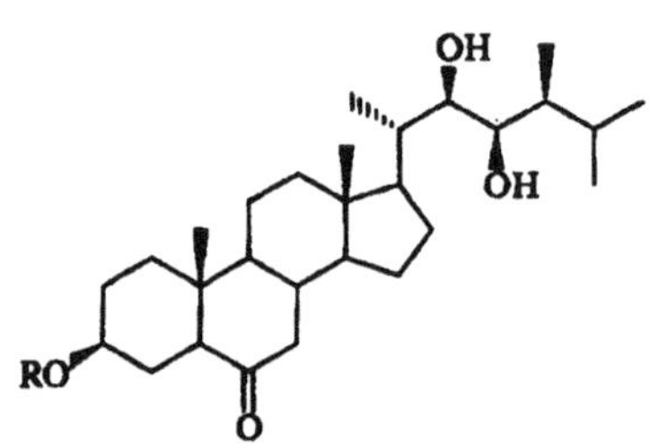

Teasterone (6) : R = H

Teasterone-3-laurate (6a) : R = $CO(CH_2)_{10}CH_3$
Teasterone-3-myristate (6b) : R = $CO(CH_2)_{12}CH_3$

Figure 1. Structures of brassinosteroids 1-6

This chapter covers the search for new brassinosteroids in the plant kingdom from 1993-1999 along with analytical and synthetic aspects. The analysis of BR's is discussed especially in regard to GC/MS and HPLC combined with fluorimetric and MS detection.

STRUCTURES

Brassinosteroids are derived from the 5α-cholestane skeleton. Their structural variations comprise substitutions at rings A and B as well as in the C-17-side chain (Fig. 2).[4,8,14]

The two vicinal hydroxy groups at ring A (2α,3α) represent a general structural feature of most active brassinosteroids, such as brassinolide (**1**) and castasterone (**2**). Biogenetic precursors, like typhasterol (**5**) and teasterone (**6**), have only one hydroxyfunction in ring A (Fig. 1). Ring B is characterized either by a seven-membered 6a-oxalactone as in (Fig. 1,(**1**) or a 6-membered ring as in (Fig. 1,(**2**). The 6-deoxobrassinosteroids have an unsubstituted ring B.

The variations in the side chain are related to the corresponding biogenetic precursors, the phytosterols. According to the cholestane side chain, the various skeletons can be classified into seven types with different substituents at C-24 and C-25, respectively (Fig. 2):

1) 24*S*-methyl
2) 24*R*-methyl
3) 24-methylene
4) 24*S*-ethyl
5) 24-ethylidene
6) 24-methylene-25-methyl
7) without substituent at C-24

NATURAL OCCURRENCE

Table 1 summarizes the natural occurrence of new brassinosteroid structures detected since 1993 (Fig. 3).[15-32] All brassinosteroids described so far have a vicinal *R,R*-configurated diol functionality at C-22 and C-23. An interesting exception represents the not yet fully characterized isomeric brassinosteroids with an oxo function at C-23 (ξ-epi-23-dehydrobrassinolide).[20] During the past few years, brassinosteroids have been detected with only one oxygen function in ring A (Table 1, Fig. 3). These compounds are biosynthetic intermediates, both in the late and the early C-6-oxidation pathway.[18, 19, 33-35] Furthermore, a brassinosteroid was found in pollen of *Lolium perenne* L. (Poaceae) whose HPLC data were

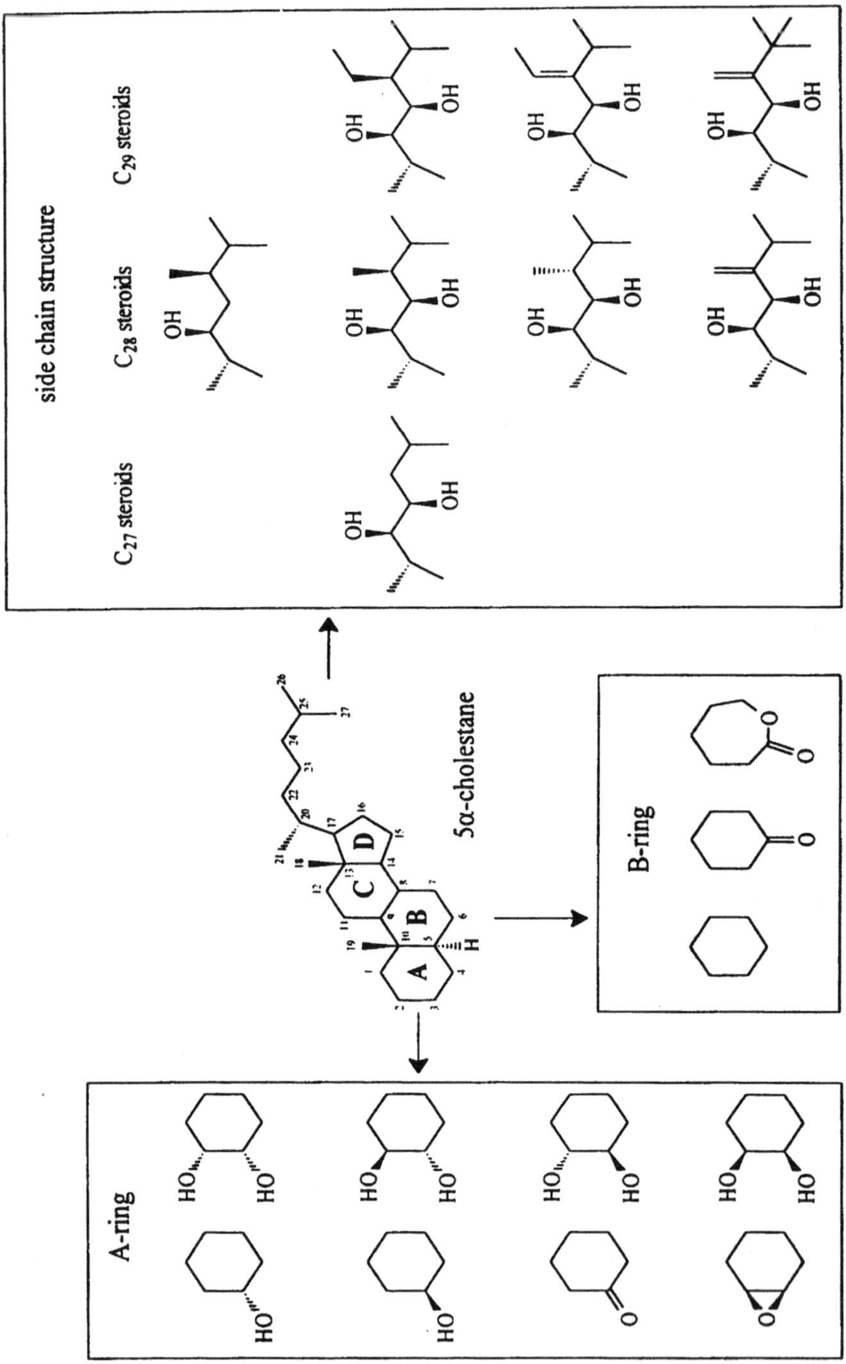

Figure 2. Structural variations of brassinosteroids

identical to those of 25-methylcastasterone.[36] In addition to free brassinosteroids, two fatty acid conjugates also have been detected in the 1990's. Thus, teasterone-3-O-laurate (**6a**) and teasterone-3-O-myristate (**6b**) (Fig. 1) are known to occur in the anthers of *Lilium longiflorum*.[24, 25]

Table 1. Natural occurrence of new brassinosteroids (1993-1999).

Plant source	Plant family	Plant part(s)	Brassinosteroid(s)	Reference
Apium graveolens L.	Apiaceae	seeds	**9**	[15]
Arabidopsis thaliana (L.) Heynh.	Brassicaceae	shoots	**13**	[16]
		seeds, siliques	**13, 14**	[17]
Catharanthus roseus G. Don	Apocynaceae	cultured cells	**10**	[18]
		cultured cells	**12, 13, 14**	[19]
Cryptomeria japonica D. Don.	Taxodiaceae	pollen, anthers	**11, 19-22**	[20]
Cupressus arizonica Greene	Cupressaceae	pollen	**11, 12, 13**	[21]
Distylium racemosum Sieb. et. Zucc.	Hamamelidaceae	leaves	**11**	[22]
Erythronium japonicum Decne	Liliaceae	pollen, anthers	**11**	[23]
Lilium longiflorum Thunb.	Liliaceae	anthers	**11**	[22]
		anthers	**6a**	[24]
		pollen	**6a, 6b**	[25]
Lychnis viscaria L.	Caryophyllaceae	seeds	**16**	[26]
Ornithopus sativus Brot.	Fabaceae	shoots	**17, 18**	[27]
Oryza sativa L.	Poaceae	bran	**7, 8**	[28]
Pisum sativum L.	Fabaceae	seeds	**9**	[29]
Raphanus sativus L.	Brassicaceae	seeds	**7**	[30]
Secale cereale L.	Poaceae	seeds	**15**	[31]
Triticum aestivum L.	Poaceae	grain	**11**	[32]

DISTRIBUTION IN THE PLANT KINGDOM

The known brassinosteroids have been found in a large variety of higher plants and have been detected in more than 50 species.[2-6,9,11,14,37] Most described so far have been isolated from angiosperms, especially dicotyledons (Fig. 4),although the occurrence of brassinosteroids is also known in some gymnosperms. Further natural occurrences are known from one pteridophyte (*Equisetum arvense*)[38] and one chlorophyte, the alga *Hydrodictyon reticulatum*.[39] The most widely distributed brassinosteroid is castasterone (**2**),[40] followed by brassinolide (**1**).[1] The trihydroxybrassinosteroids typhasterol (**5**)[41, 42] and teasterone (**6**),[43] as well as 6-deoxocastasterone and 28-norcastasterone, are also of common occurrence. However, compounds of the 24-*epi* series, such as 24-epibrassinolide (**3**)[44] and 24-epicastasterone (**4**),[39] represent relatively rare brassinosteroids (Fig. 1).

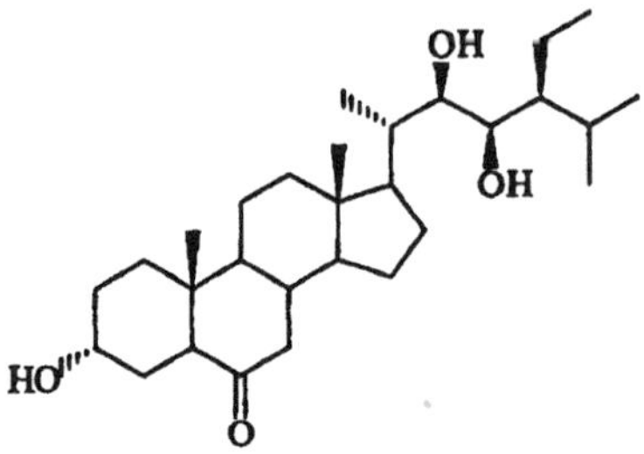

28-Homotyphasterol (7)

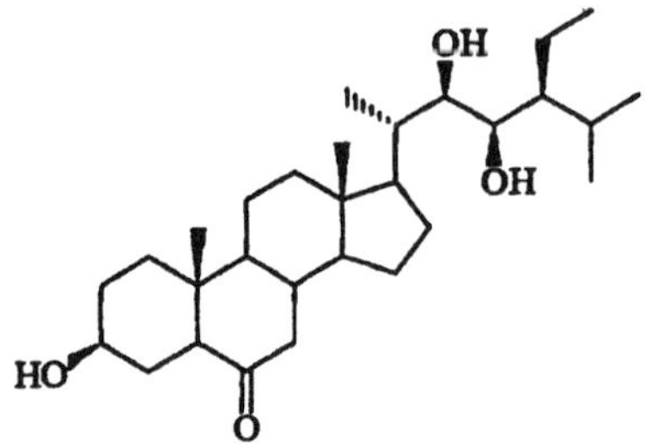

28-Homoteasterone (8)

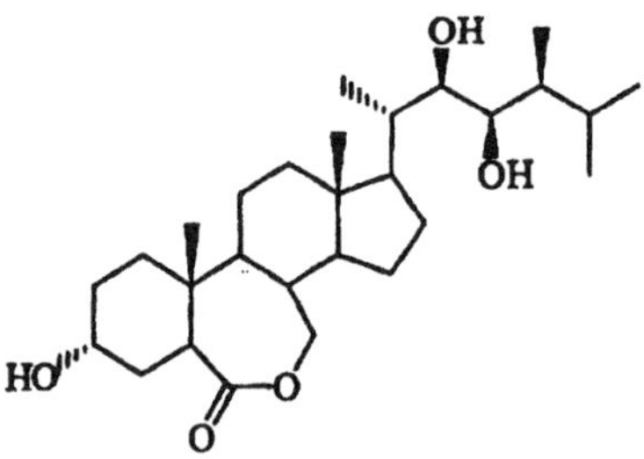

2-Deoxybrassinolide (9)

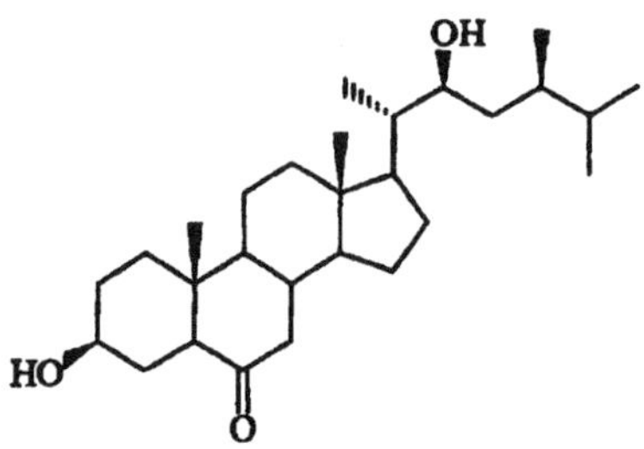

Cathasterone (10)

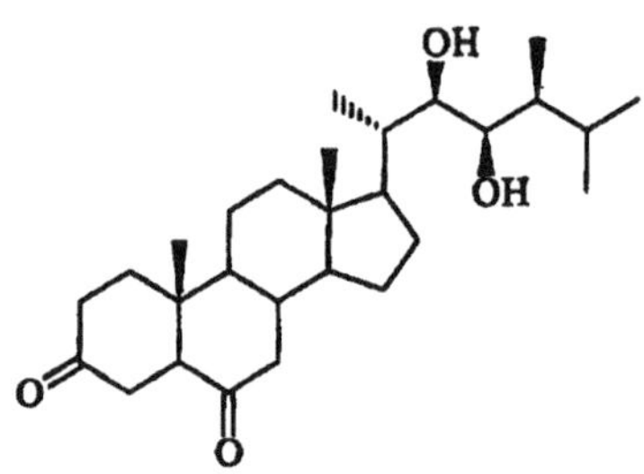

3-Dehydroteasterone (11)

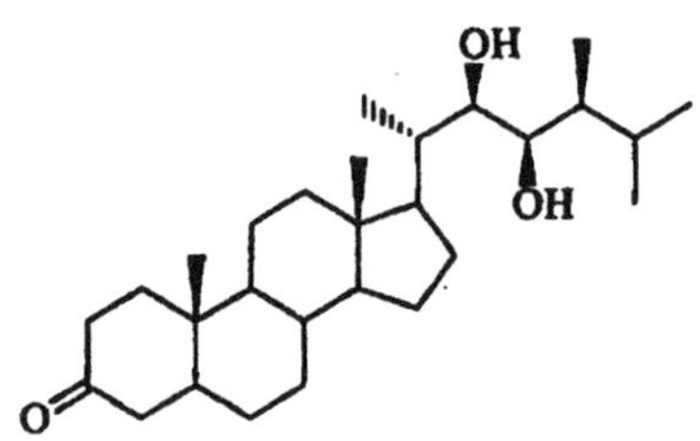

3-Dehydro-6-deoxoteasterone (12)

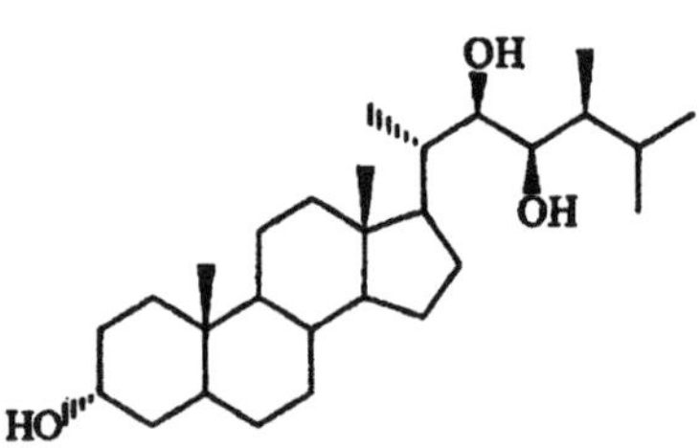

6-Deoxotyphasterol (13)

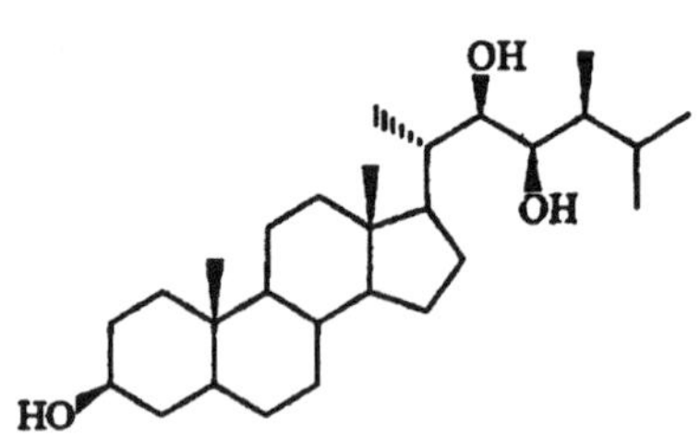

6-Deoxoteasterone (14)

Figure 3. Structures of new brassinosteroids (7-22) identified since 1993

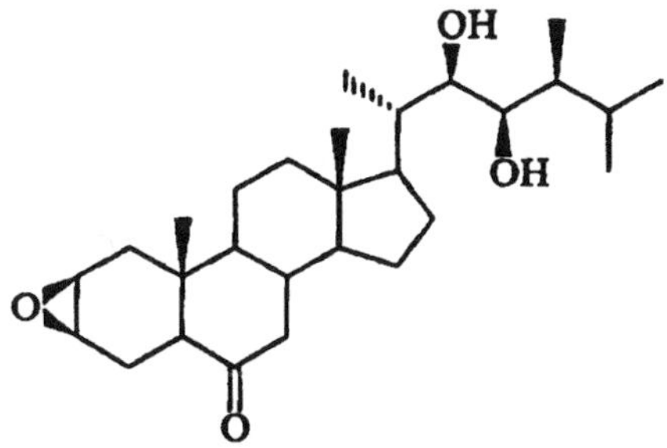

Secasterone (15)

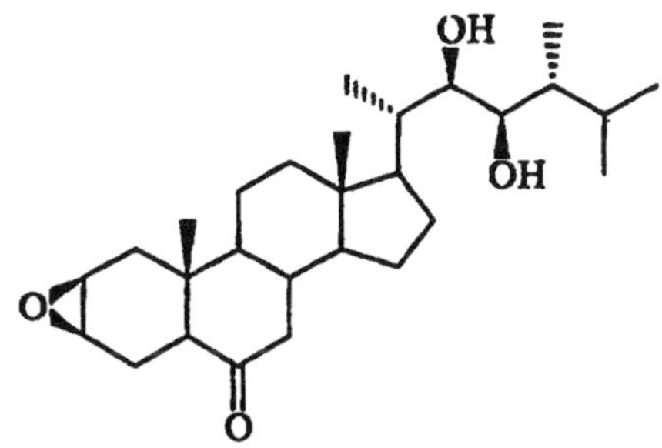

24-Episecasterone (16)

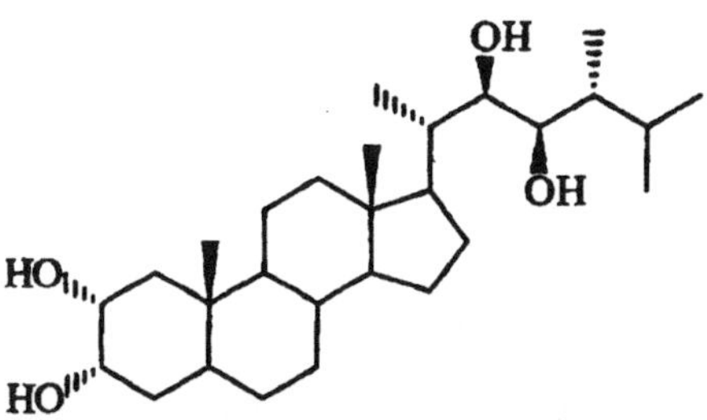

6-Deoxo-24-epicastasterone (17)

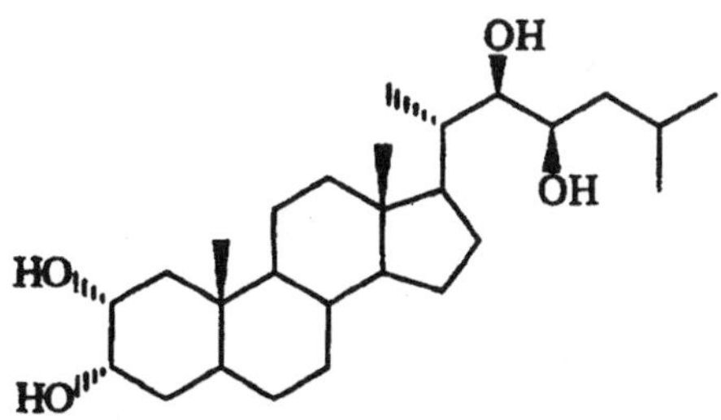

6-Deoxo-28-norcastasterone (18)

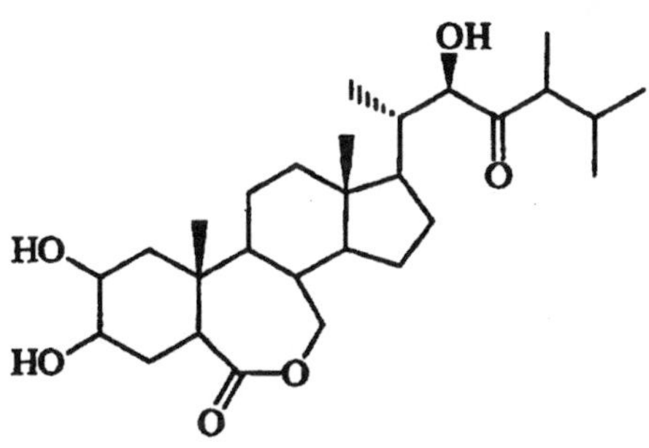

Epi-23-dehydrobrassinolides (19-22)

Figure 3. *Continued.*

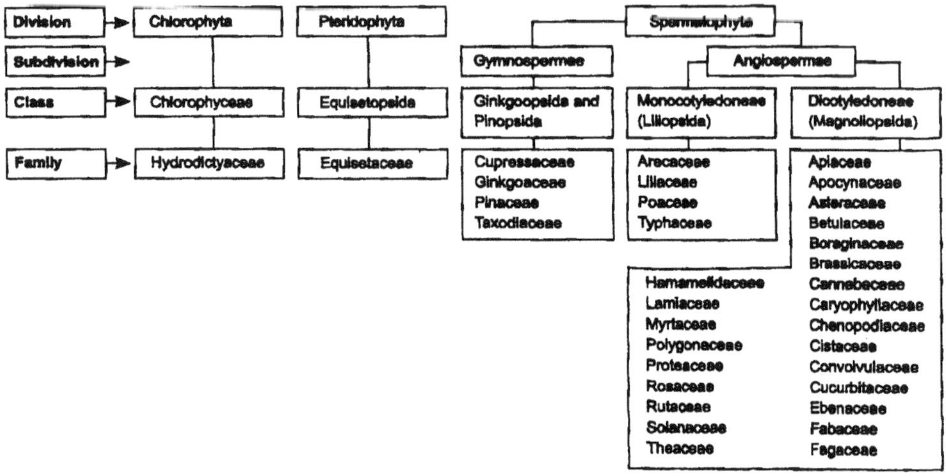

Figure 4. Distribution of brassinosteroids in the plant kingdom (from ref. 11)

ISOLATION AND PURIFICATION

Brassinosteroids can be extracted and purified via a sequence of partitioning and chromatographic processes. The extraction of BR's from plant material is carried out with methanol or methanol/ethyl acetate mixtures, followed by partitioning between water and chloroform (or ethyl acetate). Subsequently, a partitioning between 80% methanol and *n*-hexane is effective to remove nonpolar compounds. The 80% methanol fraction containing the BR's can be further purified by successive chromatography on silica gel, Sephadex LH-20 and diethylaminopropyl (DEA) ion exchange chromatography, followed by preparative HPLC (RP-18, ODS). This procedure can be modified and varied with respect to extraction, partitioning, and chromatography.[45] It is recommended to use a sensitive bioassay for monitoring the brassinosteroid-containing fractions during the several chromatographic steps. The most useful bioassay for brassinosteroids is the rice-lamina inclination test.[45, 46] This bioassay was originally developed for gibberellins and IAA. To screen for BR's, seeds of dwarf rice are germinated in water, placed on agar, incubated in the dark, and irradiated with red light for 1-2 times a day. After about 7 days, the leaf segments are excised, floated on distilled water, and incubated in a solution with probes containing BR's. The internal angles between the laminae and sheaths can be correlated with BR-content. However, it is also possible to detect known brassinosteroids without a bioassay. In these cases, fractions containing possible BR's are used for further purification processes. A typical bioassay-guided isolation and purification procedure for brassinosteroids is given in Figure 5.

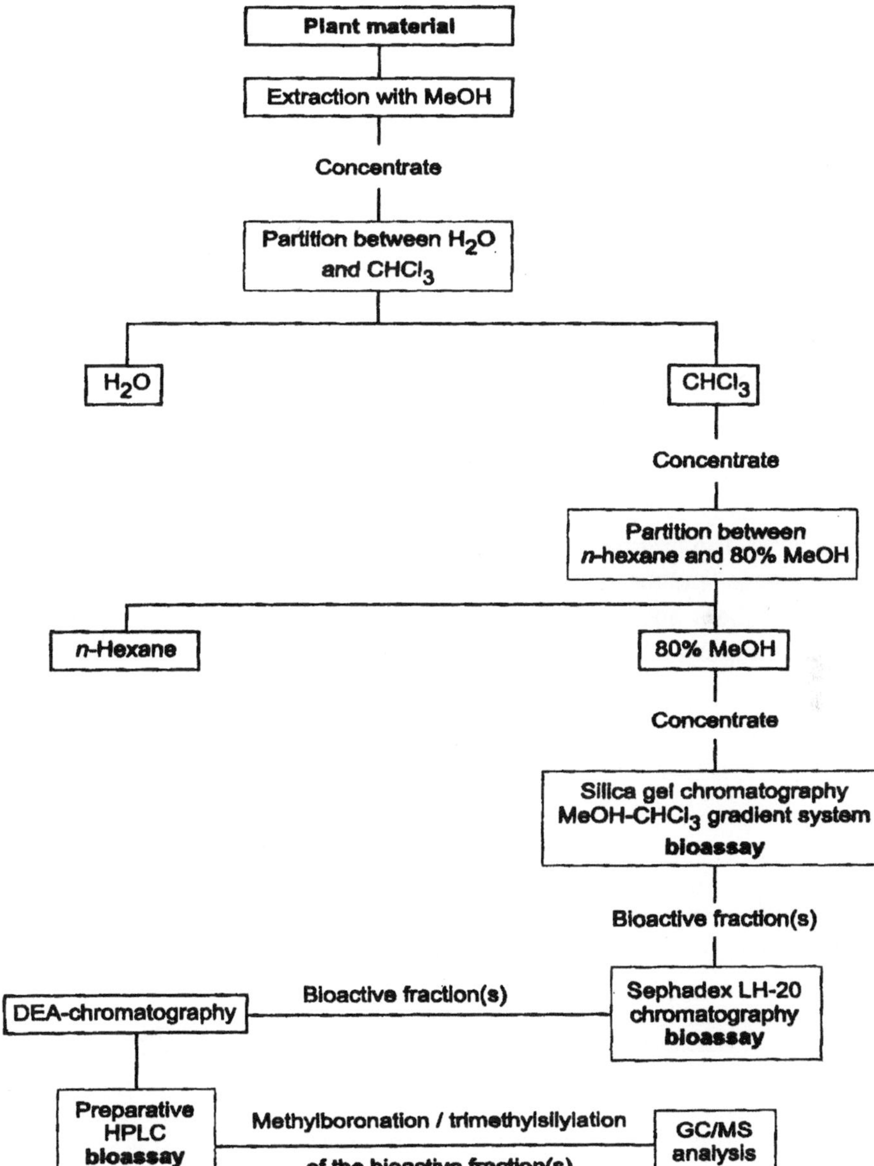

Figure 5. Typical bioassay-guided isolation and purification procedure of brassinosteroids

ANALYSIS OF BRASSINOSTEROIDS

GC-MS Analysis

The structural identification of brassinosteroids requires sensitive analytical methods since these compounds occur in trace amounts in plant material. Combination of gas chromatography and mass spectrometry (GC-MS) and GC-MS-SIM (selected ion monitoring) are the most used microanalytical methods. For this analysis, brassinosteroids must be converted into volatile derivatives. Compounds with two sets of vicinal diol functions (*e.g.* brassinolide (**1**) or castasterone (**2**)) form bismethylboronates (BMB's) by using methylboronic acid and pyridine (Fig. 6).[47-49] 2-Deoxybrassinosteroids, such as typhasterol (**5**) and teasterone (**6**), are derivatized by methylboronation of the side chain, followed by a trimethylsilylation of the 3-hydroxy group (Fig. 6).[50] The electron impact (EI) mass spectra of the BMB's of brassinosteroids and the methylboronate-trimethylsilyl (MB-TMS) derivatives yield much information with respect to the structures. Additionally, a molecular ion and characteristic key ions appear allowing an assignment of both the side chain structure and the basic steroidal skeleton.[37,48-50] The fragmentation pattern of BR's is characterized mainly by side chain cleavages leading to complementary key fragments representing both the type of side chain and the steroid skeleton. Examples of such fragments are illustrated in Figure 7.

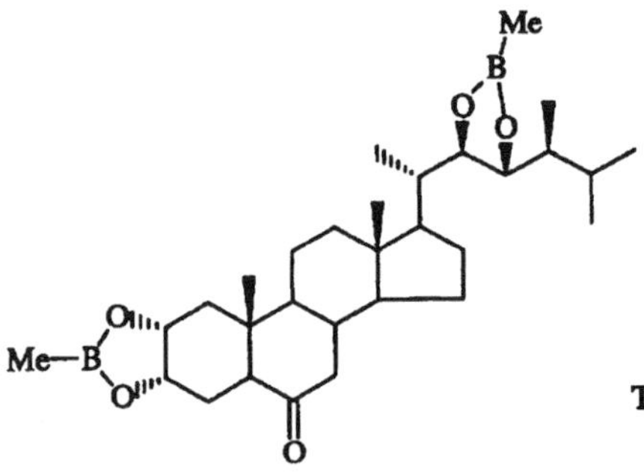

Castasterone-BMB

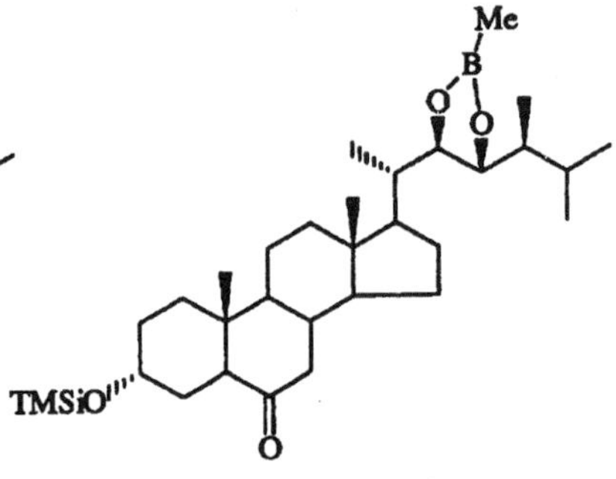

Typhasterol-MB-TMS

Figure 6. Derivatives of brassinosteroids used for gas chromatography-mass spectrometry (GC-MS)

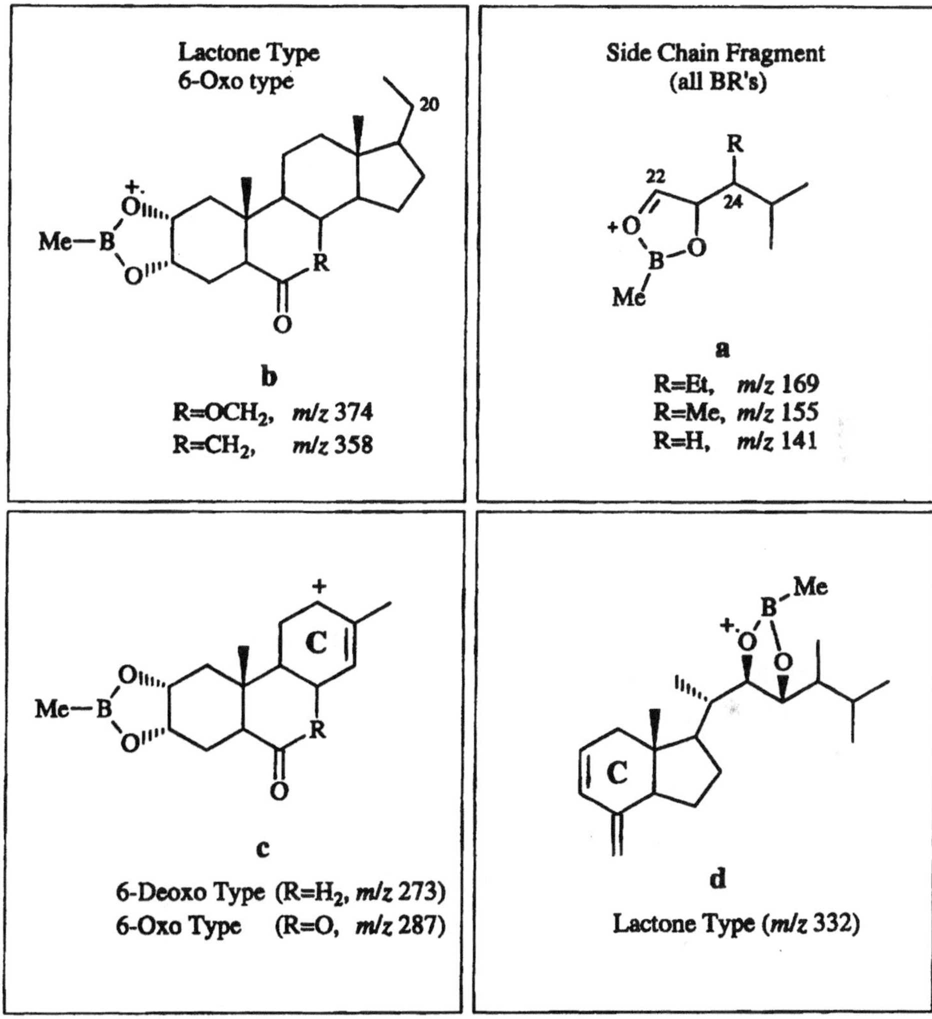

Figure 7. Important key ions in the EI-mass spectra of (*bis*)methylboronated brassinosteroids (BR's)

A saturated side chain is characterized by the key ion of type **a**, formed by an α-cleavage between C-20/C-22 (Fig. 7). The substitution at C-24 is indicated by the corresponding mass shift. Thus, an ion at *m/z* 155 (*e.g.* **1**, **2**) is typical for a methyl at C-24; a mass shift to *m/z* 169 [28-homobrassinosteroids (24*S*-ethyl), *e.g.* **7**, **8**] indicates an additional CH_2-unit in the side chain; and *m/z* 141 suggests a 28-norbrassinosteroid[51] (*e.g.* **18**). A side chain double bond between C-24/C-28, as in the case of dolichosterone[52] and 28-homodolichosterone, leads to a mass shift of 2 mass units towards lower masses (*m/z* 153 and 167, respectively) compared with the corresponding saturated analogs. Recently, four isomers of ξ-epi-23-dehydrobrassinolide with undefined stereochemistry were published.[20] In this case, the methylboronation of the side chain leads to two bismethylboronate derivatives caused by two different enol forms. While one product shows a side chain fragment at *m/z* 181 (cleavage C-17/C-20), the other one forms an ion at *m/z* 153, which originates by an α-cleavage between C-20/C-22.[20] The side chain cleavages also lead to ions comprising the rings A,B,C, and D and the C-atoms 20 and 21 (Fig. 7). The formation of the **b**-type ion, being complementary to the ion of type **a**, can be accompanied by a hydrogen transfer in each direction. The fission C17/C-20 (loss of the side chain) leads to an ion representing the A,B,C, and D rings. The saturated brassinosteroids show another key ion, comprising the rings A,B,C, and D and a part of the side chain (C-20 to C-23), which originates by an α-cleavage of the bond C-23/C-24. Thus, the formed fragments at $[M-57]^+$(*e.g.* of **18**), $[M-71]^+$ (*e.g.* of **2**), and $[M-85]^+$-ions (*e.g.* **7**), respectively, indicate the type of substitution at C-24.

Another type of fragmentation, originating by cleavages in the D ring, yields further possibilities for differentiating among the several types of brassinosteroids. Thus, a significant ion comprising the rings A,B, and C is ion **c**, which results from a hydrogen transfer from the charged species. The **c**-type ion (Fig. 7) is typical of both 6-deoxo and 6-oxobrassinosteroids with two hydroxy groups on ring A (*m/z* 273 and *m/z* 287, respectively). In the case of the 6-deoxocompounds, a further important key ion at *m/z* 288 is apparently originated by cleavages through ring D. M/z 205, comprising the A and B rings, is typical for 6-deoxobrassinosteroid-BMB derivatives.[51] An important key ion, characteristic of lactone-type brassinosteroids such as brassinolide (**1**), is the ion at m/z 332 (**d**) formed by cleavages through ring B (C-9/C-10 and O/C-7). It indicates a mass shift corresponding to the side chain structure (substitution at C-24).[49] Consequently, the ion at *m/z* 177, containing rings C and D and the carbons C-20, C-21, also suggests a lactone-type brassinosteroid (Fig. 7).

The EI mass spectra of the methylboronate-trimethylsilylether (MB-TMS) of the 2-deoxybrassinosteroids **5** and **6** show a typical fragmentation pattern involving rings A and B. Besides the prominent $(M-Me]^+$ - and $(M-TMSiOH]^+$ ions, the MB-TMS derivatives of typhasterol (**5**), teasterone (**6**), 28-homotyphasterol (**7**), and 28-homoteasterone (**8**) display an abundant $(M-C_2H_5]^+$-ion.[11,45,50] The presence of a 3-hydroxy-6a-oxalactone structure such as 2-deoxybrassinolide (**9**), is indicated by a highly specific key ion at *m/z* 156. This conjugated oxonium ion, formed by cleavages through rings A and B, was established by high-resolution mass measurements (Fig. 8).[11,53] The same holds true for the ions at *m/z* 211 and 195.

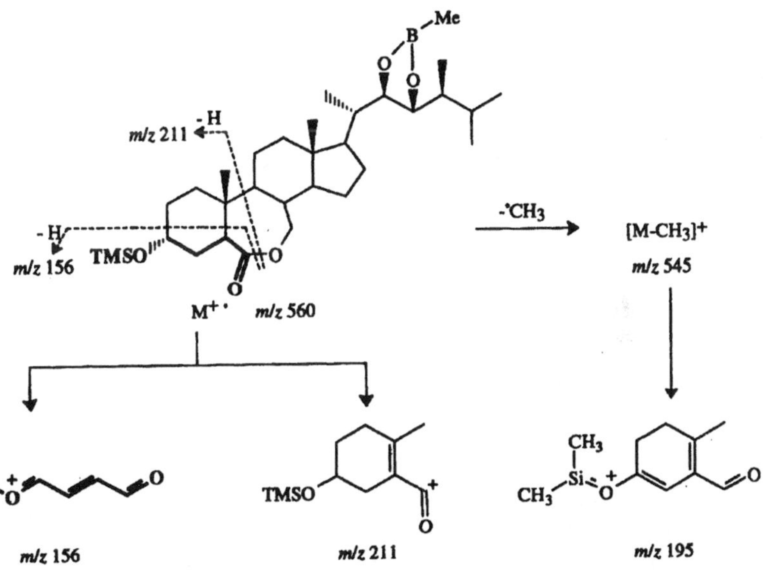

Figure 8. Ring AB fragmentation of 2-deoxybrassinolide (9) -MB-TMS

BR's with an oxo- or epoxy-function at ring A, such as 3-dehydroteasterone (**11**), secasterone (**15**), and 3-dehydro-6-deoxoteasterone (**12**), form methylboronates (MB). They show the same side chain fragmentation as the other 6-oxo - and deoxobrassinosteroids (Fig. 7). The ions comprising the steroidal ring system and the $[M-71]^+$-ion appear with the corresponding mass shift.[16,19,21,22, 31, 32]

The derivatized brassinosteroids exhibit sharp peaks in the capillary GC, and most of them can be separated from each other.[45.49] Both the Kovats retention indices, which use a hydrocarbon mixture,[29] as well as relative retention times with respect to an internal standard,[27,31] can be used to obtain reproducible retention data for the derivatized brassinosteroids.

HPLC Analysis

Brassinosteroids have no suitable chromophore for HPLC detection. Therefore, they have to be derivatized with a pre-labelling reagent. Gamoh *et al.*

have developed several applications of HPLC combined with ultraviolet (UV)[54] and fluorimetric detection[55-58] for determining brassinosteroids by using different boronic acid derivatization reagents (9-phenanthreneboronic acid,[55] dansylaminophenylboronic acid,[56] 1-naphthaleneboronic acid,[54] 1-cyanoisoindole-2-*m*-phenylboronic acid,[57] and *m*-aminophenylboronic acid[58]). Depending on the reagent used for fluorescence detection, the detection limits range from 25 to 100 pg per injection.[45] Compared with other boronating reagents, dansylaminophenylboronic acid (DABA) seems to be the most promising because the brassinosteroid dansylaminophenylboronates can be detected at longer wavelengths (excitation 345 nm/emission 515 nm).[45,56] By using the DABA-derivatives, it is also possible to separate the 24-epimeric brassinosteroids, brassinolide (**1**) and 24-epibrassinolide (**3**), as well as castasterone (**2**) and 24-epicastasterone (**4**).[58]

HPLC-MS Analysis

Only a few reports have focused on the application of HPLC/MS techniques to structural investigations of brassinosteroids.[45] Thus, combined HPLC-*frit*FABMS (fast atom bombardment mass spectrometry), which uses negative ionization, was first applied in the structure elucidation of teasterone 3-O-myristate (**6a**) from *Lilium longiflorum* anthers.[24] The HPLC-APCIMS (atmospheric pressure chemical ionization mass spectrometry) was described as a useful tool for the identification of castasterone (**2**) and teasterone (**6**) as their naphthalene boronate derivatives from *Cannabis sativa* seeds.[59,60]

CHEMICAL SYNTHESIS

Nearly all syntheses of brassinosteroids are partial syntheses. In most cases, the starting compounds are Δ^{22}-unsaturated sterols like stigmasterol, ergosterol, and campesterol or even the scarce 22-dehydrocampesterol. During the early 1980's, three different pathways leading to brassinolide (**1**) were published by American and Japanese chemists.[61-63] In the 1990's, some improved syntheses of castasterone (**2**) and brassinolide (**1**) were published.[64-66] Development in brassinosteroid synthesis has been the subject of a series of reviews.[9, 67-69] The typical structural features of the A/B rings can be introduced by classical reactions of steroid chemistry. However, the construction of the side chain moiety involves either a C_{22}-aldehyde as a key intermediate, or the Δ^{22} double bond of a suitable sterol precursor is directly used for introduction of the 22,23-diol functions (Figs. 9, 10).

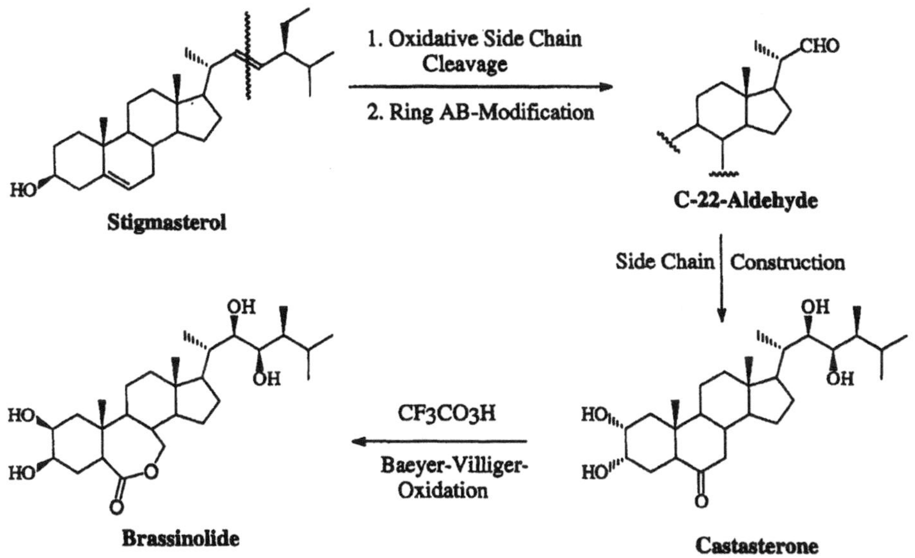

Figure 9. General synthetic routes for the synthesis of castasterone and brassinolide via a C-22-aldehyde intermediate.

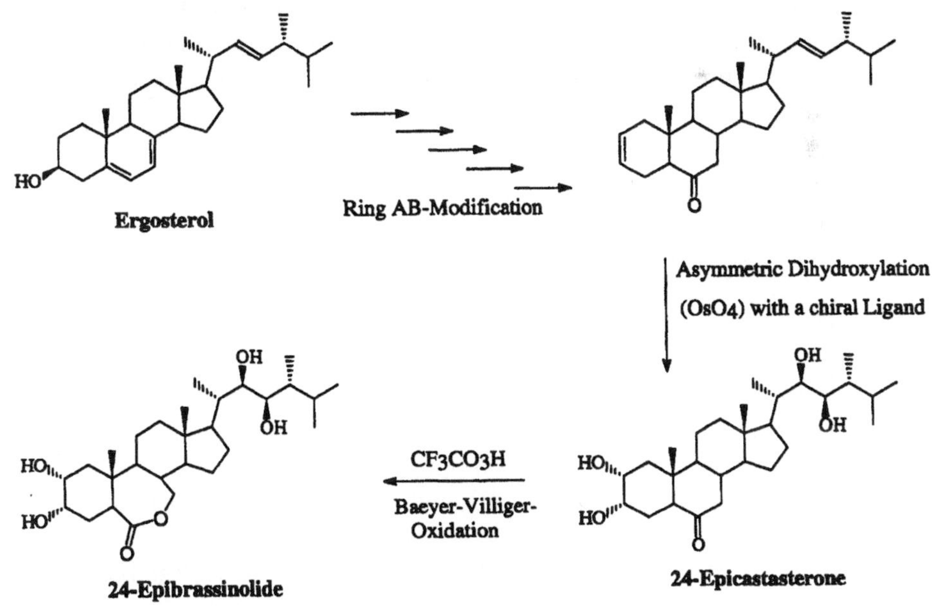

Figure 10. Typical synthetic strategy for the synthesis of 24-epicastasteroneand 24-epibrassinolide starting from ergosterol

Brassinolide (**1**) represents a preferred target for synthetic effort because of its high biological activity. The improved synthesis of **1** by Watanabe *et al.*[66] is one of the most practicable and effective procedures. In the course of this sequence, asymmetric dihydrocylation with OsO_4 is the common procedure to introduce both vicinal hydroxyfunctions (ring A, side chain). Stereoselectively leads both to the 2α,3α-diol and, in presence of the Sharpless catalyst 1,4-bis(9-O-dihydroquinidinyl)phthalazine as the chiral ligand, preferentially to castasterone (**2**) with the desired (22*R*,23*R*) configuration. In such a manner, castasterone (**2**) was obtained from stigmasterol by 13 steps. Direct Baeyer-Villiger oxidation of **2** with trifluoroperoxy acetic acid as a useful standard reagent yielded brassinolide (**1**).

24-Epibrassinolide (**3**) can be synthesized from ergosterol without side chain reconstruction (Fig. 10). Based on earlier studies from other laboratories, McMorris and Patil[70] published an improved synthesis of **3**. In that case, the starting compound ergosterol was transformed *via* solvolysis of its mesylate to the corresponding 3α,5-cyclo-6β-hydroxy compound, followed by oxidation with CrO_3-pyridine to yield a cyclo-6-ketone. Finally, the desired (22*R*,23*R*) diastereomer 24-epicastasterone (**4**) was obtained by reaction of the intermediate (22*E*,24*R*)-5α-ergosta-2,22-dien-6-one with OsO_4 in the presence of a chiral ligand (dihydroquinidine 4-chlorobenzoate). In analogy to the brassinolide synthesis, direct Baeyer-Villiger oxidation led to 24-epibrassinolide (**3**), resulting in the synthesis of the target compound **3** from ergosterol in only 7 steps.

SUMMARY

A brief review of structures, occurrence, analysis, and synthesis of BR's has been given. With regard to analysis, the EI mass spectra of derivatized BR's (BMB, MB, and MB-TMS) specifically yield valuable structural information. There is a set of ions characteristic for the substructures of the single brassinosteroid. Additionally, all results from the gas chromatographic behavior of the 24-epimers indicate that members of the 24*S* series elute generally earlier than the 24*R* analogs.[3,15,27,31,47-49,71] Combining the several types of key fragments in the EI mass spectra of the BMB, MB, and MB-TMS derivatives, it is possible to deduce the general basic brassinosteroid structure.

Based on structural proposals obtained from mass spectral data, one can synthesize suitable reference compounds for comparing both the EI mass spectra and the GC retention data of the brassinosteroid found in the natural source with the synthetic compound. In this way, we identified 24-epi-6-deoxocastasterone (**17**),[27] secasterone (**15**),[31,72] and 24-episecasterone (**16**),[26] as well as 2-

deoxybrassinolide (**9**),[15, 73] as new brassinosteroids.

During the last decade, new microanalytical techniques for BR's have been developed.[45] HPLC is combined with fluorimetric detection by using several boronating reagents for the BR's to achieve highly sensitive methods.[45,74] For the mass spectrometric detection, the frit-FAB (fast atom bombardment),[24] APCI (atmospheric pressure chemical ionization), and the ESI (electrospray ionization) are used.[45,59] Thus, in addition to molecular and biochemical studies, further development of microanalytical methods will play an important role in solving biosynthetic, analytical, and physiological problems of brassinosteroids in the future.

ACKNOWLEDGEMENTS

We gratefully acknowledge Prof. Takao Yokota (Teikyo University, Utsunomiya) and Dr. Suguru Takatsuto (Joetsu University, Niigata) for fruitful collaboration. We are also indebted to Mrs. Christine Kuhnt for the GC/MS-measurements and Mrs Helga Menzky (IPB Halle) for skillful technical assistance, as well as the Deutsche Forschungsgemeinschaft for financial support.

REFERENCES

1. GROVE, M.D., SPENCER, G.F., ROWEDDER, W.K., MANDAVA, N., WORLEY, J.F., WARTHEN, Jr., J.D., STEFFENS, G.L., FLIPPEN-ANDERSON, J.L., COOK, Jr., J.K. 1979. Brassinolide, a plant growth-promoting steroid isolated from *Brassica napus* pollen. Nature 281: 216-217.
2. SAKURAI, A., FUJIOKA, S. 1993. The current status of physiology and biochemistry of brassinosteroids. Plant Growth Regul. 13: 147-159.
3. ADAM, G., PORZEL, A., SCHMIDT, J., SCHNEIDER, B., VOIGT, B. 1996. New developments in brassinosteroid research. pp. 495-549 *in* Studies in Natural Products Chemistry (Atta-ur-Rahman, ed.), Amsterdam, Elsevier Science.
4. YOKOTA, T. 1997. Structure, biosynthesis and function of brassinosteroids. Trends Plant Sci. 2: 137-143.
5. SASSE, J.M. 1997. Recent progress in brassinosteroid research. Physiol. Plant. 100: 696-701.
6. FUJIOKA, S., SAKURAI, A. 1997. Brassinosteroids. Nat. Prod. Rep. 14: 1-10.
7. CLOUSE, S.D., SASSE, J.M. 1998. Brassinosteroids: Essential regulators of plant growth and development. Annu. Rev. Plant Physiol. Plant Mol. Biol. 49:

427-451.
8. CUTLER, H.G., YOKOTA, T., ADAM, G. (eds.) 1991. Brassinosteroids - Chemistry, Bioactivity, and Applications. American Chemical Society, Washington, DC.
9. KHRIPACH, V.A., ZHABINSKII, V.N., DE GROOT, A.E. 1999. Brassinosteroids - A New Class of Plant Hormones. Academic Press, San Diego.
10. SAKURAI, A., YOKOTA, T., CLOUSE, S.D. (eds.) 1999. Brassinosteroids, Steroidal Plant Hormones. Springer-edition, Tokyo.
11. ADAM, G., SCHMIDT, J., SCHNEIDER, B. 1999. Brassinosteroids. pp. 1-46 *in* Progress in the Chemistry of Natural Products 78, Springer-edition, Wien-New York.
12. CLOUSE, S.D., FELDMANN, K. A. 1999. Molecular genetics of brassinosteroid action *in* ref. 10, pp. 163-189.
13. SZEKERES, M., KONCZ, C. 1998. Biochemical and genetic analysis of brassinosteroid metabolism and function in *Arabidopsis*. Plant Physiol. Biochem. 36: 145-155.
14. FUJIOKA, S. 1999. Natural occurrence of brassinosteroids in the plant kingdom *in* ref. 10, pp. 21-45.
15. SCHMIDT, J., VOIGT, B., ADAM, G. 1995. 2-Deoxybrassinolide - a naturally occurring brassinosteroid from *Apium graveolens*. Phytochemistry 40: 1041-1043.
16. FUJIOKA, S., CHOI, Y.-H., TAKATSUTO, S., YOKOTA, T., LI, J., CHORY, J., SAKURAI, A. 1996. Identification of castasterone, 6-deoxocastasterone, typhasterol and 6-deoxotyphasterol from shoots of *Arabidopsis thaliana*. Plant Cell Physiol. 37: 1201-1203.
17. FUJIOKA, S., NOGUCHI, T., YOKOTA, T., TAKATSUTO, S., YOSHIDA, S. 1998. Brassinosteroids in *Arabidopsis thaliana*. Phytochemistry 48: 595-599.
18. FUJIOKA, S., INOUE, T., TAKATSUTO, S., YANAGISAWA, T., YOKOTA, T., SAKURAI, A. 1995. Identification of a new brassinosteroid, cathasterone, in cultured cells of *Catharanthus roseus* as a biosynthetic precursor of teasterone. Biosci. Biotechnol. Biochem. 59: 1543-1547.
19. CHOI, Y.-H., FUJIOKA, S., NOMURA, T., HARADA, A., YOKOTA, T., TAKATSUTO, S., SAKURAI, A. 1997. An alternative brassinolide biosynthetic pathway via late C-6 oxidation. Phytochemistry 44: 609-613.
20. YOKOTA, T., HIGUCHI, K., TAKAHASHI, N., KAMURO, Y., WATANABE, T., TAKATSUTO, S., 1998. Identification of brassinosteroids

with epimeric substituents and/or the 23-oxo group in pollen and anthers of *Japanese cedar*. Biosci. Biotechnol. Biochem. 62: 526-531.

21. GRIFFITHS, P.G., SASSE, J.M., YOKOTA, T., CAMERON, D.W. 1995. 6-Deoxotyphasterol and 3-dehydro-6-deoxoteasterone, possible precursors to brassinosteroids in the pollen of *Cupressus arizonica*.Biosci. Biotechnol. Biochem. 59: 956-959.
22. ABE, H., HONJO, C., KYOKAWA, Y., ASAKAWA, S., NATSUME, M., NARUSHIMA, M. 1994. 3-Oxoteasterone and the epimerization of teasterone: identification in lily anthers and *Distylium racemosum* leaves and its biotransformation into typhasterol. Biosci. Biotechnol. Biochem. 58: 986-989.
23. YASUTA, E., TERAHATA, T., NAKAYAMA, M., ABE, H., TAKATSUTO, S., YOKOTA, T. 1995. Free and conjugated brassinosteroids in the pollen and anthers of *Erythronium japonicum* Decne. Biosci. Biotechnol. Biochem. 59: 2156-2158.
24. ASAKAWA, S., ABE, H., KYOKAWA, Y., NAKAMURA, S., NATSUME, M. 1994. Teasterone 3-myristate: a new type of brassinosteroid derivative in *Lilium longiflorum* anthers. Biosci. Biotechnol. Biochem. 58: 219-220.
25. ASAKAWA, S., ABE, H., NISHIKAWA, N., NATSUME, M., KOSHIOKA, M. 1996. Purification and identification of new acyl-conjugated teasterones in lily pollen. Biosci. Biotechnol. Biochem. 60: 1416-1420.
26. FRIEBE, A., VOLZ, A., SCHMIDT, J., VOIGT, B., ADAM, G., SCHNABL, H. 1999. 24-Epi-secasterone and 24-epi-castasterone from *Lychnis viscaria* seeds. Phytochemistry 52: 1607-1610.
27. SPENGLER, B., SCHMIDT, J., VOIGT, B., ADAM, G. 1995. 6-Deoxo-28-norcastasterone and 6-deoxo-24-epicastasterone - two new brassinosteroids from *Ornithopus sativus*. Phytochemistry 40: 907-910.
28. ABE, H., TAKATSUTO, S., NAKAYAMA, M., YOKOTA, T. 1995. Homotyphasterol, a new natural brassinosteroid from rice (*Oryza sativa* L.) bran. Biosci. Biotechnol. Biochem. 59: 176-178.
29. YOKOTA, T., MATSUOKA, T., KOARAI, T., NAKAYAMA, M. 1996. 2-Deoxybrassinolide, a brassinosteroid from *Pisum sativum* seed. Phytochemistry 42: 509-511.
30. SCHMIDT, J., YOKOTA, T., SPENGLER, B., ADAM, G. 1993. 28-Homoteasterone - a new naturally occurring brassinosteroid from seeds of *Raphanus sativus*. Phytochemistry 34: 391-392.
31. SCHMIDT, J., SPENGLER, B., YOKOTA, T., NAKAYAMA, M., TAKATSUTO, S., VOIGT, B., ADAM, G. 1995. Secasterone, the first naturally occurring 2,3-epoxybrassinosteroid from *Secale cereale*.

Phytochemistry 38: 1095-1097.

32. YOKOTA, T., NAKAYAMA, M., WAKISAKI, T., SCHMIDT, J., ADAM, G. 1994. 3-Dehydroteasterone, a 3,6-diketobrassinosteroid as a possible biosynthetic intermediate of brassinolide from wheat grain. Biosci. Biotechnol. Biochem. 58: 1183-1186.
33. SUZUKI, H., INOUE, T., FUJIOKA, S., TAKATSUTO, S., YANAGISAWA, T., YOKOTA, T., MUROFUSHI, N., SAKURAI, A. 1994. Possible involvement of 3-dehydroteasterone in the conversion of teasterone to typhasterol in cultured cells of *Catharanthus roseus*. Biosci. Biotech. Biochem. 58: 1186-1188.
34. FUJIOKA, S., SAKURAI, A. 1997. Biosynthesis and metabolism of brassinosteroids. Physiol. Plant. 100: 710-715.
35. SAKURAI, A. 1999. Biosynthesis *in* ref. 10, pp. 91-111.
36. TAYLOR, P.E., SPUCK, K., SMITH, P.M., SASSE, J.M., YOKOTA, T., GRIFFITHS, P.G.,. CAMERON, D.W. 1993. Detection of brassinosteroids in pollen of *Lolium perenne* by immunocytochemistry. Planta 189: 91-100.
37. TAKATSUTO, S. 1994. Brassinosteroids: Distribution in plants, bioassay and microanalysis by gas chromatography-mass spectrometry. J. Chromatogr. A 658: 3-15.
38. TAKATSUTO, S., ABE, H., GAMOH, K. 1990. Evidence for brassinosteroids in strobilus of *Equisetum arvense* L.. Agric. Biol. Chem. 54: 1057-1059.
39. YOKOTA, T., KIM, S.-K., FUKUI, Y., TAKAHASHI, N., TAKEUCHI, Y., TAKEMATSU, T. 1987. Brassinosteroids and sterols from a green alga, *Hydrodictyon reticulatum*: Configuration at C-24. Phytochemistry 26: 503-506.
40. YOKOTA, T., ARIMA, M., TAKAHASHI, N. 1982. Castasterone, a new phytosterol with plant-hormone potency, from chestnut insect gall. Tetrahedron Letters 23: 1275-1278.
41. YOKOTA, T., ARIMA, M., TAKAHASHI, N., TAKATSUTO, S., IKEKAWA, N., TAKEMATSU, T. 1983. 2-Deoxycastasterone, a new brassinolide-related bioactive steroid from pinus pollen. Agric. Biol. Chem. 47: 2419-2420.
42. SCHNEIDER, J.A., YOSHIHARA, K., NAKANISHI, K., KATO, N. 1983. Typhasterol (2-deoxycastasterone): A new plant growth regulator from cat-tail pollen. Tetrahedron Letters 24: 3859-3860.
43. ABE, H., MORISHITA, T., UCHIYAMA, M., TAKATSUTO, S., IKEKAWA, N. 1984. A new brassinolide-related steroid in leaves of *Thea sinensis*. Agric. Biol. Chem. 48: 2171-2172.

44. IKEKAWA, N., NISHIYAMA, F., FUJIMOTO, Y. 1988. Identification of 24-epibrassinolide in bee pollen of the broad bean, *Vicia faba* L.. Chem. Pharm. Bull. 36: 405-407.
45. TAKATSUTO, S., YOKOTA, T. 1999. Biochemical analysis of natural brassinosteroids *in* ref. 10, pp. 47-68.
46. WADA, K., MARUMO, S., ABE, H., MORISHITA, T., NAKAMURA, K., UCHIYAMA, M., MORI, K. 1984. A rice lamina inclination test - a micro-quantitative bioassay for brassinosteroids. Agric. Biol. Chem. 48: 719-726.
47. IKEKAWA, N., TAKATSUTO, S., KITSUWA, T., SAITO, H., MORISHITA, T., ABE, H. 1984. Analysis of natural brassinosteroids by gas chromatography and gas chromatography-mass spectrometry. J. Chromatogr. 290: 289-302.
48. TAKATSUTO, S., YING, B., MORISAKI, M., IKEKAWA, N. 1982. Microanalysis of brassinolide and its analogues by gas chromatography and gas chromatography-mass spectrometry. J. Chromatogr. 239: 233-241.
49. IKEKAWA, N., TAKATSUTO, S. 1984. Microanalysis of brassinosteroids in plants by gas chromatography/mass spectrometry. Mass Spectroscopy (Japan) 32: 55-70.
50. TAKATSUTO, S., IKEKAWA, N. 1986. Analysis of 2-deoxybrassinosteroids by gas chromatography-mass spectrometry. Chem. Pharm. Bull. 34: 3435-3439.
51. ABE, H., MORISHITA, T., UCHIYAMA, M., TAKATSUTO, S., IKEKAWA, N., IKEDA, M., SASSA, T., KITSUWA, T., MARUMO, S. 1983. Occurrence of three new brassinosteroids: Brassinone, (24*S*)-24-ethylbrassinone and 28-norbrassinone, in higher plants. Experientia 39: 351-353.
52. ABE, H., NAKAMURA, K., MORISHITA, T., UCHIYAMA, M., TAKATSUTO, S., IKEKAWA, N. 1984. Endogenous brassinosteroids of the rice plant: Castasterone and dolichosterone. Agric. Biol. Chem. 48: 1103-1104.
53. SCHMIDT, J., SPENGLER, B., VOIGT, B., ADAM, G. 1996. New brassinosteroids from european cultivated plants, p. 61 *in* Proceedings 23rd Annual Meeting Plant Growth Regulator Society of America, Calgary.
54. GAMOH, K., KITSUWA, T., TAKATSUTO, S., FUJIMOTO, Y., IKEKAWA, N. 1988. Determination of trace brassinosteroids by high performance liquid chromatography. Anal. Sci. 4: 533-535.
55. GAMOH, K., OMOTE, N., OKAMOTO, N., TAKATSUTO, S. 1989. High-performance liquid chromatography of brassinosteroids in plants with derivatization using 9-phenanthreneboronic acid. J. Chromatogr. 469: 424-428.

56. GAMOH, K., OKAMOTO, N., TAKATSUTO, S., TEIJIMA, I. 1990. Determination of traces of natural brassinosteroids as dansylaminophenylboronates by liquid chromatography with fluorimetric detection. Anal. Chim. Acta 228: 101-105.
57. GAMOH, K., TAKATSUTO, S. 1989. A boronic acid derivative as a highly sensitive fluorescence derivatization reagent for brassinosteroids in liquid chromatography. Anal. Chim. Acta 222: 201-204.
58. GAMOH, K., TAKATSUTO, S., IKEKAWA, N. 1992. Effective separation of C-24 epimeric brassinosteroids by liquid chromatography. Anal. Chim. Acta 256: 319-322.
59. GAMOH, K., ABE, H., SHIMADA, K., TAKATSUTO, S. 1996. Liquid chromatography/mass spectrometry with atmospheric pressure chemical ionization of free brassinosteroids. Rap. Commun. Mass Spectrom. 10: 903-906.
60. GAMOH, K., PRESCOTT, M.C., GOAD, L.J., TAKATSUTO, S. 1996. Analysis of brassinosteroids by liquid chromatography-mass spectrometry. Bunseki Kagaku 45: 523-527.
61. FUNG, S., SIDDALL, J.B. 1980. Stereoselective synthesis of brassinolide: A plant promoting steroidal lactone. J. Amer. Chem. Soc. 102: 6581-6582.
62. ISHIGURO, M., TAKATSUTO, S., MORISAKI, M., IKEKAWA, N. 1980. Synthesis of brassinolide, a steroidal lactone with plant growth promoting activity. Chem. Comm. 20: 962-964.
63. THOMPSON, M.J., MANDAVA, N.B., MEUDT, W.J., LUSBY, W.R., SPAULDING, D.W. 1981. Synthesis and biological activity of brassinolide and its 22β,23β-isomer: Novel plant growth steroids. Steroids 38: 567-580.
64. BACK, T.G., BLAZECKA, P.G., KRISHNA, M.V. 1993. A new synthesis of castasterone and brassinolide from stigmasterol: A concise and stereoselective elaboration of the side chain from a C-22 aldehyde. Can. J. Chem. 71: 156-163.
65. McMORRIS, T.C., CHAVEZ, R. G., PATIL, P. A. 1996. Improved synthesis of brassinolide. J. Chem. Soc. Perkin Trans 1 1996: 295-302.
66. WATANABE, T., TAKATSUTO, S., FUJIOKA, S., SAKURAI, A. 1997. Improved synthesis of castasterone and brassinolide. J. Chem. Research (S), 360-361.
67. BACK, T.G. 1995. Stereoselective synthesis of brassinosteroids. pp. 321-364 *in* Studies in Natural Products Chemistry (Atta-ur-Rahman, ed.), Elsevier Science, Amsterdam.
68. KOVGANKO, N.V., ANANICH, S.K. 1997. Advances in the chemical synthesis of brassinosteroids. Chem. Nat. Comp. 33: 389-416.

69. McMORRIS, T.C. 1999. Chemical synthesis of brassinosteroids *in* ref. 10, pp. 69-90.
70. McMORRIS, T.C., PATIL, P. A. 1993. Improved synthesis of 24-epibrassinolide from ergosterol. J. Org. Chem. 58: 2338-2339.
71. SCHMIDT, J., SPENGLER, B., YOKOTA, T. ADAM, G. 1993. The co-occurrence of castasterone and 24-epicastasterone in seeds of *Ornithopus sativus*. Phytochemistry 32: 1614-1615.
72. VOIGT, B, TAKATSUTO, S., YOKOTA, T., ADAM, G. 1995. Synthesis of secasterone and further epimeric 2,3-epoxybrassinosteroids. J. Chem. Soc. Perkin Trans 1, 1995: 1495-1498.
73. VOIGT, B, SCHMIDT, J., ADAM, G. 1996. Synthesis of 24-epiteasterone, 24-epityphasterol and their B-homo-6a-oxalactones from ergosterol. Tetrahedron 52: 1997-2004.
74. GAMOH, K., TAKATSUTO, S. 1994. Liquid chromatographic assay of brassinosteroids in plants. J. Chromatogr. A 658: 17-25.

Chapter Thirteen

RECENT ADVANCES IN MOLECULAR GENETIC STUDIES OF THE FUNCTIONS OF BRASSINOLIDE, A STEROID HORMONE IN PLANTS

Zhi-Yong Wang[1] and Joanne Chory

Howard Hughes Medical Institute and Plant Biology Laboratory
The Salk Institute
10010 N. Torrey Pines Rd.
La Jolla, CA 92037

[1]To whom correspondence should be addressed.

Evolution of Metabolic Pathways – J.T. Romeo et al. (Editors)

INTRODUCTION

Twenty years ago, the first steroidal compound with growth promoting activity in plants, named brassinolide (BL), was identified in rape pollen.[1] Since then, over 40 structurally related compounds called brassinosteroids (BRs) have been found naturally in various plant species ranging from algae to higher plants.[2,3] Among these, BL remains the most active BR in eliciting biological responses. Physiological studies during the past 20 years have documented a diverse range of BR-induced responses.[4,5] Among these, the most exciting observation is perhaps the increase of crop yield as a result of BR application in the field. BR-induced morphological changes include increased cell elongation, expansion and division, enhanced stem elongation and xylem differentiation, induced leaf bending, unrolling and epinasty, and inhibition (or promotion depending on the test system and concentration of BL) of root growth. At the physiological level, BR treatment increases photosynthetic activity and enhances resistance to pathogens, temperature, drought, and salinity stresses.[4] The biochemical changes that underlie these morphological and physiological effects include activation of proton pump ATPases and invertases, and induction of specific gene expression.[4,5] The effects of BRs often vary in different systems, showing complex interactions with other phytohormones. In general, the effects of BRs are synergistic with auxin, additive with gibberellins, and can be inhibited by ABA.[4] These physiological studies demonstrate the potent biological activity of exogenously applied BRs. However, they fail to provide a definitive proof for the requirement and function of endogenous BRs in normal plant growth and development. The numerous observed physiological responses and complex interactions with other growth regulators have been exciting, but have somewhat obscured the role of BRs as plant hormones.

Recent molecular genetic studies of BR-deficient and BR-insensitive mutants have established a hormonal role in normal plant growth and development.[6,7] The cloning of the corresponding genes has advanced our insight into the physiology and biochemistry of plant steroid hormone action, and provided molecular tools for further dissecting the BR signal transduction pathways.[8] This chapter reviews the current understanding of the functions of the genes involved in BR metabolism and response, and discusses new efforts toward genetic identification of additional components involved in BR action.

BIOSYNTHETIC PATHWAYS AND BIOSYNTHETIC MUTANTS

Like animal steroid hormones, all BRs are composed of a typical steroidal skeleton with various substitutions required for biological activity.[3] A BR biosynthetic pathway has been proposed based on extensive feeding experiments that use radiolabeled compounds (Fig. 1).[3,9] Such information on biosynthesis has been instrumental in characterizing mutants,

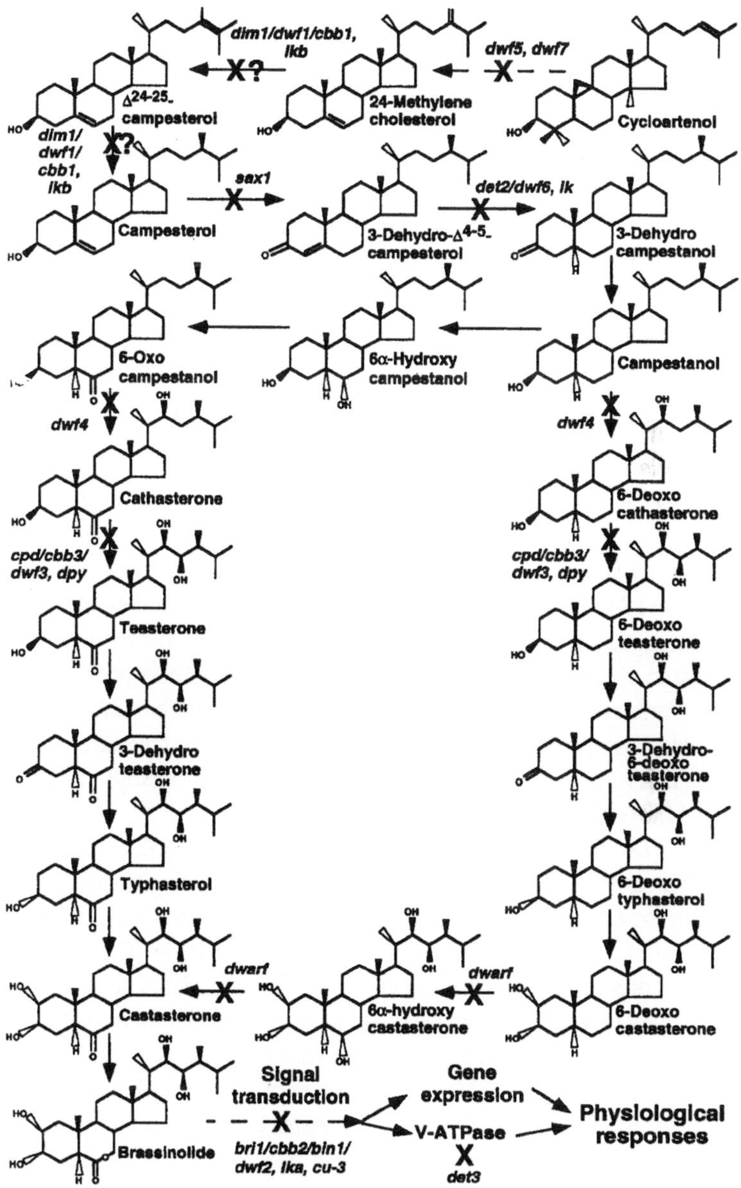

Figure 1. The brassinolide biosynthetic pathways. The branched pathways of early C-6 oxidation (left) and late C-6 oxidation (right) are shown. Known mutations that block either biosynthesis of or response to brassinosteroids are indicated (X). Question marks indicate uncertainty about exact biochemical lesions and broken arrows denote multiple steps. This diagram is adapted from Li and Chory (1999).[7]

which, in turn, have confirmed and refined the proposed biosynthetic pathway (Fig. 1). So far, at least eight independent BR biosynthetic mutants have been identified in *Arabidopsis*, and five corresponding genes have been cloned.[7,10] BR deficient mutants also have been isolated from other species, such as pea and tomato.[5,11-14]

The *DET2* gene encodes a steroid 5a-reductase that is structurally and functionally conserved in *Arabidopsis* and humans. Cloning of the *Arabidopsis DET2* gene has revealed its sequence homology to mammalian steroid 5a-reductases and suggested that DET2 is involved in the biosynthesis of an important plant steroid. Indeed, all the phenotypes of *det2* can be rescued by application of BL, demonstrating that the phenotypes of *det2* are due to a defect in BL biosynthesis.[15] The conservation between *Arabidopsis* DET2 and human steroid 5a-reductase goes beyond sequence similarity, as the DET2 protein can catalyze the 5a-reduction of several human steroids, such as testosterone and progesterone. Moreover, expression of human steroid 5a-reductases can complement the *Arabidopsis det2* mutant.[16] Such a structural and functional conservation between this steroid biosynthetic enzyme in humans and *Arabidopsis* suggests that steroid biosynthesis is highly conserved throughout evolution.

Further characterization of *det2* mutants and the DET2 protein has elucidated the precise role of DET2 in BR biosynthesis. The *det2* null mutant accumulates only about 10% of most BR intermediates, and its phenotypes can be rescued by all intermediates downstream of campestanol in the BR biosynthetic pathway. The *det2* phenotypes are not rescued by campesterol. Furthermore, *det2* mutants fail to convert deuterium-labeled campesterol to deuterium-labeled campestanol.[17] More detailed studies show that *det2* is defective in the conversion of (24R)-24-methylcholest-4-ent-3-one to (24R)-24-methyl-5a-cholestan-3-one in brassinosteroid biosynthesis.[18] (Fig. 1).

The *CPD/CBB3/DWF3* gene encodes a cytochrome P450 (CYP90A), a steroid hydroxylase that converts cathasterone to teasterone by catalyzing hydroxylation at position C23.[19] Feeding experiments suggest that the tomato *dpy* mutant has a defect in a homolog of *CPD*.[5] The *DWF4* gene encodes another cytochrome P450 (CYP90B) that converts campestanol to cathasterone by catalyzing hydroxylation at C22.[20] The tomato *DWARF* gene also encodes a cytochrome P450 that catalyses the C-6 oxidation of 6-deoxocastasterone (6-deoxoCS) to castasterone (CS), the immediate precursor of BL.[13,14] The *DWF7* gene encodes a D7 sterol C-5 desaturase involved in the production of 24-methylenecholesterol in the sterol biosynthetic pathway.[21] The *dim1/dwf1/cbb1* mutants are deficient in the conversion of 24-methylenecholesterol to campesterol.[22] Similar BR biosynthetic defects have been shown for the pea *lkb* mutant.[11,12] The *dim1* mutants also have altered profiles of phytosterols, which may contribute to some aspects of their phenotypes. The *DIM1/DWF1/CBB1* gene encodes a protein that contains a domain conserved among several FAD-dependent oxidoreductases. Its biochemical activity has not been defined.[23] Recently, another mutant called *sax1* was shown to be deficient in brassinosteroid synthesis.[10,24] Feeding experiments that used BR biosynthetic intermediates suggested that the *sax1* mutation blocked BR synthesis at a step from campesterol to (22S, 24R)-22-hydroxyergost-4-en-3-one, a substrate of DET2.[10]

DEFICIENT MUTANT PHENOTYPES AND FUNCTION IN DEVELOPMENT

All BR biosynthetic mutants of *Arabidopsis* have similar phenotypes characterized by dwarfism, dark green leaves, reduced fertility when grown in light, de-etiolation in dark, defects in xylem cell differentiation, and altered expression of stress-, pathogenesis- and light-regulated genes. These phenotypic changes are consistent with previous observations of the effect of exogenous BRs on related processes.[4,5,7,25] Some of the physiological and cellular alterations have been correlated to changes in gene expression.[19,22,26] For example, BR-regulated genes coding for cell wall-modifying enzymes [5] and tubulins [27] may contribute to the BR effect on cell elongation. Other BR actions may involve more direct effects on cytoplasmic and plasma membrane activities. In particular, the vacuolar H^+-ATPase activity has been found to be important for BR regulation of cell elongation.[28]

Brassinosteroids and Light Regulation of Plant Development

The phenotypes of de-etiolation (or constitutive photomorphogenesis) in the dark, observed in all BR-deficient mutants, suggest an important role of BRs in light regulation of plant development. In fact, several of the first BR-deficient mutants, such as *det2* and *cpd*, were originally isolated as de-etiolated/constitutive-photomorphogenesis mutants that show characteristics of light-grown plants when grown in the dark. The finding that these mutants are deficient in BR biosynthesis brought renewed attention to BR research.

Because light regulates plant development, wild-type plants can have different morphologies depending on the light conditions under which they grow. *Arabidopsis* seedlings grown in the dark have elongated hypocotyls, closed cotyledons, apical hooks, and undifferentiated plastids; in contrast, light-grown seedlings have short hypocotyls, open cotyledons, express light induced genes, have no apical hook, and develop chloroplasts. Mutants with phenotypes of light-grown plants when grown in the dark are believed to represent components involved in light control of plant development.[29]

A dark-grown BR-deficient mutant displays nearly all the morphological features of a light-grown wild type plant, including a short hypocotyl, open cotyledons, lack of an apical hook, and development of primary leaves. In some BR-deficient mutants, such as *det2* and *cpd*, the expression of light induced genes is de-repressed. Although this has not been true for the *dim1* mutant, it is not known whether other alterations in *dim1*, such as that of its sterol profile, contribute to the difference of this mutant from other more-specific BR mutants. Unlike other de-etiolated mutants, such as *det1* and *cop1*, BR-deficient mutants do not show chloroplast differentiation. It appears that BR deficiency initiates a large part of a developmental program that is triggered by light in wild type plants.

Although it has also been argued that the de-etiolated phenotype of brassinosteroid mutants is a secondary consequence of dwarfism,[22] there is much evidence for direct cross talk between light and BR actions. Differences in BR biosynthesis have been noticed between light- and dark-grown *Arabidopsis*. In the branched pathways of BR synthesis, early C-6 oxidation has been shown to be predominant in dark-grown *Arabidopsis*, while the late C-6 oxidation pathway is predominant in light-grown plants.[17,20] However, the significance of this alternation in biosynthetic pathways is not clear. Quantitative comparisons between active BR levels in light- and dark-grown plants have not been reported, and it is not known whether light affects the levels of active BRs.

Light may also modulate BR sensitivity. Dual regulation of a common signaling component would indicate a cross talk between the two signal transduction pathways. Recently, a mutation of the C-subunit of the V-type H^+-pumping ATPase was found to be responsible for the BR-insensitive and de-etiolation phenotypes of the *det3* mutant.[28] Interestingly, genes encoding several other subunits of the V-ATPase are down regulated by light.[30] The fact that the BR responsive *TCH4* gene is up-regulated by dark treatments,[31] and the light induced genes are de-repressed in dark grown BR mutants[19,32] may suggest that the interaction between light and BR signals occurs before the signals alter gene expression. Further molecular dissection of the light- and BR- response pathways will elucidate the precise role of BRs in photomorphogenesis and skotomorphogenesis.

Cross Talk with Other Phytohormone Responses

Interactions between BRs and other hormones have been studied at the physiological level. Although the results vary in different systems, it is known that the effects of BRs are generally synergistic with that of auxin, independent and additive with that of gibberellins, and antagonistic with that of ABA.[4]

An effect of a BR biosynthetic mutation on responses to other plant hormones was demonstrated recently by the identification of the *sax1* mutant. The *sax1* mutant was identified in a screen for mutants hypersensitive to auxin in a root elongation response.[24] Adult *sax1* plants have phenotypes of dwarfism, delayed development, and reduced fertility. In addition to increased sensitivity to auxin, *sax1* also has altered sensitivity to other hormones, including largely increased sensitivity of root elongation and stomata aperture to ABA, and reduced sensitivity of hypocotyl growth to gibberellins and ethylene. Further study demonstrated that *sax1* is defective in BR-biosynthesis. Almost all of the phenotypes of *sax1*, including phenotypes of altered sensitivity to auxin and ABA, can be restored to wild type by exogenous application of brassinosteroid.[10,24] These results provide genetic evidence for interactions between BRs and other phytohormones.

However, brassinosteroid interactions appear to be complex, and results from different studies have not been consistent. Whereas the increased sensitivity to ABA in the BR-deficient *sax1* mutant is consistent with an antagonistic relationship between BRs and ABA, the increased sensitivity to auxin is in opposition with the known synergistic relationship between BRs and auxin.[4] The *bri1* mutant, which is insensitive to brassinosteroids, has an increased response to ABA, unaffected response to auxin, cytokinin, and ethylene, and a slightly reduced response to gibberellins in root growth.[33] Similarly, the biosynthetic mutants *cbb1* and *cbb3* respond like wild-type plants to auxin, cytokinin, gibberellins, and ethylene.[26] Obviously, the effects of BR-deficiency on responses to other hormones need to be more carefully evaluated in additional BR mutants. The cross talk between BRs and other phytohormones will remain an exciting and challenging subject of research.

REGULATION OF ACTIVE BRASSINOSTEROID LEVELS

Regulation of Biosynthetic Genes

Hormone levels are usually regulated by developmental and environmental cues. Although the effect of developmental stage or environmental conditions on endogenous BR levels has not been well studied, expression information is available for several of the biosynthetic genes. Whereas the expression of *DET2* RNA is constitutive and ubiquitous in both light and dark,[15] the expression of the *CPD* gene shows developmental and tissue-specific regulation and is controlled by BR itself.[34]

The *CPD* RNA accumulates to high levels in flowers and leaves, is low in roots and stems, and is down-regulated by cytokinin treatment.[19] A *CPD* promoter-*uidA* reporter gene fusion directs GUS enzyme expression in cotyledon and leaf primodia of young seedlings, in leaf buds and young rosette leaves of adult plants, and in sepals and cauline leaves of inflorescence.[34] Except for young leaves where BRs promote leaf expansion, the rest of the expression pattern of the *CPD* gene does not correlate with the location where BRs are found to accumulate (such as in pollen and seeds),[4] nor with the sites where most BR dependent cell elongation occurs (such as stems and hypocotyls). If the expression of *CPD* gene correlates with BR biosynthesis, these results suggest that BL, or some of its precursors, is synthesized in a tissue specific manner and is transported to the sites of BR accumulation and responses. However, efficient transport of active BRs was not supported by the formation of dwarf and wild-type sectors in the transposon-tagged tomato *dwarf* mutant in which somatic excision of the transposon led to sectors of wild-type tissue. An alternative interpretation of the discrepancy between CPD expression and BR action would take into account the fact that BRs negatively regulate *CPD* gene expression.[34] As such, the absence of *CPD* gene expression in tissues of high BR levels or BR sensitivity may be a result of feedback inhibition of the *CPD* gene by BR signaling, while the high level of expression of the *CPD* gene could be a result of lack of BR feedback inhibition. In this case, additional regulatory steps in the biosynthetic pathway, or in the response pathway, must exist to account for the lack of BR accumulation or response, respectively, in tissues with high levels of *CPD* gene expression.

Expression of both the endogenous *CPD* gene and the *CPD* promoter-*uidA* reporter gene is inhibited by treatment with BRs, demonstrating a negative feedback regulation mechanism.[34] Such inhibition is sensitive to cycloheximide and, therefore, appears to involve *de novo* synthesis of a negative regulator. The activities of various BRs in inhibiting *CPD* expression correlates well with their activities determined in other biological assays, suggesting that feedback inhibition may share at least part of the signaling pathways with other BR responses.[34] Consistent with this observation, it was shown recently that *bri1* mutants accumulate a high level of 6-oxobrassinosteroids, including brassinolide, castasterone, typhasterol, and teasterone. The level of BR accumulation correlated with the severity of phenotypes in different *bri1* alleles.[35] These results indicate that the feedback inhibition of BR biosynthesis is mediated by a BRI1-dependent BR-signaling event, possibly leading to the suppression of the *CPD* gene.

Feedback regulation of biosynthetic genes by end products plays an important role in hormone homeostasis in animals. For example, sterol deprivation leads to the release of sterol regulatory element-binding proteins (SREBPs) from cell membranes into the nucleus and subsequent activation of genes involved in sterol biosynthesis and uptake; whereas high sterol levels lead to the suppression of these genes.[36] The biological significance of the feedback regulation of *CPD* could be evaluated by characterizing plants that express the *CPD* gene from a constitutive promoter.[19]

Metabolic Inactivation

Metabolism is as important as biosynthesis in controlling the steady level of active hormones. However, much less is known about how BRs are inactivated in plants. Recent studies demonstrate that sulfonation and hydroxylation of BRs play an important role in the turnover of this hormone.[37,38]

In animal systems, sulfotransferases play an important role in the modulation of the biological activity of a number of compounds including steroid hormones.[39] Genes encoding sulfotransferases have been isolated recently from *Arabidopsis* and *Brassica napus*.[37,40] It is estimated, based on the incomplete genome sequence (70%), that *Arabidopsis* contains about 15 sulfotransferase genes (L. Varin, personal communication). The *Brassica BNST3* gene encodes a sulfotransferase that can catalyze the *in vitro* O-sulfonation of BRs at the 22-hydroxyl group. The sulfonation of 24-epibrassinolide by BNST3 abolishes its biological activity in a bean second internode elongation assay, suggesting that plants may modulate the biological activity of steroids by sulfonation. The up-regulation of the *BNST3* gene expression by salicylic acid suggests a cross talk between plant-pathogen signaling and BR action. The BNST3 sulfotransferase has a strong substrate preference for 24-epicathasterone, an early precursor with little biological activity, but can not sulfonate BL.[37] It is yet to be determined whether other sulfotransferases have preference for brassinolide. The *in vivo* functions of sulfotransferases in regulating BR activities are yet to be demonstrated.

Hydroxylation has been implicated in the metabolic inactivation of BRs, and a cytochrome P450 has been proposed to be involved.[41,42] Recently, a gene encoding a cytochrome P450 steroid hydroxylase was identified in our lab.[38] Activation of this gene, by a T-DNA insert containing multiple copies of the cauliflower mosaic virus 35S enhancers, reduces BL levels and causes a dwarf phenotype in a photoreceptor mutant background. Feeding experiments that use radiolabeled BL show that plants overexpressing this gene convert BL to 26-hydroxybrassinolide at a higher rate than wild type plants, demonstrating that the gene is involved in hydroxylation of brassinolide.[38]

Glycosyl and acyl conjugates of BRs have been detected in plants and proposed to be products of BR inactivation.[5] However, direct evidence for *in vivo* functions of these modifications is lacking. Neither mutants with defects in BR glycosylation or acylation, nor enzymes involved in these processes have been identified.

RESPONSE MUTANTS AND SIGNAL TRANSDUCTION

Whereas hormone deficient mutants often identify biosynthesis enzymes, mutants with altered sensitivity to a hormone should represent components involved in perception and transduction of the hormonal signals, or in key processes of downstream responses.[43] BR-insensitive mutants have been identified in *Arabidopsis* (*bri1*),[8,26,33,35] pea (*lka*),[11] and tomato (*cu-3*).[25] In addition, the *Arabidopsis det3* mutant has been shown to have greatly reduced sensitivity to BL.[19] The *BRI1* gene encodes a putative receptor kinase, which is likely the receptor for brassinosteroids.[8] The *DET3* gene encodes a regulatory subunit of the vacuolar-type H^+-ATPase (V-ATPase), which is likely to mediate BL-induced cell elongation.[28]

BRI1, the Putative Brassinosteroid Receptor

Genetic screens performed in several labs have led to the identification of over 25 BR insensitive mutants that turned out to be alleles of the same *bri1* locus.[7] The *bri1-1* mutant was isolated on the basis of its resistance to BRs in a root growth inhibition assay.[33] The *cbb2* (*bri1-2*)[26] and additional *bri1* alleles[8,35] were identified based on the dwarf phenotype that can not be rescued by BR treatment. All *bri1* mutants have essentially the same phenotypes as strong BR deficient mutants, such as *cpd* and *dwarf4*, but can not be rescued by BL, which indicates that *BRI1* is a specific and essential component of the BR response pathways.[7]

The *BRI1* gene was cloned by chromosomal walking and found to encode a putative leucine rich repeat (LRR) receptor kinase.[8] The BRI1 protein is predicted to contain an N-terminal extra-cellular LRR domain, a single transmembrane domain, and a carboxyl-terminal cytoplasmic domain homologous to serine/threonine kinases. The BRI1 protein is localized on the plasma membrane, as shown by expressing a BRI1-GFP fusion protein in transgenic *Arabidopsis* (D. Friedrichsen and J. Chory, unpublished results). Therefore, BRI1 most likely functions as a cell surface receptor that transduces a signal into the cytoplasm through protein phosphorylation. The difficulty in recovering additional loci causing BR insensitivity suggests that downstream signal transduction components are either encoded by redundant genes, or essential for the viability of plants.

The extracellular domain of BRI1 most likely functions as a ligand-binding domain, which possibly interacts with BR directly or through a BR binding protein. LRR domains are often shown to be involved in protein-protein interactions, or in binding to peptide hormones.[44] However, several additional sequence motifs were noticed in the extra-cellular domain. Most notably, there is a 70 amino acid stretch buried between the twenty first and twenty second LRR repeats, which is not homologous to any known protein. Two pieces of evidence suggest that this 70-amino acid island plays an important role in BR signaling. First, such an insert in the LRR domain is unique to BRI1 and a limited number of homologous proteins, and is absent in most other LRR receptor kinases, such as CLV1 and Xa21, that are involved in developmental regulation and plant-pathogen interactions, respectively.[45,46] Second, mutations in several *bri1* alleles cluster in this domain. Four of the seven independent missense mutations in the extracellular domain are substitutions of three glycine residues in this 70-aa island domain[8,35] (and unpublished data). It is possible that this 70-aa island domain is involved in direct ligand binding, or in maintaining a structure for ligand binding. Another sequence motif located at the N-terminal of the LRRs, a leucine-zipper motif, may mediate homo- or heterodimerization of BRI1. Two pairs of cysteine residues, located on each end of the extracellular domain, may form intramolecular disulfide bridges or between dimer partners. Disulfide-bound linked heteromeric complexes have been demonstrated for the CLV1 LRR receptor kinase.[47] Mutation of one cysteine residue of the N-terminal pair is responsible for a weak *bri1* phenotype.[35]

The importance of the cytoplasmic kinase domain is evident from the fact that most *bri1* alleles have mutations in this domain[8] (and Friedrichsen *et al.* unpublished results). Serine/threonine kinase activity has been demonstrated for recombinant BRI1 proteins expressed in *E. coli* and mammalian cells (Li and Chory, unpublished data). Mutant BRI1 proteins corresponding to several *bri1* alleles have undetectable or greatly reduced kinase activity. Therefore, BRI1 must mediate the BR response by transducing a signal from the cell surface through phosphorylating cytoplasmic proteins. Although the possibility has not been excluded that BRI1 potentiates BR signal transduction but does not transduce the BR signal, the most attractive model is that BRI1 is activated by BR and transduces the signal into the cell. Demonstration of a BR-induced, BRI1-mediated phosphorylation event would be a critical test for this model.

The kinase function of BRI1 must be regulated, whatever the signal it perceives. The mechanism of receptor kinase activation has been well-studied in animal systems, and the most common principle appears to be a ligand-induced receptor complex formation and subsequent transphosphorylation. For example, ligand binding to the extracellular domain of receptor tyrosine kinases induces dimerization and subsequent phosphorylation of the kinase domain.[48] In the system of transforming-growth-factor b (TGFb) receptors, which is one of the best-characterized receptor serine/threonine kinase systems, two types of receptors, type I and type II, each exist as homodimers in the absence of ligand. Ligand binding causes association of the type I and type II receptor dimers into a tetrameric complex, and subsequent phosphorylation of the type I receptors by the constitutively active kinase of the type II receptors. Activated type I receptor kinase then phosphorylates the SMAD proteins, which then move into the nucleus to activate gene expression.[49]

We tested whether homo-dimerization of the BRI1 kinase could activate its kinase activity and cause a BR response. Mutations that cause constitutive dimerization have been identified in the extracellular domain of the receptor protein tyrosine kinase (RPTP),[50] whereas wild type RPTP exists as a monomer in the absence of its ligand.[51] We constructed chimeric proteins containing dimer or monomer forms of the extracellular domain and transmembrane domain of RPTP fused to the kinase domain of BRI1. Each fusion construct was inserted downstream of the strong cauliflower mosaic virus 35S promoter and transformed into wild type and *det2 Arabidopsis*. If BRI1 kinase could be activated through dimerization, the dimerized chimeric proteins would be constitutively active and could trigger BR responses and rescue BR deficient phenotypes of *det2* mutants. In contrast, the monomeric form would not do so. However, neither rescue of the *det2* phenotype nor constitutive BR response in wild type plants was observed when these constructs were transformed into *Arabidopsis* (Z.-Y. Wang, C. Joazeiro, and J. Chory, unpublished results). In contrast, some transgenic plants, transformed with either wild-type or mutant chimeras, showed *bri1* phenotypes. These *bri1*-like phenotypes were correlated with co-suppression of the endogenous *BRI1* gene, whereas some plants overexpressing the transgene encoding dimerizing RPTP-BRI1 fusion proteins showed no phenotype. These results suggest that simply dimerizing BRI1 kinase is not sufficient to cause activation of the BR signaling pathway. BRI1 is either not activated through homo-dimerization, or a particular conformation of the dimer is required.

BRI1, therefore, may function as a heteromeric complex that requires the extracellular domain of BRI1 for assembly and function. Multiple genes encoding LRR-receptor kinases have been found in the *Arabidopsis* genome, and some of them may function as partners of BRI1. CLV1 and ERACTA may represent tissue specific partners of BRI1, as mutations of these genes cause organ specific defects related to cell elongation.[45,52]

The cell wall-associated receptor kinases[53] are also potential candidates for BRI1 interacting partners, as regulation of cell elongation should involve communication between the cell wall and cytoplasm in order to balance the synthesis of cell wall and the increase of cell volume. Isolation of BRI1 interacting proteins by using molecular and immuno-biochemical tools will shed light on the network of signal transduction mediated by BRI1.

BRI1 represents the first known membrane-bound receptor mediating a steroid hormone response in any system. The signal transduction mechanism implicated by BRI1 for brassinosteroids is in contrast to the classical steroid nuclear receptors but is analogous to the non-genomic action of steroids observed in some animal systems. The classical steroid receptors found in animals are nuclear receptors that upon steroid binding directly participate in regulation of gene expression as transcription factors.[54] Such nuclear steroid receptors have not been found in plants. On the other hand, non-genomic actions of steroid hormones, which are fast and independent of gene expression, have been observed in many animal experimental systems since the 1960s.[55] Such non-genomic responses to steroids have been shown to be mediated by putative receptors on the cell surface, and involve generation of second messengers, opening and closing of ion channels, and activation of protein kinases.[55] The first receptor mediating such a non-genomic steroid action was identified only recently.[56] In this case, progesterone, a steroid hormone, was shown to bind directly to the receptor for the peptide hormone oxytocin, which is a member of the G-protein coupled receptor family. Binding of progesterone to the oxytocin receptor reduced two consequences of oxytocin signaling: production of the second messenger inositol-1, 4, 5 triphosphate, and an increase in intracellular calcium concentration. More recently, a membrane receptor mediating endocytosis was shown to mediate cell up-take of vitamin D, a steroid-like hormone, in its vitamin D-binding protein-bound state.[57] Neither of these membrane bound steroid receptors of animals share homology with BRI1. It would be interesting to see how much functional similarity there is between BR responses in plants and the non-genomic action of animal steroid hormones.

Vacuolar H^+-ATPase Required for Brassinosteroid Responses

Like *det2*, *det3* was isolated as a de-etiolated mutant with light-grown phenotypes when grown in the dark, and a dwarf phenotype when grown in the light.[58] Unlike *det2*, however, the hypocotyl phenotypes of *det3* can not be rescued by treatment with BRs, and *det3* shows greatly reduced, if any, responses to BL.[19] Thus, *det3* has been considered defective in BR sensitivity. The *DET3* gene was recently cloned by using chromosomal walking.[28] *DET3* encodes the C-subunit of the V-ATPase, which is localized to both the vacuolar membrane and the plasma membrane.[59] The V-ATPase consists of a V_0 subcomplex imbedded in the membrane and a V_1 subcomplex assembled onto the V_0 subcomplex. Subunit C is known to mediate the assembly of the holoenzyme.[59] The *det3* contains a mutation in an intron of the gene, which leads to reduced *DET3* RNA and protein levels. Compared to wild type plants, the membrane fractions of *det3* have a normal level of V_0 subcomplex but much reduced levels of V_1 subcomplex, which accumulate in the soluble fraction. These results indicate that normal V-ATPase function is required for BR induced cell elongation. It is yet to be seen whether BR deficient mutants or *bri1* mutants have altered assembly and/or activity of V-ATPase.

Although the finding that *det3* is defective in V-ATPase assembly was unforeseen, it is not surprising in light of previous reports that ATPase activity is activated during BR-induced cell elongation. Cerana *et al.* showed that BR treatments increase ATPase activity and acid secretion in Azuki bean epicotyls and maize roots.[60,61] Dicyclohexyl carbodiimide, an inhibitor for membrane-bound ATPase (including plasma membrane H^+-ATPase and V-ATPase), affected only BR-mediated elongation in cucumber hypocotyls but did not inhibit the GA-mediated hypocotyl growth in the same test.[62] Together with these physiological observations, the finding of the defect in V-ATPase activity in *det3* suggests that regulation of V-ATPase by BR might be an integral part of BR responses. We hypothesize that there are two BR signaling pathways. One pathway leads to regulation of nuclear gene expression, including those encoding cell wall modifying enzymes. The other pathway is independent of gene expression and leads to direct regulation of cellular activities, such as V-ATPase that mediates multiple transport processes across the membrane and affects turgor pressure that drives cell elongation and expansion.[59]

NEW MUTANTS

The structure and subcellular localization of BRI1 suggest a signal transduction pathway leading from the cell surface to the nucleus and other cytoplasmic machinery. Yet, extensive genetic screens for BR insensitive mutants have yielded only alleles of this single locus.[8] Additional components of the BR pathway may be encoded by redundant genes, or by genes essential for viability. In any case, new genetic approaches are required to uncover additional genes involved in BR responses.

One modification to the original genetic screens could be to look for subtle BR phenotypes. If a function is encoded by redundant genes, loss-of-function mutations of one gene might only give a subtle phenotype. Such a screen demands precise knowledge of the physiological responses, and tends to be obscured by by-pass mutations that affect a parallel or interacting pathway. For example, a de-etiolated mutant with weak BR-responsiveness could be due to a defect in BR response or in the photomorphogenic pathway. A deviation of this screen is to look for a specific subset of the phenotypes displayed by *bri1*; *e.g.* the altered expression of a BR-regulated gene. This kind of screen could identify components in branched downstream pathways.

Another approach would be to screen for pathway-active mutants. Previous genetic screens of BR response mutants were based on the failure of a mutant to show a response elicited in wild-type plants by BRs.[8,33] Such screens could only uncover mutations that cause inactivation of the BR pathway. Mutations that cause activation of the BR response have not been screened for. Such mutants could represent loss-of-function mutations of negative components or gain-of-function mutations of positive components in the pathway. Suppressor screens in the background of a pathway-inactive mutant, such as *det2* or *bri1*, would be an ideal way to uncover pathway-activation mutations.

det2 Activation-Tagged Suppressors

It has been predicted, based on the genome sequence of yeast, that most genes in a eukaryotic genome have redundant functions and only a small fraction of them give phenotypes when inactivated.[63] Genetic screens designed for gain-of-function mutations are particularly useful in uncovering genes that cause no phenotypes when inactivated, either due to redundancy or lethality. Over-expression of genes encoding regulatory or signal transduction components often causes phenotypes that are opposite to those caused by loss-of-function mutations.[64] Likewise, gene activation by random insertion of a strong enhancer or promoter unit into the genome (activation tagging) can produce novel phenotypes that provide clues for the function of the activated gene.[65] Such an activation tagging approach has been used successfully in the isolation of several gain-of-function mutants,[65] including a suppressor of a phytochrome B mutant that led to the identification of a BR-inactivating hydroxylase.[38]

One can expect that activation of some of the genes involved in the BR response pathway could suppress the phenotype of a *det2* mutant (Fig. 2). These genes may encode: (1) signal transduction components that when over-expressed can increase BR sensitivity or cause signal-independent activation of the pathway; (2) key regulators whose expression is normally induced by BRs; (3) biosynthetic enzymes that when overexpressed can partially compensate for the *det2* defect.

We have performed an activation-tagging screen in the *det2* background. The *det2* mutant plants were transformed with a T-DNA containing 4 copies of the 35S enhancers, and T1 plants were screened in the dark for long hypocotyl phenotype. Out of about 18,000 independent T1 plants, we identified several with longer hypocotyls than *det2*. One of these showed a nearly wild-type phenotype as an adult plant, having elongated petioles and expanded leaf blades (Fig. 3). This plant was genotyped to be homozygous for the *det2* mutation. The *det2* suppressor phenotype was semidominant and co-segregated with kanamycin resistance (T-DNA) in the T2 generation. Further genetic analysis of these *det2* suppressor mutants and cloning of the tagged genes will elucidate their functions in BR biosynthesis or signaling.

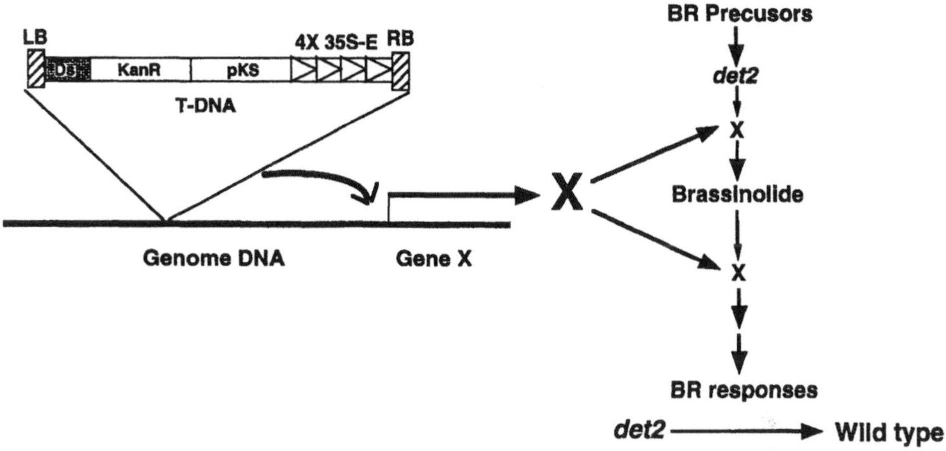

Figure 2. Diagram of *det2* suppressor screen using activation tagging. The T-DNA contains 4 copies of the 35S enhancers (4X 35S-E), pBluescript for plasmid rescue (pKS), Kanamycin resistance marker (KanR), and a transposable element (Ds). Activation by the T-DNA of a gene (X) involved in BR action may suppress the *det2* phenotypes.

Insensitive Mutants

The reduced fertility of BR mutants presents a difficulty in suppressor screens. In addition, a suppressor mutant with no phenotype in the wild-type background might also be difficult to map. A BR biosynthesis inhibitor named brassinazole (Brz) was recently shown to block specifically the BR biosynthesis at steps from campestanol to teasterone that are catalyzed by cytochrome P450s encoded by *CPD* and *DWF4*.[66] As such, wild type plants grown on medium containing brassinazole display a phenotype similar to BR-deficient mutants, namely short hypocotyls and open cotyledons in the dark, and dwarfism in the light. These phenotypes can be reversed by application of BRs together with Brz, but can not be reversed by GA, suggesting the specificity of Brz as a BR biosynthesis inhibitor.

We performed a genetic screen in which about 200,000 EMS mutagenized M2 seeds were grown in the dark on medium containing Brz, and mutants with longer hypocotyls than the wild-type background were isolated. These mutants have reduced sensitivity to Brz and are named *Brz resistant* (*bzr*) mutants. Figure 4 shows pictures of four *bzr* mutants grown in the dark on Brz medium. In addition to the long hypocotyls, some of the mutants also have apical hooks and closed cotyledons, in contrast to wild type seedlings. We believe that some of these mutant phenotypes are likely to be due to enhanced sensitivity to BRs, or BR-independent activation of the BR response pathways. Further phenotypic characterization and genetic mapping of these mutations are underway.

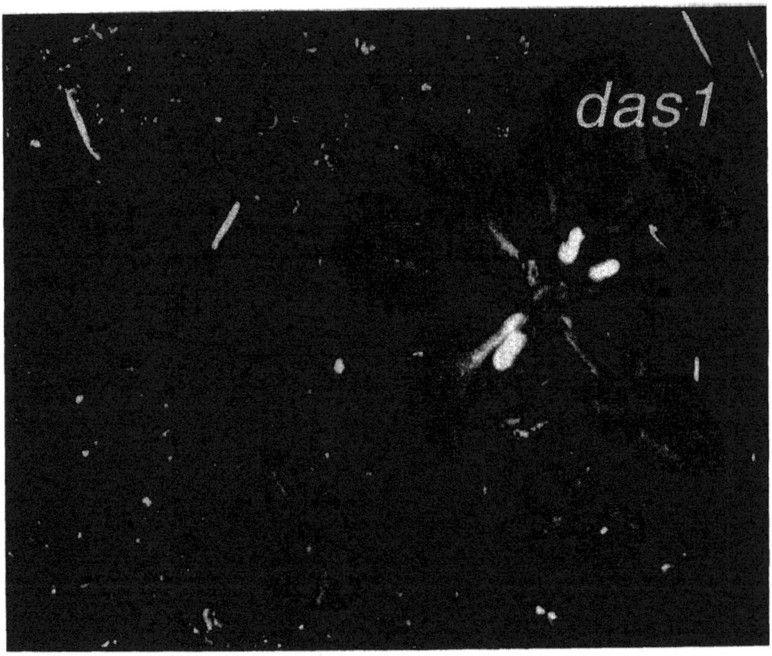

Figure 3. Phenotypes of *das1* mutant in comparison with the background of *det2*-like plants. Plants were grown in light for 4 weeks.

SUMMARY AND PERSPECTIVES

Recent molecular genetic studies have demonstrated the biological role of BRs in plant development and have led to the identification of genes encoding BR biosynthetic and metabolic enzymes, a putative BR receptor, and a cellular target of BR signaling. Some of these genes have revealed evolutionary conservation in the biosynthesis, metabolism, and signal transduction mechanisms of steroid hormones between plants and animals. Molecular studies of the regulation and function of these genes in *Arabidopsis* have started to unravel the underlying mechanism of BR homeostasis and signal transduction. The *BRI1* gene, which is required for BR responsiveness, encodes a leucine-rich-repeat receptor kinase that is localized to the plasma membrane. The BR response, thus, appears to involve signal transduction from the cell surface into the cytoplasm and nucleus, a mechanism similar to non-genomic steroid action rather than the nuclear receptor system in animal steroid responses. Our current data indicate that the signal transduction pathway from BRI1 is

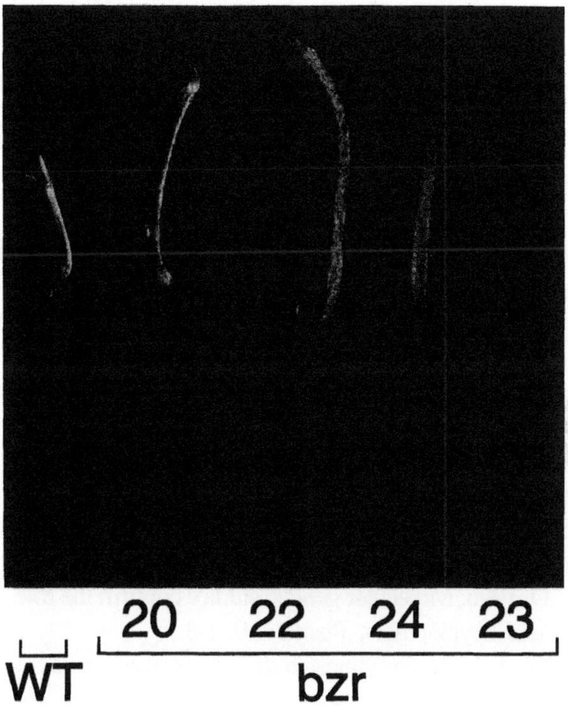

Figure 4. Phenotypes of *bzr* mutants grown in the dark for 6 days on medium containing 1/2x MS salt, 1% sucrose and 2 mM brassinazole. WT, wild type control.

branched: one branch controls rapid change of cell elongation, while the second involves changes in gene expression. Molecular tools are available to isolate BRI1-interacting proteins. We can expect more exciting discoveries in the near future from biochemical studies of the putative BR receptor BRI1. Novel genetic screens are yielding additional mutants affected in BR responses and will lead to identification of new components involved in BR action. Research along these lines will lead to a better understanding of not only how the BR signals are transduced, but also how different hormonal signals interact with each other and with environmental cues in regulating plant development, and how much evolutionary conservation there is between plant and animal steroid actions.

ACKNOWLEDGEMENTS

We thank Jianping Hu, Karen Schumacher, and Michael Neff for their helpful comments in the preparation of this manuscript, and Dionne Vifados for her technical assistance. We thank Dr. Luc Varin for sharing unpublished data. Our work on brassinosteroids is funded by grants from the USDA and the Howard Hughes Medical Institute to J. C.. Z.-Y. Wang is an NSF postdoctoral fellow and J. C. is an investigator of the Howard Hughes Medical Institute.

REFERENCES

1. GROVE, M.D., SPENSER, G.F., ROHWEDDER, W.K. 1979. Brassinolide, a plant growth-promoting steroid isolated from *Brassica napus* pollen. Nature 281: 216-217.
2. SASSE, J. 1997. Recent progress in brassinosteroid research. Physiol. Plant.100: 696-701.
3. YOKODA, T. 1997. The structure, biosynthesis and function of brassinosteroids. Trends in Plant Science 2: 137-143.
4. MANDAVA, N.B. 1988. Plant growth-promoting brassinosteroids. Annu. Rev. Plant Physiol. Plant Mol. Biol. 39: 23-52.
5. CLOUSE, S.D., SASSE. J. 1998. Brassinosteroids: Essential regulators of plant growth and development. Annu. Rev. Plant Physiol. Plant Mol. Biol. 49: 427-451.
6. CLOUSE, S.D. 1996. Molecular genetic studies confirm the role of brassinosteroids in plant growth and development. Plant J. 10: 1-8.
7. LI, J., CHORY, J. 1999. Brassinosteroid actions in plants. J. Exp. Bot. 50: 332-340.
8. LI, J., CHORY, J. 1997. A putative leucine-rich repeat receptor kinase involved in brassinosteroid signal transduction. Cell 90: 929-38.
9. FUJIOKA, S., SAKURAI, A. 1997. Biosynthesis and metabolism of brassinosteroids. Physiol. Plant. 100: 710-715.
10. EPHRITIKHINE, G., PAGANT, S., FUJIOKA, S., TAKATSUTO, S., LAPOUS, D., CABOCHE, M., KENDRICK, R., BARBIER-BRYGOO, H. 1999. The *sax1* mutation defines a new locus involved in the brassinosteroid biosynthesis pathway in *Arabidopsis thaliana*. Plant. J. 18: 315-320.
11. NOMURA, T., NAKAYAMA, M., REID, J.B., TAKEUCHI, Y., YOKOTA, T. 1997. Blockage of brassinosteroid biosynthesis and sensitivity causes dwarfism in garden pea. Plant Physiol. 113: 31-37.
12. NOMURA, T., KITASAKA, Y., TAKATSUTO, S., REID, J.B., FUKAMI, M., YOKOTA, T. 1999. Brassinosteroid/Sterol synthesis and plant growth as affected by *lka* and *lkb* mutations of Pea. Plant Physiol. 119: 1517-26.
13. BISHOP, G.J., NOMURA, T., YOKOTA, T., HARRISON, K., NOGUCHI, T., FUJIOKA, S., TAKATSUTO, S., JONES, J.D., KAMIYA, Y. 1999. The tomato DWARF enzyme catalyses C-6 oxidation in brassinosteroid biosynthesis. Proc. Natl. Acad. Sci. USA 96: 1761-6.

14. BISHOP, G.J., HARRISON, K., JONES, J.D.G. 1996. The tomato *dwarf* gene isolated by heterologous transposon tagging encodes the first member of a new cytochrome P450 family. Plant Cell 8: 959-969.
15. LI, J., NAGPAL, P., VITART, V., MCMORRIS, T. C., CHORY, J. 1996. A role for brassinosteroids in light-dependent development of *Arabidopsis*. Science 272: 398-401.
16. LI, J., BISWAS, M. G., CHAO, A., RUSSELL, D. W., CHORY, J. 1997. Conservation of function between mammalian and plant steroid 5alpha- reductases. Proc. Natl. Acad. Sci. USA 94: 3554-9.
17. FUJIOKA, S., LI, J., CHOI, Y. H., SETO, H., TAKATSUTO, S., NOGUCHI, T., WATANABE, T., KURIYAMA, H., YOKOTA, T., CHORY, J., SAKURAI, A. 1997. The *Arabidopsis deetiolated2* mutant is blocked early in brassinosteroid biosynthesis. Plant Cell 9: 1951-62.
18. NOGUCHI, T., FUJIOKA, S., TAKATSUTO, S., SAKURAI, A., YOSHIDA, S., LI, J., CHORY, J. 1999. *Arabidopsis det2* is defective in the conversion of (24R) -24-methylcholest-4-en-3-one to (24R)-24-methyl-5a-cholestan-3-one in brassinosteroid biosynthesis. Plant Physiol. 120: 833-840.
19. SZEKERES, M., NÉMETH, K., KONCZ-KÁLMÁN, Z., MATHUR, J., KAUSCHMANN, A., ALTMANN, T., RÉDEI, G.P., NAGY, F., SCHELL, J., KONCZ, C. 1996. Brassinosteroids rescue the deficiency of CYP90, a Cytochrome P450, controlling cell elongation and de-etiolation in *Arabidopsis*. Cell 85: 171-182.
20. CHOE, S., DILKES, B. P., FUJIOKA, S., TAKATSUTO, S., SAKURAI, A., FELDMANN, K. A. 1998. The *DWF4* gene of *Arabidopsis* encodes a cytochrome P450 that mediates multiple 22alpha-hydroxylation steps in brassinosteroid biosynthesis. Plant Cell 10: 231-43.
21. CHOE, S., NOGUCHI, T., FUJIOKA, S., TAKATSUTO, S., TISSIER, C.P., GREGORY, B.D., ROSS, A.S., TANAKA, A., YOSHIDA, S., TAX, F.E., FELDMANN, K.A. 1999. The *Arabidopsis dwf7/ste1* mutant is defective in the delta7 sterol C-5 desaturation step leading to brassinosteroid biosynthesis. Plant Cell 11: 207-221.
22. KLAHRE, U., NOGUCHI, T., FUJIOKA, S., TAKATSUTO, S., YOKOTA, T., NOMURA, T., YOSHIDA, S., CHUA, N.H. 1998. The *Arabidopsis DIMINUTO/DWARF1* gene encodes a protein involved in steroid synthesis. Plant Cell 10: 1677-90.
23. TAKAHASHI, T., GASCH, A., NISHIZAWA, N., CHUA, N.-H. 1995. The *DIMINUTO* gene of *Arabiodpsis* is involved in regulating cell elongation. Genes & Devel. 9: 97-107.
24. EPHRITIKHINE, G., FELLNER, M., VANNINI, C., LAPOUS, D., BARBIER-BRYGOO, H. 1999. The *sax1* dwarf mutant of *Arabidopsis thaliana* shows altered sensitivity of growth responses to abscisic acid, auxin, gibberellins and ethylene and is

partially rescued by exogenous brassinosteroid. Plant J. 18: 303-314.

25. ALTMANN, T. 1998. A tale of dwarfs and drugs: brassinosteroids to the rescue. Trends in Genetics 14: 490-495.
26. KAUSCHMANN, A., JESSOP, A., KONCZ, C., SZEKERES, M., WILLMITZER, L., ALTMANN, T. 1996. Genetic evidence for an essential role of brassinosteroids in plant development. Plant J. 9: 701-713.
27. MUNOZ, F.J., LABRADOR, E., DOPICO, B. 1998. Brassinolides promote the expression of a new *Cicer arietinum* bea-tubulin gene involve in the epicotyl elongation. Plant Mol. Biol. 37: 807-817.
28. SCHUMACHER, K., VAFEADOS, D., MCCARTHY, M., SZE, H., WILKINS, T., CHORY, J. 1999. The *Arabidopsis det3* mutant reveals a central role for the vacuolar H+ ATPase in plant growth and development. Genes Dev. In press.
29. CHORY, J., PETO, C., FEINBAUM, R., PRATT, L., AUSUBEL, F. 1989. *Arabidopsis thaliana* mutant that develops as a light-grown plant in the absence of light. Cell 58: 991-999.
30. VIERECK, R., KIRSCH, M., LOW, R., RAUSCH, T. 1996. Down regulation of plant V-type H^+-ATPase genes after light induced inhibition of growth. FEBS Lett. 384: 285-288.
31. XU, W., PURUGGANAN, M.M., POLISENSKY, D.H., ANTOSIEWICZ, D.M., FRY, S.C., BRAAM, J. 1995. *Arabidopsis TCH4*, regulated by hormones and the environment, encodes a xyloglucan endotransglycosylase. Plant Cell 7: 1555-1567.
32. CHORY, J., NAGPAL, P., PETO, C.A. 1991. Phenotypic and genetic analysis of *det2*, a new mutant that affects light regulated seedling development in *Arabidopsis*. Plant Cell 3: 445-459.
33. CLOUSE, S., LANGFORD, M., MCMORRIS, T.C. 1996. A brassinosteroid-insensitive mutant in *Arabidopsis thaliana* exhibits multiple defects in growth and development. Plant Physiol. 111: 671-678.
34. MATHUR, J., MOLNAR, G., FUJIOKA, S., TAKATSUTO, S., SAKURAI, A., YOKOTA, T., ADAM, G., VOIGT, B., NAGY, F., MAAS, C., SCHELL, J., KONCZ, C., SZEKERES, M. 1998. Transcription of the *Arabidopsis CPD* gene, encoding a steroidogenic cytochrome P450, is negatively controlled by brassinosteroids. Plant J. 14: 593-602.
35. NOGUCHI, T., FUJIOKA, S., CHOE, S., TAKATSUTO, S., YOSHIDA, S., YUAN, H., FELDMANN, K. A., TAX, F. E. 1999. Brassinosteroid-insensitive dwarf mutants of *Arabidopsis* accumulate brassinosteroids. Plant Physiol. 121: 743-752.
36. BROWN, M.S., GOLDSTEIN, J.L. 1997. The SREBP pathway: Regulation of cholesterol metabolism by proteolysis of a membrane-bound transcription factor. Cell 89: 331-340.

37. ROULEAU, M., MARSOLAIS, F., RICHARD, M., NICOLLE, L., VOIGT, B., ADAM, G., VARIN, L. 1999. Inactivation of brassinosteroid biological activity by a salicylate-inducible steroid sulfotransferase from *Brassica napus*. J. Biol. Chem. 274: 20925-30.
38. NEFF, M., NGUYEN, S., MALANCHARUVIL, E., FUJIOKA, S., NOGUCHI, T., SETO, H., TSUBUKI, M., HONDA, T., TAKATSUTO, S., YOSHIDA, S., CHORY, J. 2000. *BAS1*: A gene regulating brassinosteroid levels and light responsiveness in *Arabidopsis*. Proc. Natl. Acad. Sci. USA In press.
39. STROTT, C.A. 1996. Steroid sulfotransferases. Endocr. Rev. 17: 670-697.
40. LACOMME, C., ROBY, D. 1996. Molecular cloning of a sulfotransferase in *Arabidopsis thaliana* and regulation during development and in response to infection with pathogenic bacteria. Plant Mol. Biol. 30: 995-1008.
41. HAI, T., SCHNEIDER, B., ADAM, G. 1995. Metabolic conversion of 24-epi-brassinolide to pentahydroxylated brassinosteroid glucosides in tomato cell cultures. Phytochemistry 40: 443-448.
42. WINTER, J., SCHNEIDER, B., STRACK, D., ADAM, G. 1997. Role of a cytochrome P450-dependent monooxygenase in the hydroxylation of 24-epibrassinolide. Phytochemistry 45: 233-37.
43. MCCOURT, P. 1999. Genetic analysis of hormone signaling. Annu. Rev. Plant Physiol. Plant Mol. Biol. 50: 219-243.
44. BUCHANAN, S.G.S.C., Gay, N. J. 1996. Structural and functional diversity in the leucine-rich repeat family of proteins. Progr. Biophys. Molec. Biol. 65: 1-44.
45. CLARK, S.E., WILLIAMS, R.W., MEYEROWITZ, E.M. 1997. The *CLAVATA1* gene encodes a putative receptor kinase that controls shoot and floral meristem size in *Arabidopsis*. Cell 89: 575-585.
46. SONG, W.-Y., WANG, G.-L., CHEN, L.-L., KIM, H.-S., PI, L.-Y., HOLSTEN, T., GARDNER, J., WANG, B., ZHAO, W.-X., ZHU, L.-H., FAUQUET, C., RONALD, P. 1995. A receptor kinase-like protein encoded by the rice disease resistance gene, *Xa21*. Science 270: 1804-1806.
47. TROTOCHAUD, A.E., HAO, T., WU, G., YANG, Z., CLARK, S.E. 1999. The CLAVATA1 receptor-like kinase requires CLAVATA3 for its assembly into a signaling complex that includes KAPP and a Rho-related protein. Plant Cell 11: 393-406.
48. WELLS, A. 1999. EGF receptor. Int. J. Biochem. Cell Biol. 31: 637-643.
49. MASSAGUE, J. 1998. TGF-b signal transduction. Annu. Rev. Biochem. 67: 753-791.
50. JIANG, G., DEN HERTOG, J., SU, J., NOEL, J., SAP, J., HUNTER, T. 1999. Dimerization inhibits the activity of receptor-like protein-tyrosine phosphatase-alpha. Nature 401: 606-610.
51. BILWES, A.M., DEN HERTOG, J., HUNTER, T., NOEL, J. P. 1996. Structural basis for inhibition of receptor protein-tyrosine phosphatase-alpha by dimerization. Nature 382:

555-559.
52. TORII, K.U., MITSUKAWA, N., OOSUMI, T., MATSUURA, Y., YOKOYAMA, R., WHITTIER, R.F., KOMEDA, Y. 1996. The *Arabidopsis ERECTA* gene encodes a putative receptor protein kinase with extracellular leucine-rich repeats. The Plant Cell 8: 735-746.
53. HE, Z.H., FUJIKI, M., KOHORN, B. D. 1996. A cell wall-associated, receptor-like protein kinase. J. Biol. Chem. 271: 19789-93.
54. EVANS, R.M. 1988. The steroid and thyroid hormone receptor superfamily. Science 240: 889-895.
55. WEHLING, M. 1997. Specific, nongenomic actions of steroid hormones. Annu. Rev. Physiol. 59: 365-393.
56. GRAZZINI, E., GUILLON, G., MOUILLAC, B., ZINGG, H. H. 1998. Inhibition of oxytocin receptor function by direct binding of progesterone. Nature 392: 509-512.
57. NYKJAER, A., DRAGUN, D., WALTHER, D., VORUM, H., JACOBSEN, C., HERZ, J., MELSEN, D., CHRISTENSEN, E. I., WILLNOW, T. E. 1999. An endocytic pathway essential for renal uptake and activation of the steroid 25-(OH) vitamin D_3. Cell 96.
58. CHORY, J. 1992. A genetic model for light-regulated seedling development in *Arabidopsis*. Development 115: 337-354.
59. SZE, H., LI, X., PALMGREN, M. G. 1999. Energization of plant cell membranes by H^+-pumping ATPase: Regulation and biosynthesis. Plant Cell 11: 677-89.
60. CERANA, R., LADO, P., ANASTASIA, M., CIUFFREDA, P., ALLEVI, P. 1984. Regulating effects of brassinosteroids and of sterols on growth and H^+ secretion in maize roots. J. Plant Physiol. 114: 221-225.
61. CERANA, R., BONETTI, A., MARRE, M. T., ROMANI, G., LADO, P., MARRE, E. 1983. Effects of a brassinosteroid on growth and electrogenic proton extrusion in Azuki bean epicotyls. Physiol. Plant. 59: 23-27.
62. KATSUMI, M. 1985. Interaction of a brassinosteroid with IAA and GA3 in the elongation of cucumber hypocotyl sections. Plant Cell Physiol. 26: 615-25.
63. OLIVER, S.G., VAN DER AART, Q. J. M., AGOSTONI-CARBONNE, M. L., AIGLE, M., ALBERGHINA, L., ET AL. 1992. The complete DNA sequence of yeast chromosome III. Nature 357: 38-46.
64. CHAO, Q.M., ROTHENBERG, M., SOLANO, R., ROMAN, G., ECKER, J. 1997. Activation of the ethylene gas response pathway in *Arabidopsis* by the nuclear protein ethylene-insensitive3 and related proteins. Cell 89: 1133-1144.
65. WEIGEL, D., AHN, J., BLAZQUEZ, M., BOREVITZ, J., CHRISTENSEN, S., FANKHAUSER, C., FERRANDIZ, C., KARDAILSKY, I., NEFF, M., NGUYEN, J., SATO, S., WANG, Z., XIA, Y., DIXON, R., HARRISON, M., LAMB, C., YANOFSKY, M., CHORY, J. 2000. Activation tagging in *Arabidopsis*. Plant Phsyiol.

In press.
66. ASAMI, T., YOSHIDA, S. 1999. Brassinosteroid biosynthesis inhibitors. Trends in Plant Sci. 4: 348-353.

Chapter Fourteen

PLANT SOLUBLE SULFOTRANSFERASES: STRUCTURAL AND FUNCTIONAL SIMILARITY WITH MAMMALIAN ENZYMES

Frédéric Marsolais, Satinder Kaur Gidda, Jason Boyd, and Luc Varin

Department of Biology
Concordia University
1455 de Maisonneuve Blvd. W.
Montreal, Quebec
Canada H3G 1M8

Evolution of Metabolic Pathways – J.T. Romeo et al. (Editors)

INTRODUCTION

Sulfotransferases (STs) catalyze the transfer of a sulfonate group from 3′-phosphoadenosine 5′-phosphosulfate (PAPS) to an appropriate hydroxyl group of acceptor molecules. According to their cellular localization, they fall in two groups. Members of the first group are associated with membranes and accept as substrates macromolecules such as proteins and glycosaminoglycans.[1,2] To date, members of this family have been characterized from animal tissues only. However, the recent discovery of sulfated peptides excreted from tobacco cells grown in culture suggests that members of this group may also be present in plants.[3] The members of the second group are often described in the literature as the cytosolic STs. This terminology is misleading and should be replaced by "soluble STs", since it does not reflect accurately the subcellular distribution of the protein. The soluble STs accept as substrates small organic molecules such as flavonoids, steroids, neurotransmitters, as well as xenobiotics with diversified structures. They have been characterized from bacteria, plants, and animals,[4,5] and can be subdivided in two subgroups according to their biological function. The first subgroup is represented by enzymes involved in the detoxification of xenobiotics and/or endogenous compounds, and are generally characterized by their broad substrate specificity. Members of the second subgroup are involved in important metabolic processes, such as steroid transport or inactivation, and they exhibit high specificity for their substrates.

The evolutionary link between soluble STs from different phyla is obvious when we consider their common structural domains and the similarity in the chemical structure of their substrates. In addition, recent results on the characterization of plant STs specific for brassinosteroids[6] suggest that the evolutionary link between mammalian and plant STs at the structural level might be further extended to their similar biological functions. This chapter presents the state of our knowledge of plant soluble STs that have been characterized at the molecular level, with emphasis on their proposed biological function, and on the structural determinants involved in substrate recognition and catalysis.

FLAVONOL SULFOTRANSFERASES OF *Flaveria chloraefolia* AND *Flaveria bidentis*

The flavonol 3- and 4′-STs from *Flaveria chloraefolia* were the first plant STs for which cDNA clones were isolated and characterized.[7] They participate in a sequential pathway of polysulfation in *Flaveria* species.[8] The flavonol 3- and 4′-STs catalyze the sulfonation of flavonol aglycones and flavonol 3-sulfates, respectively. The cloning of soluble STs from plant origin allowed the establishment of the main features of the general structure of these enzymes.

Despite extensive knowledge of the biochemistry of flavonol sulfate biosynthesis, the understanding of its biological function is limited. The results of studies on the distribution of flavonol sulfates, and on the regulation of expression of flavonol STs, might provide clues about their function. The accumulation of flavonol sulfates is developmentally regulated in *F. bidentis*, being more abundant at the apex and in the first pair of extended leaves.[9] In cell culture, flavonol ST expression is induced by the synthetic auxin 2,4 D.[10] Furthermore, flavonol aglycones are known to inhibit polar auxin transport by binding to the naphthylphthalamic acid receptor, while sulfate conjugates have an antagonistic effect.[11,12] Flavonol sulfate biosynthesis may, therefore, allow auxin efflux from tissues where it is produced in large amounts.[10]

SULFOTRANSFERASES OF *Arabidopsis thaliana* AND *Brassica napus*

Recently, we initiated a functional genomics project with the objective of characterizing the biological function of all ST-coding genes from one flowering plant. This approach was made possible by the international effort to characterize the genome of *Arabidopsis thaliana* and by the production of powerful molecular tools such as T-DNA tagged mutant lines that are available to the scientific community. A search of the expressed sequence tag and genomic databases of *A. thaliana* allowed the identification of 13 genes named AtST1 to AtST13. The presence of a relatively high number of different ST-coding genes in *A. thaliana* is quite surprising and suggests that this plant accumulates a variety of sulfated metabolites other than the glucosinolates. In the future, methods optimized for the isolation and characterization of sulfated metabolites will have to be used in order to identify these unknown compounds from *A. thaliana*.[13] This number is high also when compared with the presence of only five different soluble ST-coding genes in the human genome,[14] and reflects the high metabolic plasticity of plants as compared with mammals.

Our experimental strategy is simple and involves the cloning of the ST-coding sequences in prokaryotic expression vectors. The purified recombinant enzymes are tested with a wide range of substrates including flavonoids, desulfoglucosinolates, phenolic acids, terpenoids, phytohormones, as well as fractionated methanolic extracts of *A. thaliana*. The initial biochemical characterization is followed by a study of the expression pattern of the gene during plant development and in response to stress, and by the construction of transgenic *A. thaliana* plants expressing the ST gene in the sense and antisense orientation. The growth behavior of the transgenic plants is then evaluated in order to assess the biological function of the enzyme. Recently, we also started to search for T-DNA tagged insertions in ST-coding genes in recombinant inbred lines. This approach allows the isolation of "knockout" mutants for the gene of interest by screening pooled DNA samples from the mutant lines using PCR. It has the advantage of by-passing the requirement for the production of antisense transgenic plants. Out of the 13 ST-coding genes, four have been partially characterized. The following section will briefly describe the results of our studies on AtST1 to AtST4.

AtST1

RaR047 was the first cDNA clone encoding a sulfotransferase to be characterized from *A. thaliana.*[15] In order to simplify the nomenclature of ST-coding genes, RaR047 was renamed AtST1. It was isolated from a cDNA library constructed from *A. thaliana* cultured cells inoculated with an avirulent strain of *Xanthomonas campestris*. AtST1 was found to encode a protein having 302 amino acids with an overall identity of 52% with the previously characterized flavonol STs from *Flaveria* species. No attempt was made in the original report to demonstrate the ST activity or to define the substrate specificity of AtST1.

Studies on the developmental regulation of AtST1 expression showed that it is expressed in the aerial parts of young seedlings. This pattern of expression is not unique to AtST1 and was also observed for the flavonol ST genes in *F. bidentis*.[10] The expression of AtST1 was also shown to be induced in response to salicylate and pathogen infection. Furthermore, the expression of AtST1 was induced preferentially in response to avirulent pathogens as compared with virulent strains. This pattern of expression is similar to previously characterized *Pathogenesis-Related* (*PR*) genes, and suggests that AtST1 might play a role in the plant defense response to pathogen infection.

Brassica napus Steroid Sulfotransferases

Isolation and Biochemical Characterization. Three genes coding for STs, designated BnST1 to BnST3, were isolated from a *Brassica napus* genomic library by screening with AtST1 as a probe.[6] The BnST amino acid sequences are similar to AtST1, with BnST3 being the closest relative with 87% amino acid sequence identity. The BnST proteins share approximately 40% sequence identity with the flavonol STs of *F. chloraefolia*, and 27-30% identity with the mammalian phenol, estrogen, and hydroxysteroid STs. The BnST genes comprise a family of at least twelve members. They represent orthologs of AtST1 in *B. napus*. When expressed in *Escherichia coli*, only BnST1 and BnST3 were produced as soluble proteins.

Surprisingly, the BnST1 and BnST3 enzymes were found to be active with estrogenic steroids. However, they did not exhibit any activity with the structurally related androgens, phytosterols, or ecdysteroids. Although still controversial, several reports described the presence in plants of mammalian steroids, such as estradiol, estrone, or testosterone, as well as their physiological effects on growth, development, and flowering.[16,17] The significance of estrogen sulfonation by BnST enzymes remains to be elucidated. This enzymatic activity may not be relevant *in vivo*, since estrogens are accepted by BnST3 only at relatively high substrate concentrations,[6] but it may reflect the conservation of biochemical function between steroid STs across phyla.

In sharp contrast with mammalian estrogen STs that catalyze sulfonate transfer to the phenolic hydroxyl group at position 3 of estrogens, BnST3 sulfonates the hydroxyl group at position 17 of estradiol. If we consider the alcoholic character of the hydroxyl group that is sulfonated, BnST3 is more related to the mammalian hydroxysteroid STs, which can sulfonate the 3ß-hydroxyl group of dehydroepiandrosterone or the 17ß-hydroxyl group of testosterone.[14] Furthermore, the enzyme shows strict stereospecificity for the 17-hydroxyl group of estradiol in the ß configuration.[6] Stereospecificity is a common feature of steroid STs, and has been described for the mammalian estrogen and hydroxysteroid STs.[18] It reflects constraints imposed on the architecture of the binding site, depending on the type of steroid that is accepted.

The enzymatic activity of BnST3 with brassinosteroids also has been characterized.[6] Among the substrates tested, 24-epicathasterone was accepted with the highest affinity, followed by 24-epiteasterone (Fig. 1). BnST3 does not accept 22-deoxy-24-epiteasterone, suggesting that sulfonation is taking place at the 22-hydroxyl group of the steroid side chain. Apparent kinetic parameters determined with the two best substrates suggest that only 24-epicathasterone may be sulfonated *in vivo*.[6]

An interesting feature of BnST3 substrate specificity is its ability to sulfonate 24-epibrassinolide and the synthetic enantiomer (22*S*, 23*S*)-28-homobrassinolide, while no enzymatic activity with brassinolide can be observed. An opposite orientation of the side chain hydroxyl groups relative to the 24-methyl group is, therefore, a strict requirement for catalytic activity. As far as natural brassinosteroids are concerned, 22*R*-, 23*R*-hydroxyls and a 24*R*-methyl group are strictly required for activity. Another factor influencing enzyme activity is the level of oxidation of ring B, since 6-deoxo-24-epibrassinosteroids are poor substrates. Finally, the 3-hydroxyl group in ß, found in 24-epicathasterone and 24-epiteasterone, is favored, as compared with the 2α- and 3α-hydroxyl groups of 24-epibrassinolide.

The characterization of the enzymatic activity of BnST1 has revealed that its substrate preference is different from that observed with BnST3. To date, a substrate for which BnST1 would have a high affinity has yet to be identified. This may be due to the limited number of brassinosteroids that have been tested. 2α, 3ß-Epimers of brassinosteroids are important metabolic intermediates that are subject to 3-acyl or 3-glucosyl conjugation, or to side chain degradation.[19] It would be interesting to test the enzymatic activity of BnST1 with these intermediates. Another important aspect to be considered is the relationship between the substrate specificity of AtST1 and those of the BnSTs. A single gene in *A. thaliana* corresponds to a family of homologous genes in *B. napus*. If BnST enzymes have distinct substrate specificities, does AtST1 catalyze all of their reactions, or does it have a more specialized biochemical function?

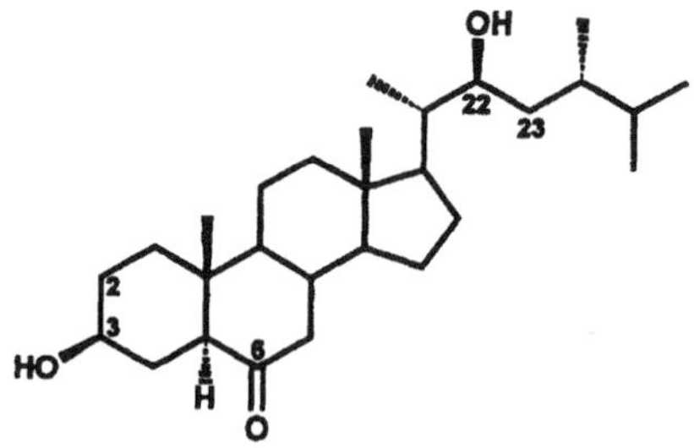

24-Epicathasterone

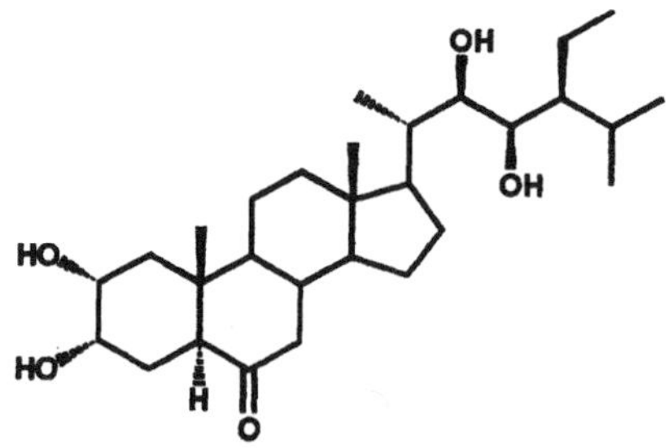

28-Homocastasterone

Figure 1. Chemical structures of representative brassinosteroids.

Occurence of Brassinosteroid Sulfate Conjugates in Plants. Sulfate conjugates of brassinosteroids have not yet been characterized from plants.

However, feeding experiments with [^{3}H]-castasterone to mung bean explants showed that two types of conjugates were formed.[20] A few were glucosides, but most conjugates were polar and non-glycosidic, since the aglycone could not be released by enzymatic hydrolysis. In a similar experiment with [^{3}H]-brassinolide, a reverse pattern of conjugate formation was observed. The major part consisted of a glycoside, 23-O-β-D-glucopyranosylbrassinolide.[21] During feeding experiments with rice seedlings, non-glycosidic water-soluble metabolites of [^{3}H]-castasterone and [^{3}H]-brassinolide were the major conjugates formed.[22] Based on their chromatographic behavior and susceptibility to solvolysis, it was proposed that they might be sulfate esters. The results of experiments with mung bean suggest that, in some species, brassinolide forms predominantly glucosyl conjugates, whereas the intermediates are metabolized into sulfate conjugates. This might explain the substrate preference of BnST3 for the precursor 24-epicathasterone, as opposed to the final product 24-epibrassinolide.[6]

Functional Conservation of Plant and Animal Steroid Sulfotransferases. Since BnST1 and BnST3 accept mammalian estrogenic steroids, the enzymatic activity of human steroid STs was tested with brassinosteroids. The human estrogen ST was essentially inactive with the steroid plant hormones as substrates, consistent with a strict requirement for a phenolic hydroxyl group at position 3 of the steroid nucleus (A.-M. Schinas, unpublished results). However, the human hydroxysteroid ST showed significant activity with brassinosteroid intermediates having a 3ß-hydroxyl group, such as 24-epicathasterone. In contrast, the activity measured with brassinosteroids having a 3α-hydroxyl group was negligible. The affinity of the enzyme for 24-epicathasterone, as measured by its apparent Km, was comparable to that observed with its natural substrate, dehydroepiandrosterone, although its specific activity was reduced approximately 10-fold. These results demonstrate a surprisingly high level of conservation of biochemical function between plant and animal steroid STs.

Regulation of BnST Gene Expression in Response to Salicylic Acid. The expression of BnST genes and their proteins was found to be induced by salicylic acid, a signal molecule in the plant defense response against pathogen infection.[6] The kinetics of induction at the level of steady-state mRNA was found to be similar to those of *PR* genes.[23] The increase in gene expression was correlated with an increase in the accumulation of BnST proteins, as detected with anti-BnST1 polyclonal antibodies. These results are in agreement with those obtained for the *Arabidopsis* homologue AtST1.[15]

Function of Brassinosteroid Sulfonation. Steroid sulfonation in mammals can serve two functions, either in hormone inactivation or in the transport of a precursor from the site of biosynthesis to a target tissue, where the precursor will be further metabolized and bioactivated.[18] Estrogen sulfonation represents a good example of a deactivation reaction. The estrogen STs have a high affinity for estradiol, comparable to that of the estrogen receptor, and their function has been characterized in specific organ systems, such as the endometrium or testicles. The sulfonation of dehydroepiandrosterone by the human hydroxysteroid ST in the adrenal gland constitutes a good example of the function in steroid transport.

Similar functions may be considered for BnST enzymes. However, the inducibility of AtST1 and BnSTs by salicylic acid, or infection with virulent or avirulent pathogens,[6,15] is consistent with a function in brassinosteroid inactivation. During pathogen infection, cell elongation promoted by brassinosteroids is undesirable, since active mechanisms are taking place to reinforce the cell wall against invasion. The function in inactivation is also supported by our results demonstrating that 24-epibrassinolide sulfate is inactive in the bean second internode bioassay.[6] In addition to brassinosteroid inactivation by sulfonation, a reduction in the biosynthesis of sterol precursors is presumed to take place during pathogen infection. Suppression of phytosterol biosynthesis in response to fungal elicitors has been demonstrated in parsley, potato, tobacco, and *Tabernaemontana divaricata*.[24-27] There is also a concomitant reduction in expression, at the level of steady-state mRNA, of the genes coding for squalene synthase and Δ24-sterol *C*-methyltransferase.[28,29] Squalene synthase acts at an important branch point where a common precursor, farnesyl pyrophosphate, is partitioned between the sterol biosynthetic pathway and the pathways leading to other terpenoids.[30]

The action of BnST3 on the precursor 24-epicathasterone, instead of the hormonally active end-product 24-epibrassinolide, is intriguing. Hydroxylation at position 22 of the sterol side chain, catalyzed by a cytochrome P450 encoded by *DWF4* (CYP90B) of *Arabidopsis*, appears to represent a rate-limiting step in brassinosteroid biosynthesis.[31] By acting on the product of this enzyme, BnST3 may block further biosynthesis of 24-epibrassinosteroids (Fig. 2).

Tissue Localization of BnST Expression. The tissue localization of expression of BnST2 and BnST3 was studied in transgenic *Arabidopsis* expressing GUS-promoter fusions. In seedlings, expression was restricted to roots and cotelydons (F. Marsolais, unpublished results), which is consistent with the results of Western blot experiments that use *B. napus* seedlings extracts. Three days after seed imbibition, expression was found only in root hairs, with intense staining near the hypocotyl junction. The following day, expression is apparent in the root, but

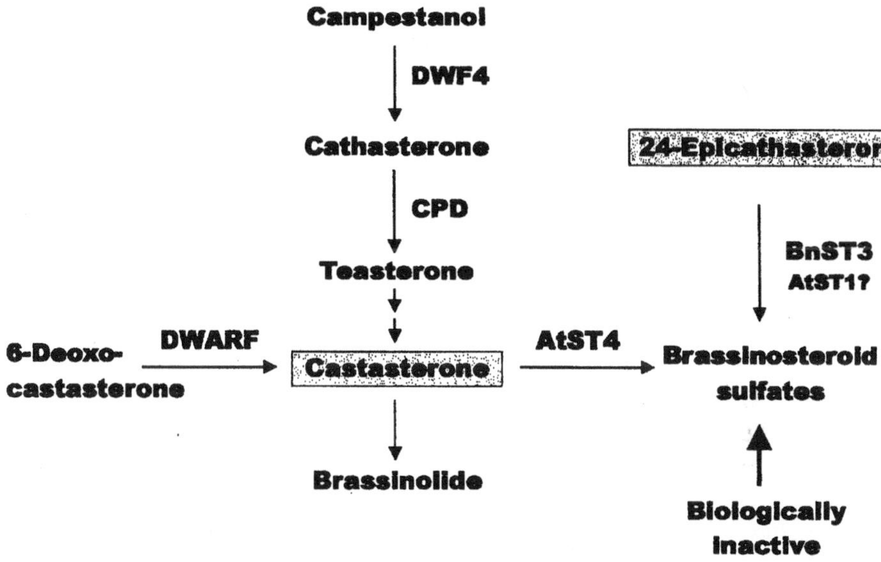

Figure 2. Proposed function of steroid STs in brassinosteroid metabolism.

the promoters direct expression to different zones of the tissue. Expression of BnST3 is found in the zone of cell division at the root tip, while BnST2 expression is restricted to the zone of cell elongation. This tissue-specific pattern of expression is reminiscent of that observed for genes coding for enzymes that catalyze two different steps in brassinosteroid biosynthesis (Fig. 2). The *Arabidopsis CPD* gene, coding for the P450 (CYP90A) responsible for the 23-hydroxylation of cathasterone, is expressed in the meristematic tissue of the stem.[32] This result is surprising in view of the function of brassinosteroids in cell elongation, and implies that brassinosteroid intermediates need to be transported before bioactivation. In contrast, the tomato *DWARF* gene, which encodes a cytochrome P450 (CYP85) responsible for the formation of castasterone from 6-deoxocastasterone (Fig. 2), is expressed in the zone of cell elongation.[33,34] It is interesting to note the similarities in the spatial patterns of expression of *CPD* and BnST3, their substrates being different epimers of the same intermediate, cathasterone and 24-epicathasterone, respectively.

Isolation and Characterization of an Ethanol-Inducible Member of the BnST Family. BnST4 was isolated from a cDNA library produced from *B. napus* seedlings treated with ethanol (F. Marsolais, unpublished results). AtST1 and BnST3 are its closest relatives, with 86-87% amino acid sequence identity.

Preliminary characterization of the recombinant BnST4 enzyme indicates that it catalyzes the sulfonation of dehydroepiandrosterone. In contrast with BnST1 and 3, estrogens are not accepted, and all brassinosteroids tested are poorer substrates than dehydroepiandrosterone.

It is known that plants respond to ethanol treatment by increasing the expression of several cytochrome P450 hydroxylase genes, some of which are involved in herbicide detoxification.[35,36] Analogous to mammalian liver, the induction of plant P450s and BnST4 by ethanol could be considered a response to chemical stress, increasing the levels of phase I and II detoxifying enzymes. The expression of alcohol dehydrogenase genes in plants has been studied in relation to oxygen stress that results in a metabolic switch towards anaerobic fermentation.[37] However, hypoxia does not result in an increase in the level of BnST protein in *B. napus* seedlings. Furthermore, induction of BnST4 does not seem to be part of a general response to chemical stress, since it is not observed after treatment with other alcohols, such as methanol, propanol, or butanol. In order to exclude any link between BnST4 expression and chemical stress, the effect of other known inducers of plant cytochrome P450s, such as herbicide safeners, phenobarbital, or heavy metals will have to be tested.[35] Alternatively, ethanol-inducible expression may be considered a response to normal metabolic processes. For example, in tissues having a high metabolic rate, such as the root tip and the reproductive organs of flowers, there is a co-occurrence of respiration and aerobic fermentation leading to the production of ethanol.[38] In these tissues, the accumulation of ethanol might lead to the induction of BnST4 and the inactivation of specific steroids. Further experiments are needed in order to evaluate this hypothesis.

AtST2

The purified recombinant AtST2 enzyme was found to exhibit strict specificity and high affinity for 12-hydroxyjasmonate (Fig. 3) (S. K. Gidda, unpublished results). The enzyme did not accept structurally related compounds, such as cucurbic acid, arachidonyl alcohol, or prostaglandins. 12-Hydroxyjasmonate is a natural metabolite, known as tuberonic acid, which was initially characterized from *Solanum tuberosum*, and has been shown to induce tuber formation in this plant.[39] Its presence has been detected in plant species other than *S. tuberosum*.[40,41] 12-Hydroxyjasmonate sulfate is also a natural metabolite and has been characterized from *Tribulus cistoides* (Zygophyllaceae).[42] However, 12-hydroxyjasmonate and its sulfated derivative have not yet been reported to occur in *A. thaliana*.

The presence of an enzyme that sulfonates 12-hydroxyjasmonate in *A. thaliana* suggests that this metabolite may play a role in the development of plants that do not produce tubers. The availability of AtST2 from *A. thaliana* provides a useful tool for studying the biological function of 12-hydroxyjasmonate in transgenic potato plants. Based on our knowledge of the function of STs in animals and plants, we expect that transgenic potato plants overexpressing AtST2 will exhibit delayed tuber formation.

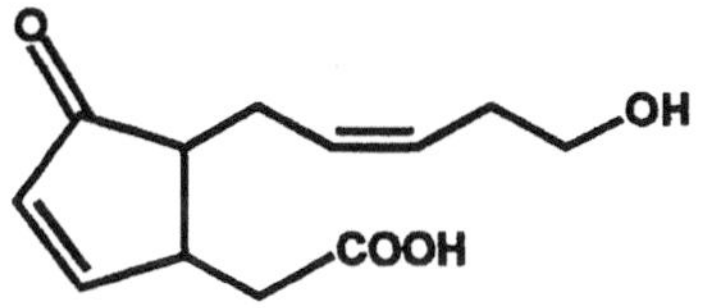

12-Hydroxyjasmonate

Figure 3. Chemical structure of the acceptor substrate of AtST2.

AtST3

The purified recombinant AtST3 enzyme was found to exhibit strict specificity for position 7 of flavonoids (S. K. Gidda, unpublished results). In contrast with the previously characterized flavonol 7-ST from *F. bidentis* that sulfonates only flavonol 3, 3′- and 3, 4′-disulfates,[43] AtST3 accepts a number of flavonols and flavone aglycones as well as their 3-monosulfated derivatives. This difference in substrate specificity may reflect differences in the flavonoids synthesized in *Arabidopsis* as compared with *Flaveria* species.

The natural occurrence of a ST exhibiting high specificity for flavonoids in *A. thaliana* is surprising considering the absence of reports on the presence of flavonoid sulfates in this plant. This discovery suggests that flavonoid sulfates might have a more widespread occurrence in the plant kingdom than was generally believed. The presence in *A. thaliana* of an enzyme involved in flavonoid sulfonation provides an elegant experimental system to test the hypothesis that flavonoid sulfates act as regulators of polar auxin transport. In order to answer this important question, we are presently producing transgenic *A. thaliana* plants to evaluate the impact of altering the expression levels of AtST3 on plant growth and development.

AtST4

The purified recombinant AtST4 enzyme was found to exhibit strict specificity for brassinosteroids having 22*R*-, 23*R*-hydroxyls, and a 24*S*-methyl or ethyl group on the steroid side chain (Fig. 1) (J. Boyd, unpublished results). In contrast with BnST3, AtST4 shows substrate preference for castasterone and 28-homocastasterone, the immediate precursors of brassinolide and 28-homobrassinolide, respectively. Together, AtST4 and AtST1 may function to inactivate brassinosteroids in *A. thaliana* through sulfonate conjugation (Fig. 2). The presence of two genes encoding stereospecific brassinosteroid STs in *A. thaliana* suggests that the brassinosteroid stereoisomers may play different roles in plant development, and that their activity requires independent regulation. If this hypothesis is true, we can expect to observe different phenotypes in transgenic plants overexpressing AtST1 or AtST4. Furthermore, plants overexpressing both ST activities should exhibit a more severe phenotype. We are presently producing transgenic *A. thaliana* that overexpress AtST1, AtST4, or both genes in order to test this hypothesis.

SEQUENCE RELATEDNESS AND PHYLOGENETIC ANALYSIS

Amino acid sequence relatedness of plant cytosolic STs allows us to classify them into three families, designated SULT3 to SULT5. According to a proposed molecular classification of soluble STs from eukaryotes, the SULT1 and SULT2 families belong to mammals and represent the phenol and hydroxysteroid STs, respectively, with the phenol STs being further divided in two sub-families, the phenol and estrogen STs.[5] According to this classification, STs belonging to the same family share at least 45% amino acid sequence identity, whereas members of sub-families are 60% or more identical in sequence.[5]

The plant SULT3 family includes the flavonol 3- and 4′-STs from *Flaveria* species, and a ST from *F. bidentis* of unknown biochemical function (FbSTX) (Fig. 4). BnST proteins from *B. napus* and their homologue AtST1 from *A. thaliana* form the SULT4 family, while the SULT5 family comprises seven other proteins from *A. thaliana*. The AtST2 gene encoding the 12-hydroxyjasmonate ST is distant from the rest of the sequences, and should be considered as the first member of a new SULT6 family (Fig. 4).

Members of the SULT3 family share at least 60% amino acid sequence identity, while members of the SULT4 family are at least 70% identical. The SULT5 family represents the less homogeneous group of sequences. The level of amino acid sequence identity between its members is generally lower than 60%.

Therefore, they may belong to new sub-families. The inclusion of enzymes with completely different substrate specificities, such as AtST3 (flavonol 7-ST) and AtST4 (steroid ST), supports this hypothesis.

The molecular classification of plant cytochrome P450s divides them in two classes.[44] Class A P450s perform plant-specific reactions of secondary metabolic pathways. The non-class A P450s are more similar to sequences from other kingdoms and belong to common metabolic pathways, such as lipid or steroid biosynthesis. Although mammalian and plant soluble STs are related phylogenetically, we do not observe the same types of relationships as those seen for cytochrome P450s. Plant STs involved in the reactions of secondary metabolism do not form a distinct phylogenetic group. Furthermore, plant steroid STs cluster with other plant STs, and not with the animal steroid STs, despite their related biochemical functions.

An interesting feature deduced from the phylogenetic analysis of plant soluble STs is the fact that flavonol STs from *Flaveria* species and the flavonol 7-ST (AtST3) from *A. thaliana* have evolved independently, as illustrated by the fact that they fall in different families (Fig. 4). This is not so surprising if we consider their different enzymatic properties, the specificity of AtST3 being broader, as compared with the highly specific flavonol 3- and 4′-STs.

GENERAL STRUCTURE OF SULFOTRANSFERASES

Structural similarities shared among members from different phyla represent a distinctive feature of soluble STs, as compared with other families of enzymes involved in conjugation reactions of plant secondary metabolism. For instance, mammalian and plant soluble STs generally share 25 to 30 % amino acid sequence identity. Soluble STs have similar lengths and hydropathy profiles, and contain four regions of conserved amino acid residues (Fig. 5).[7] Based on their conserved features, they should have similar three-dimensional structures. The bacterial STs encoded by *nodH* from *Rhizobium meliloti* and the ORF4 of the *avrD* locus from *Pseudomonas syringae* pv. *tomato* have shorter sequences of approximately 240 amino acids, with a low level of sequence similarity with soluble STs from eukaryotes. However, they do contain the sequence motifs present in the first three conserved regions.[45-47]

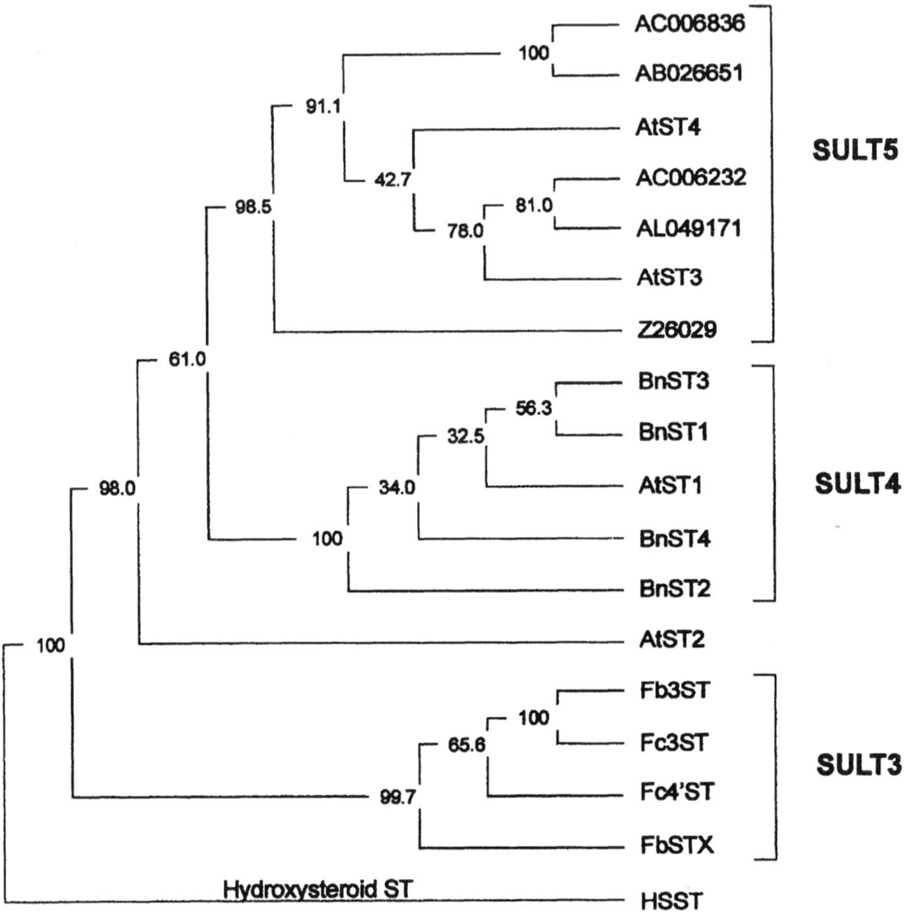

Figure 4. Phylogenetic reconstruction of plant soluble STs prepared using the PHYLIP package.[60] Numbers indicate bootstrap values of branches. SULT3 through SULT5 designate gene families defined by a minimum level of 45% amino acid sequence identity among members. *Arabidopsis* genes with unknown function are designated by their Genbank accession numbers. For ST genes with known biochemical functions, accession numbers are: HSST (X84816), FbSTX (U10277), Fc4′ST (M84136), Fc3ST (M84135), Fb3ST (U10275), AtST2 (AB010697), BnST2 (AF000306), AtST1 (Z23001), BnST1 (AF000305), BnST3 (AF000307), AtST3 (T75675), AtST4 (AC005396).

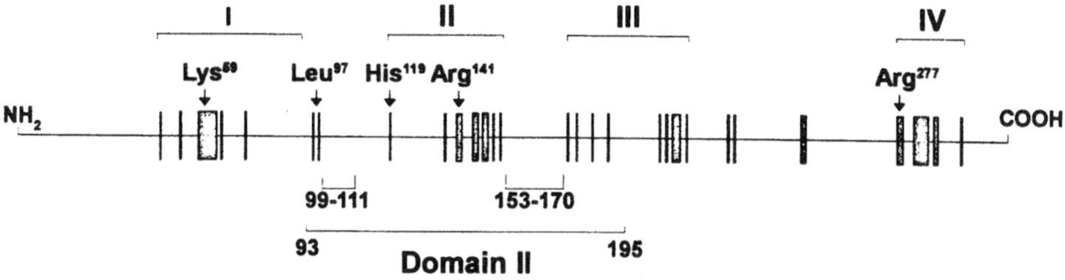

Figure 5. General structure of soluble STs. Amino acid numbers correspond to their positions in the flavonol 3-ST sequence from *F. chloraefolia*.[7] The boxes represent conserved residues. Roman numbers refer to the four conserved regions presented in ref. 7. The position of important functional residues is marked by arrows. The two subdomains of high divergence present within domain II are indicated. The amino acid sequence alignment used to build this representation can be found in ref. 55

The enzymatic reactions of phosphorylation and sulfonation are chemically similar, although STs do not utilize a Mg^{2+} ion like kinases. Such mechanistic similarities are reflected in the three-dimensional structure of soluble STs. The three-dimensional structure of mouse estrogen ST reveals striking similarity with nucleotide kinases, including a P-loop fold corresponding to the first conserved region.[48]

SPECIFIC FUNCTIONAL MOTIFS

The flavonol 3- and 4′-STs from *F. chloraefolia* represent an ideal experimental system for studying the determinants of ST substrate specificity. They represent the only cloned STs to date that are involved in a clearly defined sequential pathway of polysulfation.[8] The main differences between their reactions are the presence of a sulfate group at position 3 of the 4′-ST substrate, and the requirement for a modified orientation of the acceptor molecule for transfer of the sulfonate group to different positions on the flavonoid ring. Assuming that PAPS binding sites are equivalent between the two enzymes, one can predict that the different orientation of the flavonol will impose constraints on the architecture of the binding sites. In addition, specific amino acid side chains, which are most likely positively charged, will be present in the 4′-ST binding site in order to stabilize the sulfate group of the flavonol 3-sulfate.

The flavonol 3-ST represents also a good model for studying cosubstrate binding and the catalysis of sulfonate transfer, considering the structural similarity between plant and animal soluble STs. Amino acid residues involved in these common functions should be strictly conserved in these enzymes. Research on the flavonol 3-ST allowed the characterization of specific amino acid residues that are involved in PAPS binding and catalysis in all soluble STs from eukaryotes. The function of these residues was confirmed when the crystal structure of the mouse estrogen ST in complex with 3′-phosphoadenosine 5′-phosphate (PAP) and ß-estradiol was elucidated.[48] Based on the coordinates of this structure, three-dimensional models of the flavonol 3- and 4′-STs in complex with PAP and their respective flavonol substrate were constructed.[49] In future work, the structural models will allow the rational design of experiments aimed at characterizing the determinants of substrate specificity in these enzymes.

3′-Phosphoadenosine 5′-Phosphate Binding Site

Region I of STs and the P-loop of kinases have a similar fold, although they have no similarity in primary structure other than the presence of a strictly conserved lysine. In adenylate kinase, the side chain of the invariant lysine interacts with the ß- and γ-phosphates of ATP.[50] Similarly, in the crystal structure of the mouse estrogen ST in complex with PAP and ß-estradiol, the conserved lysine of region I, corresponding to residue 59 of the flavonol 3-ST, interacts with the 5′-phosphate of PAP (Figs. 5 and 6).[48] This assignment was confirmed by the results of comparative affinity chromatography experiments with the K59R and wild-type flavonol 3-ST using several PAPS analogs as ligands.[49]

Regions II and IV contain arginine residues (Arg141 and Arg277 of the flavonol 3-ST) that coordinate the 3′-phosphate group of PAP (Figs. 5 and 6).[48,49] In soluble STs, high affinity for PAP is dependent essentially on the interactions with the 3′-phosphate group.[49] The results of PAP-affinity chromatography and photoaffinity labeling with [^{35}S]PAPS indicated that the contribution of Arg141 to cosubstrate and product affinity is more important quantitatively than that of Arg277.[46,51]

Catalysis and Mechanism of Sulfonate Transfer

The most important residues for catalysis in the flavonol 3-ST are Lys59 and His119. Analogous to kinases, the ST reaction proceeds *via* nucleophilic substitution of a hydroxyl group from the acceptor substrate on the sulfur atom of PAPS.[48,52] In the crystal structure of estrogen ST, the residue corresponding to

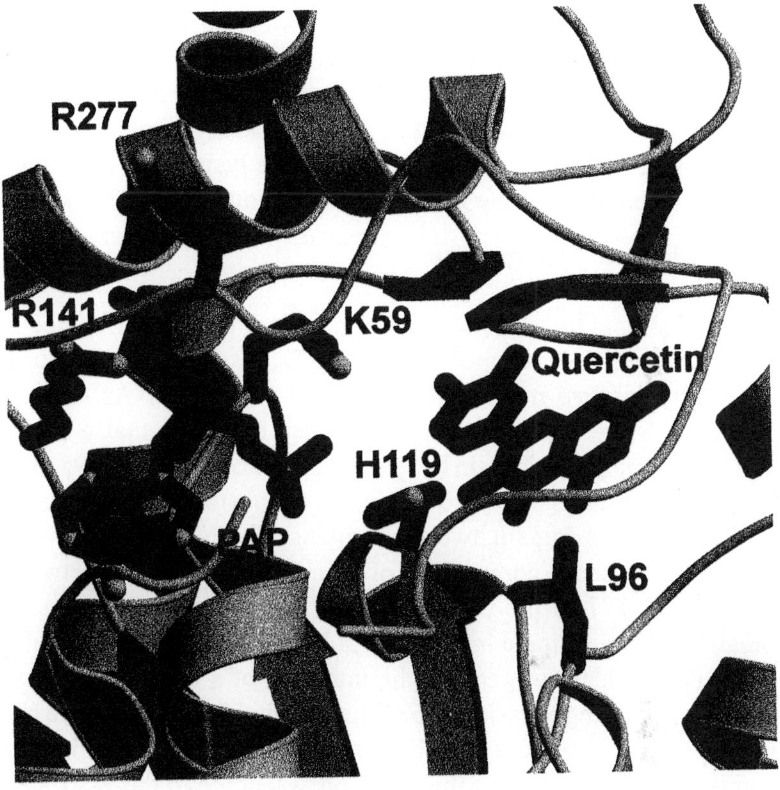

Figure 6. Schematic representation of the active site of the flavonol 3-ST from *F. chloraefolia* in complex with PAP and quercetin. The model was prepared according to ref. 49.

His119 of the flavonol 3-ST interacts with the hydroxyl group at position 3 of ß-estradiol (Figs. 5 and 6).[48] Replacement of this residue with alanine abolishes enzyme activity.[51,53,54] However, a lysine mutant at this position retains significant enzyme activity in both the flavonol 3- and mouse estrogen STs, consistent with a possible role of the histidine as a catalytic base that abstracts a proton in order to activate the nucleophile.[51,54]

The conserved lysine residue present in region I does not contribute to the affinity for PAPS but plays an important role in catalysis. Alanine and arginine mutants retain a similar or a slightly higher affinity for PAPS, as compared to the wild-type enzyme.[46] However, the K59R mutant displays a 13-fold reduction in specific activity as compared to the wild-type enzyme, while K59A is essentially inactive. In addition to the 5′-phosphate, the lysine side chain may also interact with the sulfate group.[49,54] The interaction with the sulfate may stabilize the transition state, assisting nucleophilic attack by reducing the negative charge. Alternatively, the lysine may stabilize the developing negative charge on the bridging oxygen of the 5′-phosphate. From the analysis of ^{31}P NMR spectra of flavonol 3-ST mutants in complex with PAP, the lysine also seems to determine the adequate orientation of the phosphosulfate group of PAPS for transfer.[49]

Acceptor Substrate Binding Site

To elucidate the structural aspects underlying the difference in substrate specificity between the flavonol 3- and 4′-STs, a series of hybrid proteins were constructed by *in vitro* manipulation of their cloned cDNAs. Analysis of the specificity of the resulting chimeric enzymes indicated that a segment of the flavonol STs, designated domain II, spanning amino acids 93 to 195 of the flavonol 3-ST sequence, contains all the determinants of substrate and position preferences (Fig. 5).[55] Within this domain, two subdomains of high divergence were identified by amino acid sequence comparison of animal and plant STs, between positions 99-111 and 154-171 of the flavonol 3-ST (Fig. 5). It was proposed that these subdomains may participate in acceptor substrate binding in all soluble STs.

In the mouse estrogen ST, the mutation of a tyrosine to a leucine at the position corresponding to residue 96 of the flavonol 3-ST results in a change of the substrate preference to that of the hydroxysteroid ST.[56] In the flavonol STs, the same amino acids are found at this position: a leucine in the 3-ST and a tyrosine in the 4′-ST (Fig. 5). In the three-dimensional model of the flavonol 3-ST, the leucine side chain is located near ring C of quercetin (Fig. 6). The L96Y mutation had an impact on flavonol binding, and the results of the kinetic analysis with several flavonoid acceptors suggested that the tyrosine side chain of the mutant comes in direct contact with ring B of the flavonol.[51]

In the 4′-ST, the corresponding tyrosine residue may have two different functions. By interacting with ring B, the tyrosine side chain may properly orient the flavonol for sulfonate transfer at position 4′. At the same time, a bulkier tyrosine may prevent productive binding of the aglycone in the flavonol 4′-ST catalytic site, and subsequent transfer at position 3. We have constructed the reciprocal mutant of the flavonol 4′-ST, Y106L. Preliminary characterization indicates that the mutant exhibits a new enzymatic activity with quercetin, although with low affinity (F. Marsolais, unpublished results).

Evidence supporting the involvement of the two subdomains of high divergence in the determination of substrate specificity derives from two other experimental systems. In the crystal structure of the mouse estrogen ST, it is apparent that several residues present in the two subdomains are in contact with ß-estradiol.[48,56] In addition, a single mutation in the second subdomain changes the substrate preference of the human phenol-preferring phenol ST to that of the catecholamine-preferring phenol ST.[57-59] The use of similar protein segments to form the acceptor substrate binding site is yet another feature of the evolutionary conservation between plant and animal soluble STs.

CONCLUSION

Recent studies on plant soluble STs have benefited from the *Arabidopsis* genome project. It is now possible to undertake a systematic characterization of the biochemical and biological functions of all soluble STs present in a higher plant. To date, the biochemical function of four out of thirteen ST-coding genes of *Arabidopsis* has been elucidated (AtST1 to 4). Currently, more ST genes are known from plants than individual sulfated metabolites! Future progress in the field will depend on the identification of the unknown metabolites, through the development of methods of extraction and purification compatible with the retention of the labile sulfate group. Despite the progress made in the last few years, genes coding for important plant ST enzymes, such as the choline ST implicated in resistance to salt stress, or the desulfoglucosinolate STs of the Brassicaceae, remain to be identified.[4]

ACKNOWLEDGMENTS

We thank Lars Pedersen for the structural models of the flavonol 3- and 4′-STs, and Ingrid Muzac for her help with the phylogenetic analysis.

REFERENCES

1. BOWMAN, K. G., BERTOZZI, C. R. 1999. Carbohydrate sulfotransferases: Mediators of extracellular communication. Chem. Biol. 6: R9-R22.
2. NIEHRS, C., BEISSWANGER, R., HUTTNER W. B. 1994. Protein tyrosine sulfation, 1993 - an update. Chem. Biol. Interact. 92: 257-271.
3. MATSUBAYASHI, Y., SAKAGAMI, Y. 1996. Phytosulfokine, sulfated peptides that induce the proliferation of single mesophyll cells of *Asparagus officinalis* L. Proc. Natl. Acad. Sci. USA 93:7623-7627.

4. VARIN, L., MARSOLAIS, F., RICHARD, M., ROULEAU, M. 1997. Biochemistry and molecular biology of plant sulfotransferases. FASEB J. 11: 517-525.
5. WEINSHILBOUM, R. M., OTTERNESS, D. M., AKSOY, I. A., WOOD, T. C., HER, C., RAFTOGIANIS, R. B. 1997. Sulfotransferase molecular biology: cDNAs and genes. FASEB J. 11: 3-14.
6. ROULEAU, M., MARSOLAIS, F., RICHARD, M., NICOLLE, L., VOIGT, B., ADAM, G., VARIN, L. 1999. Inactivation of brassinosteroid biological activity by a salicylate-inducible steroid sulfotransferase from *Brassica napus*. J. Biol. Chem. 274: 20925-20930.
7. VARIN, L., DELUCA, V., IBRAHIM, R. K., BRISSON, N. 1992. Molecular characterization of two plant flavonol sulfotransferases. Proc. Natl. Acad. Sci. USA 89: 1286-1290.
8. VARIN, L. 1992. Flavonoid sulfation. Rec. Adv. Phytochemistry. 26: 233-254.
9. HANNOUFA, A., VARIN, L., IBRAHIM, R. K. 1991. Spatial distribution of flavonoid conjugates in relation to glucosyltransferase and sulfotransferase activities in *Flaveria bidentis*. Plant Physiol. 97: 259-263.
10. ANANVORANICH, S., VARIN, L., GULICK, P., IBRAHIM, R. K. 1994. Cloning and regulation of flavonol 3-sulfotransferase in cell suspension cultures of *Flaveria bidentis*. Plant Physiol. 106: 485-491.
11. JACOBS, M., RUBERY, P. H. 1988. Naturally occurring auxin transport regulators. Science 241: 346-349.
12. FAULKNER, I. J., RUBERY, P. H. 1992. Flavonoids and flavonol sulfates as probes of auxin-transport regulation in *Cucurbita pepo* hypocotyl segments and vesicles. Planta 186: 618-625.
13. BARRON, D., VARIN, L., IBRAHIM, R. K., HARBORNE, J. B., WILLIAMS, C. A. 1988. Sulphated flavonoids - an update. Phytochemistry 27: 2375-2395.
14. FALANY, C. N. 1997. Enzymology of human cytosolic sulfotransferases. FASEB J. 11: 206-216.
15. LACOMME, C., ROBY, D. 1996. Molecular cloning of a sulfotransferase in *Arabidopsis thaliana* and regulation during development and in response to infection with pathogenic bacteria. Plant Mol. Biol. 30: 995-1008.
16. GEUNS, J. M. C. 1978. Steroid hormones and plant growth and development. Phytochemistry 17: 1-14.
17. SIMONS, R. G., GRINWICH, D. L. 1989. Immunoreactive detection of four mammalian steroids in plants. Can. J. Bot. 67: 288-296.
18. STROTT, C. A. 1996. Steroid sulfotransferases. Endocr. Rev. 17: 670-697.

19. ADAM, G., SCHNEIDER, B. 1999. Uptake, transport and metabolism. pp.113-136 *in* Brassinosteroids: Steroidal Plant Hormones (A. Sakurai, T. Yokota, S. D. Clouse, eds.) Springer-Verlag, Tokyo.
20. YOKOTA, T., OGINO, Y., SUZUKI, H., TAKAHASHI, N., SAIMOTO, H., FUJIOKA, S., SAKURAI, A. 1991. Metabolism and biosynthesis of brassinosteroids. pp. 86-96 *in* Brassinosteroids: Chemistry, Bioactivity and Applications (H. C. Cutler, T. Yokota and G. Adam, eds.) ACS Symp. Ser. 474, Amer. Chem. Soc., Washington, DC.
21. SUZUKI, H., KIM, S.-K., TAKAHASHI, N., YOKOTA, Y. 1993. Metabolism of castasterone and brassinolide in mung bean explant. Phytochemistry 33: 1361-1367.
22. YOKOTA, T., HIGUCHI, K., KOSAKA, Y., TAKAHASHI, N. 1992. Transport and metabolism of brassinosteroids in rice. pp. 298-305 *in* Progress in Plant Growth Regulation (C. M. Karssen, L. C. van Loon, D. Vreugdenhil, eds.), Kluwer, Dordrecht.
23. WARD, E. R., UKNES, S. J., WILLIAMS, S. C., DINCHER, S. S., WIEDERHOLD, D. L., ALEXANDER, D. C., AHL-GOY, P. METRAUX, J.-P., RYALS, J. A. 1991. Coordinate gene activity in response to agents that induce systemic acquired resistance. Plant Cell 3: 1085-1094.
24. HAUDENSCHILD, C., HARTMANN, M.-A. 1995. Inhibition of sterol biosynthesis during elicitor-induced accumulation of furanocoumarins in parsley cell suspension cultures. Phytochemistry 40: 1117-1124.
25. BRINDLE, P. A., KUHN, P. J., THRELFALL D. R. 1988. Biosynthesis and metabolism of sesquiterpenoid phytoalexins and triterpenoids in potato cell suspension culture. Phytochemistry 27: 133-150.
26. VOGELI, U., CHAPPELL, J. 1988. Induction of sesquiterpene cyclase and suppression of squalene synthase activities in plant cell cultures treated with fungal elicitor. Plant Physiol. 88: 1291-1296.
27. VAN DER HEIJDEN, R., THRELFALL, D. R., VERPOORTE, R., WHITEHEAD, I. M. 1989. Regulation and enzymology of pentacyclic triterpenoid phytoalexin biosynthesis in cell suspension cultures of *Tabernaemontana divaricata*. Phytochemistry 28: 2981-2988.
28. DEVARENNE, T. P., SHIN, D. H., BACK, K., YIN, S., CHAPPELL, J. 1998. Molecular characterization of tobacco squalene synthase and response to fungal elicitor. Arch. Biochem. Biophys. 349: 205-215.
29. SHI, J., GONZALES, R. A., BHATTACHARYYA, M. K. 1996. Identification and characterization of an S-adenosyl-L-methionine: Δ24-sterol-*C*-methyltransferase cDNA from soybean. J. Biol. Chem. 271: 9384-9389.

30. NEWMAN, J. D., CHAPPELL, J. 1999. Isoprenoid biosynthesis in plants: carbon partitioning within the cytoplasmic pathway. Crit. Rev. Biochem. Mol. Biol. 34: 95-106.
31. CHOE, S., DILKES, B. P., FUJIOKA, S., TAKATSUTO, S., SAKURAI, A., FELDMANN, K. A. 1998. The *DWF4* gene of Arabidopsis encodes a cytochrome P450 that mediates multiple 22α-hydroxylation steps in brassinosteroid biosynthesis. Plant Cell 10: 231-243.
32. MATHUR, J., MOLNAR, G., FUJIOKA, S., TAKATSUTO, S., SAKURAI, A., YOKOTA, T., ADAM, G., VOIGT, B., NAGY, F., MAAS, C., SCHELL, J., KONCZ, C., SZEKERES, M. 1998. Transcription of the Arabidopsis *CPD* gene, encoding a steroidogenic cytochrome P450, is negatively controlled by brassinosteroids. Plant J. 14: 593-602.
33. BISHOP, G. J., NOMURA, T., YOKOTA, T., HARRISON, K., NOGUCHI, T., FUJIOKA, S., TAKATSUTO, S., JONES, J. D., KAMIYA, Y. 1999. The tomato DWARF enzyme catalyses C-6 oxidation in brassinosteroid biosynthesis. Proc. Natl. Acad. Sci. USA 96: 1761-1766.
34. LENTON, J. 1998. Plant hormones on the move! Trends Plant Sci. 3: 457-458.
35. REICHHART, D., SALAUN, J.-P., BENVENISTE, I., DURST, F. 1979. Induction by manganese, ethanol, phenobarbital, and herbicides of microsomal cytochrome *P*-450 in higher plant tissues. Arch. Biochem. Biophys. 196: 301-303.
36. POTTER, S., MORELAND, D. E., KREUZ, K., WARD, E. 1995. Induction of cytochrome P450 genes by ethanol in maize. Drug Metab. Drug Interact. 12: 317-327.
37. FREELING, M. 1973. Simultaneous induction by anaerobiosis or 2,4 D of multiple enzymes specified by two unlinked genes: differential *Adh*1-*Adh*2 expression in maize. Mol. Gen. Genet. 127: 215-227.
38. TAGEDE, M., KUHLMEIER, C. 1997. Aerobic fermentation during tobacco pollen development. Plant Mol. Biol. 35: 343-354.
39. YOSHIHARA, T., OMER, E. A., KOSHINO, H., SAKAMURA, S., KIKUTA, Y., KODA, Y. 1989. Structure of tuber inducing stimulus from potato leaves (*Solanum tuberosum* L.). Agric. Biol. Chem. 53: 2835-2837.
40. HELDER, H., MIERSCH, O., VREUGDENHIL, D., SEMBDNER, G. 1993. Occurence of hydroxylated jasmonic acids in the leaflets of *Solanum demissum* plants grown under long- and short-day conditions. Physiol. Plant. 88: 647-653.
41. KODA, Y. 1992. The role of jasmonic acid and related compounds in the regulation of plant development. Int. Rev. Cytol. 135: 155-199.

42. ACHENBACH, H., HUBNER, H., BRANDT, W., REITER, M. 1994. Cardioactive steroid saponins and other constituents from the aerial parts of *Tribulus cistoides*. Phytochemistry 35: 1527-1543.
43. VARIN, L., IBRAHIM, R. K. 1991. Partial purification and some properties of flavonol 7-sulfotransferase from *Flaveria bidentis*. Plant Physiol. 90: 1254-1258.
44. DURST, F., NELSON, D. R. 1995. Diversity and evolution of plant P450 and P450-reductases. Drug Metab. Drug Interact. 12: 189-206.
45. ROCHE, P., DEBELLE, F., MAILLET, F., LEROUGE, P., FAUCHER, C., TRUCHET, G., DENARIE, J., PROME, J. C. 1991. Molecular basis of host specificity in *Rhizobium meliloti*: *nodH* and *nodPQ* genes encode the sulfation of lipo-oligosaccharide signals. Cell 67: 1131-1143.
46. MARSOLAIS, F., VARIN, L. 1995. Identification of amino acid residues critical for catalysis and cosubstrate binding in the flavonol 3-sulfotransferase. J. Biol. Chem. 270: 30458-30463.
47. KAKUTA, Y., PEDERSEN, L. G., PEDERSEN, L. C., NEGISHI, M. 1998. Conserved structural motifs in the sulfotransferase family. Trends Biochem. Sci. 23: 129-130.
48. KAKUTA, Y., PEDERSEN, L. G., CARTER, C. W., NEGISHI, M., PEDERSEN, L. C. 1997. Crystal structure of estrogen sulphotransferase. Nat. Struct. Biol. 4: 904-908.
49. MARSOLAIS, F., LAVIOLETTE, M., KAKUTA, Y., NEGISHI, M., PEDERSEN, L. C., AUGER, M., VARIN, L. 1999. 3′-Phosphoadenosine 5′-phosphosulfate binding site of flavonol 3-sulfotransferase studied by affinity chromatography and ^{31}P NMR. Biochemistry 38: 4066-4071.
50. MULLER, C. W., SCHULZ, G. E. 1992. Structure of the complex between adenylate kinase from *Escherichia coli* and the inhibitor Ap_5A refined at 1.9 Å resolution. J. Mol. Biol. 224: 159-177.
51. MARSOLAIS, F., VARIN, L. 1997 Mutational analysis of domain II of flavonol 3-sulfotransferase. Eur. J. Biochem. 247: 1056-1062.
52. MARSOLAIS, F., VARIN, L. 1998. Recent developments in the study of the structure-function relationship of flavonol sulfotransferases. Chem. Biol. Interact. 109: 117-122.
53. HOMMA, H., OGAWA, K., HIRONO, K., MORIOKA, Y., HIROTA, M., TANAHASHI, I., MATSUI, M. 1996. Site-directed mutagenesis of rat hepatic hydroxysteroid sulfotransferases. Biochim. Biophys. Acta 1296: 159-166.
54. KAKUTA, Y., PETROTCHENKO, E. V., PEDERSEN, L. C., NEGISHI, M. 1998. The sulfuryl transfer mechanism. Crystal structure of a vanadate

complex of estrogen sulfotransferase and mutational analysis. J. Biol. Chem. 273: 27325-27330.

55. VARIN, L., MARSOLAIS, F., BRISSON, N. 1995. Chimeric flavonol sulfotransferases define a domain responsible for substrate and position specificities. J. Biol. Chem. 270: 12498-12502.
56. PETROTCHENKO, E. V., DOERFLEIN, M. E., KAKUTA, Y., PEDERSEN, L. C., NEGISHI, M. 1999. Substrate gating confers steroid specificity to estrogen sulfotransferase. J. Biol. Chem. 274: 30019-30022.
57. SAKAKIBARA, Y., TAKAMI, Y., NAKAYAMA, T., SUIKO, M., LIU, M.-C. 1998. Localization and functional analysis of the substrate specificity/catalytic domains of human M-form and P-form phenol sulfotransferases. J. Biol. Chem. 273: 6242-6247.
58. BRIX, L. A., DUGGLEBY, R. G., GAEDICK, A., McMANUS, M. E. 1999. Structural characterization of human aryl sulphotransferases. Biochem. J. 337: 337-343.
59. DAJANI, R., HOOD, A. M., COUGHTRIE, M. W. 1998. A single amino acid, glu146, governs the substrate specificity of a human dopamine sulfotransferase, SULT1A3. Mol. Pharmacol. 54: 942-948.
60. FELSENSTEIN, J. 1993. PHYLIP (Phylogeny Inference Package) version 3.5c. Distribution by the author. Department of Genetics, University of Washington, Seattle, WA.

INDEX

FSC
www.fsc.org
MIX
Papier | Fördert
gute Waldnutzung
FSC® C083411

Kolibri 360 GmbH
Ferdinand-Jühlke-Straße 7
99095 Erfurt, Deutschland
produktsicherheit@kolibri360.de